D0867326

THE FRONTIERS COLLECTION

THE FRONTIERS COLLECTION

Series Editors:
A.C. Elitzur M.P. Silverman J. Tuszynski R. Vaas H.D. Zeh

The books in this collection are devoted to challenging and open problems at the forefront of modern science, including related philosophical debates. In contrast to typical research monographs, however, they strive to present their topics in a manner accessible also to scientifically literate non-specialists wishing to gain insight into the deeper implications and fascinating questions involved. Taken as a whole, the series reflects the need for a fundamental and interdisciplinary approach to modern science. Furthermore, it is intended to encourage active scientists in all areas to ponder over important and perhaps controversial issues beyond their own speciality. Extending from quantum physics and relativity to entropy, consciousness and complex systems – the Frontiers Collection will inspire readers to push back the frontiers of their own knowledge.

Other Recent Titles

The Thermodynamic Machinery of Life
By M. Kurzynski

The Emerging Physics of Consciousness
Edited by J. A. Tuszynski

Weak Links
Stabilizers of Complex Systems from Proteins to Social Networks
By P. Csermely

Mind, Matter and the Implicate Order
By P.T.I. Pylkkänen

Quantum Mechanics at the Crossroads
New Perspectives from History, Philosophy and Physics
Edited by J. Evans, A.S. Thorndike

Particle Metaphysics
A Critical Account of Subatomic Reality
By B. Falkenburg

The Physical Basis of the Direction of Time
By H.D. Zeh

Asymmetry: The Foundation of Information
By S.J. Muller

Mindful Universe
Quantum Mechanics and the Participating Observer
By H. Stapp

For a complete list of titles in The Frontiers Collection, see back of book

Maximilian Schlosshauer

DECOHERENCE AND THE QUANTUM-TO-CLASSICAL TRANSITION

With 72 Figures

Springer

Dr. Maximilian Schlosshauer
The University of Melbourne
Department of Physics
Melbourne, Victoria, 3010
Australia
e-mail: m.schlosshauer@unimelb.edu.au

Cover figure: Image courtesy of the Scientific Computing and Imaging Institute, University of Utah (www.sci.utah.edu).

Library of Congress Control Number: 2007930038
MSC 1612-3018

ISSN 1612-3018
ISBN 978-3-540-35773-4 Springer Berlin Heidelberg New York

Springer is a part of Springer Science+Business Media

springer.com

Typesetting: Data supplied by the author
Production: LE-TEX Jelonek, Schmidt & Vöckler GbR, Leipzig
Cover design: KünkelLopka, Werbeagentur GmbH, Heidelberg

Printed on acid-free paper SPIN 11428633 57/3180/YL - 5 4 3 2 1 0

For Kari

Preface

Over the course of the past decade, decoherence has become a ubiquitous scientific term popular in all kinds of research, from fundamental theories of quantum physics to applications in nanoengineering. Decoherence has been hailed as the solution to long-standing foundational problems dating back to the beginnings of quantum mechanics. It has been cursed as the key obstacle to next-generation technologies, such as quantum computers (another seemingly omnipresent field of research). And while decoherence has been directly observed in various experiments, its scope and meaning have often been misunderstood and misrepresented. Decoherence makes a fantastic subject of research, as it touches upon many different facets of physics, from philosophically inclined questions of interpretation all the way to down-to-earth problems in experimental settings and engineering applications.

This book will introduce the reader, in an accessible and self-contained manner, to these various fascinating aspects of decoherence. It will focus in particular on the relation of decoherence to the so-called quantum-to-classical transition, i.e., the question of how decoherence may explain the emergence of the classical appearance of the macroscopic world around us from the underlying quantum substrate.

The scope of this book is relatively broad in order to familiarize the reader with the many facets of decoherence, in both the theoretical and experimental domains. Throughout the book, I have sought to maintain a healthy balance between the conceptual ideas associated with the decoherence program on the one hand and the formal and mathematical details on the other hand. This book will establish a proper understanding of decoherence as a pure quantum phenomenon and will emphasize the importance of the correct interpretation of the consequences and achievements of decoherence.

One beautiful thing about learning about decoherence is that, as vast as its implications and applications are, the basic ideas and formal structures are actually quite clear and simple. As a general rule, I will wherever possible avoid muddling important general insights with complicated mathematical exercises. A basic knowledge of the formalism of quantum mechanics should suffice to follow most, if not all, explanations and derivations in this book. While certain sections inevitably contain somewhat lengthy mathematical considerations (the derivation of master equations in Chaps. 4 and 5 is probably the most striking example), readers less interested in these formal structures underlying the decoherence program should be able to just glance over these sections—or even skip them altogether—without significantly compromising their understanding of other parts of the book. At the same time, the more advanced material included in this book will be useful to the working physicist who may already have some knowledge of decoherence and is looking for a self-contained and detailed reference. Philosophers of physics interested in the foundations of quantum mechanics should also find plenty of interesting material throughout this book (especially in Chaps. 1, 2, 8, and 9).

The book is organized as follows. In Chap. 1, we will take a first "bird's-eye look" at decoherence by introducing some of the basic ideas and concepts. We will emphasize the importance of considering "open" quantum systems in addressing some of the long-standing issues of quantum theory, and contemplate why it may have taken over half a century for this realization and the ideas of the decoherence program to take hold.

The core chapter of the book is Chapter 2, in which we will introduce and discuss in detail the key conceptual ideas and formal descriptions of decoherence. First, we will analyze fundamental concepts of quantum mechanics, such as quantum states (and their differences to classical states), the superposition principle, quantum entanglement, and density matrices. A proper grasp of these topics will turn out to be very important for the development of a solid understanding of decoherence. We will then illustrate and discuss different components of what has become known as the "quantum measurement problem." This problem encapsulates many of the fundamental conceptual difficulties that have to this date prevented us from arriving at a commonly agreed-upon understanding of the physical *meaning* of the formalism of quantum mechanics and of how this formalism relates to the perceived world around us. The measurement problem is also intimately related to decoherence, since decoherence has direct implications for the different components of the problem.

We will then illustrate basic concepts of decoherence in the context of the well-known double-slit experiment. This approach will allow the reader to develop a rather natural understanding of decoherence as a consequence of environmental "monitoring" and quantum entanglement. It will also establish a modern view of Bohr's famous "complementarity principle." We will formalize decoherence in terms of system–environment entanglement and reduced density matrices and discuss the two main consequences of decoherence, the environment-induced suppression of quantum interference and the selection of preferred "pointer" states through the interaction with the environment.

After the reader has thus become familiar with the ideas and formalism of decoherence, the subsequent chapters can either be read in order, or the reader may focus on particular chapters of interest. Each chapter is designed to present a fairly self-contained discussion of a particular aspect of decoherence.

In Chap. 3, we will consider a very important model that describes decoherence of quantum objects due to collisions with environmental particles such as photons and air molecules. This scattering-induced decoherence is ubiquitous in nature and of paramount importance in describing the quantum-to-classical transition on macroscopic everyday-world scales.

Next, in Chap. 4, we will introduce the master-equation formalism that provides us with a general method for determining the dynamics of decoherence models in many cases of physical interest. We will spend some time deriving the important Born–Markov master equation that will allow us to treat many decoherence problems in a fairly straightforward and intuitive fashion.

In Chap. 5, we will then show how a large class of system–environment models can be reduced to a few "canonical" decoherence models. We will then analyze these models in detail. In particular, we will discuss so-called quantum Brownian motion, which can be viewed as the quantum approximation to the familiar classical Newtonian trajectories in phase space. We will also introduce the famous spin–boson model which has recently received additional attention in the context of quantum computing.

After so much theoretical material, the reader will certainly be longing for a break. Thus, in Chap. 6, we will describe some fascinating experiments that have made it possible to directly observe in the laboratory the gradual action of decoherence and therefore the transition from the quantum world to the classical domain.

In Chap. 7, we will shift gears somewhat and enter the field of quantum computing that has attracted so much interest over the past decade. We will explain the crucial role that decoherence plays in this field. We will then describe how the effects of decoherence can be mitigated through sophisticated (but ultimately easy to understand) methods such as quantum error correction, decoherence-free subspaces, and environment engineering.

Chapter 8 will discuss the implications of decoherence for several of the main interpretations of quantum mechanics. We will describe how decoherence may enhance, redefine, or challenge the most common interpretations, such as the orthodox and Copenhagen interpretations, relative-state interpretations, physical collapse models, modal interpretations, and Bohmian mechanics.

Finally, in Chap. 9, we will discuss the role of the observer in quantum theory and the question of decoherence processes in the brain. We will explain why this question is of interest in the first place and then review some explicit model calculations that demonstrate the efficiency of decoherence in the brain. The implications of these results will be discussed, in particular with respect to a "subjective" observer-based resolution of the measurement problem.

A brief remark on notation. I have set $\hbar \equiv 1$ throughout most of the book except in situations where explicit numerical estimates play a role. In this way, I hope to have kept the notation as clear as possible without compromising the reader's ability to derive and reproduce numerical values where needed.

There are many people who have contributed to making this book possible. First and foremost, I would like to thank my Ph.D. advisor, Arthur Fine, for giving me both the freedom and guidance to study the field of decoherence. He suggested to me that I write up some "personal notes" on decoherence so that he and I would better understand this area of research (which was, at the time, new to both of us). These notes evolved into a review article on decoherence [1], which in turn motivated this book. In this context, I am deeply indebted to H. Dieter Zeh for many helpful discussions and for bringing the idea for this book to the attention of Angela Lahee, editor at Springer, who has since lent her patient, encouraging, and helpful support to every aspect in the production of this book.

I thank Michael Nielsen and Gerard Milburn for their hospitality at the University of Queensland where parts of this book were written. I would also like to express my gratitude to Stephen Adler for comments on Sect. 8.4, to Erich Joos for feedback on Chap. 3, to Gerard Milburn for introducing me to quantum-electromechanical systems, and to Wojciech Zurek for many valuable comments on the manuscript and for inspiring discussions. Most importantly, though, I would like to thank my wife Kari for all her patience and all-around inspiration during the long process of writing this book.

Melbourne, Australia
June 2007

Maximilian Schlosshauer

Contents

1 **Introducing Decoherence** ... 1

2 **The Basic Formalism and Interpretation of Decoherence** .. 13
2.1 The Concept and Interpretation of Quantum States ... 14
2.1.1 Classical Versus Quantum States ... 14
2.1.2 The Probabilistic Nature of Quantum States ... 16
2.1.3 The Ontological Status of Quantum States ... 18
2.2 The Superposition Principle ... 20
2.2.1 The Interpretation of Superpositions ... 20
2.2.2 Experimental Verification of Superpositions ... 21
2.2.3 The Scope of the Superposition Principle ... 26
2.3 Quantum Entanglement ... 28
2.3.1 Quantum Versus Classical Correlations ... 30
2.3.2 Quantification of Entanglement and Distinguishability 32
2.4 The Concept and Interpretation of Density Matrices ... 33
2.4.1 Pure-State Density Matrices and the Trace Operation . 34
2.4.2 Mixed-State Density Matrices ... 36
2.4.3 Quantifying the Degree of "Mixedness" ... 39
2.4.4 The Basis Ambiguity of Mixed-State Density Matrices 41
2.4.5 Mixed-State Density Matrices Versus Physical Ensembles ... 43
2.4.6 Reduced Density Matrices ... 44
2.5 The Measurement Problem and the Quantum-to-Classical Transition ... 49
2.5.1 The Von Neumann Scheme for Ideal Quantum Measurement ... 50
2.5.2 The Problem of the Preferred Basis ... 53
2.5.3 The Problem of the Nonobservability of Interference .. 55
2.5.4 The Problem of Outcomes ... 57
2.6 Which-Path Information and Environmental Monitoring ... 60
2.6.1 The Double-Slit Experiment, Which-Path Information, and Complementarity ... 60
2.6.2 The Description of the Double-Slit Experiment in Terms of Entanglement ... 63
2.6.3 The Environment as a Which-Path Monitor ... 65

2.7 Decoherence and the Local Damping of Interference 68
2.8 Environment-Induced Superselection 71
2.8.1 Pointer States in the Quantum-Measurement Limit ... 76
2.8.2 Pointer States in the Quantum Limit of Decoherence .. 81
2.8.3 General Methods for Determining the Pointer States .. 81
2.8.4 Selection of Quasiclassical Properties 83
2.9 Redundant Encoding of Information in the Environment and "Quantum Darwinism" 85
2.10 A Simple Model for Decoherence 88
2.11 Decoherence Versus Dissipation 93
2.12 Decoherence Versus Classical Noise 95
2.13 Virtual Decoherence and Quantum "Erasure" 98
2.14 Resolution into Subsystems 101
2.15 Formal Tools and Their Interpretation 103
2.15.1 The Schmidt Decomposition 104
2.15.2 The Wigner Representation 106
2.15.3 "Purifying" the Environment 109
2.15.4 The Operator-Sum Formalism 110
2.16 Summary 112

3 Decoherence Is Everywhere: Localization Due to Environmental Scattering 115
3.1 The Scattering Model 119
3.2 Calculating the Decoherence Factor 122
3.3 Full Versus Partial Which-Path Resolution 128
3.3.1 The Short-Wavelength Limit 128
3.3.2 The Long-Wavelength Limit 130
3.4 Decoherence Due to Scattering of Thermal Photons and Air Molecules 132
3.4.1 Photon Scattering 132
3.4.2 Scattering of Air Molecules 136
3.4.3 Comparison with Experiments 138
3.5 Illustrating the Dynamics of Decoherence 139
3.6 Summary 150

4 Master-Equation Formulations of Decoherence 153
4.1 General Formalism 154
4.2 The Born–Markov Master Equation 155
4.2.1 Structure of the Born–Markov Master Equation 156
4.2.2 Derivation of the Born–Markov Master Equation 158
4.3 Master Equations in the Lindblad Form 165
4.4 Non-Markovian Dynamics 169

5 **A World of Spins and Oscillators: Canonical Models for Decoherence** 171
5.1 Mapping onto Canonical Models 173
5.1.1 Mapping of the Central System 173
5.1.2 Mapping of the Environment 174
5.2 Quantum Brownian Motion 178
5.2.1 Derivation of the Born–Markov Master Equation 178
5.2.2 Harmonic Oscillator as the Central System 182
5.2.3 Ohmic Decoherence and Dissipation 188
5.2.4 The Caldeira–Leggett Master Equation 191
5.2.5 Dynamics of Quantum Brownian Motion 194
5.2.6 Limitations of the Quantum Brownian Motion and Caldeira–Leggett Models 203
5.2.7 Exact Master Equation 206
5.3 The Spin–Boson Model 207
5.3.1 Simplified Spin–Boson Model Without Tunneling 208
5.3.2 Born–Markov Master Equation for the Spin–Boson Model 218
5.4 Spin-Environment Models 222
5.4.1 A Simple Dynamical Spin–Spin Model 223
5.4.2 Spin-Environment Models in the Weak-Coupling Limit: Mapping to Oscillator Environments 228
5.4.3 Beyond Markov: Solving General Spin-Environment Models 237
5.5 Summary 237

6 **Of Buckey Balls and SQUIDs: Observing Decoherence in Action** 243
6.1 The First Milestone: Atoms in a Cavity 244
6.1.1 Atom–Field Interactions and Rabi Oscillations 246
6.1.2 Creating the Cat State 247
6.1.3 Observing the Gradual Action of Decoherence 251
6.1.4 Bringing Schrödinger Cats Back to Life 255
6.2 Interferometry with C_{70} Molecules 258
6.2.1 The Double-Slit Experiment with Electrons 258
6.2.2 Experimental Setup 259
6.2.3 Confirming the Wave Nature of Massive Molecules 262
6.2.4 Which-Path Information and Decoherence 263
6.2.5 Decoherence Due to Emission of Thermal Radiation 265
6.2.6 Beyond Buckey Balls 267
6.3 SQUIDs and Other Superconducting Qubits 270
6.3.1 Superconductivity and Supercurrents 271
6.3.2 Basic Physics of SQUIDs 272
6.3.3 Superposition States and Coherent Oscillations in SQUIDs 275

6.3.4 Observing and Quantifying Decoherence 279
6.4 Other Experimental Domains 282
6.4.1 Decoherence in Bose–Einstein Condensates 282
6.4.2 Decoherence in Quantum-Electromechanical Systems . 284
6.5 Outlook 289

7 Decoherence and Quantum Computing 293
7.1 A Brief Overview of Quantum Computing 294
7.1.1 The Power of Quantum Computing 294
7.1.2 Reading Out a Quantum Computer 297
7.1.3 Simulating Physical Systems 298
7.1.4 Examples of Famous Quantum Algorithms 300
7.1.5 Physical Realizations of Quantum Computers 300
7.2 Decoherence Versus Controllability in Quantum Computers . . 301
7.3 Decoherence Versus Classical Fluctuations 302
7.4 Quantum Error Correction 304
7.4.1 Classical Versus Quantum Error Correction 305
7.4.2 Representing the Influence of Decoherence by Discrete Errors 307
7.4.3 “Undoing” Decoherence in a Quantum Computer 311
7.4.4 When Does an Error-Correcting Code Exist? 314
7.4.5 Importance of Redundant Encoding and the Three-Bit Code for Phase Errors 315
7.4.6 Apparatus-Induced Decoherence and Fault Tolerance . 320
7.5 Quantum Computation on Decoherence-Free Subspaces 321
7.5.1 What Does a Decoherence-Free Subspace Look Like? . 322
7.5.2 Experimental Realizations of Decoherence-Free Subspaces 325
7.5.3 Environment Engineering and Dynamical Decoupling . 326
7.6 Summary and Outlook 327

8 The Role of Decoherence in Interpretations of Quantum Mechanics 329
8.1 The Standard and Copenhagen Interpretations 330
8.1.1 The Problem of Outcomes 331
8.1.2 Observables, Measurements, and Environment-Induced Superselection 333
8.1.3 The Concept of Classicality in the Copenhagen Interpretation 335
8.2 Relative-State Interpretations 336
8.2.1 Everett Branches and the Preferred-Basis Problem . . . 337
8.2.2 Probabilities in Relative-State Interpretations 339
8.2.3 The “Existential Interpretation” 343
8.3 Modal Interpretations 344

8.3.1 Property Assignment Based on Environment-Induced Superselection . . . 345
8.3.2 Property Assignment Based on Instantaneous Schmidt Decompositions . . . 345
8.3.3 Property Assignment Based on Decompositions of the Decohered Density Matrix . . . 346
8.4 Physical Collapse Theories . . . 347
8.4.1 The Preferred-Basis Problem . . . 349
8.4.2 Simultaneous Presence of Decoherence and Spontaneous Localization . . . 350
8.4.3 The Tails Problem . . . 351
8.4.4 Connecting Decoherence and Collapse Models . . . 352
8.4.5 Experimental Tests of Collapse Models . . . 353
8.5 Bohmian Mechanics . . . 354
8.5.1 Particles as Fundamental Entities . . . 355
8.5.2 Bohmian Trajectories and Decoherence . . . 356
8.6 Summary . . . 357

9 Observations, the Quantum Brain, and Decoherence . . . 359
9.1 The Role of the Observer in Quantum Mechanics . . . 359
9.2 Quantum Observers and the Von Neumann Chain . . . 361
9.3 Decoherence in the Brain: The Brain as a Quantum Computer? . . . 365
9.3.1 Decoherence Timescales for Superposition States in Neurons . . . 368
9.3.2 Decoherence Timescales for Superposition States in Microtubules . . . 371
9.4 "Subjective" Resolutions of the Measurement Problem . . . 375

Appendix: The Interaction Picture . . . 379

References . . . 383

Index . . . 409

1 Introducing Decoherence

The "paradox" is only a conflict between reality and your feeling of what reality "ought to be."

Richard P. Feynman

What qualifies as a good experiment in physics? Clearly, the typical goal of the experimenter is to study a particular aspect of a phenomenon of interest in such a way that disturbances of this aspect by undesired influences are minimized. For example, suppose we are interested in the dynamical laws of a body. We may then devise an experiment in which we make appropriate measurements on the motion of a ball rolling down an inclined plane. And, of course, we will try our best to reduce any sources of noise, such as friction, to arrive at the "exact" laws of motion.

This strategy has led to the idealization of isolated systems in physics, that is, the fundamental idea that we always ought to be able to sufficiently shield our system of interest from unwanted environmental disturbances in such a way as to find out the objective, "true," underlying nature of our system under study. In fact, this idea has proven extremely successful and fruitful in the history of physics, and an experiment is considered "good" if it yields maximum information about the phenomenon of interest with a minimum of environmental noise.

When quantum effects were discovered and quantum theory was formulated in the early decades of the twentieth century, it caused an enormous paradigm shift in our view of physics in particular and of nature in general. Yet, the idealized and ubiquitous notion of isolated systems remained a guiding principle of physics and was adopted in quantum mechanics without much further scrutiny.

This might not come as a surprise. The experimental evidence at the time hardly seemed to necessitate a reevaluation of the isolated-systems notion. For example, the agreement between the experimental data and the theoretical predictions for the discrete spectrum of the hydrogen atom was spectacular, and those predictions were based on an application of quantum theory to the model of a completely isolated atom.

More generally, the systems for which quantum effects were observed in the early days of quantum mechanics typically resided in the microscopic domain in which the isolated-system approximation indeed held to a good degree in many cases. From our modern perspective, we now know that the observation of quantum effects in these early experiments was made possible precisely because the systems under study did not interact much with their environments.

To be sure, the pioneers of quantum theory already understood that certain peculiar quantum features, while observable on microscopic scales, seemed strikingly absent from the world of our everyday experience. But the connection between the loss of "quantumness" and environmental interactions was not realized for a long time.

At the same time, it was also recognized early that the predictions of quantum theory imply that, by coupling a microscopic system to a macroscopic system, these quantum features should be transferable to the classical-appearing objects around us, in obvious contradiction to our experience. Of course, no other example has illustrated this problem of the *quantum-to-classical transition* more poignantly and drastically than Schrödinger's infamous cat [2], which appears, by the verdict of quantum theory, to be doomed into a netherworldy superposition of being alive and dead.

In this paradox, Schrödinger imagined a cat confined to a box (see Fig. 1.1). Inside the box, the decay of an unstable atom serves as a trigger for the hammer to break a vial containing poison. The release of the poison will then kill the cat. According to the laws of quantum mechanics, the atom is at all times described by a superposition of "decayed" and "not decayed." The feature of quantum entanglement (see below) then implies that this superposition spreads to the total system containing the cat, hammer, and poison, which must then be described by a superposition of two states which seem mutually exclusive according to our experience. One state corresponds to the atom not yet decayed, the hammer untriggered, the vial unharmed, and thus the cat alive. The other state represents a situation in which the atom has

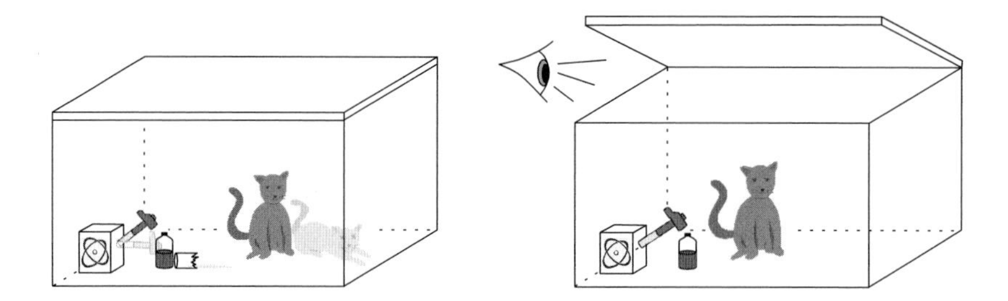

Fig. 1.1. Illustration of Schrödinger's cat paradox. *Left:* The laws of quantum mechanics force the composite system containing the unstable atom, hammer, poison, and cat into a superposition of two classically mutually exclusive states, one of which contains an alive cat, the other one a dead cat. *Right:* According to the standard interpretation of quantum mechanics, an observer opening the box and looking at the cat then "collapses" the superposition onto either one of the two components (here the "alive" component). However, this begs the question of the state of the cat before the box was opened.

decayed, the hammer has broken the vial, and the poison thus released has killed the cat.

The second part of the paradox is established by the appearance of an external observer. When the observer opens the box, standard quantum theory predicts that she will "collapse" the superposition onto one of its two component states. Thus it is ensured that the observer will perceive only either one of these states, in agreement with our experience. The observer would therefore seem to suddenly decide the fate of the cat by simply looking at the unfortunate animal. The paradox, then, consists of the simple question: What was the state of the cat *before* the observer opened the box? Alive or dead, both alive and dead, or neither? Has this question any meaning at all? In a similar vein, as Einstein is reported to have asked sarcastically [3], is the moon there when nobody looks?

In a certain sense, the seeds of a recognition of the importance of the *openness* of quantum systems for their ability to lose their "quantumness" and become effectively classical are already planted into examples such as the cat paradox. The observer seems to be required to literally *open* the box and look at the cat to ensure that it is "classical" (i.e., either alive *or* dead).

But in Schrödinger's setup the role of the observer is simply derived from the collapse (or "projection") postulate of quantum mechanics. The observer is not considered in physical terms as a macroscopic quantum system interacting with the cat. Furthermore, the isolated-system assumption implicitly underlies the remainder of the cat setup. Schrödinger considered the cat as interacting with the poison—which in turn is linked to the unstable atom via the hammer that breaks the vial containing the poison, thereby establishing the coupling between the microscopic and the macroscopic domains—but as otherwise isolated from the rest of the universe.

In the 1930s, when Schrödinger presented his cat paradox, it was considered a mere *Gedankenexperiment* (i.e., a thought experiment). Quantum phenomena, such as interference effects, had at that time been observed only in the microscopic domain. It was thus not only argued that quantum mechanics is unnecessary for a description of the macroscopic world of our experience, but moreover that quantum mechanics should be banned from this realm altogether. An example of the latter stance is the Copenhagen interpretation of quantum mechanics, which postulates a fundamental dualism between a microscopic "quantum" domain and a macroscopic "classical" realm.

Today, our view has changed drastically. On the one hand, quantum effects have been observed in the laboratory far beyond the microscopic domain. Researchers have created mesoscopic and macroscopic "Schrödinger kittens" such as superpositions of microampere currents flowing in opposite directions and interference patterns for massive molecules composed of dozens of carbon atoms (see Chap. 6).

On the other hand, over the past three or so decades it has been slowly realized that the isolated-system assumption—which, as we have described above, had proved so fruitful in classical physics and had simply been taken over to quantum physics—had in fact been the crucial obstacle to an understanding of the quantum-to-classical transition. It was recognized that the openness of quantum systems, i.e., their interaction with the environment, is *essential* to explaining how quantum systems (and thus, assuming that quantum mechanics is a universal physical theory, all systems) become effectively classical: How their "quantumness" seems to slip out of view as we go to larger scales, finally arriving in the world of our experience where cats are either alive or dead but are never observed to be in a superposition of such two classically distinct properties.

How is it possible that the importance of interactions with the environment for the emergence of the classical world from the underlying quantum domain had been overlooked for so long? We already discussed one likely reason, namely, the deeply ingrained notion that physics is about (idealized) isolated systems. Related to this issue, we can unearth another important reason. States in classical physics, for which the isolated-systems notion was originally conceived, are *local*.

As an illustration, consider a small grain of pollen immersed in a liquid. The collisions with the molecules in the liquid may then cause the particle to move in a random zig-zag path (which is the well-known effect of Brownian motion). Thus the environment perturbs the motion of the particle, i.e., it acts as classical "noise." Nevertheless, it is clear that the grain of pollen retains its "individuality" at all times, even in the presence of the liquid. In other words, all physical properties of the particle always remain contained within the particle itself. Thus we can imagine removing the liquid, and we would expect to find the same particle as before.

This notion of the locality of states is deeply ingrained—in fact, so deeply that we take it for granted, and anyone who has not come into contact with the puzzling features of quantum theory is unlikely ever to challenge this notion. Indeed, the locality of states provides the physical justification for the isolated-system idealization. In this picture, isolating the system from its surroundings would only diminish environmental noise effects (i.e., perturbations of the system by its environment) but not alter the actual "nature" of the system. Accordingly, it may be easy to see why the isolated-system view had been deemed innocuous for such a long time.

Quantum mechanics has forced us to radically reevaluate our notion of the locality of states. While interactions continue to be local also in the quantum theory (i.e., quantum mechanics is still a *local theory*), the *states* that can be generated by these local interactions are distinctly *nonlocal*. The key concept is *quantum entanglement*, where two systems (which may be well-separated in space) are described by a quantum state that, loosely speaking, cannot be "broken down" into two separated quantum states for each individual system (see Fig. 1.2). Entangled states encapsulate *quantum correlations* between the two systems. Such correlations often embody entirely new physical properties

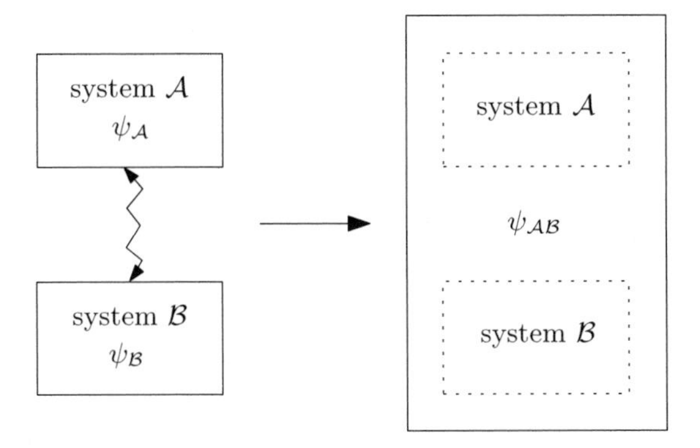

Fig. 1.2. Illustration of the principles of quantum entanglement. Consider two systems $\mathcal{A}$ and $\mathcal{B}$ initially described by their individual quantum states $\psi_{\mathcal{A}}$ and $\psi_{\mathcal{B}}$. If these two systems interact in certain ways, they can no longer be described by such individual quantum states. They have become entangled with each other, and only the composite system $\mathcal{AB}$ can be assigned a quantum state $\psi_{\mathcal{AB}}$.

for the composite system that are not present in any of the two individual subsystems. We may say that these subsystems have lost their individuality, in the sense that physical properties are now at least partially encapsulated in the nonlocal quantum correlations and therefore cannot be attributed to only one of the subsystems. Broadly speaking, we may thus conclude that quantum entanglement represents a situation where the quantum-mechanical whole is different from the sum of its parts.

Thus, in the quantum picture, the (local) interactions between a system and its environment now have the power to play a much greater role than in classical physics. They will typically lead to entanglement between the two interacting partners and thus change the nature of the object itself, in the sense of fundamentally altering what we may observe at the level of the system.

Thus environmental interactions in the quantum setting no longer amount to a mere "perturbation" of the system of interest that ought to be minimized in order to properly describe the physics of this system, or that could simply be neglected if these interactions are sufficiently "weak" (for instance, in the case of the influence of air molecules on the Newtonian trajectory of a billiard ball). Instead, the coupling to the environment now *defines* the observable physical properties of the system. At the same time, quantum *coherence*, a measure for the "quantumness" of the system, is delocalized into the entangled system–environment state, which effectively removes it from our observation. This process is usually irreversible in practice and constitutes a key component in explaining how the classical world of our experience emerges from the underlying quantum substrate. (We will make these ideas much more precise in the next chapter.)

Thus there are two main, and intimately related, consequences of environmental interactions (and thus of quantum entanglement) for a quantum system:

1. The effectively irreversible disappearance of quantum coherence, the source of quantum phenomena such as interference effects, from the system.
2. The dynamical "definition" of the observable properties of the system, i.e., the selection of a set of robust preferred states (or, formally, observables) for the system.

These consequences are subsumed under the heading of *environment-induced decoherence*, or *decoherence* for short, the subject of this book. The motivation for the term "decoherence" should be obvious from the first consequence listed above. The second consequence is sometimes referred to in the literature by its own name, *environment-induced superselection.* This term is motivated by the analogy with so-called superselection rules in physics, which (usually axiomatically) restrict the space of allowable superpositions, i.e., prohibit the existence of certain observables. Similarly, the interaction with the environment strongly limits, for all practical purposes, which physical quantities can actually be observed on a given quantum system. These restrictions arise dynamically from the interaction of the system with its environment and explain why in our everyday world we only observe a few "classical," robust quantities, such as position and momentum.

Readers may still wonder why such a long time elapsed between the completion of the main formalism of quantum mechanics in the 1930s—including the realization of the problem of the quantum-to-classical transition as exemplified by Schrödinger's cat—and the first groundbreaking work on the importance of environmental interactions and decoherence for the quantum-to-classical transition in the 1970s and 1980s [4–9]. After all, quantum entanglement, including its counterintuitive consequences for our well-established notion of locality, was already well known in the 1930s. In 1935, Einstein, Podolsky, and Rosen (EPR) [10] had written a seminal paper in which the authors pointed out the puzzling "holistic" nonlocal character of the quantum world implied by entanglement. Why, then, did it take another forty years for researchers to recognize the crucial importance of environmental interactions and entanglement for the explanation of how the classical world emerges from the quantum domain?

There are at least two possible, complementary answers to this question. First, even after Schrödinger had presented his cat paradox (which is based on entangling a microscopic and a macroscopic system), for quite some time entanglement may have seemed like a rather unusual state of affairs. It was often regarded as a phenomenon that would need to be created in the laboratory by means of an elaborate setup but would play no significant role in other areas, and certainly not in the macroscopic world of our experience.

This assessment turned out to be incorrect. In fact, it has been shown that entanglement is ubiquitous *especially* in the macroscopic domain. Furthermore, the focus was frequently put on *usable* entanglement, i.e., on quantum correlations that can be manipulated and exploited in a controlled way such as to implement certain "quantum protocols" (such as quantum cryptography). This stance is evident in the slogan, often heard in the quantum-information and quantum-computing community, of "entanglement as a resource." While entanglement can indeed be used in such a way, it usually comes in the form of *uncontrolled* and thus "unusable" environmental entanglement.

A second reason for the relatively late discovery of the importance of environmental interactions and entanglement for the emergence of classicality may be found in the fact that the extreme effectiveness of even minimal environments in bringing about the quantum-to-classical transition was not realized for a long time. For example, in classical physics, it is perfectly legitimate to neglect the influence of incident light (i.e., photons) on the motional state of a macroscopic body in virtually all cases (see Fig. 1.3). The amount of momentum transferred to the body by the photons is usually negligibly small in comparison with the amount that would be needed to perturb the motion of the body. Additionally, the incident photons are often distributed isotropically, such that the net momentum transfer averages out to zero.

However, when we switch from the classical to the quantum picture, the situation changes completely (see again Fig. 1.3). Now the scattering interaction between the photons and the body will lead to the formation of quantum

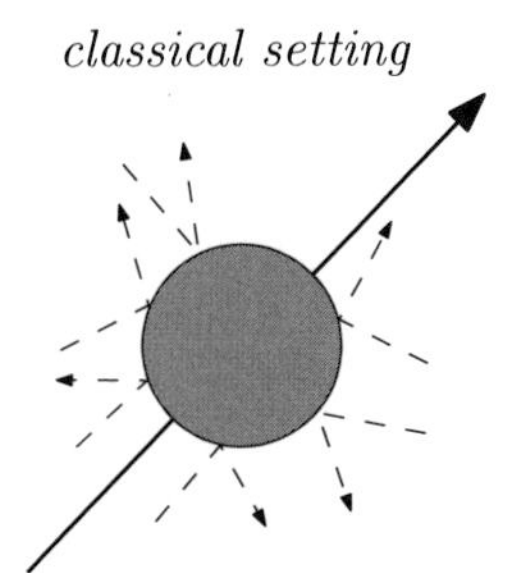

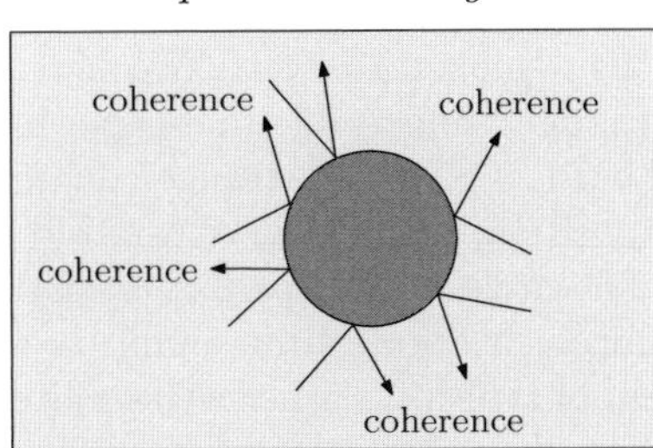

Fig. 1.3. The different influence of the environment on the system in the classical and quantum settings, illustrated for the case of a macroscopic body immersed into light (photons) incident from all directions. *Left:* In the classical case, light scattering off the body will not change the motion of the body, even though the environment interacts strongly with the system. *Right:* In the quantum setting, the interaction leads to an entangled object–photon state $\psi_{\text{object+photons}}$. Coherence becomes delocalized from the object, making quantum effects such as interference patterns unobservable at the level of the system.

correlations between the two partners, which "carry away" coherence from the body, thereby diminishing the degree of "quantumness" that we may observe at the level of the body (again, we will make these rather figurative terms more precise later). The amount of momentum transferred to the body does not play any role in this creation of entanglement—in fact, we may simply assume that no momentum at all is transferred from each individual photon to the system. Thus the photons do not need to lead to classical perturbations of the motional state of the body, and nonetheless they will exert a strongly decohering influence on the body. Although decoherence *may* occur simultaneously with the classical process of dissipation (i.e., the loss of energy from the system), decoherence is a distinct and purely quantum-mechanical effect without any analog in classical physics.

As we will see in this book, the decoherence of a macroscopic object induced by the scattering of environmental particles is astonishingly strong and fast. If we tried to prepare our object in a nonclassical quantum state such as a coherent superposition of two different well-separated positions (a situation we may refer to, sloppily speaking, as the body being in "two places at the same time"), environmental entanglement with photons scattering off the body will lead to a complete delocalization of the coherence between the two components in the superposition within a tiny fraction of a second. In this way, for all practical purposes of observations at the level of the object, the behavior of the object will in virtually all cases become effectively classical much faster than we could ever resolve. Even more surprisingly, we do not even need the presence of an environment of incident light for this to happen. Not only will the omnipresent thermal radiation completely suffice to accomplish the task, but even cosmic background radiation, a relict from the Big Bang that permeates the universe, will lead to rapid decoherence of macroscopic bodies (see Chap. 3).

This demonstrates that decoherence is not only extremely effective, but also virtually impossible to escape. In turn, this explains why it is so exceedingly hard to observe "strange" quantum effects, such as the Schrödinger-cat superposition, in the everyday world of our experience. In turn, now coming back full-circle to the opening discussion of this chapter, it may also indicate why for such a long time quantum mechanics was usually discussed in the context of typically rather well-isolated microscopic systems. Unless a system is extremely well shielded from its environment (which is virtually impossible for macroscopic objects), decoherence will usually make it prohibitively difficult to observe the peculiar quantum effects of interest.

The question of how the appearance of a classical world can be brought into agreement with, or even be derived from, the predictions of quantum mechanics extended into the macroscopic domain is one of the greatest foundational problems of quantum mechanics. It is almost ironic that this question has found an (at least partial) answer not by forcing preconceived classical concepts onto quantum mechanics in order to simply exorcise the "strange" implications of quantum mechanics for the macroscopic realm, but from within quantum mechanics itself. After all, it is the distinctly *quantum-mechanical* phenomenon of entanglement—that was initially only regarded as leading to such "paradoxes" as Schrödinger's cat and EPR—which ultimately enforces the observed *classical* behavior of systems.

We may therefore understand our deeply rooted notion of "classicality" as an *emergent concept*, as something that is not contradicted but rather justified by the peculiar features of quantum mechanics. To what extent it may be possible to derive *all* classical concepts as an effective consequence of environmental interactions and decoherence—and thus to deny these concepts a fundamental role in physical theories—is certainly an open question that continues to fuel a lively debate among researchers interested in the foundations of quantum mechanics. In this book, we will show that decoherence provides us with an explanation of the appearance of the classical world around us that is "as good as it gets," if quantum theory is assumed to be universally valid—and there exists no compelling experimental evidence to the contrary—and classicality is to be explained from within this theory (which constitutes a highly desirable goal).

There certainly exist some open conceptual questions related to the "ultimate physical reality" underlying the *effective classicality* induced by decoherence. The particular form of these questions tends to vary according to one's interpretive stance toward quantum mechanics. But such questions are unavoidable if the emergence of classicality is to be completely derived from quantum mechanics. After all, at some global level all the "strange" nonclassical phenomena of quantum theory, such as superposition states, nonlocal quantum correlations, etc., must still persist, since they have not been exorcised by postulates but have rather been used to *derive* the classical world of our experience.

Before digressing too far into these interpretive issues (which we will come back to in many places of this book, especially in Chap. 8), let us emphasize that, over the past years, decoherence has been studied extensively in many experiments (see Chap. 6). In various such experiments, it has become possible to control the interactions of the system of interest with the environment in such a way as to observe the gradual action of decoherence and thus the step-by-step transition between the quantum regime and the classical domain. For example, quantum interference patterns produced by large fullerene molecules sent through diffraction gratings were observed to decay gradually as the density of surrounding gas molecules, and thus the rate of scattering events between the fullerenes and the environmental particles, was increased. The characteristic time of such decoherence processes has been measured with great precision and has been found to be in stunning agreement with theoretical predictions. These experiments are so remarkable because they allow us to study different quantitative degrees of decoherence. Instead of simply attributing the *already existing* appearance of a fully classical macroscopic world around us to the extremely effective and fast action of decoherence, we are now in a position to directly measure how the continuous interaction with the environment gradually degrades our ability to observe quantum phenomena in the mesoscopic and macroscopic domains.

Many practitioners of quantum mechanics, especially researchers in experimental disciplines of physics and in the area of quantum computing, have now taken a very practical attitude toward decoherence. For example, decoherence is often viewed as a form of "quantum noise" that hampers the experimental implementation of devices (e.g., by introducing "errors" into quantum computations). More and more realistic decoherence models have been developed, and their predictions have been compared to experimentally observed decoherence phenomena. For all practical purposes of the working physicist, decoherence provides a complete and self-contained framework for a qualitative and quantitative description of the quantum-to-classical transition. On the other hand, it is hardly surprising that decoherence, as it applies to philosophically charged notions such as the "classical world of our experience," leads to some interesting questions of a conceptual nature.

As decoherence is simply a consequence of the application of the standard formalism of quantum mechanics to the interaction between a system and its environment, decoherence is neither an extraneous theory distinct from quantum mechanics itself nor something that we could freely choose to include or neglect. Decoherence is a ubiquitous effect in nature, with far-reaching and fascinating consequences that must be taken into account in order to arrive at a realistic description of physical systems.

The volume of literature on decoherence has grown to enormous proportions over the past years. For readers who are interested in exploring the groundbreaking early papers that introduced the key ideas and concepts of decoherence, we shall now give a brief survey of some of the main references (see [11] for further historical remarks on the development of the decoherence program). The first paper on what was later to become known as "decoherence" was written by H. Dieter Zeh at the University of Heidelberg and published in 1970 in *Foundations of Physics* [4]. There, Zeh pointed out that realistic macroscopic quantum systems are never closed and interact strongly with their environments. Hence, if the Schrödinger equation is assumed to be universally valid, such systems will typically be found in states that are quantum-correlated with the environment, leading to a "dynamical decoupling" of wave-function components and to the inability to describe the dynamics of the system itself by the Schrödinger equation. Zeh suggested that this could explain the observed fragility of quantum states of macroscopic systems and the emergence of superselection rules. In a subsequent paper [6], he and Olaf Kübler investigated the dynamics of entanglement and emphasized the dynamical robustness of coherent states. However, the term "decoherence" was not introduced until the late 1980s. In fact, throughout much of the 1970s decoherence did not attract much attention at all (Zeh recently called it the "dark ages of decoherence" [11]).

A milestone in the development of the decoherence program was reached in 1981–82 when Wojciech Zurek (a postdoc of John Wheeler's in 1981 and a Tolman Fellow at Caltech in 1982) published two defining papers on decoherence in *Physical Review D* [8,9]. Zurek clearly pointed out the importance of the so-called "preferred-basis problem" (to be discussed in the next chapter), which is central to decoherence and the problem of the quantum-to-classical transition. Zurek developed the concept of environment-induced superselection that is a cornerstone of the decoherence program, and he defined a precise framework for determining the environment-superselected preferred states (which he called "pointer states"). He emphasized the importance of the preservation of quantum correlations as the key criterion for the selection of the preferred states. He also showed how environment-induced superselection effectively remedies the preferred-basis problem and how it explains the fact that position is observed to be the ubiquitous preferred quantity in the everyday world. Zurek's work started to put decoherence into the spotlight and marked the beginning of the end of the "dark ages" of decoherence. Another important contribution to the development of the decoherence program was made in 1984 when Zurek derived a quite general and simple (and still widely used) expression from which typical decoherence timescales could be evaluated [12]. Although advertised in a series of talks at the time and also described in a Los Alamos report, Zurek's work was "officially" published (as a part of the proceedings of a 1984 conference) only in 1986.

In 1985, Erich Joos (a student of Zeh's) and Zeh coauthored a seminal paper in *Zeitschrift für Physik* [7] that presented a detailed model for decoherence induced by the scattering of environmental particles. The article also included the first explicit numerical estimates of decoherence timescales for objects of various sizes and physical nature immersed into different types of environments. Both the papers by Zurek and by Joos and Zeh made it clear that decoherence constitutes an extremely fast and efficient process, especially on macroscopic scales.

In 1991, Zurek's *Physics Today* article [13] introduced decoherence to a broader audience. The paper helped establish decoherence as a mainstay of physics, and decoherence finally began to attract widespread attention from physicists, material scientists, and philosophers alike. (Joos once called it a "historical accident" [14, p. 13] that the implications of decoherence for fundamental problems of quantum mechanics had been overlooked for so long.) Ongoing investigations of the role of the environment continue to lead to new developments (such as so-called "quantum Darwinism," the "environment as a witness" program, or the connection between quantum entanglement and probabilities—which are all research projects initiated and currently worked on by Zurek and collaborators), yielding a more complete picture of the quantum origin of the classical everyday world we are accustomed to. The ever-growing interest in quantum computing has further boosted research on decoherence, opening up new theoretical and experimental avenues toward an active control, shielding, and mitigation of decoherence processes. Thus decoherence, with its manifold aspects, implications, and applications, is far from fully explored to date, and we can be certain to continue to see many exciting developments in the coming years.

In addition to this book, the interested reader may find reviews of the decoherence program in the articles by Paz and Zurek [15], Zurek [16], and in the book by Joos et al. [17]. The more technical book by Breuer and Petruccione [18] also contains material on decoherence. In the next chapter, we will make more precise, both conceptually and formally, the various notions and ideas which constitute the foundations of decoherence and which have been introduced, albeit rather handwavingly, in this preliminary discussion.

2 The Basic Formalism and Interpretation of Decoherence

In the first part of this chapter, we will introduce the reader to some of the formal and interpretive structures of quantum mechanics that underlie the decoherence program. We will begin by discussing quantum states (Sect. 2.1). Much of what makes quantum mechanics different from classical physics can be centered around the concept and interpretation of quantum states. In classical physics, the "state of the system" simply denotes a catalog of values of physical quantities such as position, momentum, temperature, etc. In quantum physics, systems are described by abstract state vectors that in general can be related to observed physical properties only through the additional—but at best vaguely defined—concept of quantum measurement.

In Sect. 2.2, we will then discuss the superposition principle of quantum mechanics. If the concept of abstract quantum states had been the first significant difference between classical and quantum mechanics, the superposition principle widens the gap even further. Decoherence is to a large extent centered around the issue of quantum superpositions: It is the key mechanism for telling us which superpositions we may be able to observe in nature, and it provides an explanation for why most superpositions, allowed in principle by the superposition principle, are in fact not seen in the world around us.

Section 2.3 will introduce another peculiar quantum-mechanical concept important to decoherence, namely, that of quantum entanglement. Entanglement has become the key ingredient in harnessing the power of quantum mechanics in technological applications such as quantum computing. It is also the basic mechanism underlying decoherence. Section 2.4 will then discuss the formalism and interpretation of density matrices, which are important tools in the formal description of decoherence. In Sect. 2.5, we will introduce in some detail the measurement problem of quantum mechanics and the related problem of the quantum-to-classical transition. Decoherence is intimately related to these problems and provides, as we shall see in this chapter and also in Chap. 8, at least partial solutions.

In Sect. 2.6, we will discuss, using the example of the double-slit experiment, a consequence of entanglement that is crucial to decoherence, namely, the encoding of "which-path" information in the partner entangled with a given system. Our account will pave the way toward an understanding of decoherence as a consequence of environmental "monitoring" of a system via system–environment entanglement.

In the subsequent Sects. 2.7, 2.8, and 2.9, we will then discuss in detail the three main consequences of this environmental monitoring: The suppression of interference effects at the level of the system; the selection of quasiclassical preferred states, which are the states least sensitive to entanglement with the environment; and the robust and redundant encoding of information about these preferred states in the environment. As already pointed out in Chap. 1, decoherence in the narrow sense is often associated only with the first consequence (the decay of interference), whereas the second consequence (the selection of preferred states) is often denoted by the separate term "environment-induced superselection." However, these two phenomena have a common origin and are inextricably connected (since the damping of interference must refer to a particular set of states between which interference becomes suppressed). In this bigger picture, the "decoherence program" should certainly be thought of as incorporating all three aforementioned consequences of environmental monitoring.

In Sect. 2.10, we will explore these ideas in the context of a concrete, simple model for decoherence. The remaining sections will be concerned with the discussion of some important further topics, thus completing the picture. We will emphasize fundamental differences between decoherence and dissipation (Sect. 2.11) and between decoherence and classical noise (Sect. 2.12). In Sect. 2.13, we will describe instances in which decoherence may be reversible and analyze the related process of quantum "erasure." In Sect. 2.14, we will discuss the problem of the resolution of the universe into subsystems, which constitutes an important conceptual issue of the decoherence program. Finally, in Sect. 2.15, we will outline a few additional formal tools, namely, the Schmidt decomposition theorem (Sect. 2.15.1), the Wigner representation (Sect. 2.15.2), the purification of nonpure states (Sect. 2.15.3), and the Kraus operator-sum formalism (Sect. 2.15.4).

2.1 The Concept and Interpretation of Quantum States

2.1.1 Classical Versus Quantum States

In classical physics, the notion of the "state" of a physical system is quite intuitive. We typically focus on certain measurable quantities of interest, for example, the position and momentum of a moving body, and subsequently assign mathematical symbols to these quantities (such as "x" and "p"). The motional state of the body is then simply specified by assigning numerical values to these symbols. In other words, there exists a one-to-one correspondence between the physical properties of the object (and thus the entities of the physical world) and their formal and mathematical representation in the theory.

To be sure, we may certainly think of some cases in classical physics where this direct correspondence is not always established as easily as in the example of Newtonian mechanics used here. For instance, historically it turned out to be rather difficult to relate the formal definition of temperature in the theory of thermodynamics to the underlying molecular processes leading to the physical notion of temperature. However, reference to other physical quantities and phenomena (e.g., by relating a collective thermodynamic variable to the behavior of the microscopic constituents in a bulk of matter) usually allowed one to resolve this identification problem at least at some level.

With the advent of quantum theory in the early twentieth century, this straightforward bijectivism between the physical world and its mathematical representation in the theory came to a sudden end. Instead of describing the state of a physical system by means of intuitive symbols that corresponded directly to the "objectively existing" physical properties of our experience, in quantum mechanics we have at our disposal only an abstract *quantum state* that is defined as a vector (or, more generally, as a ray) in a similarly abstract Hilbert vector space.

The conceptual leap associated with this abstraction is hard to overestimate. In fact, the discussions regarding the "interpretation of quantum mechanics" that have occupied countless physicists and philosophers since the early years of quantum theory are to a large part rooted precisely in the question of how to relate the abstract quantum state to the "physical reality out there."

Textbook quantum mechanics tells us that the connection with the familiar physical quantities of our experience is only made in an indirect manner, namely, through measurements of physical quantities, that is, of *observables* (the "real-world" part) represented by Hermitian operators in a Hilbert space (the "formal-theory" part). According to the commonly used collapse (or projection) postulate, measurements then instantaneously change the quantum state into one of the eigenstates of the operator representing the measured observable, where the probability of each of these eigenstates is given by the Born rule [19] (see footnote 9 on p. 35).

The eigenstates of the measured operator represent the different definite "values" that the corresponding physical quantity may assume in a measurement. The notion associated with these eigenstates is therefore similar to that encountered in classical physics discussed above (of course, the measured observable may be distinctly nonclassical, as in the case of spin). Thus, to a certain extent, the measurement allows us to revert to a one-to-one correspondence between the mathematical formalism and the "objectively existing physical properties" of the system, i.e., to the concept familiar from classical physics.

However, an additional and distinctly quantum-mechanical caveat is lurking here. Due to the fact that many observables are mutually incompatible (formally expressed by the noncommutativity of the representing operators), a quantum state will in general be a simultaneous eigenstate of only a very small set of operator-observables. Accordingly, we may ascribe only a limited number of definite physical properties to a quantum system, and additional measurements will in general alter ("disturb") the state of the system—unless, of course, we measure, by virtue of sheer luck or prior knowledge, an operator-observable with an eigenstate that happens to coincide with the quantum state of the system before the measurement.

This fragility of quantum states is also reflected in the famous *no-cloning theorem* (see Sect. 7.4.1), which states that it is in general impossible to duplicate an unknown quantum state [20, 21]. Another way of expressing this fact is to say that it is impossible to uniquely determine an unknown quantum state of an individual system by means of measurements performed on that system only [22–26].

Needless to say, this situation is in stark contrast with classical physics. Here we can enlarge our "catalog" of physical properties of the system (and therefore specify its state more completely) by performing an arbitrary number of measurements of additional physical quantities, in any given order. Furthermore, many independent observers may carry out such measurements (and agree on the results) without running into any risk of disturbing the state of the system, even though they may have been initially completely ignorant about this state. Of course, this *objective preexistence* of classical states has led to our deeply ingrained notion of "classical reality."

We note that in the following we shall often—unless an ambiguity may arise—refer to (Hermitian) operators representing (physical) observables simply as "observables," instead of using the more precise term "operator-observable." Nonetheless the reader should bear in mind that the identification between formal operators and physical quantities is introduced axiomatically into the quantum theory and is therefore nontrivial (although the concept of observables may also be viewed as emergent and therefore derivable from the quantum-state formalism alone, thus matching the spirit of John Bell's quest for "beables" [27]; see Sect. 2.2 of [17] for an example of this approach). The danger of falling into a naïve realism about such operators by describing measurements of physical quantities in an oversimplified manner as "measurements of operators" has been pointed out before [28].

2.1.2 The Probabilistic Nature of Quantum States

In view of the properties of quantum states discussed in the preceding section, it has often been argued that these states represent only "potentialities" for the various observed "classical" states. At the same time, however, it is important to emphasize that (according to our current knowledge) quantum states represent a *complete description* of a quantum system, i.e., the quantum state encapsulates *all there is to say* about the physical state of the system. Yet, in general quantum states do not tell us which particular outcome will be obtained in a measurement but only the *probabilities* of the various possible outcomes. This seemingly intrinsic probabilistic character of quantum mechanics is one of the central features distinguishing this theory from classical physics. In an experimental situation, the probabilistic aspect is represented by the fact that, if we measure the same physical quantity on a collection of systems all prepared in exactly the same quantum state, we will in general obtain a set of different outcomes.

As is well known, throughout his life Einstein remained reluctant to accept this apparent intrinsic randomness of nature [29], as captured in his famous slogan that "God does not play dice with the universe." This discomfort has been shared by others, who have sought a way out by suggesting that quantum mechanics does not constitute a complete theory, in the sense that the quantum state does not suffice to completely specify the physical state of a system. Thus, an ensemble of "identically prepared systems" would not actually represent a collection of *physically* identical systems. While each system in the collection would be described by the same quantum state, the "complete state" of each system, which determines the outcomes of all possible measurements on each individual system, would not be the same for every member in the ensemble.

Such approaches are usually referred to as *hidden-variables theories.* They attempt to restore determinism at the fundamental level by augmenting the quantum state through the introduction of postulated "hidden variables" whose values determine the particular outcome of any measurement performed on the system. Therefore only the combination of the quantum state together with the values of the relevant hidden variables specifies the complete physical state of the system. The ensemble of systems, all of which are prepared in the same quantum state, would then be described by a range of different values of the hidden variables. This would give a completely deterministic account of the observation of different measurement outcomes. Quantum mechanics would then only *appear* probabilistic to us, since we do not have any means of knowing the specific values of the hidden variables (hence the term "hidden"). Probabilities in quantum mechanics would therefore be purely epistemic: Just as in classical statistical mechanics, they would be only an expression of our subjective lack of knowledge, but they would not be representative of an objective indeterminism at the fundamental physical level.

However, it turns out that it is in fact extremely difficult to construct a hidden-variables approach that does not violate the experimentally well-confirmed predictions of standard quantum mechanics. Most famously, in a seminal paper of 1964 John Bell presented an inequality that would be obeyed by any *local* hidden-variables theory but violated by standard quantum mechanics [30] (see also Bell's later refined accounts in [27, 31]). Numerous experiments have shown a clear violation of Bell's inequality, thereby almost definitely[1] ruling out local hidden-variables theories. These results show that any hidden-variables theory would need to be highly nonlocal in order to work. Probably the most famous representative of such a nonlocal hidden-variables theory is the de Broglie–Bohm approach [34–37] (see Sect. 8.5). Here, the hidden variable is chosen to be position, i.e., each particle in the system has a definite position at all times.

Furthermore, Kochen and Specker [38] demonstrated that any hidden-variables theory must necessarily be *contextual*: The particular value of a physical quantity ascribed to a quantum system (by means of the hidden variables) will in general be dependent on the measurement context, i.e., in which particular manner this value is eventually measured. For example, depending on which other observables are co-measured on the system, the system would possess different values of a particular observable. Contextuality therefore forces us to relinquish the key idea motivating the hidden-variables program in the first place: That the physical world is independent of any measurements performed on it, and that a measurement simply reveals the preexisting value (determined by the values of the hidden variables) of a physical quantity. It it thus clear that any postulated hidden-variables concept would necessitate a sophisticated, highly nonlocal, and contextual mechanism to work as suggested.

2.1.3 The Ontological Status of Quantum States

Another dispute has centered around the question of the ontological status of quantum states (see [39] for an accessible short historical overview). In the early days of quantum mechanics, Schrödinger attempted to identify narrow wave packets in real space with actual physical particles (see the beginning of Chap. 3). This strategy encountered two core problems. First of all, initially localized wave packets generally tend to spread out very rapidly over large regions of space, which is irreconcilable with the concept of particles, which, by definition, are localized in space. (As we shall see in Chap. 3, this spreading becomes effectively suppressed by the interaction with the environment, a process Schrödinger had of course been unaware of at the time.) Second, the wave function describing the quantum state of $N > 1$ particles in three-dimensional real space resides in a 3^N-dimensional Hilbert space and not anymore in the familiar three-dimensional space of our experience.

Subsequently, Born and Pauli took a different approach and formulated their famous interpretation of quantum states as representing a *probability amplitude*, i.e., as specifying the probabilities of the outcomes of all possi-

[1] There remains a small, albeit ever diminishing, amount of leeway in interpreting the experimental results. The main such "loophole" is due to imperfect detectors with efficiencies of less than 100% [32]. Another issue is the so-called "communication loophole," which is now commonly considered to have been closed through an experiment by Weihs and coworkers [33].

ble measurements that could be performed on the system. The act of measurement was then assumed to play the fundamental role of dynamically "actualizing" these potential properties. Wheeler [40, p. 182] (see also [41]) poignantly summarized this view as "no elementary [i.e., quantum] phenomenon is a phenomenon until it is a recorded (observed) phenomenon." Similar ideas are apparent in Heisenberg's statement that "the particle trajectory is created by our act of observing it"[2] [42, p. 185], and in Pauli's letter to Born in which he suggest that "the appearance of a definite position x_0 during an observation (...) is then regarded as a creation existing outside the laws of nature"[3] [43]. Thus the widely accepted Born–Pauli interpretation of the wave function is neither purely ontological (in the sense of viewing quantum states as directly representing physical reality) nor simply epistemic (i.e., as representing but the lack of our subjective knowledge).

However, this interpretation leaves unanswered the fundamental question of the ontological role of the "actualization by observation" and the physical explanation of this process in dynamical terms. In this sense, the problem of the interpretation of the quantum state is simply replaced by the problem of the interpretation of measurement. Note also that the problem of the "actualization" of physical properties is not solved by the assumption of a wave-function collapse that reduces the quantum state to an eigenstate of the measured observable. Such a collapse does not *per se* alter the ontological status of this state, and thus the state of the system after the collapse cannot be regarded as more "physically real" than before the collapse.

The opposite pole to a "realist" interpretation of quantum states is represented by the aforementioned purely epistemic view of quantum states. This interpretation regards quantum states as describing only our knowledge but not the objective physical state of a system (see, e.g., [44] and also Ballentine's "ensemble interpretation" [45, 46] for examples of such a stance). This view, while obviously difficult to refute empirically, stubbornly refuses to address the fundamental question of the physical reality underlying our observations and thus our "knowledge."

In this book, we shall adopt the widely accepted notion that a quantum-state vector (expressed, for example, as a ket $|\psi\rangle$ in the standard Dirac notation) provides a complete description of the physical state of an individual system. To reflect the "completeness" of such quantum states, they are commonly called *pure*. (By contrast, so-called *mixed states* are simply classical ensembles of pure states; we will discuss them in Sect. 2.4.2.) Our aim will be to explain the nature of our observations from within the quantum formalism, instead of taking the cop-out route of introducing *ad hoc* as-of-yet undiscov-

[2]The original German quote reads: *"Die Bahn entsteht erst dadurch, daß wir sie beobachten."*

[3]The original text reads: *"Das Erscheinen eines bestimmten Ortes* x_0 *bei der Beobachtung (...) wird dann als außerhalb der Naturgesetze stehende Schöpfung aufgefaßt."*

ered "hidden variables" or by taking our "knowledge" as the fundamental entity described by the quantum formalism.

2.2 The Superposition Principle

The superposition principle lies at the heart of quantum mechanics, and it is one of the features of quantum mechanics that most distinctly marks the departure from classical concepts. Formally, the superposition principle is rooted in the linearity of the Hilbert space. Since quantum states are represented by vectors in a Hilbert space, we may form linear combinations of such vectors. The superposition principle then states that such a linear combination of vectors again corresponds to a new quantum state. This means that, if the kets $|\psi_n\rangle$ represent a set of quantum states, then the superposition

$$|\Psi\rangle = \sum_n c_n |\psi_n\rangle \tag{2.1}$$

also corresponds to a possible and equally admissible quantum state, with the c_n denoting arbitrary complex coefficients. Thus, by virtue of the egalitarian nature of the superposition principle, the superposition state $|\Psi\rangle$ corresponds to some possible (physical) state of the system in the same way as the component states $|\psi_n\rangle$ do.

For example, suppose we are able to prepare a spin-$\frac{1}{2}$ particle in one of the two states $|0\rangle$ ("spin up") or $|1\rangle$ ("spin down"). Then the superposition principle tells us that $|\Psi\rangle = (|0\rangle + |1\rangle)/\sqrt{2}$ is an equally admissible quantum state, and that we could therefore (at least in principle) also prepare our system in this superposition state.

2.2.1 The Interpretation of Superpositions

To understand why the superposition principle has such counterintuitive consequences, the correct interpretation of a quantum-mechanical superposition is crucial. Such a superposition state does not simply represent a classical ensemble of its components (such an ensemble is often referred to as a *proper mixture* [47–49]), i.e., a situation in which the quantum system *actually is* in only *one* of the component states $|\psi_n\rangle$, but we simply do not know in which (we may call this the "naïve ensemble interpretation"). Instead, each of the components $|\psi_n\rangle$ in (2.1) is simultaneously present in the quantum state. This situation is referred to the existence of *coherence* between these components. To clearly emphasize the distinction from the classical case, a quantum-mechanical superposition of the form (2.1) is therefore often referred to as a *coherent superposition*. (In this book, the term "superposition" will always, unless explicitly stated otherwise, refer to a quantum, i.e., coherent, superposition.) Such a superposition of the component states defines a new (physical) state of an individual system and not merely a statistical distribution of the component states.

We may give a general argument against the interpretation of superpositions as classical ensembles of their component states, i.e., as an ensemble of more fundamentally determined states. The proof is by contradiction. Suppose that such an ensemble view could indeed be attached to a superposition $|\Psi\rangle = \sum_n c_n |\psi_n\rangle$ of states $|\psi_n\rangle$. By virtue of measurements, we could then obtain additional knowledge that would allow us to single out a subensemble consisting of the states compatible with the results obtained in the measurement. Because the time evolution has been completely deterministic according to the Schrödinger equation,

$$\mathrm{i}\frac{\mathrm{d}}{\mathrm{d}t} |\Psi(t)\rangle = \hat{H} |\Psi(t)\rangle , \tag{2.2}$$

this would enable us to backtrack this particular subensemble in time. This, in turn, would allow us to specify the initial (premeasurement) state of the system more completely (this process is often referred to as "postselection" [50]), which implies that this state must physically differ from the initially prepared superposition, establishing a contradiction.

2.2.2 Experimental Verification of Superpositions

How can we experimentally demonstrate the existence of coherence and thereby show that a superposition is indeed different from an ensemble (i.e., a proper mixture) of its component states? There are two general methods: A repeated direct projective measurement onto the superposition state, or an indirect confirmation of the presence of all components in the superposition by means of an interference experiment. Let us illustrate these two approaches using some simple examples.

Direct Measurements

The most direct way of confirming the existence of a superposition state of the form (2.1) would be to carry out a projective measurement of the observable $\hat{O} = |\Psi\rangle\langle\Psi|$ on every member of an ensemble of identically prepared systems. If we obtain an outcome equal to one in each measurement and the ensemble is sufficiently large, we can conclude that the systems must indeed have been prepared in the superposition (2.1).

An example for a setup that realizes such a measurement is the *Stern–Gerlach* apparatus (see Fig. 2.1). A typical Stern–Gerlach experiment goes as follows. Silver atoms are heated in an oven and then pass through collimating slits and a magnetic field inhomogeneous in the z direction (for simplicity, we shall neglect field components in other directions). Each silver atom has 47 electrons, 46 of which are contained in a spherically symmetric electron

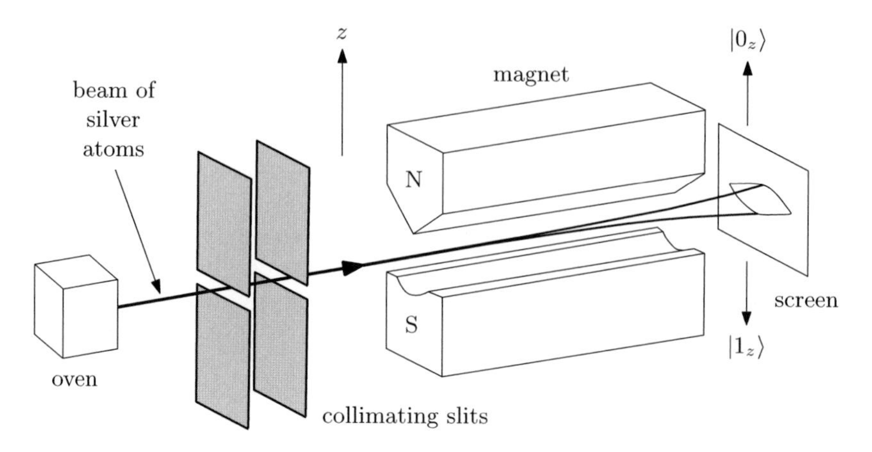

Fig. 2.1. Schematic illustration of the Stern–Gerlach experiment. Silver atoms with a magnetic moment equal to the spin of a single electron are emitted from an oven and subsequently pass through an inhomogeneous magnetic field in the z direction (the inhomogeneity is created, for example, by the shape of the pole pieces of the magnet). Depending on the spin state, the path of the atom is deflected along two possible trajectories, thereby indicating the quantized value of the spin along the z axis. After many atoms have traversed the apparatus, two distinct bands become observable on the detection screen.

cloud. Thus only the spin of the 47th electron yields a net magnetic moment of the atom as a whole. We therefore deal with a comparably massive atom characterized by a magnetic moment equal to the spin magnetic moment of a single electron.

Since the electron can take only two equal but opposite values of spin along the z axis, namely, spin "up" or spin "down," the silver atom has two possible values $\pm\mu_z$ of magnetic moment. When a silver atom passes through the inhomogeneous magnetic field, it experiences a force in the z direction due to the interaction of the magnetic moment with the field. This force is proportional to the value of the magnetic moment. Thus the trajectory of the silver atom will be deflected in either the $+z$ or the $-z$ direction, depending on whether the 47th electron in the atom is in the spin-up or the spin-down state along the z axis (we shall denote these states by $|0_z\rangle$ and $|1_z\rangle$, respectively, which are the eigenstates of the Pauli z-spin operator $\hat{\sigma}_z$). When the beam of atoms hits a detector screen, two spatially separated spots will appear, corresponding to the two distinct trajectories that the atoms may take. Alternatively, we may place a particle detector in one of the paths that will then tell us whether a particular atom took this path. Either way, this setup corresponds to a measurement of the z-component of electron spin.

Suppose now we prepare the incoming atoms in a quantum state corresponding to "spin up" of the electron along the z axis, i.e., in the eigenstate

$|0_z\rangle$ of the z-spin operator $\hat{\sigma}_z$. As expected, the pattern on the screen will experimentally confirm this preparation: All atoms will be found to impinge in the upper region of the screen. At the same time, $|0_z\rangle$ can be equally rewritten as a linear combination of "spin up" and "spin down" along the orthogonal x axis, $|0_z\rangle = (|0_x\rangle + |1_x\rangle)/\sqrt{2}$, and thus the experiment can be viewed as confirming the preparation of this superposition. If the superposition did represent a classical ensemble of the states $|0_x\rangle$ and $|1_x\rangle$, i.e., if each electron actually *was* in either of the two states, a single spot in the center of the screen would appear, since the inhomogeneity of the magnetic field is oriented along the z axis only and would therefore not induce any splitting of the beam of atoms. Of course, this is not the behavior observed in the experiment. Conversely, we may choose to rotate the orientation of the magnetic field by 90 degrees into the x direction. Now the 50–50 splitting of the beam would be observed for all atoms prepared in the initial state $|0_z\rangle$.

Therefore the superposition state $|0_z\rangle = (|0_x\rangle + |1_x\rangle)/\sqrt{2}$ corresponds to a (physical) state of an individual system in which the components $|0_x\rangle$ and $|1_x\rangle$ are simultaneously present. The superposition thus does not just manifest itself in form of interference fringes (see below) between the components $|0_x\rangle$ and $|1_x\rangle$. Instead, there always exists an orientation of the magnetic field in the Stern–Gerlach apparatus such that the trajectory of the atom can be "predicted with certainty" [10, 39].

Interference Experiments

The example of spin measurements in a Stern–Gerlach apparatus represents a rather unique case, since it is typically very difficult to directly measure the projective observable corresponding to a superposition state. Instead, all we usually have available are devices that perform measurements in a particular basis corresponding to the *components* of the superposition. For example, there are measuring devices (namely, our own eyes!) that can detect whether Schrödinger's cat is alive or dead, i.e., that perform measurements in the $\{|\text{"alive"}\rangle, |\text{"dead"}\rangle\}$ basis. But there exists no obvious procedure that would correspond to a measurement of observables in the *conjugate basis* $\{(|\text{"alive"}\rangle \pm |\text{"dead"}\rangle)/\sqrt{2}\}$. This, of course, is the reason why Schrödinger-cat superposition states appear so utterly counterintuitive: We simply have usually no means (i.e., no measuring devices, and in particular no human senses) that would allow us to observe such states directly.

To give a more down-to-earth example of this problem, let us consider the famous double-slit experiment in which we let electrons pass individually, i.e., one at a time, through a double slit. At the level of the slits, the electron is described by a superposition $|\Psi\rangle = (|\psi_1\rangle + |\psi_2\rangle)/\sqrt{2}$ of the components $|\psi_1\rangle$ and $|\psi_2\rangle$ corresponding to passage through slit 1 and 2, respectively. To directly confirm the existence of this superposition, we would need to perform a projective measurement onto the state $|\Psi\rangle$. However, the wave function $\Psi(x) \equiv \langle x|\Psi\rangle$ is spatially *delocalized* over each plane parallel to the slits, while the only type of measurement we have available is the measurement of the position of the particles behind the slits. Thus in order to measure the projective observable $|\Psi\rangle\langle\Psi|$, we would need to exactly refocus the two partial waves $\psi_1(x) \equiv \langle x|\psi_1\rangle$ and $\psi_2(x) \equiv \langle x|\psi_2\rangle$ onto a single point in space. Obviously, this is virtually impossible in practice.

In such cases—where one has available only a measurement procedure for the component states of the superposition—one typically resorts to demonstrating *interference effects* between the components in the superposition. The basic idea consists of designing the experiment in such a way as to induce a spatial or temporal variation in the expansion coefficients c_n that define the superposition state $|\Psi\rangle = \sum_n c_n |\psi_n\rangle$ [see (2.1)], while this variation would be absent if the system was instead described by a classical ensemble of the component states $|\psi_n\rangle$. We then carry out measurements in the component basis $\{|\psi_n\rangle\}$ on an ensemble of identically prepared systems. Since $|c_n|^2$ specifies the probability of finding the component state $|\psi_n\rangle$, we can infer that the systems had indeed been prepared in the superposition state $|\Psi\rangle = \sum_n c_n |\psi_n\rangle$ if we observe the variation of these probabilities with position or time. Let us mention two important examples.

In the double-slit experiment, we measure the *spatial* variation of the density pattern on the distant screen, i.e., the (position-space) density distribution $\varrho(x)$ of the particles at the level of the screen. The well-known result is that we obtain an interference pattern, which is distinctly different from the classical pattern that would be expected if we assumed that each electron passes through either one of the slits (see Fig. 2.2). The density $\varrho(x)$ is not described by the sum of the squared wave functions describing the addition of

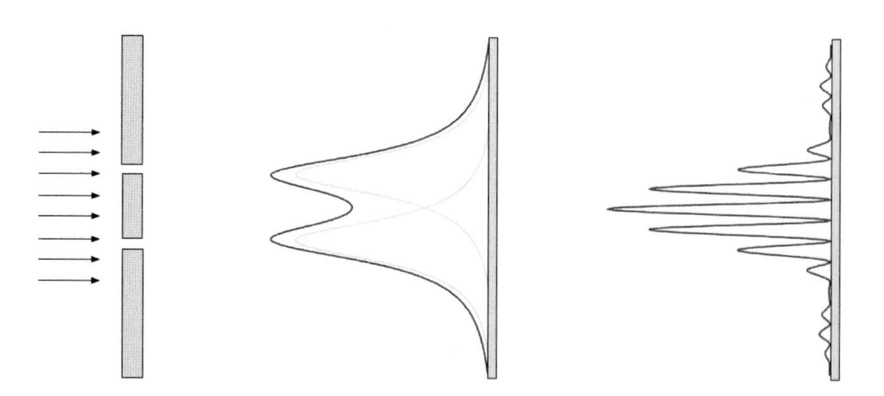

Fig. 2.2. The double-slit experiment with particles. *Left:* Particles are incident on a double slit. *Center:* The resulting "classical" density pattern obtained on a distant detection screen. This pattern corresponds to a simple addition of the contributions from each individual slit. *Right:* The interference pattern obtained in the quantum setting (with the envelope modulated by diffraction).

individual passages through a single slit (i.e., $\varrho(x) \propto |\psi_1(x)|^2 + |\psi_2(x)|^2$), corresponding to a classical distribution of component states $\psi_1(x)$ and $\psi_2(x)$, but by the square of the sum of the individual wave functions,

$$\varrho(x) = \frac{1}{2} |\psi_1(x) + \psi_2(x)|^2 = \frac{1}{2} |\psi_1(x)|^2 + \frac{1}{2} |\psi_2(x)|^2 + \mathrm{Re}\{\psi_1(x)\psi_2^*(x)\}, \tag{2.3}$$

where the last term is responsible for the characteristic interference pattern on the screen.

Thus this experiment clearly shows that, within the standard quantum-mechanical formalism,[4] the individual electron cannot be described by either one of the wave functions describing the passage through a particular slit, but only by a superposition of these wave functions, $\Psi(x) = (\psi_1(x) + \psi_2(x))/\sqrt{2}$. Formally, this finding is explained by the fact that quantum states represent probability *amplitudes* rather than actual probabilities. This implies that a superposition state describes a linear combination of probability amplitudes rather than of probabilities, leading to interference terms in the probability distribution (2.3) corresponding to the superposition. Note also that, since it is completely feasible to carry out the double-slit experiment such that only a single particle is present in the apparatus at any one time (see Sect. 6.2), the interference pattern cannot be due to any interactions between different electrons.

Another example of the indirect verification of the existence of a superposition via the observation of interference effects—this time, in terms of *temporal* (instead of spatial) variations—is given by the technique of Ramsey interferometry [51]. Ramsey interferometry is directly based on the principle of quantum-coherent Rabi oscillations in two-level systems, which we will explain in more detail in Sect. 6.1.1. Suppose we have a two-level atom described by a ground state $|g\rangle$ and an excited state $|e\rangle$, and we would like to demonstrate the existence of the superposition state $|\psi\rangle = (|g\rangle + |e\rangle)/\sqrt{2}$ of the atom. However, our experiment only allows us to measure the energy state of the atom, i.e., we can only perform measurements in the $\{|g\rangle, |e\rangle\}$ basis but not in the conjugate basis $\{(|g\rangle \pm |e\rangle)/\sqrt{2}\}$, which would be required for a direct projective measurement of the superposition.

We can again use the method of performing an interference experiment to accomplish our goal. First, we prepare our atom in the ground state $|g\rangle$ by means of a projective measurement of the observable $\hat{O} = |g\rangle\langle g|$. Then, by applying a laser pulse of a particular duration (see Sect. 6.1.1 for details), we can transform this state into the coherent superposition

[4] There exist alternative formulations of quantum mechanics which permit a different view. For example, in Bohmian mechanics the positions of all particles are taken to be determinate at all times (see Sect. 8.5), and thus each particle follows a definite trajectory through either one of the two slits. While the wave function is still delocalized over the slits, it only defines the various *possible* paths that a particle may take.

$$|\Psi\rangle = \frac{1}{\sqrt{2}} \left(|g\rangle - \mathrm{i} \, |e\rangle \right) . \tag{2.4}$$

The state then continues to evolve freely,

$$|\Psi\rangle = \frac{1}{\sqrt{2}} \left(|g\rangle - \mathrm{i}e^{-\mathrm{i}\phi(t)} \, |e\rangle \right) , \tag{2.5}$$

where $\phi(t)$ denotes the phase shift induced by the unitary evolution after a time t has passed since the application of the laser pulse (we omit here any global phase factors in $|\Psi\rangle$). Equation (2.5) represents the superposition whose existence we would now like to experimentally verify. To do so, we apply a second laser pulse of the same duration as the first pulse, which changes the state (2.5) into

$$|\Psi\rangle = \sin\left(\phi(t)/2\right) |g\rangle - \cos\left(\phi(t)/2\right) |e\rangle . \tag{2.6}$$

Here the phase shift $\phi(t)$ can be adjusted by changing the time t between the two laser pulses. Thus we have introduced a temporal variation into the coefficients in the superposition. If we now measure the atom in the $\{|g\rangle, |e\rangle\}$ basis, we will find the atom in the ground or excited state with probabilities that explicitly depend on the value of $\phi(t)$.

On the other hand, suppose that during the time between the laser pulses the atom had *not* been described by the superposition (2.5), but instead by a classical ensemble of the components $|g\rangle$ and $|e\rangle$. As it is evident from (2.5), during the time between the pulses the probabilities of finding the atom in the ground or excited state are identical, i.e., the phase shift $\phi(t)$ present in the superposition (2.5) would be completely irrelevant to the corresponding classical ensemble derived from (2.5). Consequently, if between the application of the two pulses the atom *was* described by an ensemble (a proper, *incoherent* mixture) of the component states instead of the superposition (2.5), then after the application of the second pulse and the subsequent measurement of the atom in the $\{|g\rangle, |e\rangle\}$ basis, the ground-state and excited-state probabilities would be independent of the phase shift $\phi(t)$. By finding that the measurement statistics indeed depend on $\phi(t)$, i.e., on the time t between the pulses, we are therefore able to confirm the existence of coherent superpositions of the form (2.5).

The Ramsey technique is by no means limited to the example of an atom controlled by a laser pulse. The atom can be replaced by any two-level system that can be addressed by some suitable control field. For example, Sect. 6.3.4 will describe an application of the Ramsey method to the measurement of decoherence effects in superconducting systems.

2.2.3 The Scope of the Superposition Principle

Historically, quantum theory, and thus the superposition principle, were applied to microscopic phenomena only, for example, in explanations of the discrete spectrum of the hydrogen atom or in descriptions of interference experiments with microscopic particles. The spectacular agreement between theoretical predictions and experimental observations explicitly demonstrated the validity and power of the superposition principle in the microscopic domain. Since superpositions seemed to be limited to the world of tiny physical entities (such as electrons) far removed from the range of our direct experience, it was not too difficult to convince oneself that such superpositions were simply a peculiar feature of the microworld.

As already noted in our introduction in Chap. 1, the reason for the initial restriction to the microscopic domain was twofold. First of all, only on microscopic scales had the explanation of certain phenomena observed at the beginning of the twentieth century demanded a departure from classical physics. By contrast, the classical framework seemed to continue to provide a satisfactory explanation of our observations in the macroscopic everyday world of our experience. Moreover, however, the application of the superposition principle to macroscopic systems seemed to immediately lead to counterintuitive consequences that appeared to blatantly contradict our experience, as exemplified by the Schrödinger-cat paradox discussed in Chap. 1. After all, macroscopic systems are always observed to reside in a few "classical" macroscopic states, defined by having a small number of determinate and robust properties such as position and momentum. How could one possibly reconcile this observation with the vastness of the quantum-mechanical Hilbert space and with the superposition principle, which would seem to allow for arbitrary superpositions?

Accordingly, quantum mechanics was not only deemed unnecessary for a description of the macroworld, but in fact often banned *a priori* from the macroscopic realm. For example, the hugely influential Copenhagen interpretation of quantum mechanics postulated a fundamental dualism (the so-called *Heisenberg cut*) between the microscopic regime, in which quantum theory, and in particular the superposition principle, was assumed to hold, and the macroscopic domain, which was supposed to be entirely described in irreducibly classical terms (see Sect. 8.1).

However, over the past decade a rapidly growing number of sophisticated experiments have demonstrated the validity of the superposition principle on larger and larger scales (see Chap. 6), leaving behind the "unproblematic" territory of superpositions involving only microscopic entities such as electrons and photons that had been studied by the founders of quantum theory. Furthermore, these experiments have shown that any observed disappearance of quantum coherence and interference can be attributed to interactions with the environment, that is, to decoherence. These results suggest that there may indeed exist no fundamental limit to the validity of the superposition principle on macroscopic scales, i.e., that the "breakdown" of this principle observed in the macroscopic world of our experience is of a purely apparent nature and simply a consequence of decoherence. On the other hand, as we shall make more clear below, decoherence does not actually *destroy* the superposition, it simply *extends* it to include the environment, which (as we shall show, too) precludes the observation of coherence at the level of the system.

This may indeed be enough to "save the phenomena," but many people do find, understandably, the idea of the persistent existence of "global Schrödinger cats" (that now include the environment) worrisome on some philosophical level. Accordingly, there exist proposals for exorcising such "cat states" by means of postulated nonlinear corrections to the Schrödinger equation, such that superpositions are, at some stage, physically and objectively reduced to their components states (we will discuss such approaches in Sect. 8.4). Since these proposals postulate deviations from standard unitary dynamics, they are, at least in principle, experimentally falsifiable. However, while *conceptually* the effect of such an "objective" reduction mechanism is fundamentally different from a merely apparent reduction in form of a nonobservability of coherence phenomena due to environmental interactions, it would in fact be extremely difficult to *experimentally* distinguish these two effects (see also Sect. 8.4.5).

2.3 Quantum Entanglement

Let us now turn to quantum entanglement, which is the key process underlying decoherence. First of all, let us define entanglement. Suppose we are given a quantum system $\mathcal{S}$, described by a state vector $|\Psi\rangle$, that is composed of two subsystems $\mathcal{S}_1$ and $\mathcal{S}_2$ ($\mathcal{S}$ is therefore called a *bipartite* quantum system). The state vector $|\Psi\rangle$ of $\mathcal{S}$ is called *entangled* with respect to $\mathcal{S}_1$ and $\mathcal{S}_2$ if it cannot be written as a tensor product of state vectors of these two subsystems, i.e., if there do not exist any state vectors $|\psi\rangle_1$ of $\mathcal{S}_1$ and $|\phi\rangle_2$ of $\mathcal{S}_2$ such that $|\Psi\rangle = |\psi\rangle_1 \otimes |\phi\rangle_2$.[5] Put in terms of entangled systems instead of entangled states, $\mathcal{S}_1$ and $\mathcal{S}_2$ are called entangled if the state of the composite system $\mathcal{S}$ cannot be expressed in the tensor-product form $|\Psi\rangle = |\psi\rangle_1 \otimes |\phi\rangle_2$, with $|\psi\rangle_1$ and $|\phi\rangle_2$ denoting some state vectors of $\mathcal{S}_1$ and $\mathcal{S}_2$, respectively. Note that in the remainder of this book, we shall refrain from explicitly writing out the tensor-product symbol "$\otimes$" for tensor products of quantum states. We shall, however, retain the symbol "$\otimes$" in denoting products of operators pertaining to different Hilbert spaces and in referring to products of Hilbert spaces.

[5] A brief remark on notation. When dealing with multiple systems, we will often denote which system a ket is associated with by using a subscript at the end of the ket symbol, such as in "$|\psi\rangle_1$" to indicate that this ket refers to system $\mathcal{S}_1$. Sometimes, on the other hand, it will be possible to label the quantum state in direct reference to the system, e.g., by writing "$|a\rangle$" to refer to the state of a system $\mathcal{A}$. In such (and other similarly unambiguous) cases we will omit the subscript for the sake of notational clarity.

As an example for pure entangled states, consider two spin-$\frac{1}{2}$ particles described by the mutually orthogonal basis states $|0\rangle_i$ and $|1\rangle_i$, $i = 1, 2$, of their respective two-dimensional Hilbert spaces. The states $|0\rangle_i$ and $|1\rangle_i$ correspond to particle i having its spin pointing "up" and " down," respectively, along some given axis in space. There are several different pure entangled quantum states for the composite system consisting of the two spin-$\frac{1}{2}$ particles that are maximally entangled (we will explain the notion of "maximum entanglement" below). These states are commonly referred to as *Bell states* and given by

$$|\Phi^{\pm}\rangle = \frac{1}{\sqrt{2}} \left(|0\rangle_1 |0\rangle_2 \pm |1\rangle_1 |1\rangle_2 \right), \tag{2.7a}$$

$$|\Psi^{\pm}\rangle = \frac{1}{\sqrt{2}} \left(|0\rangle_1 |1\rangle_2 \pm |1\rangle_1 |0\rangle_2 \right). \tag{2.7b}$$

It is quite easy to explicitly confirm that it is indeed impossible to write these Bell states as a tensor product of states of the two spin-$\frac{1}{2}$ subsystems. Because in (2.7) the states $|0\rangle_1$ and $|1\rangle_1$ are one-to-one correlated with the states $|0\rangle_2$ and $|1\rangle_2$, one often says that $|0\rangle_2$ and $|1\rangle_2$ are *relative states* of $\mathcal{S}_2$ with respect to the states $|0\rangle_1$ and $|1\rangle_1$ of $\mathcal{S}_1$ (and vice versa).

What does our definition of entanglement *mean*? If $|\Psi\rangle = |\psi\rangle_1 |\phi\rangle_2$, i.e., if $\mathcal{S}_1$ and $\mathcal{S}_2$ are *not* entangled, then we may regard the two subsystems $\mathcal{S}_1$ and $\mathcal{S}_2$ as individual entities. Each subsystem possesses its own quantum state, which constitutes a complete description of the physical state of the subsystem, and there exist no physical properties that could be measured only on the composite system $\mathcal{S}$ but could not be derived from measurements on the individual subsystems (such composite, or "global," properties would be embodied in *quantum correlations* between the subsystems). In other words, the subsystems, although considered as part of a larger composite system, completely retain their individuality. The whole (the composite system $\mathcal{S}$) is therefore simply the sum of the parts (the subsystems $\mathcal{S}_1$ and $\mathcal{S}_2$). This, of course, is the situation familiar from classical physics. Conversely, if there exist no subsystem states $|\psi\rangle_1$ of $\mathcal{S}_1$ and $|\phi\rangle_2$ of $\mathcal{S}_2$ such that $|\Psi\rangle = |\psi\rangle_1 |\phi\rangle_2$, all of the aforementioned "classical" features of separability fail to hold. Now the subsystems $\mathcal{S}_1$ and $\mathcal{S}_2$ cannot be attributed quantum states of their own; instead, they can only be described by a global composite quantum state (see also Fig. 1.2).

The term "entanglement" ("Verschränkung" in German) was first coined by Schrödinger in 1935 [2, 52, 53] who immediately emphasized the nonclassical implications of entanglement [52, p. 555]:

> When two systems, of which we know the states by their respective representatives, enter into temporary physical interaction due to known forces between them, and when after a time of mutual influence the systems separate again, then they can no longer be described in the same way as before, viz. by endowing each of them

> with a representative of its own. I would not call that *one* but rather *the* characteristic trait of quantum mechanics, the one that enforces its entire departure from classical lines of thought.

The peculiar features of entanglement have now been confirmed in many experiments, for instance, by using entangled pairs of photons separated by distances of many kilometers.[6] Entanglement is also at the heart of the "second quantum revolution" [55] that is concerned with quantum technologies such as quantum computers and quantum cryptography. Quantum computers would be able to solve certain computational problems faster than any classical computer could ever do (see Chap. 7), and quantum cryptography allows for completely secure communication. While it remains to be seen whether and when a reasonably complex quantum computer may be experimentally realizable, quantum cryptography already represents a comparably mature field with existing commercial applications.

2.3.1 Quantum Versus Classical Correlations

It is important to note that quantum correlations (i.e., entanglement) are fundamentally different from classical correlations.[7] In classical physics, correlations often arise due to certain conservation laws. For example, a particle at rest may decay into two identical fragments that, due to the conservation of total momentum, will then fly apart at the same speed but in opposite direction. If we measure the momentum of one of the fragments, we can therefore immediately infer that the momentum of the other particle must be equal in magnitude but of opposite sign. This is a purely classical (statistical) one-to-one correlation between the two particles: The momentum of each fragment "exists" independently of the measurement performed on the first fragment, and the inference of the momentum of the second particle follows directly from the conservation of momentum.

Now let us switch to the case of quantum entanglement. Suppose a pair of spin-$\frac{1}{2}$ particles is described by the Bell state [see (2.7b)]

$$|\Psi^+\rangle = \frac{1}{\sqrt{2}} \left(|0\rangle_1 |1\rangle_2 + |1\rangle_1 |0\rangle_2 \right). \tag{2.8}$$

The analogy with the case of classical correlations holds to the extent that, whenever we measure the first particle and find "spin up," we can immediately

[6]For prospects of extending such demonstrations into space to achieve separations of many thousands of kilometers for proof-of-principle experiments on entanglement, the Bell inequalities, and theories of wave-function collapse, see [54].

[7]Schrödinger himself seems to sometimes not have made this difference sufficiently clear in his early papers on entanglement. For instance, two of these papers bear the common title "Discussion of probability relations between separated systems" [52, 53], which fails to reflect the peculiar quantum nature of entanglement that goes far beyond (classical) probability relations.

infer that the second particle would, upon measurement, always be found in the "spin down" state.[8]

Thus we may say that, by the act of the first measurement, quantum correlations are transformed into classical (purely statistical) correlations. Note, however, that in the quantum setting the outcome of the measurement on the *first* particle is completely random. We will get "spin up" or "spin down" with equal probabilities, but we have no means of predicting which particular outcome will be obtained. It would therefore appear that information about the outcome of the measurement on the first particle is seemingly instantaneously transmitted to the second particle, which may be spatially separated by an arbitrary distance. (This observation is the basis of the EPR "paradox" [10].)

This would, at a first glance, seem to violate the principle of special relativity that no (classical) signal can travel faster than at the speed of light. However, there is in fact no such violation. Although the outcome of the measurement on the second particle is instantaneously fixed by the outcome of the measurement on the first particle, the complete randomness of the latter outcome means that no useful information can be transmitted between the two partners. (A more rigorous proof of this so-called *no-signaling theorem* can be given in terms of the statistics of measurements performed on the two subsystems; see, e.g., [56].) This is a rather remarkable result. Quantum mechanics *per se* is a nonrelativistic theory that does not contain any explicit axiom that would introduce fundamental limitations on the speed of transmission of information, that is to say, quantum mechanics does not "know" special relativity. Yet, its probabilistic structure nonetheless indirectly imposes no-signaling constraints such as the one described above.

A good way of thinking about such apparent instantaneous "spooky action at a distance" (as Einstein put it) is to realize that the entangled state must have been prepared *locally* by having let the two subsystems interact at some point in the past. When these two subsystems are then separated from each other, the quantum state is simply "spread out" (i.e., delocalized) over a larger spatial region. There is no information that would need to be physically transmitted between the two subsystems: The entangled state is already the most complete description for the individual *sub*systems and encapsulates all possible information about these systems.

This viewpoint is also reflected in Schrödinger's earliest papers on entanglement [2] of 1935 (quoted from the English translation by J. D. Trimmer [57]):

[8] To use the famous phrase of the original EPR paper [10], we can "predict with certainty" the outcome of the measurement on system 2. Intuitively, we would therefore conclude that the second system has been in the "down" state already *before* any explicit measurement on this system has confirmed the prediction. In essence, this viewpoint corresponds to EPR's (in)famous "criterion of reality."

> That a portion of the knowledge should float in the form of disjunctive conditional statements between the two systems can certainly not happen if we bring up the two from opposite ends of the world and juxtapose them without interaction. For then indeed the two "know" nothing about each other. A measurement on one cannot possibly furnish any grasp of what is to be expected of the other. Any "entanglement of predictions" that takes place can obviously only go back to the fact that the two bodies at some earlier time formed in a true sense one system, that is were interacting, and have left behind traces on each other.

The feature of entanglement seems to suggests that nature is fundamentally *nonlocal*: The outcome of the *local* measurement on the second particle is determined by quantum correlations encoded only in the *global* entangled quantum state of the composite system.

2.3.2 Quantification of Entanglement and Distinguishability

In our introduction of the four Bell states (2.7), we noted that these states are "maximally" entangled, suggesting the existence of different quantitative degrees of entanglement. This raises the question of how to quantify entanglement, i.e., of how to measure "how much" a given state is entangled. This question is important to decoherence. As we shall see, broadly speaking, the higher the degree of entanglement between the system of interest and its environment, typically the stronger will be the decohering effect of the environment.

Let us consider the situation of a bipartite entangled state of the form

$$|\Psi\rangle = \frac{1}{\sqrt{2}} \left(|\psi_1\rangle_1 |\phi_1\rangle_2 \pm |\psi_2\rangle_1 |\phi_2\rangle_2 \right), \tag{2.9}$$

where $|\psi_i\rangle_1$ and $|\phi_i\rangle_2$, $i = 1, 2$, are now arbitrary and not necessarily mutually orthogonal states of the subsystems $\mathcal{S}_1$ and $\mathcal{S}_2$. A useful intuitive way of quantifying the entanglement present in this state is to consider the following question: How much can the observer learn about one system by measuring the other system?

Let us have a look back at the Bell states (2.7). These states are maximally entangled, because a projective measurement on the system $\mathcal{S}_2$ in the $\{|0\rangle_2, |1\rangle_2\}$ basis immediately tells us in which of the states $|0\rangle_1$ and $|1\rangle_1$ we will find the system $\mathcal{S}_1$ in a subsequent measurement. This fact relies on two properties. First, the "relative" states $|0\rangle_2$ and $|1\rangle_2$ of $\mathcal{S}_2$ are one-to-one correlated with the states $|0\rangle_1$ and $|1\rangle_1$ of $\mathcal{S}_1$. Second, the states $|0\rangle_2$ and $|1\rangle_2$ are mutually orthogonal, i.e., they are perfectly distinguishable. We may think of these states as corresponding to some pointer of an apparatus that indicates the (relative) state of system $\mathcal{S}_1$. (In fact, this is precisely the basic idea underlying the von Neumann measurement scheme, which we shall discuss in Sect. 2.5.1 below.)

Now consider the state (2.9) and the limiting case in which the states $|\phi_1\rangle_2$ and $|\phi_2\rangle_2$ of $\mathcal{S}_2$ are far from being orthogonal, i.e., in which they have large overlap. It will therefore be very difficult to distinguish these two states in a projective measurement performed on $\mathcal{S}_2$. In turn, this implies that it also will be difficult to infer the corresponding relative state of $\mathcal{S}_1$ ($|\psi_1\rangle_1$ or $|\psi_2\rangle_1$) from this measurement. Broadly speaking, we may say that the system $\mathcal{S}_2$ encodes very little distinguishing "information" about $\mathcal{S}_1$ with respect to the states $|\psi_1\rangle_1$ and $|\psi_2\rangle_1$.

In the extreme situation of the states $|\phi_1\rangle_2$ and $|\phi_2\rangle_2$ being equal, $|\phi_1\rangle_2 = |\phi_2\rangle_2 \equiv |\phi\rangle_2$, we can write (2.9) as a product state,

$$|\Psi\rangle = \frac{1}{\sqrt{2}} \left(|\psi_1\rangle_1 \pm |\psi_2\rangle_1 \right) |\phi\rangle_2 \,. \tag{2.10}$$

Therefore, this state is no longer entangled and does not contain any quantum correlations between the two systems. If we now measure system $\mathcal{S}_2$ (in any basis!), we will not be able to infer anything about which of the states $|\psi_1\rangle_1$ and $|\psi_2\rangle_1$ we would expect to find upon a subsequent projective measurement performed on $\mathcal{S}_1$. The two subsystems are now in separate pure states, each of which completely specifies the physical state (or, put more epistemically, "all that can be known," e.g., in terms of measurement statistics) about each individual system, and the feature of "quantum holism" has disappeared. Finally, we can phrase the above argument in an analogous manner in regards to the correlation between the degree of entanglement and the overlap of the states $|\psi_1\rangle_1$ and $|\psi_2\rangle_1$ of $\mathcal{S}_1$.

As we shall soon see, this distinguishability of the states of one system correlated with the states of another system lies at the heart of an understanding of the conceptual basis of decoherence. Here, the subsystem $\mathcal{S}_2$ corresponds to an environment that encodes, via quantum correlations of the form (2.9), "information" about $\mathcal{S}_1$. Following our above argument, the amount of this information increases with the amount of system–environment entanglement, and thus also with the distinguishability of the relative states of the environment correlated with the different component states of the system. The larger the amount of this information about the system learned by the environment becomes, the more the system loses its individuality (in the sense of the inability to assign an individual quantum state to it). As a consequence, quantum coherence initially localized within the system will become a "shared property" of the composite system–environment state and can no longer be observed at the level of the system, leading to decoherence. We will make these ideas more precise in Sects. 2.6 and 2.7 below.

2.4 The Concept and Interpretation of Density Matrices

Density matrices, especially so-called reduced density matrices, play an important role in the formal description of decoherence. The main reason for this role can be traced back to the fact that entanglement between the system and the environment makes it impossible to assign an individual quantum state vector to the system (see the previous Sect. 2.3), and thus we typically cannot describe the system of interest in terms of such pure quantum states. As we shall see, reduced density matrices then provide an elegant method for representing the *measurement statistics* for the system. In the following, we will spend some time discussing the various aspects of pure-state, mixed-state, and reduced density matrices.

2.4.1 Pure-State Density Matrices and the Trace Operation

Let us begin with the familiar concept of quantum state vectors. As we have explained above, a quantum state vector $|\psi\rangle$ encapsulates maximum knowledge about the state of a physical system. We can also define the *density operator* $\hat{\rho}$ corresponding to such a pure state $|\psi\rangle$ as

$$\hat{\rho} \equiv |\psi\rangle\langle\psi|, \tag{2.11}$$

which is simply the projection operator onto the state $|\psi\rangle$. In this book, we shall interchangeably use the term *density matrix* for the density operator $\hat{\rho}$. Strictly speaking, the density matrix refers to the *matrix representation* of the density operator in a particular basis, but the use of "density matrix" for the operator $\hat{\rho}$ is so established in the literature that we shall follow this widely accepted terminology. In cases where we would like to explicitly refer to the matrix representation of the density operator, we will make the distinction clear by omitting the operator-indicating caret from $\hat{\rho}$, that is, by denoting such actual density *matrices* by ρ.

If we express $|\psi\rangle$ as a superposition of basis states $|\psi_i\rangle$,

$$|\psi\rangle = \sum_i c_i \, |\psi_i\rangle \,, \tag{2.12}$$

the corresponding density matrix written in this basis $\{|\psi_i\rangle\}$ reads

$$\hat{\rho} = |\psi\rangle\langle\psi| = \sum_{ij} c_i \, c_j^* |\psi_i\rangle\langle\psi_j|. \tag{2.13}$$

The terms $i \neq j$ on the right-hand side of this equation embody the quantum coherence between the different components $|\psi_i\rangle$. Accordingly, they are usually referred to as *interference terms*, or *off-diagonal terms* (since these terms correspond to the off-diagonal elements in the matrix representation of $\hat{\rho}$ in the basis $\{|\psi_i\rangle\}$).

However, it is important to keep in mind that such interference terms are always to be understood with respect to a particular basis $\{|\psi_i\rangle\}$, i.e., coherence and interference is present *between certain components* $|\psi_i\rangle$. There always exists a basis in which the density matrix becomes diagonal, and thus there will be no interference terms in this basis. Thus our association between interference (coherence) and the "quantumness" of a system must not lead us to the erroneous conclusion that the absence of interference terms from the density matrix written in some basis necessarily implies that the system "does not have quantum properties" or "behaves classically." The basis in which the density matrix takes this diagonal form may not at all correspond to the familiar determinate quantities of our experience (see also Sect. 2.15.1). Expressed in a different basis, interference terms will in general reappear, showing the persistent quantum coherence between these basis states.

Let us now introduce the *trace operation*, denoted by "Tr." This operation always acts on some operator $\hat{A}$ and is implemented in the following way. Choose an orthonormal basis $\{|\phi_i\rangle\}$ of the Hilbert space of the system, and perform the operation

$$\mathrm{Tr}(\hat{A}) \equiv \sum_i \langle\phi_i| \hat{A} |\phi_i\rangle . \tag{2.14}$$

It is easy to show that this operation is linear,

$$\mathrm{Tr}(\hat{A} + \hat{B}) = \mathrm{Tr}(\hat{A}) + \mathrm{Tr}(\hat{B}), \tag{2.15}$$

and that it is independent of the particular choice of the orthonormal basis $\{|\phi_i\rangle\}$. Thus we can use any arbitrary set of orthonormal basis vectors to compute the trace.

What is the reason for introducing the trace operation? To answer this question, let us consider the operator $\hat{A} = \hat{\rho}\hat{O}$, formed by the product of the pure-state density matrix (2.11) and a Hermitian operator $\hat{O}$ representing some observable that is measured on the system. Let us choose the eigenstates $|o_i\rangle$ of $\hat{O}$, with corresponding eigenvalues o_i (we shall assume a discrete spectrum here), as the orthonormal basis for evaluating the trace (2.14) of $\hat{\rho}\hat{O}$. This yields

$$\mathrm{Tr}(\hat{\rho}\hat{O}) = \sum_i \langle o_i| \left(|\psi\rangle\langle\psi|\right) \hat{O} |o_i\rangle = \sum_i o_i \left|\langle o_i|\psi\rangle\right|^2 . \tag{2.16}$$

But the term $|\langle o_i|\psi\rangle|^2$ is simply the Born probability of the outcome o_i in a measurement represented by $\hat{O}$.[9] Thus $\mathrm{Tr}(\hat{\rho}\hat{O})$ represents an average over

---

[9]The Born rule [19] can be stated as follows. Suppose a system is described by a pure state $|\psi\rangle$. Suppose further that an observable represented by a Hermitian operator $\hat{O}$, with eigenstates $|o_i\rangle$ and corresponding eigenvalues o_i (again assuming a discrete spectrum for $\hat{O}$), is measured on the system. Then:

(i) The states $|o_i\rangle$ and corresponding (eigen)values o_i are the only possible outcomes of the measurement.
(ii) The probability of finding the system in the state $|o_i\rangle$ after the measurement (or, put differently, of measuring the value o_i) is given by $|\langle o_i|\psi\rangle|^2$.

all possible outcomes o_i of this measurement, weighted by the corresponding Born probabilities. But this is precisely the definition of the *expectation value* $\langle \hat{O} \rangle$ of the observable $\hat{O}$. This connection between the (mathematical) procedure of the trace of the operator $\hat{\rho}\hat{O}$ and the (physical) concept of an expectation value of a measurement is known as the *trace rule*,[10]

$$\langle \hat{O} \rangle = \mathrm{Tr}\,(\hat{\rho}\hat{O}). \tag{2.17}$$

If we choose $\hat{O} = \hat{I}$, we get the expected result

$$\mathrm{Tr}\,\hat{\rho} = 1. \tag{2.18}$$

This relation simply reflects the fact that pure states are normalized, i.e., that $|\langle \psi|\psi \rangle|^2 = 1$. Before proceeding, we emphasize again that the concept and interpretation of the trace fundamentally relies on the Born rule. The importance of this point will become clear later.

2.4.2 Mixed-State Density Matrices

So far, using density matrices to describe the state of a system has not really yielded any advantage. If our system is in a completely known quantum state, then the descriptions of a system in terms of a (pure) quantum state $|\psi\rangle$ or of the corresponding (pure-state) density matrix (2.11) are completely equivalent, both formally and physically. However, we may also describe our system by a *mixed state*. A mixed state expresses insufficient information about the state of the system, in the sense that the system is (before the measurement) in one of the pure states $|\psi_i\rangle$ (which do not need to be orthogonal) but the observer simply does not know in which. Therefore we can only ascribe probabilities $p_i \geq 0$ to each of the states $|\psi_i\rangle$.[11]

Such a situation typically arises if the physical procedure used to prepare a quantum state contains a probabilistic element (albeit with known probabilities). For instance, the preparation device may be able to prepare one of two possible states $|\psi_1\rangle$ and $|\psi_2\rangle$ (Fig. 2.3). Which particular state is prepared is decided by a spin measurement on a spin-$\frac{1}{2}$ particle in an unknown quantum state, which will yield "spin up" and "spin down" with equal probabilities

---

Statement (i) is, in essence, the collapse postulate of quantum mechanics and hence often separated out from the Born rule, reducing the content of the Born rule to the prescription (ii) for calculating the actual values of the probabilities.

[10] Of course, the trace rule is completely equivalent to computing pure-state expectation values via $\langle \hat{O} \rangle = \langle \psi | \hat{O} | \psi \rangle$, as the reader can immediately confirm by evaluating the trace (2.16) using a set of orthonormal basis states that contains the state $|\psi\rangle$ of the system.

[11] For simplicity, we shall consider here the case of a finite-dimensional Hilbert space. The generalization to the case of infinitely many dimensions is rather straightforward (see, for example, [58]).

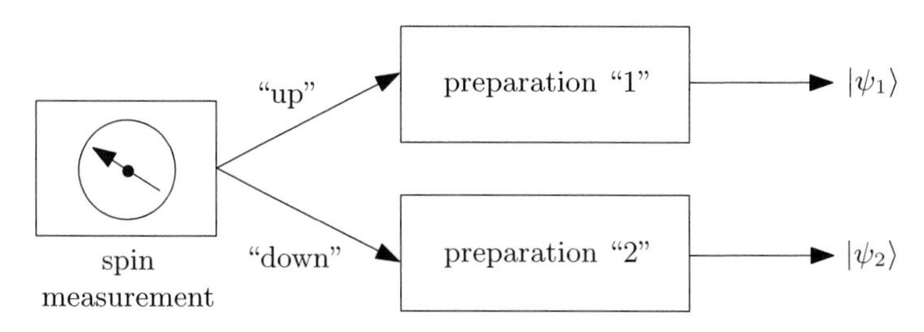

Fig. 2.3. Example for a probabilistic state-preparation procedure. Depending on the outcome of a spin measurement on a spin-$\frac{1}{2}$ particle, the device prepares a given system in either the quantum state $|\psi_1\rangle$ or the quantum state $|\psi_2\rangle$. The observer, ignorant of the particular outcome of the spin measurement, will assign equal (and classical!) probabilities to the states $|\psi_1\rangle$ and $|\psi_2\rangle$. To express this subjective ignorance, she will describe the prepared system by the mixed-state density matrix $\hat{\rho} = \frac{1}{2}|\psi_1\rangle\langle\psi_1| + \frac{1}{2}|\psi_2\rangle\langle\psi_2|$.

(such a measurement could be implemented, for example, using the Stern–Gerlach device described in Sect. 2.2.2). Then, depending on the outcome of this measurement, the device prepares (completely deterministically!) the system of interest in either the state $|\psi_1\rangle$ or the state $|\psi_2\rangle$. However, the observer does not inquire about the outcome of the spin measurement. She will therefore only know that either $|\psi_1\rangle$ or $|\psi_2\rangle$ has been prepared but not which of these two states. Of course, this is a somewhat artificial example, because the probabilistic character of the preparation procedure is here introduced deliberately. But it is easy to imagine other, more realistic situations in which the probabilistic element is inherent to the physical device used for state preparation. For example, the device may be imperfect and thus produce a range of possible pure states.

Regardless of the physical origin of the probabilistic element, the mixed state, i.e., the resulting set of possible pure states $|\psi_i\rangle$ with associated probabilities p_i, represents a *classical ensemble.* By this we mean to convey the notion that the origin of the probabilities is purely classical. These probabilities simply express the subjective ignorance of the observer about the quantum state of the system, while physically the system has indeed been prepared in a pure and thus completely known state (albeit not known to the observer). In principle, we could always completely follow or backtrack every step of the state-preparation procedure to determine which of the pure states $|\psi_i\rangle$ has been produced in each run of the procedure. The probabilities p_i simply express our practical decision not to inquire into the finer details of the preparation procedure, but there is nothing fundamental about these probabilities. Therefore they simply correspond to a "coarse-graining" approach as in classical statistical mechanics, where we cannot in practice (but could in principle) follow, say, the deterministic path of each individual molecule in a gas.

How would the observer describe the statistics of measurements performed on a system described by a mixed state? Clearly, since she is no longer able to assign a pure quantum state $|\psi\rangle$ to the system, she cannot compute expectation values of observables $\hat{O}$ via the usual rules $\langle \hat{O} \rangle = \langle \psi | \hat{O} | \psi \rangle$ or $\langle \hat{O} \rangle = \mathrm{Tr}\,(\hat{\rho}\hat{O})$ [see (2.17)] with a pure-state density matrix $\hat{\rho} = |\psi\rangle\langle\psi|$. However, these rules are easily generalized by combining the classical probability concept (arising from the ignorance of the observer about the prepared pure state of the system) with the intrinsic quantum-mechanical probabilities (arising from the probabilistic "collapse" of the quantum state into an eigenstate of the measured observable). The intuitive idea consists of simply weighting the expectation values $\langle \psi_i | \hat{O} | \psi_i \rangle$ for each of the possible pure states $|\psi_i\rangle$ contained in the mixed state by their respective classical probabilities p_i of the $|\psi_i\rangle$ and sum the results over the entire ensemble, i.e.,

$$\langle \hat{O} \rangle = \sum_i p_i \langle \psi_i | \hat{O} | \psi_i \rangle . \tag{2.19}$$

This expression still contains two separate statistical elements, namely, the classical probabilities p_i as well as the quantum expectation values $\langle \psi_i | \hat{O} | \psi_i \rangle$ for each of the pure states $|\psi_i\rangle$. It turns out we can introduce a *mixed-state density matrix* that encapsulates both of these parts and thus completely encodes all statistical properties of the system. This density matrix is given by

$$\hat{\rho} = \sum_i p_i |\psi_i\rangle\langle\psi_i|, \tag{2.20}$$

with $p_i \geq 0$ and $\sum_i p_i = 1$. We can view this density matrix as a classical probability distribution of pure-state density matrices $\hat{\rho}_i = |\psi_i\rangle\langle\psi_i|$.

Since the classical probability concept is already built into the mixed-state density matrix (2.20), expectation values of observables $\hat{O}$ can now be computed in exactly the same way as for pure states by using the trace rule (2.17), i.e.,

$$\langle \hat{O} \rangle = \mathrm{Tr}\,(\hat{\rho}\hat{O}). \tag{2.21}$$

As it should be, this is equivalent to the method (2.19) of computing the expectation values in the pure component states $|\psi_i\rangle$ and weighting them by the mixed-state probabilities p_i.

Just as the pure-state density matrix, the mixed-state density matrix also obeys the normalization condition (2.18),

$$\mathrm{Tr}\,\hat{\rho} = 1. \tag{2.22}$$

This follows directly from the normalization (2.18) of each constituent pure-state density matrix $\hat{\rho}_i = |\psi_i\rangle\langle\psi_i|$ and the fact that $\sum_i p_i = 1$,

$$\mathrm{Tr}\,\hat{\rho} = \sum_i p_i \mathrm{Tr}\,(|\psi_i\rangle\langle\psi_i|) \overset{(2.18)}{=} \sum_i p_i = 1. \tag{2.23}$$

As discussed in Sect. 2.3, entanglement between two subsystems corresponds to the inability of writing the composite pure state $|\Psi\rangle$ as a tensor product $|\Psi\rangle = |\psi\rangle_1 |\phi\rangle_2$ of pure subsystem states $|\psi\rangle_1$ and $|\phi\rangle_2$. We may now restate this situation equivalently in the language of density matrices. The two subsystems are entangled with each other if the composite system is described by a (pure-state or mixed-state) density matrix $\hat{\rho}$ that cannot be written in the tensor-product form $\hat{\rho} = \hat{\rho}_1 \otimes \hat{\rho}_2$, where $\hat{\rho}_1$ and $\hat{\rho}_2$ are density matrices pertaining to the two subsystems. This criterion follows directly from the above definition of entanglement in terms of state vectors. Thus, whenever the total density matrix factorizes into subsystem density matrices, $\hat{\rho} = \hat{\rho}_1 \otimes \hat{\rho}_2 \otimes \cdots$, there exist no quantum correlations between these subsystems.

Finally, at the risk of overstating an important point, we emphasize that a mixed state must be clearly distinguished from a pure-state superposition of the form

$$|\psi\rangle = \sum_i \sqrt{p_i}\,|\psi_i\rangle\,. \tag{2.24}$$

As discussed in Sect. 2.2.1, here all component states $|\psi_i\rangle$ are simultaneously present, which can (at least in principle) always be experimentally verified. There is no *a priori* probabilistic element contained in this superposition: $|\psi\rangle$ is a pure state and therefore encapsulates maximum knowledge about the system. The density matrix corresponding to the superposition (2.24) is

$$\begin{aligned}\hat{\rho} = |\psi\rangle\langle\psi| &= \sum_{ij} \sqrt{p_i p_j} |\psi_i\rangle\langle\psi_j| \\ &= \sum_i p_i |\psi_i\rangle\langle\psi_i| + \sum_{i\neq j} \sqrt{p_i p_j} |\psi_i\rangle\langle\psi_j|.\end{aligned} \tag{2.25}$$

The presence of the off-diagonal terms $i \neq j$ (that represent interference between different states $|\psi_i\rangle$) clearly distinguishes this pure-state density matrix from the mixed-state density matrix (2.20).

2.4.3 Quantifying the Degree of "Mixedness"

Since a pure-state density matrix $\hat{\rho} = |\psi\rangle\langle\psi|$ is simply the projection operator onto the pure state $|\psi\rangle$, it immediately follows that

$$\hat{\rho}^2 = \hat{\rho}. \tag{2.26}$$

This projection-operator property $\hat{\rho}^2 = \hat{\rho}$ in fact constitutes a necessary and sufficient condition for the system to be in a pure state, as we can see from the following argument. Suppose the system is not in a pure state but is instead described by the mixed-state density matrix (2.20), with at least two non-zero probabilities p_i. For this density matrix, we obtain

$$\hat{\rho}^2 = \sum_{ij} p_i p_j |\psi_i\rangle\langle\psi_j|\langle\psi_i|\psi_j\rangle. \tag{2.27}$$

First, in general $\langle\psi_i|\psi_j\rangle \neq 0$ for $i \neq j$, but even if $\langle\psi_i|\psi_j\rangle = 0$ for $i \neq j$, we nonetheless end up with

$$\hat{\rho}^2 = \sum_i p_i^2 |\psi_i\rangle\langle\psi_i|. \tag{2.28}$$

This expression differs from $\hat{\rho}$, see (2.20), since the value of each p_i is less than one. Thus, if we are given some arbitrary density matrix $\hat{\rho}$, we can easily determine whether the system is in a pure state by simply checking whether (2.26) holds, i.e., whether $\hat{\rho}^2 = \hat{\rho}$.

We can go one step further and define measures that not only tell us whether a given density matrix is pure, but also quantify the degree of "mixedness" of this density matrix in the following sense. Consider again the mixed-state density matrix as in (2.20),

$$\hat{\rho} = \sum_i p_i |\psi_i\rangle\langle\psi_i|. \tag{2.29}$$

As we know by now, the case of a pure-state density matrix corresponds to all p_i being equal to zero except for one, say $p_1 = 1$. That is, there is no "ignorance" about the state of the system in this situation. Suppose now we decrease the magnitude of p_1. The normalization condition for probabilities, $\sum_i p_i = 1$, tells us that now at least one other $p_i \neq p_1$ must attain a non-zero value, say, $p_2 > 0$. Now there is a degree of "ignorance" present, as we do not know with certainty in which of the pure states $|\psi_1\rangle$ and $|\psi_2\rangle$ the system has been prepared. Nonetheless, if, for example, $p_1 = 90\%$ and $p_2 = 10\%$, we can still be reasonably confident that the state $|\psi_1\rangle$ has indeed been prepared. Accordingly, the degree of "mixedness" is low. The opposite extreme of the pure state is attained if the density matrix is proportional to the identity operator, which corresponds to an equal-weight ensemble of mutually orthogonal pure states which form a basis of the Hilbert space of the system. In this case the density matrix contains no information whatsoever about which pure state has been prepared and therefore expresses a maximum degree of ignorance ("mixedness").

Let us now introduce two measures that quantify the degree of mixedness in a more general way. A simple and commonly used measure is the so-called *purity* of the density matrix, defined as

$$\varsigma \equiv \mathrm{Tr}\,(\hat{\rho}^2). \tag{2.30}$$

Why is this a sensible measure? If $\hat{\rho}$ represents a pure-state density matrix, it follows from (2.18) and (2.26) that $\varsigma = 1$. On the other hand, assuming that the states $\{|\psi_i\rangle\}$ form an orthonormal basis of the N-dimensional Hilbert space of the system, the mixed-state density matrix (2.20) has purity [see (2.28)]

$$\varsigma = \sum_{i=1}^{N} p_i^2. \tag{2.31}$$

The sum is bounded from below by the value $1/N$, which is attained precisely if the density matrix is maximally mixed, i.e., if $p_i = 1/N$ for all $i = 1, \ldots, N$.

Another commonly used measure for quantifying the (im)purity of the density matrix is the so-called *von Neumann entropy.* This measure was first introduced in 1927 by the mathematician John von Neumann [59], who subsequently developed it further in his monumental book on quantum mechanics [60] (see also [61] for some interesting historical remarks). The von Neumann entropy can be viewed as a generalization of the notion of entropy in classical statistical mechanics to the case of quantum-mechanical density operators. It is given by

$$S(\hat{\rho}) \equiv -\mathrm{Tr}\,(\hat{\rho}\log_2\hat{\rho}) \equiv -\sum_i \lambda_i \log_2 \lambda_i, \tag{2.32}$$

where the λ_i denote the eigenvalues of the density matrix $\hat{\rho}$. By convention, the case $\lambda_i = 0$ is handled by defining $0\log_2(0) \equiv 0$, such that states absent from a mixture do not enter into the value of the entropy (as it should be).

Let us now again determine some explicit values for this measure. If $\hat{\rho}$ is pure, then all $\lambda_i = 0$ except for one (which must therefore take on a value of one), and thus $S(\hat{\rho}) = 0$. Thus a pure state is characterized by a value of zero of the von Neumann entropy. On the other hand, for a maximally mixed state that corresponds to complete ignorance about which of the mutually exclusive (and thus orthogonal) pure states $|\psi_i\rangle$, $i = 1, \ldots, N$, has been prepared, we have $\lambda_i = p_i = 1/N$ and thus

$$S(\hat{\rho}) = \log_2(N), \tag{2.33}$$

which is the maximum value that $S(\hat{\rho})$ can take. For the intermediate case of non-maximally mixed states, the value is correspondingly lower. For example, if the preparation device prepares only a subset of orthogonal states $|\psi_i\rangle$, $i = 1, \ldots, M < N$, with equal probabilities, we obtain $S(\hat{\rho}) = \log_2(M)$. These values show that the definition of the von Neumann entropy matches our intuition about entropy in the classical (i.e., thermodynamic) setting, where entropy is a measure of the amount of information—or, conversely, of ignorance—about the state of the system, typically quantified by the number of different possible states available to the system.

2.4.4 The Basis Ambiguity of Mixed-State Density Matrices

In Sect. 2.4.2 above we discussed how the notion of a mixed state is based on a classical probability concept. Accordingly, one also says that a mixed-state density matrix (2.20) represents an *ignorance-interpretable* (proper) mixture of pure states [47–49],[12] in order to express the fact that a mixed-state density matrix of the form (2.20) can, to some extent, be interpreted as a classical probability distribution of pure quantum states $|\psi_i\rangle$. However, this is only true if we actually *know* that the system has indeed been prepared in one of the states $|\psi_i\rangle$, but we simply do not possess more specific information about which of these states has been prepared. On the other hand, if we are simply confronted with the density matrix (2.20) but are given no further information (e.g., about the preparation procedure), we cannot infer that the system actually is in one of the states $|\psi_i\rangle$. This is so because any nonpure density matrix can be written in many different ways, which shows that any partition into a particular ensemble of quantum states is arbitrary. In other words, the mixed-state density matrix alone does not suffice to uniquely reconstruct a classical probability distribution of pure states.

To illustrate this fact, let us consider the mixed-state density matrix

$$\hat{\rho} = \frac{1}{2}|0_z\rangle\langle 0_z| + \frac{1}{2}|1_z\rangle\langle 1_z|. \tag{2.34}$$

We would use such a density matrix to describe a system that has been prepared in either one of the eigenstates $|0_z\rangle$ and $|1_z\rangle$ of the Pauli spin operator $\hat{\sigma}_z$ (with equal likelihoods), but we simply do not know in which and must therefore resort to a description in terms of a classical probability distribution of these states.

On the other hand, the states $|0_z\rangle$ and $|1_z\rangle$ can also be rewritten in terms of the eigenstates $|0_x\rangle$ and $|1_x\rangle$ of the Pauli spin operator $\hat{\sigma}_x$ for spin along along the x axis,

$$\begin{aligned} |0_z\rangle &= \frac{1}{\sqrt{2}}\left(|0_x\rangle + |1_x\rangle\right), \\ |1_z\rangle &= \frac{1}{\sqrt{2}}\left(|0_x\rangle - |1_x\rangle\right). \end{aligned} \tag{2.35}$$

Using these expressions, the density matrix (2.34) can be equivalently written as

$$\hat{\rho} = \frac{1}{2}|0_x\rangle\langle 0_x| + \frac{1}{2}|1_x\rangle\langle 1_x|. \tag{2.36}$$

In fact, $\hat{\rho}$ can be rewritten in this form in infinitely many ways in terms of spin eigenstates along *any* axis. Therefore, unless we know the physical axis along which the spin state of the system has been prepared, the density matrix alone provides us only with information about the probabilities of different sets of pure states (here these sets would be represented by $\{|0_z\rangle, |1_z\rangle\}$, $\{|0_x\rangle, |1_x\rangle\}$, etc.), but not about which particular set of states has been prepared.

[12] This classification will become important when introducing reduced density matrices (see Sect. 2.4.6), which will turn out to be *not* ignorance-interpretable.

2.4.5 Mixed-State Density Matrices Versus Physical Ensembles

An important remark is in order here. A mixed-state density matrix of the form (2.20), $\hat{\rho} = \sum_i p_i |\psi_i\rangle\langle\psi_i|$, is also often used to describe a *physical ensemble* of identical systems $\mathcal{S}_k$, $k = 1 \dots N$ (that is, a collection of "copies" of a particular system of interest), each of which is prepared, in a deterministic manner, in one of the pure states $|\psi_i\rangle$, $i = 1, \dots, M \leq N$. The quantity p_i now denotes the fraction of systems that have been prepared in the state $|\psi_i\rangle$. This "ensemble" approach is frequently found in textbooks when motivating the introduction of the density-matrix concept (see, e.g., [58]).

However, there is an important distinction to be made between (physical) ensembles and (statistical) mixtures. In the case of ensembles, there is no insufficiency of information whatsoever that would justify the description by a mixed-state density matrix (unless, of course, each individual system is *not* prepared in a pure state). Instead, the ensemble of all individual systems $\mathcal{S}_k$ must rather be described by a Hilbert space $\mathcal{H}$ that is the tensor product of the N constituent state-spaces $\mathcal{H}_k$ of each system $\mathcal{S}_k$,

$$\mathcal{H} = \bigotimes_{k=1}^{N} \mathcal{H}_k. \tag{2.37}$$

Similarly, the total system $\mathcal{S}$ is described by a *pure* (tensor) product state $|\Psi\rangle$ of the individual states $|\psi_k\rangle \in \{|\psi_i\rangle\}$,

$$|\Psi\rangle = \left(\prod_{j=1}^{Np_1} |\psi_1\rangle\right)\left(\prod_{j=1}^{Np_2} |\psi_2\rangle\right)\cdots\left(\prod_{j=1}^{Np_M} |\psi_M\rangle\right). \tag{2.38}$$

The corresponding N-system pure density matrix $\hat{\rho} = |\Psi\rangle\langle\Psi|$ is therefore manifestly different from the expression (2.20) for the mixed-state density matrix that describes a probability distribution of *single-system* pure states $|\psi_i\rangle$.

Only in the limited and purely statistical sense of an *ensemble average* can the description of ensembles of physical systems by a single-system mixed-state density matrix be justified. This is the case if we restrict ourselves to measurements on *single* systems in the ensemble and consider the average statistics for the entire ensemble. Specifically, if we measure a single-system observable $\hat{O}$ on every member in the ensemble described by the N-system pure state (2.38) and average the results over the ensemble, we obtain

$$\langle\hat{O}\rangle = \frac{1}{N}\sum_{i=1}^{M} Np_i \langle\psi_i|\,\hat{O}\,|\psi_i\rangle, \tag{2.39}$$

where we have for simplicity assumed that the $|\psi_i\rangle$ form an orthonormal set of basis states for each individual system. If we instead compute the expectation value of $\hat{O}$ for the mixed state $\hat{\rho}$ [see (2.20)] by using the trace rule, we get the same result,

$$\langle \hat{O} \rangle = \sum_{i=1}^{M} p_i \mathrm{Tr}(\hat{\rho}\hat{O}) = \sum_{i=1}^{M} p_i \langle \psi_i | \hat{O} | \psi_i \rangle . \tag{2.40}$$

Thus, for single-system observables, in the ensemble average (and only in this average!) the statistics generated by the pure-state density matrix corresponding to the N-system pure state $|\Psi\rangle$ will be identical to those obtained from the single-system mixed-state density matrix (2.20). By contrast, the proper pure-state N-system description (2.38) of the ensemble allows one to compute the statistics for any arbitrary observable of the ensemble of systems and for each individual system.

2.4.6 Reduced Density Matrices

Reduced density matrices play an important role in the formal description of decoherence and date back to the early years of quantum mechanics [60, 62, 63] (see also [64] for some interesting remarks). The basic motivation underlying the concept of reduced density matrices is the description of a quantum system $\mathcal{A}$ that is quantum-correlated (i.e., entangled) with another system $\mathcal{B}$. In this case, the quantum state of the total combined system $\mathcal{AB}$ may well be pure (and therefore in principle be completely known). However, suppose that the observer only has access to the first system $\mathcal{A}$, i.e., she can perform measurements only on $\mathcal{A}$ but not on $\mathcal{B}$. Everything that can be known about the state of the composite system must therefore be derived from such *local measurements* on $\mathcal{A}$ only, which will yield the possible measurement outcomes for this system and their probability distribution. The key question is then: What is the suitable mathematical object in the quantum-mechanical formalism that contains, exhaustively and correctly, all information (i.e., all measurement statistics) that can be extracted by the observer of system $\mathcal{A}$?

It turns out that this object is the *reduced density matrix* given by

$$\hat{\rho}_{\mathcal{A}} \equiv \mathrm{Tr}_{\mathcal{B}}\, \hat{\rho}. \tag{2.41}$$

Here the subscript "$\mathcal{B}$" means that the trace is to be performed using an orthonormal basis of the Hilbert space $\mathcal{H}_{\mathcal{B}}$ of $\mathcal{B}$ only. Accordingly, the operation "$\mathrm{Tr}_{\mathcal{B}}$" is also referred to as the *partial trace* over $\mathcal{B}$ and may be interpreted as an "averaging" over the degrees of freedom of the unobserved system $\mathcal{B}$ (this interpretation will become more clear later). Equation (2.41) therefore implies that the measurement statistics for all observables pertaining only to system $\mathcal{A}$ are completely contained in the reduced density matrix $\hat{\rho}_{\mathcal{A}}$ obtained by "tracing out" the degrees of freedom of $\mathcal{B}$ (which is quantum-correlated with $\mathcal{A}$).

Relevance to Decoherence

Before deriving (2.41), let us first discuss the importance of the concept of reduced density matrices for the description of decoherence. Recall that decoherence arises from interactions between two systems, namely, the "system of interest" and its environment. Typically, such interactions will then lead to an entangled state for the system–environment combination. Usually, the observer will perform measurements only on the system of interest, whereas the environment is typically either inaccessible, cannot be completely measured, or is simply of no interest. As an example, we may consider the environment of photons scattering off an object. In practice, it will usually be impossible to intercept all of these scattered photons, and thus the observer will only be able to measure observables that pertain to the system and a fraction of the environmental photons.

This is where the reduced density matrix comes into play. By tracing over (all, or a fraction of) the degrees of freedom of the environment of the system–environment density matrix, we obtain a complete and exhaustive description of the measurement statistics for our system of interest in terms of the reduced density matrix of the system. All influences of the environment on local measurements performed on the system will automatically be encapsulated in this reduced density matrix. Since the system is entangled with its environment, no individual quantum state can be attributed to the system itself. Therefore, the reduced density matrix is all we have available to describe the statistics of measurements on the system,[13] and this reduced density matrix is necessarily nonpure due to the presence of system–environment entanglement.

Derivation of the Reduced Density Matrix

Let us now derive the expression (2.41) for the reduced density matrix. To do so, let us consider an entangled state of two systems $\mathcal{A}$ and $\mathcal{B}$ of the form

$$|\Psi\rangle = \frac{1}{\sqrt{2}} \left(|a_1\rangle |b_1\rangle + |a_2\rangle |b_2\rangle \right), \tag{2.42}$$

where $|a_i\rangle$ and $|b_i\rangle$, $i = 1, 2$, are arbitrary normalized (but not necessarily orthogonal) states of $\mathcal{A}$ and $\mathcal{B}$, respectively. From (2.11), the corresponding pure-state density matrix is

$$\hat{\rho} = |\Psi\rangle\langle\Psi| = \frac{1}{2} \sum_{ij=1}^{2} |a_i\rangle\langle a_j| \otimes |b_i\rangle\langle b_j|. \tag{2.43}$$

[13] The so-called "envariance" program, recently developed Zurek [16, 65–67] and outlined in Sect. 8.2.2, uses symmetries of entangled states to provide a framework that allows one to discuss (given a set of assumptions) the state of, and measurements on, a system entangled with another system without the use of reduced density matrices (and thus without presuming the Born rule).

Also, let $\{|\psi_k\rangle\}$ and $\{|\phi_l\rangle\}$ be orthonormal bases of the Hilbert spaces $\mathcal{H}_\mathcal{A}$ and $\mathcal{H}_\mathcal{B}$ of $\mathcal{A}$ and $\mathcal{B}$.

Let us now consider observables acting on system $\mathcal{A}$ only, which can be written as $\hat{O} = \hat{O}_\mathcal{A} \otimes \hat{I}_\mathcal{B}$, where $\hat{I}_\mathcal{B}$ is the identity operator in the Hilbert space of $\mathcal{B}$. Following our question posed earlier, we would like to investigate whether there exists a mathematical object which is more simple than the bipartite density matrix (2.43) but that will nonetheless allow us to compute the expectation values of all such $\mathcal{A}$-observables.

As we know, the expectation value $\langle \hat{O} \rangle$ of any observable can be computed using the standard trace rule (2.17), $\langle \hat{O} \rangle = \mathrm{Tr}\,(\hat{\rho}\hat{O})$. Since now $\hat{O} = \hat{O}_\mathcal{A} \otimes \hat{I}_\mathcal{B}$, the part of the trace operation pertaining to system $\mathcal{B}$ can immediately be carried out. Explicitly, we then obtain

$$\begin{aligned}
\langle \hat{O} \rangle &= \mathrm{Tr}\,(\hat{\rho}\hat{O}) \\
&= \sum_{kl} \langle \phi_l| \langle \psi_k| \hat{\rho} \left(\hat{O}_\mathcal{A} \otimes \hat{I}_\mathcal{B} \right) |\psi_k\rangle |\phi_l\rangle \\
&= \sum_{k} \langle \psi_k| \left(\sum_{l=1} \langle \phi_l| \hat{\rho} |\phi_l\rangle \right) \hat{O}_\mathcal{A} |\psi_k\rangle \\
&= \sum_{k} \langle \psi_k| (\mathrm{Tr}_\mathcal{B}\, \hat{\rho})\, \hat{O}_\mathcal{A} |\psi_k\rangle \\
&\equiv \sum_{k} \langle \psi_k| \hat{\rho}_\mathcal{A} \hat{O}_\mathcal{A} |\psi_k\rangle \\
&= \mathrm{Tr}_\mathcal{A} \left(\hat{\rho}_\mathcal{A} \hat{O}_\mathcal{A} \right),
\end{aligned} \tag{2.44}$$

where $\hat{\rho}_\mathcal{A}$ is precisely the reduced density matrix introduced in (2.41).

The concept of reduced density matrices can be generalized from bipartite entangled states to any pure state $|\psi\rangle$ describing entanglement between N subsystems. Consider an observable $\hat{O}$ that pertains only to system i,

$$\hat{O} = \hat{I}_1 \otimes \hat{I}_2 \otimes \cdots \otimes \hat{I}_{i-1} \otimes \hat{O}_i \otimes \hat{I}_{i+1} \otimes \cdots \otimes \hat{I}_N. \tag{2.45}$$

The measurement statistics of $\hat{O}$ generated by applying the trace rule will then be identical regardless of whether we use the pure-state density matrix $\hat{\rho} = |\Psi\rangle\langle\Psi|$ of the composite system containing all N subsystems, or the reduced density matrix

$$\hat{\rho}_i = \mathrm{Tr}_{1,\ldots,i-1,i+1,\ldots,N}\,(\hat{\rho}) \tag{2.46}$$

of the ith subsystem (obtained by tracing over all subsystems except for the ith subsystem). This is so because it is easy to show that

$$\langle \hat{O} \rangle = \mathrm{Tr}(\hat{\rho}\hat{O}) = \mathrm{Tr}_i(\hat{\rho}_i\hat{O}_i). \tag{2.47}$$

Local Measurability of Interference and Distinguishability

For our bipartite state (2.42), we can easily evaluate $\hat{\rho}_{\mathcal{A}}$ [see (2.44)] from the expression (2.43) for the full density matrix $\hat{\rho}$. Expanding the states $|b_i\rangle$, $i = 1, 2$, of $\mathcal{B}$ in terms of the set $\{|\phi_l\rangle\}$ of orthonormal basis vectors of $\mathcal{H}_{\mathcal{B}}$,

$$|b_i\rangle = \sum_l c_l^{(i)} |\phi_l\rangle , \tag{2.48}$$

we obtain

$$\begin{aligned}
\hat{\rho}_{\mathcal{A}} &= \mathrm{Tr}_{\mathcal{A}} \left(\frac{1}{2} \sum_{ij=1}^{2} |a_i\rangle\langle a_j| \otimes |b_i\rangle\langle b_j| \right) \\
&= \frac{1}{2} \sum_{ij=1}^{2} |a_i\rangle\langle a_j| \sum_k \langle \phi_k| \left(\sum_{ll'} c_l^{(i)} \left(c_{l'}^{(j)} \right)^* |\phi_l\rangle\langle \phi_{l'}| \right) |\phi_k\rangle \\
&= \frac{1}{2} \sum_{ij=1}^{2} |a_i\rangle\langle a_j| \sum_k c_k^{(i)} \left(c_k^{(j)} \right)^* \\
&= \frac{1}{2} \sum_{ij=1}^{2} |a_i\rangle\langle a_j| \langle b_j|b_i\rangle \\
&= \frac{1}{2} \left(|a_1\rangle\langle a_1| + |a_2\rangle\langle a_2| + |a_1\rangle\langle a_2|\langle b_2|b_1\rangle + |a_2\rangle\langle a_1|\langle b_1|b_2\rangle \right) .
\end{aligned} \tag{2.49}$$

This result can be easily generalized to the case in which the composite system $\mathcal{AB}$ is not described by the two-component state (2.42) but instead by the more general N-component state $|\Psi\rangle = \frac{1}{\sqrt{N}} \sum_{n=1}^{N} |a_n\rangle |b_n\rangle$ with $N > 2$. Then the resulting reduced density matrix takes the form

$$\hat{\rho}_{\mathcal{A}} = \frac{1}{N} \sum_{ij=1}^{N} |a_i\rangle\langle a_j| \langle b_j|b_i\rangle . \tag{2.50}$$

We see from (2.49) that the influence of the system $\mathcal{B}$ on the measurement statistics is now effectively subsumed in the overlap $\langle b_2|b_1\rangle = \langle b_1|b_2\rangle^*$ of the $\mathcal{B}$-states $|b_1\rangle$ and $|b_2\rangle$ that multiplies the off-diagonal terms $|a_1\rangle\langle a_2|$ and $|a_2\rangle\langle a_1|$ in the reduced density matrix. As we have mentioned in Sect. 2.4.2, these off-diagonal terms correspond to interference between the states $|a_1\rangle$ and $|a_2\rangle$.

Thus we obtain a very important result: The amount of overlap of the relative states $|b_1\rangle$ and $|b_2\rangle$ of $\mathcal{B}$ that are one-to-one correlated with the states $|a_1\rangle$ and $|a_2\rangle$ of $\mathcal{A}$ [see (2.42)] quantifies the degree of interference in the $\{|a_1\rangle, |a_2\rangle\}$ basis that can be measured on $\mathcal{A}$. This immediately connects with our earlier observation in Sect. 2.3.2. There, we had introduced an intuitive notion of the degree of entanglement present in a bipartite pure state by referring to the amount of information encoded in one subsystem about the corresponding (relative) states of the other subsystem. We had related this measure to the distinguishability of the relevant states of one of the subsystems, quantified by the mutual overlap of these states. We thus see the intimate connection between the degree of entanglement, the amount of distinguishability, and the extent to which interference can be observed by performing measurements on only one of the subsystems.

For example, in the limiting case of vanishing overlap (and thus perfect distinguishability) of the states $|b_1\rangle$ and $|b_2\rangle$, the reduced density matrix (2.49) becomes diagonal in the $\{|a_1\rangle, |a_2\rangle\}$ basis,

$$\hat{\rho}_{\mathcal{A}} = \frac{1}{2}\left(|a_1\rangle\langle a_1| + |a_2\rangle\langle a_2|\right). \tag{2.51}$$

Since the off-diagonal (interference) terms $|a_1\rangle\langle a_2|$ and $|a_2\rangle\langle a_1|$ are now absent, there is no *local* observable $\hat{O} = \hat{O}_{\mathcal{A}} \otimes \hat{I}_{\mathcal{B}}$ that would allow us to measure interference between the states $|a_1\rangle$ and $|a_2\rangle$.

Reduced Density Matrices Versus Ensembles

An important remark is in order here. Evidently, the reduced density matrix (2.51) is *formally* identical to the density matrix that would be obtained if system $\mathcal{A}$ were described by a proper mixture. Recall that such a proper mixture would correspond to a situation in which the system is in either one of the two pure states $|a_1\rangle$ and $|a_2\rangle$ with equal probabilities, as opposed to the global entangled superposition (2.42) in which both components $|a_1\rangle$ and $|a_2\rangle$ are present (which could, at least in principle, always be confirmed by suitable interference experiments; see Sect. 2.2.1).

The formal identity between the reduced density matrix (2.51) (arising from a tracing-out of the degrees of freedom of system $\mathcal{B}$ that is entangled with system $\mathcal{A}$) and a mixed-state density matrix implies that a measurement of an observable that only pertains to system $\mathcal{A}$ cannot discriminate between the two cases, pure vs. mixed state.[14] However, it is of crucial importance to understand that this formal identity must not be interpreted as implying that the state of the system can be viewed as mixed too (see also the discussions by d'Espagnat [47–49]). In general, density matrices are only a calculational tool for computing the probability distribution of a set of possible outcomes of measurements, but they do not specify the *state* of the system. Since the two systems $\mathcal{A}$ and $\mathcal{B}$ are entangled and the total composite system is still described by the superposition (2.42), it follows from the

[14] One can show (see, e.g., pp. 208–210 of [68]) that this inability to operationally discern mixed from pure states is in fact not just a consequence of the restriction to local measurements, but also holds for any observable of the *composite* system that factorizes into the form $\hat{O} = \hat{O}_{\mathcal{A}} \otimes \hat{O}_{\mathcal{B}}$, where $\hat{O}_{\mathcal{A}}$ and $\hat{O}_{\mathcal{B}}$ do not commute with the projection operators $|a_i\rangle\langle a_i|$ and $|b_i\rangle\langle b_i|$, $i = 1, 2$, in the Hilbert spaces of $\mathcal{A}$ and $\mathcal{B}$, respectively.

standard rules of quantum mechanics that no individual definite state can be attributed to either one of the subsystems. Reduced density matrices of entangled subsystems therefore represent *improper mixtures* [47–49].

The fact that the reduced density matrix may be formally similar to a mixed-state density matrix thus cannot be used to argue that somehow, magically, a definite subsystem state—i.e., $|a_1\rangle$ or $|a_2\rangle$ with equal probabilities, as it would appear from the reduced density matrix (2.51)—has been obtained from the global entangled superposition state (2.42) by means of the (physical) interaction with system $\mathcal{B}$ and the (formal) trace operation. This observation will turn out to be important when discussing the implications of decoherence for the quantum measurement problem (see the next Sect. 2.5). We will also come back to this issue in Chap. 8 (see especially Sect. 8.1).

2.5 The Measurement Problem and the Quantum-to-Classical Transition

In this section, we shall describe the (in)famous *measurement problem* of quantum mechanics that we have already referred to in several places in the text. The choice of the term "measurement problem" has purely historical reasons: Certain foundational issues associated with the measurement problem were first illustrated in the context of a quantum-mechanical description of a measuring apparatus interacting with a system.

However, one may regard the term "measurement problem" as implying too a narrow scope, chiefly for the following two reasons. First, as we shall see below, the measurement problem is composed of three distinct issues, so it would make sense to rather speak of measurement *problems*. Second, quantum measurement and the arising foundational problems are but a special case of the more general problem of the *quantum-to-classical transition*, i.e., the question of how effectively classical systems and properties around us emerge from the underlying quantum domain.

On the one hand, then, the problem of the quantum-to-classical transition has a much broader scope than the issue of quantum measurement in the literal sense. On the other hand, however, many interactions between physical systems can be viewed as measurement-like interactions. For example, light scattering off an object carries away information about the position of the object, and it is in this sense that we thus may view these incident photons as a "measuring device." Such ubiquitous measurement-like interactions lie at the heart of the explanation of the quantum-to-classical transition by means of decoherence. Measurement, in the more general sense, thus retains its paramount importance also in the broader context of the quantum-to-classical transition, which in turn motivates us not to abandon the term "measurement problem" altogether in favor of the more general "problem of the quantum-to-classical transition."

As indicated above, the measurement problem (and the problem of the quantum-to-classical transition) is composed of three parts, all of which we shall describe in more detail in the following:

1. *The problem of the preferred basis* (Sect. 2.5.2). What singles out the preferred physical quantities in nature—e.g., why are physical systems usually observed to be in definite positions rather than in superpositions of positions?
2. *The problem of the nonobservability of interference* (Sect. 2.5.3). Why is it so difficult to observe quantum interference effects, especially on macroscopic scales?
3. *The problem of outcomes* (Sect. 2.5.4). Why do measurements have outcomes at all, and what selects a particular outcome among the different possibilities described by the quantum probability distribution?

Familiarity with these problems will turn out to be important for a proper understanding of the scope, achievements, and implications of decoherence. To anticipate, it is fair to conclude that decoherence has essentially resolved the first two problems. Since these problems and their resolution can be formulated in purely operational terms within the standard formalism of quantum mechanics, the role played by decoherence in addressing these two issues is rather undisputed.

By contrast, the success of decoherence in tackling the third issue—the problem of outcomes—remains a matter of debate, in particular, because this issue is almost inextricably linked to the choice of a specific interpretation of quantum mechanics (which mostly boils down to a matter of personal preference). In fact, most of the overly optimistic or pessimistic statements about the ability of decoherence to solve "the" measurement problem can be traced back to a misunderstanding of the scope that a standard quantum effect such as decoherence may have in resolving the more interpretive problem of outcomes.

2.5.1 The Von Neumann Scheme for Ideal Quantum Measurement

Starting from two separate (nonentangled) systems, how can an entangled composite state come about? How is it possible that the two subsystems lose their individuality to become a quantum-mechanical whole? Quantum entanglement can be viewed as arising from the kinematical concept of the superposition principle combined with the dynamical feature of the linearity of the Schrödinger time evolution. The resulting process is often represented in terms of a *von Neumann measurement*, a scheme devised by von Neumann during the early years of quantum mechanics and discussed in his seminal book of 1932 [60].

Von Neumann's goal was to describe the act of quantum measurement in entirely quantum-mechanical terms as the physical interaction between the measured system and the measuring apparatus, treating not only the system but also the apparatus (and, ultimately, the observer; see Sect. 9.2) as quantum-mechanical objects. It is worth noting that this approach represented a radical departure from the Copenhagen interpretation that had postulated the existence of intrinsically classical measurement apparatuses which were regarded as not subject to the laws of quantum mechanics (see Sect. 8.1).

Despite its name, the scope of the von Neumann measurement scheme goes far beyond the context of quantum measurement in the actual sense. In fact, the von Neumann scheme is the easiest way to understand how quantum entanglement arises. It also illustrates nicely the aforementioned three components of the measurement problem, namely, the problem of the preferred basis, the nonobservability of interference effects, and the problem of outcomes. Finally, it will allow us to introduce the basic formalism of decoherence, by regarding decoherence as a consequence of a von Neumann–type measurement interaction between the system and its environment.

After these introductory remarks, let us now formulate the von Neumann measurement scheme. Its typical ingredients are a (typically microscopic) system $\mathcal{S}$, described by a Hilbert space $\mathcal{H}_\mathcal{S}$ with basis vectors $\{|s_i\rangle\}$, and a (usually macroscopic) measuring apparatus $\mathcal{A}$, formally represented by basis vectors $\{|a_i\rangle\}$ in a Hilbert space $\mathcal{H}_\mathcal{A}$. Strictly speaking, whether the system and the apparatus are microscopic or macroscopic has no bearing on the following general argument. However, it typically reasonable, from a physical point of view, to associate microscopicity with the system, since we would like to ensure that the system can be easily prepared in a superposition state. On the other hand, the physical realization of a measuring apparatus typically involves a macroscopic system with a large number of degrees of freedom, and from our experience we would expect such an apparatus to behave according to the laws of classical physics (although the von Neumann scheme deliberately treats the apparatus in quantum-mechanical terms).

The purpose of the apparatus is now to measure the state of the system $\mathcal{S}$. We can think of the apparatus as having some kind of pointer that moves to the position "i," represented by the state $|a_i\rangle$, if the system is measured to be in the state $|s_i\rangle$ (see Fig. 2.4). Assuming that, before the measurement takes place, the apparatus starts out in some initial "ready" state $|a_\mathrm{r}\rangle$, the dynamical measurement interaction between the system and the apparatus will then be of the form

$$|s_i\rangle |a_\mathrm{r}\rangle \longrightarrow |s_i\rangle |a_i\rangle \tag{2.52}$$

for all i. Here the initial and final states reside in the tensor-product Hilbert space $\mathcal{H}_\mathcal{S} \otimes \mathcal{H}_\mathcal{A}$ describing the total $\mathcal{SA}$ system. We see that the measurement has established a one-to-one correspondence between the state of the system and the state of the apparatus: The latter perfectly mirrors the former. Also note that, in writing the right-hand side of (2.52), we have tacitly assumed

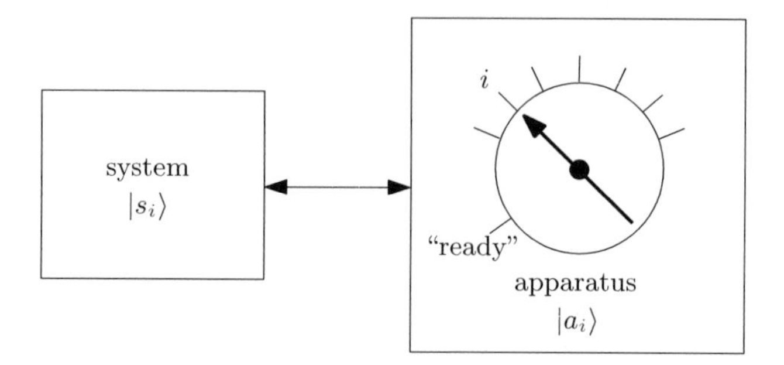

Fig. 2.4. Von Neumann scheme for ideal quantum measurement. Both system and apparatus are treated as quantum systems. The system–apparatus interaction is such that the system's being in state $|s_i\rangle$ causes the apparatus pointer to move from the initial "ready" position on the dial to position "i," represented by a quantum state $|a_i\rangle$ of the apparatus.

that the measurement interaction does not change the state of the system. Because of these assumptions, the measurement scheme (2.52) is often called *ideal.* Such measurements which do not disturb the state of the system are also known as *quantum nondemolition measurements* [69].

Now we come to the key point. Thus far, the interaction (2.52) has not led to any entanglement: The final system–apparatus state is still separable. However, let us consider what happens if the system starts out in a superposition of the basis states $|s_i\rangle$,

$$|\psi\rangle = \sum_i c_i |s_i\rangle . \tag{2.53}$$

In this case, the linearity of the Schrödinger equation implies that the system–apparatus combination $\mathcal{SA}$ will evolve according to

$$|\psi\rangle |a_\mathrm{r}\rangle = \left(\sum_i c_i |s_i\rangle \right) |a_\mathrm{r}\rangle \longrightarrow |\Psi\rangle = \sum_i c_i |s_i\rangle |a_i\rangle . \tag{2.54}$$

This evolution represents the (ideal) von Neumann quantum-measurement scheme.

Inspection of the right-hand side of (2.54) shows that the final state of the system–apparatus combination is in general described by an entangled state, i.e., we can no longer attribute an individual state vector to the system or the apparatus. Entanglement has thus been created *dynamically.* The superposition initially present only in the system has been *amplified* to the level of the (typically macroscopic) apparatus, in the sense that the final superposition involves both the system and the apparatus.

The crucial difficulty is now that it is not at all obvious how one is to regard the dynamical evolution described by (2.54) as representing measurement in the usual sense. This is so because the final state on the right-hand side of (2.54) is, for at least two reasons to be discussed in the following, not sufficient to directly conclude that the measurement has actually been completed. To emphasize this fact, the scheme (2.54) is frequently referred to as *premeasurement*.

2.5.2 The Problem of the Preferred Basis

The first problem is that there exists a basis ambiguity regarding the expansion of the final composite state on the right-hand side of (2.54). We can express this state in many different ways, implying that in the von Neumann scheme stated above the observable that was supposedly measured is not uniquely defined by this final state. In fact, given *any* set of states describing our system $\mathcal{S}$, there exists a corresponding set of apparatus states such that the final composite state takes the form (2.54),

$$|\psi\rangle |a_r\rangle \longrightarrow |\Psi\rangle = \sum_i c_i |s_i\rangle |a_i\rangle = \sum_i c'_i |s'_i\rangle |a'_i\rangle = \dots . \tag{2.55}$$

However, this freedom in the choice of the basis is in practice to some degree limited by two constraints. First of all, we will typically require the states $|a_i\rangle$ of the apparatus to be mutually orthogonal. This ensures that these states correspond to classically distinct outcomes of the measurement, such that the possible states $|s_i\rangle$ of the system can be reliably distinguished. Second, for an arbitrary choice of the set of apparatus states $|a_i\rangle$, the relative states $|s_i\rangle$ for the system may fail to be mutually orthogonal. In this case the system observables corresponding to the states $|s_i\rangle$ (i.e., the observables with eigenstates $|s_i\rangle$) will not be Hermitian, which is usually an undesired property.[15]

In view of these arguments, we may therefore require the apparatus states $|a_i\rangle$ to be mutually orthogonal, i.e., $\langle a_i|a_j\rangle = 0$ for $i \neq j$. In this case, it follows from the so-called Schmidt theorem (see Sect. 2.15.1 below) that the decomposition

$$|\Psi\rangle = \sum_i c_i |s_i\rangle |a_i\rangle , \tag{2.56}$$

with c_i real and $\sum c_i^2 = 1$, is unique, *provided* all the coefficients c_i are different from one another.

Let us consider a simple example showing that the decomposition (2.56) is in general *not* unique if this condition on the coefficients does not hold.

[15] However, non-Hermitian observables are not *a priori* forbidden and arise in certain experimental settings. For instance, in quantum optics one often performs measurements that have coherent states as their outcomes. Coherent states, however, form an overcomplete set of states and can therefore not be represented by Hermitian observables. See also the discussion by Zurek in [16].

Suppose both the system $\mathcal{S}$ and the apparatus $\mathcal{A}$ are quantum-mechanical two-state systems represented by spin-$\frac{1}{2}$ particles, described by basis states $|0_z\rangle$ and $|1_z\rangle$. Suppose further that the states $|0_z\rangle_{\mathcal{A}}$ and $|1_z\rangle_{\mathcal{A}}$ of $\mathcal{A}$ act as "pointers" for the spin states $|0_z\rangle_{\mathcal{S}}$ and $|1_z\rangle_{\mathcal{S}}$ of $\mathcal{S}$. That is, the von Neumann interaction (2.52) is here described by the dynamics

$$\begin{aligned} |0_z\rangle_{\mathcal{S}} \, |\text{“ready”}\rangle_{\mathcal{A}} &\longrightarrow |0_z\rangle_{\mathcal{S}} \, |0_z\rangle_{\mathcal{A}} \, , \\ |1_z\rangle_{\mathcal{S}} \, |\text{“ready”}\rangle_{\mathcal{A}} &\longrightarrow |1_z\rangle_{\mathcal{S}} \, |1_z\rangle_{\mathcal{A}} \, . \end{aligned} \tag{2.57}$$

Assuming the system $\mathcal{S}$ starts out in the superposition $(|0_z\rangle_{\mathcal{S}} + |1_z\rangle_{\mathcal{S}})/\sqrt{2}$, it follows from (2.57) and the linearity of the Schrödinger equation [see also (2.54)] that the final composite entangled spin–apparatus state will be

$$|\Psi\rangle = \frac{1}{\sqrt{2}} \left(|0_z\rangle_{\mathcal{S}} \, |0_z\rangle_{\mathcal{A}} + |1_z\rangle_{\mathcal{S}} \, |1_z\rangle_{\mathcal{A}} \right) . \tag{2.58}$$

Let us now use (2.35) to rewrite the z-spin states $|0_z\rangle_i$ and $|1_z\rangle_i$, $i = \mathcal{S}, \mathcal{A}$, in terms of the eigenstates $|0_x\rangle_i$ and $|1_x\rangle_i$ of the Pauli spin operator $\hat{\sigma}_x$. Expressed in the new basis $\{|0_x\rangle_i \, , |1_x\rangle_i\}$, the state (2.58) reads

$$|\Psi\rangle = \frac{1}{\sqrt{2}} \left(|0_x\rangle_{\mathcal{S}} \, |0_x\rangle_{\mathcal{A}} + |1_x\rangle_{\mathcal{S}} \, |1_x\rangle_{\mathcal{A}} \right) . \tag{2.59}$$

Recall that we regarded $\mathcal{A}$ as a measuring device for the spin of the system $\mathcal{S}$. Equations (2.58) and (2.59) then imply that the apparatus $\mathcal{A}$ has formed one-to-one correlations with *both* the z-spin and x-spin states of $\mathcal{S}$. If we interpret, in the spirit of the von Neumann scheme (i.e., without the assumption of any subsequent wave-function collapse), this formation of system–apparatus correlations as a complete measurement, this state of affairs seems to imply the following. Once $\mathcal{A}$ has measured the spin of $\mathcal{S}$ along the z axis [see the final state (2.58)], $\mathcal{A}$ may be considered as having measured also the spin of $\mathcal{S}$ along the x axis, where the latter "measurement" is represented by the equivalent form (2.59) of (2.58).

Thus our device $\mathcal{A}$ would appear to have simultaneously measured two *noncommuting* observables of the system, namely, $\hat{\sigma}_x$ and $\hat{\sigma}_z$, in apparent contradiction with the laws of quantum mechanics. What is more, it is easy to show that the state (2.58) can in fact be rewritten in infinitely many equivalent ways using the spin-$\frac{1}{2}$ basis along *any* arbitrary axis. Thus it would appear that, once our apparatus $\mathcal{A}$ has "measured" the spin of $\mathcal{S}$ along the z axis—in the von Neumann sense of the formation of correlations of the type (2.58)—it has also "measured" spin along any other spatial direction.

Such a situation of simultaneous measurement of a set of noncommuting observables is not only forbidden by quantum mechanics, but also contradicts our experience that measuring devices seem to be *designed* to measure only very particular quantities. In our example, the apparatus $\mathcal{A}$ may be realized in form of a Stern–Gerlach device (see Sect. 2.2.2), with the states $|0_z\rangle_{\mathcal{A}}$

and $|1_z\rangle_{\mathcal{A}}$ corresponding to the two separated paths through the apparatus (with the magnetic field aligned along the z axis) that distinguish the spin states $|0_z\rangle_{\mathcal{S}}$ and $|1_z\rangle_{\mathcal{S}}$ of the system $\mathcal{S}$. This Stern–Gerlach apparatus is therefore set up to measure spin only along the z axis, but not along any other axis. Performing a measurement along an axis different from z would require a physical rotation of the magnetic field and would thus correspond to a *physically different* setup.

The existence of such a "preferred observable" (or of a "preferred basis") is thus not explained by the final system–apparatus state arrived at through a von Neumann measurement. As we have seen, the form of this state will in general not uniquely fix the observable of the system that is recorded by the apparatus via the formation of quantum correlations. This *problem of the preferred basis* was first cleanly separated out from the problem of wave-function collapse and the intimately related problem of outcomes (see Sect. 2.5.4 below) by Zurek [8], who emphasized the distinct and important role played by the preferred-basis problem in any account of quantum measurement. Before the problem of outcomes may play any role, we ought to solve this preferred-basis problem, since it does not make sense to even inquire about specific outcomes if the set of possible outcomes is not clearly defined.

Zurek also recognized [8, 9] that the preferred-basis problem plays a key role in the problem of the quantum-to-classical transition, well beyond the narrow context of quantum measurement. Here, we encounter the core question of why we perceive systems, especially macroscopic ones, in only a tiny subset of the physical quantities in principle allowed by the superposition principle. Most notably, for example, macroscopic systems are always found in definite spatial positions but not in superpositions thereof. What singles out position as the preferred quantity? As first suggested by Zeh [4] and spelled out in detail by Zurek [8, 9], to answer these questions and to overcome the preferred-basis problem one must consider the system and the apparatus as open quantum systems, i.e., as interacting with their environment. We will discuss this approach in detail in Sect. 2.8.

2.5.3 The Problem of the Nonobservability of Interference

In many experiments, we can observe interference patterns indicative of the presence of quantum superpositions of component states in a particular basis (see Sect. 2.2.2). However, as we go to larger scales, such interference patterns are typically observed to vanish. For example, if we carry out the double-slit experiment with microscopic particles such as electrons, interference fringes appear on the distant detecting screen (see Fig. 6.6 in Chap. 6). However, as we perform the experiment with atoms or molecules, the interference pattern usually disappears rapidly.

In traditional textbook accounts, this inability to observe interference patterns for mesoscopic and macroscopic objects in the double-slit experiment has typically been explained by an analogy with classical light-wave interference. In the latter setting, it is well known from basic optics that the separation between the diffracting slits must be on the order of the wavelength of the light in order for an interference pattern to be observed. The analogous argument is then applied to the quantum case of the double-slit experiment with matter. Since the (de Broglie) wavelength of particles such as atoms is extremely short, it is simply impossible in practice to manufacture slits whose width and spacing would be of similar magnitude as the wavelength of the particles. In other words, in this picture our inability to observe interference patterns for massive particles would be rooted in the insufficient "resolution" of the experimental device, preventing us from unlocking the quantum nature of the particles.

However, this is only one part of the story. For example, it *is* possible to observe spatial interference patterns for mesoscopic molecules in experimental setups that are similar in spirit to the double-slit experiment but circumvent the obstacle of having to manufacture microscopic slits (see Sect. 6.2 for details on these experiments). Yet, when certain experimental parameters *unrelated to the diffraction process* are changed (for instance, the density of air surrounding the diffracted molecules), the interference pattern is observed to decay. Thus, although the diffraction stage of the experiment clearly allows for the creation of the spatial superposition (which could then be observed in form of an interference pattern), there are other factors that prevent us from observing the pattern. This clearly indicates that our difficulties in "seeing interferences" cannot solely be due to the problem of generating the superposition in the first place.

Furthermore, there exist many interference experiments involving various physical systems that do not at all fall into the category of diffraction experiments with matter. For instance, as already briefly mentioned earlier, there are experiments that allow us to generate superpositions of electrical currents flowing in opposite directions, which would lead to a temporal interference pattern in form of a current oscillating back and forth between the two directions (such experiments will be described in more detail in Sect. 6.3). Yet, despite the fact that the experimental conditions allow for the generation of the superposition state, observation of interference often fails or at least requires extremely refined experimental conditions. Clearly, this problem can no longer be described by the simple analogy between the diffraction of classical light waves and quantum "probability waves."

A forteriori, from the final state (2.54) of the von Neumann measurement scheme it follows that superpositions involving macroscopic measurement devices should be ubiquitous in nature. Why, then, do we not seem to observe interferences between different pointer positions of the apparatus in the everyday world around us? Why is it so difficult to observe *any* interference effects in the mesoscopic and macroscopic regime, although the von Neumann scheme clearly suggests that superpositions should be easily amplified from the microlevel to the macrolevel?

This, in essence, constitutes the problem of the nonobservability of interference. Shortly, we will see how decoherence provides a very elegant and general answer to this problem, by explaining the observed decay of interference patterns (or the complete inability to experimentally observe such patterns in the first place) as a result of interactions with environmental degrees of freedom.

2.5.4 The Problem of Outcomes

Our experience tells us that every measurement results in a definitive value of the measured quantity. In fact, the very definition of terms such as "outcome," "value," "quantity," etc., inherently relies on this definiteness. On the other hand, the final state (2.54) obtained via the von Neumann scheme represents a superposition of system–apparatus states. From our discussion in Sect. 2.2.1 we know that such a superposition is fundamentally different from a classical ensemble of states, i.e., from a situation in which the system–apparatus combination actually is in only one of the component states $|s_i\rangle\,|a_i\rangle$ but we simply do not know in which (see also the analysis in Sect. 2.4.2 above). Therefore, unless we supply some additional physical process (say, some collapse mechanism) or provide a suitable interpretation of such a superposition, it is not clear how to account, given the final composite state, for the definite pointer positions that are observed as the result of an actual measurement.

This problem can be further broken down into two distinct aspects. First, we are faced with the question of why we do not perceive the pointer of the apparatus in a superposition of different pointer positions $|a_i\rangle$ at the conclusion of the measurement (whatever it would actually *mean* to observe such a superposition), i.e., why measurements seem to have outcomes at all. And second, we may ask what "selects" a specific outcome. That is, why do we observe, in each run of the experiment that realizes the measurement, a *particular* pointer position i (and thus a particular pointer state $|a_i\rangle$), as opposed to one of the other possible states $|a_{j\neq i}\rangle$? We shall refer to both issues jointly as the *problem of outcomes*.

The problem of outcomes directly underlies the Schrödinger-cat paradox described in Chap. 1. Recall that the first part of the paradox is concerned with the fact that quantum mechanics seems to predict that the final composite atom–cat state is described by a superposition of classically mutually exclusive states (for simplicity, we shall here refrain from explicitly including the hammer and the poison in our discussion). This part can be understood as simply arising from a von Neumann–type measurement-like interaction between the atom and the cat. Here the atom corresponds to the microscopic system $\mathcal{S}$, while the cat represents the macroscopic apparatus (in the sense that its vitality is an indicator—a "pointer"—of the state of the atom). Following (2.52), the "measurement" scheme therefore reads

$$|\text{“atom not decayed”}\rangle\,|c_\text{r}\rangle \longrightarrow |\text{“atom not decayed”}\rangle\,|\text{“cat alive”}\rangle\,, \tag{2.60a}$$
$$|\text{“atom decayed”}\rangle\,|c_\text{r}\rangle \longrightarrow |\text{“atom decayed”}\rangle\,|\text{“cat dead”}\rangle\,. \tag{2.60b}$$

Here $|c_\text{r}\rangle$ denotes the initial state of the cat. Now, according to quantum mechanics, an unstable atom is at all times described by a superposition of the decayed state and the undecayed state of the atom,

$$|\psi\rangle = \alpha\,|\text{“atom not decayed”}\rangle + \beta\,|\text{“atom decayed”}\rangle\,, \tag{2.61}$$

where α and β are time-dependent coefficients with $|\alpha|^2 + |\beta|^2 = 1$. Just as in the general von Neumann scheme (2.54), the fact that the Schrödinger equation is linear implies that the final composite atom–cat system is then described by an entangled state of the form

$$\begin{aligned}|\psi\rangle\,|c_\text{r}\rangle \longrightarrow\; &\alpha\,|\text{“atom not decayed”}\rangle\,|\text{“cat alive”}\rangle \\ &+ \beta\,|\text{“atom decayed”}\rangle\,|\text{“cat dead”}\rangle\,.\end{aligned} \tag{2.62}$$

No individual quantum state can now be attributed to the cat, and it would thus appear that “the cat is neither alive nor dead,” as the situation described by the final state on the right-hand side of (2.62) is often interpreted.[16] Note, though, that it is not the cat itself that is described by a superposition of the states “alive” and “dead.” Rather, it is the *composite* atom–cat state that is represented by a superposition of atom–cat states (again, we have omitted the hammer and the poison from the picture).

As explained in Chap. 1, the second part of the Schrödinger-cat paradox refers to the fact that the superposition (2.62) persists until, at least according to the standard interpretation of quantum mechanics (see Sect. 8.1), the box is opened and its contents are directly observed (leading to the postulated “collapse of the wave function”). This poses the question of how and why the fate of the cat could possibly be left to the intervention (von Neumann's *“erster Eingriff”*) of an external observer. Of course, this is in essence nothing else than the problem of outcomes: How can we account for the observer's experience of a definite state of the cat (i.e., of a cat that is either alive or dead) given the superposition state (2.62)?

Remedying the problem of outcomes, i.e., solving the apparent conflict between the paramount and experimentally confirmed role of the superposition principle and the observation of single definite outcomes in measurements, has been one of the core motivations behind any interpretation of quantum mechanics. The standard (or “orthodox”) interpretation of quantum mechanics (see Sect. 8.1 for details) prescribes that an observable corresponding to a physical quantity has a definite value if and only if the system is in an eigenstate of this observable. On the other hand, if the system is described by a superposition of such eigenstates, it is considered meaningless to speak

[16] Indeed, it is difficult to properly put such a state of affairs into words, as it is simply not part of our experience.

of the (physical) state of the system as having any definite value of the observable at all. This rule of orthodox quantum mechanics is often referred to as the *eigenvalue–eigenstate link* (sometimes also called, more to the point, the *value–eigenstate link*).

However, the eigenvalue–eigenstate link is not necessitated by the structure of quantum mechanics or by any empirical constraints [68]. Furthermore, the concept of an "exact" eigenvalue–eigenstate link leads to difficulties of its own. For instance, outcomes of measurements are typically registered by pointers localized in position space. But these pointers are never perfectly localized, i.e., they cannot be described by exact eigenstates of the position operator, since such eigenstates are unphysical (they correspond to an infinite spread in momentum and therefore to an infinite amount of energy to be contained in the system). Therefore the states corresponding to different pointer positions cannot be exactly mutually orthogonal.

The concept of classical "values" that can be ascribed through the eigenvalue–eigenstate link based on observables and the existence of exact eigenstates of these observables has therefore frequently been either weakened or altogether abandoned. Either the quantum formalism and the concept of measurement have then been reinterpreted in certain ways, or actual modifications of quantum mechanics itself have been introduced. In the former category, relative-state and modal interpretations aim to interpret the final composite system–apparatus state arising in the von Neumann scheme (2.54) in such a way as to explain the existence, or at least the subjective perception, of outcomes in spite of the fact that the quantum state has the form of a superposition. In the latter category, physical collapse models (already briefly mentioned in Sect. 2.2.3) postulate the existence of some fundamental mechanism in nature that breaks the unitarity of the Schrödinger evolution and leads to an "objective" reduction of the wave function onto one of its components. These various interpretations and their relation to decoherence will be discussed in more detail in Chap. 8.

Generally, the problem of outcomes is rooted in the question of what actualizes a particular result in a probabilistic theory. In classical probabilistic theories, answering this question does not, at least in principle, pose any difficulties. Here, the notion of probability is simply a consequence of a convenient coarse-graining procedure that may simplify the treatment of certain problems. At the fundamental level of the physical system, however, the particular outcome is completely specified by the underlying, deterministic laws of physics (however complicated they may look in practice for, say, a system composed of billions of atoms). Thus, in principle, we could always rid classical physics of any probabilistic aspect.

By contrast, as we have discussed in detail in Sect. 2.1, quantum mechanics appears to possess an intrinsically probabilistic character. Pure quantum states already represent a complete description of the (physical) state of the system, and their evolution is given by the deterministic Schrödinger equation. Yet, there exists no fundamental mechanism that would determine which particular outcome is realized in each measurement instance. Therefore, the problem of outcomes is fundamental to quantum mechanics itself. Accordingly, the best hope we can have for decoherence to help us solve this problem is by explaining why only one of the possible outcomes is actually observed in a measurement (rather than a superposition of outcomes). However, the question of why a *particular* outcome appears to the observer rather than another one of the possible outcomes, none of which is formally singled out in any way in the final von Neumann state (2.54), pertains to fundamental issues in the interpretation of quantum mechanics outside of the scope of decoherence. We will come back to this topic in Chap. 8.

2.6 Which-Path Information and Environmental Monitoring

Having laid out the key elements of the formalism and interpretation of quantum mechanics relevant to decoherence—namely, quantum states, the superposition principle, entanglement, density matrices, and the measurement problem—we are now in an excellent position to finally approach our actual subject of interest, namely, decoherence. We shall go about this task by first revisiting the well-known double-slit experiment, which will provide us with a very intuitive and accessible illustration of the basic mechanism of decoherence.

2.6.1 The Double-Slit Experiment, Which-Path Information, and Complementarity

Let us have a look back at Fig. 2.2, where we sketched the usual double-slit setup. Particles (such as electrons) approaching from the left are incident on a screen with two slits. After passage through the slits, they hit a distant detector screen, leaving a permanent spot. As predicted by quantum mechanics and confirmed by experiment, there exist two different limiting regimes (see also our discussion in Sect. 2.2.2):

1. *The "wave" scenario* (illustrated on the right of Fig. 2.2). If we refrain from measuring through which slit each particle has passed, the particle density observed at the level of the detecting screen corresponds to an interference pattern given by $\varrho(x) = \frac{1}{2}\left|\psi_1(x) + \psi_2(x)\right|^2$, i.e., by the probability density corresponding to a *quantum-mechanical superposition* of the partial waves $\psi_1(x)$ and $\psi_2(x)$ representing passage through slit 1 and 2, respectively.
2. *The "particle" scenario* (shown in the center of Fig. 2.2). If we place a detector at one of the slits to find out whether the particle has passed

through this slit, the interference pattern disappears. Now the density pattern on the distant screen is simply equal to a *classical addition* of the pattern created by all particles that have traversed slit 1 (that is, the pattern that would be obtained if slit 2 was covered) and the pattern created by all particles that have passed through slit 2: $\varrho(x) = \frac{1}{2} |\psi_1(x)|^2 + \frac{1}{2} |\psi_2(x)|^2$.

The standard explanation of the second case (the "particle" scenario) goes as follows. According to quantum mechanics, the state of the particle at the level of the slits is given by the superposition $\Psi(x) = (\psi_1(x) + \psi_2(x)) / \sqrt{2}$ of the partial waves $\psi_1(x)$ and $\psi_2(x)$. This superposition is spatially spread out over the region encompassing the two slits. If we now measure the position of the particle at one of the slits and indeed find the particle to be present at this slit, we localize $\Psi(x)$ to the corresponding spatial region. That is, since $\psi_1(x)$ and $\psi_2(x)$ have negligible overlap at the level of the slits, we may say that, using the standard collapse postulate of quantum mechanics, the measurement has collapsed the superposition $\Psi(x)$ onto either one of the component states $\psi_1(x)$ or $\psi_2(x)$. Thus we can no longer obtain interference between these partial waves, and hence the interference pattern on the distant screen disappears.

In other words, whenever we try to obtain *which-path ("Welcher-Weg") information* about the particle in order to see which of the two slits the particle has traversed—i.e., whenever we attempt to make sense of the quantum-mechanical superposition of the two paths that would seem to describe a counterintuitive simultaneous passage through *both* slits—it seems that we cannot help but destroy the ability of the particle to exhibit the quantum property of (spatial) interference. Thus there exists, to use Niels Bohr's famous term [70], a *complementarity* between which-path information (the "particle" aspect") and interference (the "wave" aspect). Depending on the experimental setup (namely, depending on whether we measure the path of the particle or not), we seem to observe either "particle-like" or "wave-like" behavior. This so-called "wave–particle duality" has been considered a cornerstone of quantum theory and has been the subject of countless discussions among physicists and philosophers of physics alike.

The complementarity principle and its application was the subject of a famous debate between Einstein and Bohr at the Fifth Solvay Conference in Brussels in 1927 [71]. At the conference, Einstein had challenged Bohr with the following thought experiment involving the standard double-slit setup. He argued that, based on the law of momentum conservation, the passage of the particle through the top (bottom) slit should result in a recoil of the screen containing the slits in the upward (downward) direction. Suppose now we could measure the direction of the recoil of the screen (see Fig. 2.5). This measurement would allows us to infer, at least in principle, the path of the particle, thus providing us with which-path information. Einstein argued that, since the interference pattern is solely due to the interference between

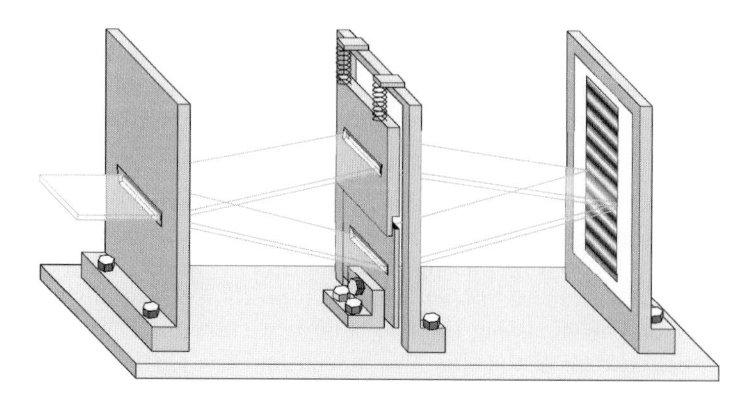

Fig. 2.5. Einstein's thought experiment for obtaining which-path information in the double-slit experiment. Particles leaving the collimating slit on the left pass through a double slit (center) and are registered on the detection screen on the right. The passage of the particle through the slit transfers momentum to the slit, which should in principle be measurable as a recoil of the screen containing the slits. While the bottom slit is kept fixed, the upper slit is suspended by springs. A particle passing through the upper slit would therefore induce a tiny oscillatory motion of the suspended slit that could (at least in principle) be detected, which would allow us to infer which slit the particle has traversed. The illustration is based on original drawings of Bohr [72] and is reproduced from [73] by permission from Macmillan Publishers Ltd: Nature, copyright 2001.

the two wave packets emerging from the slits, the determination of the direction of the recoil at the stage of the screen cannot have any effect on the subsequent evolution, and thus an interference pattern should be observable, in apparent contradiction with Bohr's principle of complementarity.

Bohr countered the challenge in the following way. The recoil imparted on the slits by an individual particle will typically be extremely small. To resolve this tiny change in momentum, we must know the initial momentum of the screen containing the slits within a range at least as small as the to-be-detected recoil. According to the uncertainty principle, measuring the momentum of the screen with such high accuracy implies a large uncertainty in the position of the screen. Bohr then showed that already due to the measurement of the initial momentum of the screen (before the incident particle even reaches the slits), the resulting indeterminacy of the position of the screen translates into a range of possible positions of the interference fringes on the detecting screen that would be on the order of the characteristic spacing between these fringes. Thus, if we average the interference pattern over this range of positions, the pattern is "washed out": The position of a maximum in the pattern for one of the possible positions coincides with the position of a minimum in the pattern for another position of the screen, and hence the net interference pattern disappears.

Bohr's central claim is therefore that obtaining which-path information implies an inevitable disturbance of the system, which is indeed true in the Bohr–Einstein example of gaining which-path information from a measurement of the recoil of the screen. However, as first shown by Wootters and Zurek [74] and further investigated by Scully and Drühl [75], in certain situations it is also possible to gather which-path information in such a way that there is no significant change in the spatial wave function of the particles, moderating the effect of the position–momentum uncertainty principle pointed out by Bohr. This will be discussed in the next section.

2.6.2 The Description of the Double-Slit Experiment in Terms of Entanglement

Thus far, the complementarity between obtaining which-path information and observing an interference pattern has been introduced as a discontinuous either–or distinction. However, we may now ask whether we could retain parts of the interference pattern by gathering only *some* which-path information, e.g., by performing an imprecise measurement of which slit the particle has traversed [74].

It turns out that the answer to this question is in the affirmative. However, to discuss and explain this feature of quantum mechanics, the description in terms of a wave-function collapse (as used above) will no longer be suitable. By postulate, the collapse is a discontinuous, irreversible process and therefore cannot account for smooth, reversible changes in the amount of which-path information and the degree of interference. Instead, we shall pursue a purely quantum-mechanical account in terms of the von Neumann measurement scheme (Sect. 2.5.1) and entanglement. This description will then also become the basis for our description of the process of decoherence, where the environment will assume the role of the which-path detector. The connection between the observability of interference in the double-slit experiment and entanglement was first discussed by Wootters and Zurek [74].

Let us denote the quantum states of the particle corresponding to passage through slit 1 and 2 by $|\psi_1\rangle$ and $|\psi_2\rangle$, respectively. As before, we place a detector at each of the two slits, with both detectors initially in the "ready" state. We can prepare our particle in, say, the state $|\psi_1\rangle$ by covering slit 2, and by placing the particle source directly behind slit 1 such that the particle will be guaranteed to pass through this slit. Consequently, the detector associated with slit 1 will trigger, while the detector at slit 2 will remain in the untriggered "ready" state. In the following we shall refer to the two individual detectors jointly as "the detector." We denote the joint "ready" state of the (composite) detector by $|\text{“ready”}\rangle$, and the quantum state of this detector system after preparation of the state $|\psi_1\rangle$ (as described above) by $|1\rangle$, indicating the passage of the particle through slit 1. Thus, in this case, the evolution of the state of the composite particle–detector system will be of the form

$$|\psi_1\rangle\,|\text{“ready”}\rangle \longrightarrow |\psi_1\rangle\,|1\rangle\,. \tag{2.63}$$

Repeating the above argument with the role of the slits reversed yields the evolution

$$|\psi_2\rangle\,|\text{“ready”}\rangle \longrightarrow |\psi_2\rangle\,|2\rangle\,. \tag{2.64}$$

Equations (2.63) and (2.64) correspond to the general description (2.52) of a measurement interaction.

Now, if both slits are open, the particle must be described by a superposition $|\psi\rangle = (|\psi_1\rangle + |\psi_2\rangle)/\sqrt{2}$ of the two components $|\psi_1\rangle$ and $|\psi_2\rangle$. Using (2.63) and (2.64), we therefore obtain a dynamical evolution of the von Neumann type (2.54), leading to an entangled composite particle–detector state,

$$\frac{1}{\sqrt{2}}\left(|\psi_1\rangle + |\psi_2\rangle\right)|\text{“ready”}\rangle \longrightarrow \frac{1}{\sqrt{2}}\left(|\psi_1\rangle\,|1\rangle + |\psi_2\rangle\,|2\rangle\right). \tag{2.65}$$

Once again, the detector states $|1\rangle$ and $|2\rangle$ act as “pointers” for the relative states $|\psi_1\rangle$ and $|\psi_2\rangle$ of the system.

What can we now learn about the particle by performing a measurement on it, for instance by letting it impinge on the detection screen, which corresponds to a measurement of position? As we know from Sect. 2.4.6, the quantity of interest is now the reduced density matrix for the particle, which for the pure state on the right-hand side of (2.65) is given by [see (2.49)]

$$\hat{\rho}_{\text{particle}} = \frac{1}{2}\left\{|\psi_1\rangle\langle\psi_1| + |\psi_2\rangle\langle\psi_2| + |\psi_1\rangle\langle\psi_2|\langle 2|1\rangle + |\psi_2\rangle\langle\psi_1|\langle 1|2\rangle\right\}. \tag{2.66}$$

This density matrix corresponds to a particle density $\varrho(x)$ at the detecting screen given by

$$\begin{aligned}\varrho(x) &\equiv \rho_{\text{particle}}(x,x)\\ &\equiv \langle x|\,\hat{\rho}_{\text{particle}}\,|x\rangle\\ &= \frac{1}{2}\left|\psi_1(x)\right|^2 + \frac{1}{2}\left|\psi_2(x)\right|^2 + \operatorname{Re}\left\{\psi_1(x)\psi_2^*(x)\langle 2|1\rangle\right\},\end{aligned} \tag{2.67}$$

where $\psi_i(x) = \langle x|\psi_i\rangle$, $i = 1, 2$. The last term describes the well-known interference pattern, and we see that the visibility of this interference pattern is quantified by the overlap $\langle 2|1\rangle$. This observation directly connects with our previous discussion in Sect. 2.4.6.

In particular, the limiting case of perfect distinguishability of the detector states $|1\rangle$ and $|2\rangle$, $\langle 2|1\rangle = 0$, corresponds to the “particle” regime,

$$\varrho(x) = \frac{1}{2}\left|\psi_1(x)\right|^2 + \frac{1}{2}\left|\psi_2(x)\right|^2. \tag{2.68}$$

Conversely, if $|1\rangle$ and $|2\rangle$ are completely unable to resolve the path of the particle, $\langle 2|1\rangle = 1$ (disregarding phase factors), the “wave” scenario of full interference pattern applies,

$$\varrho(x) = \frac{1}{2}\left|\psi_1(x)\right|^2 + \frac{1}{2}\left|\psi_2(x)\right|^2 + \mathrm{Re}\left\{\psi_1(x)\psi_2^*(x)\right\}. \tag{2.69}$$

The situation in which the detector obtains some but not full which-path information is formally represented by an overlap of the two detector states $|1\rangle$ and $|2\rangle$ that is nonzero but less than one. From (2.66) we see that we will then be able to observe an interference pattern, but the pattern will decay progressively as the overlap $\langle 2|1\rangle$ *decreases* (i.e., as the detector states $|1\rangle$ and $|2\rangle$ become more distinguishable) and thus the amount of which-path information obtained by the detector *increases*.

Thus we have shown that it is indeed possible to simultaneously observe an interference pattern and obtain some information about the path of the particle through the slits, provided this information remains incomplete. As soon as the information acquired by the detector is sufficient to enable us to infer with certainty which path the particle has taken, the interference pattern disappears. In their analysis of this fundamental trade-off between which-path information and interference and thus of the intimate connection between complementarity and entanglement, Wootters and Zurek [74] pointed out that one can obtain a fairly large amount of which-path information while retaining a visible interference pattern. Specifically, they showed that 90% certainty about the which-path question still allows for roughly 50% contrast of the interference pattern.

In summary, the degree to which an interference pattern can be observed is simply determined by the available which-path information encoded in some system entangled with the object of interest, and this amount can be changed without necessarily influencing the spatial wave function of the object itself. Thus complementarity can be regarded as a *consequence* of quantum entanglement.

2.6.3 The Environment as a Which-Path Monitor

Let us consider this book. It is immersed into a large environment of air molecules, light and thermal photons, even cosmic neutrinos and radioactive background radiation. In every second, a huge number of these particles will collide with and scatter off the book. Each of these collision will deflect the particle to some extent, depending on the position and orientation of the book. Let us look at a particular air molecule. It starts from some initial position with a certain velocity, scatters off the book, and then flies away along a trajectory deflected by a certain angle with respect to the incoming path.

Suppose now the book was oriented at a different angle. Now the air molecule would in general fly off along a different direction than in the first case. Thus, these two distinct paths of the scattered molecule would allow us to *distinguish* the existence of two different spatial orientations of the book. In other words, the scattered particles—the air molecules—carry away information about the position and orientation of the scattering object, which is here represented by the book. Using our now-familiar terminology, they therefore encode which-path information about the system. We remark that the term "which-path information" should here (as in the remainder of the text) be understood in the more general sense of "which-state information." The former term is historically motivated by the example of the double-slit experiment in which the information of interest concerns the path of the particle through the slits. Nonetheless, we shall often use the well-established term "which-path information" even in cases where the relevant information does not actually pertain to the trajectory of a particle.

A single air molecule colliding with the book may not carry away much information about the orientation of the book. After all, the scattering takes place only in a tiny region compared to the size of the book. However, as mentioned above, there are millions of such molecules scattering off the surface of the book in any given moment. The which-path (or, maybe more appropriately, "which-orientation") information encoded in all these molecules taken together will certainly be completely sufficient to distinguish two different positions and orientations of the book, even if they are very similar.

Let us express our argument schematically in terms of the quantum formalism. Suppose we focus on all N particles that will scatter off the book in the span of one second. Let us represent the quantum state of all N environmental particles before the collision by $|E_0\rangle$. After the collision, each particle will fly off along a certain trajectory with some velocity. For another orientation of the book, the particle may scatter into a slightly different direction. Now suppose all N particles have scattered off the book. Depending on the orientation of the book, we denote the post-scattering state of the environment by $|E_1\rangle$ and $|E_2\rangle$, respectively.

Following our above argument, these two states are clearly distinguishable, since the information contained in these two states is sufficient to discriminate between the two orientations of the book. Therefore the overlap between $|E_1\rangle$ and $|E_2\rangle$ will be negligibly small.[17] We clearly see the connection with the which-path formalism introduced in the previous Sect. 2.6.2: The environmental states $|E_1\rangle$ and $|E_2\rangle$ simply correspond to the states of a which-path (or, in this case, which-orientation) detector.

The crucial point is that these environmental which-path detectors are present everywhere in nature. Every object interacts with its environment, which in turn will obtain information about certain physical properties of the system. We may be able to shield our book from air molecules to some extent by placing it in a good vacuum, we may block out visible photons (light), we

[17]We can think of $|E_1\rangle$ and $|E_2\rangle$ as products of states pertaining to each individual molecule, $|E_1\rangle = \prod_{i=1}^{N} |e_1\rangle_i$ and $|E_2\rangle = \prod_{i=1}^{N} |e_2\rangle_i$. While the overlap of each individual pair of molecular states will be close to one, $(\langle e_1|e_2\rangle)_i \approx 1$, the product states $|E_1\rangle$ and $|E_2\rangle$ of a large number N of these individual states will be approximately orthogonal.

may even try to reduce the influence of thermal photons by cooling the book. Still, as we shall see in the next Chap. 3, particles from other sources remain that will continue to “measure” the position of the system. This crucial role of the environment as a ubiquitous “measuring device” which continuously performs effective measurements (in the von Neumann sense) on the system was first clearly recognized and discussed by Zurek in the early 1980s [8, 76].

We emphasize that this monitoring process does not require a human observer of any sort. The fact that the which-path information encoded in the environment could *in principle* be read out is sufficient for the interference pattern to disappear, i.e., for the particles to “lose their wave nature.” This lends a more precise, observer-independent meaning to Heisenberg's statement (already quoted in Sect. 2.1.3) that “the particle trajectory is created by our act of observing it” [42, p. 185]. This positivist attitude had resulted in much criticism directed at the quantum theory, as famously represented by Einstein's rhetorical question (mentioned in Chap. 1) of whether “the moon exists only when I look at it” [3].

We can now go through a formal argument analogous to that of Sect. 2.6.2. Suppose the system is described by a coherent superposition of two quantum states $|\psi_1\rangle$ and $|\psi_2\rangle$ representing localization around two different positions x_1 and x_2 (in the case of the double-slit experiment, the corresponding position-space wave functions $\psi_1(x)$ and $\psi_2(x)$ would represent the partial waves at the slits). Before the scattering of environmental particles takes place, the combined system–environment state has the product form

$$|\psi\rangle |E_0\rangle = \frac{1}{\sqrt{2}} (|\psi_1\rangle + |\psi_2\rangle) |E_0\rangle . \tag{2.70}$$

Following our above discussion, the system–environment interaction dynamics are given by

$$|\psi_1\rangle |E_0\rangle \longrightarrow |\psi_1\rangle |E_1\rangle , \tag{2.71a}$$

$$|\psi_2\rangle |E_0\rangle \longrightarrow |\psi_2\rangle |E_2\rangle . \tag{2.71b}$$

That is, the state of the environment evolves into $|E_1\rangle$ or $|E_2\rangle$ depending on the state of the system. Then the linearity of the Schrödinger equation implies the usual von Neumann evolution

$$\frac{1}{\sqrt{2}} (|\psi_1\rangle + |\psi_2\rangle) |E_0\rangle \longrightarrow \frac{1}{\sqrt{2}} (|\psi_1\rangle |E_1\rangle + |\psi_2\rangle |E_2\rangle) . \tag{2.72}$$

We see that the relative states $|\psi_1\rangle$ and $|\psi_2\rangle$ of the system have become entangled with the environmental states $|E_1\rangle$ and $|E_2\rangle$ that encode which-path information. The superposition initially confined to the system has now spread to the larger, composite system–environment state. Correspondingly, coherence between the components $|\psi_1\rangle$ and $|\psi_2\rangle$ is no longer a property of the system alone: It has become a shared property of the global system–environment state. One therefore often says that coherence has been “delocalized into the larger system” [77, p. 5], which now includes the environment.

The dynamical system–environment evolution described by (2.72) is the basic formal representation of the decoherence process, and we shall now discuss its consequences.

2.7 Decoherence and the Local Damping of Interference

The reduced density matrix $\hat{\rho}_{\text{particle}}$ of the system for the state (2.72) is given by [see (2.49) and (2.66)]

$$\hat{\rho}_{\text{particle}} = \frac{1}{2} \left\{ |\psi_1\rangle\langle\psi_1| + |\psi_2\rangle\langle\psi_2| + |\psi_1\rangle\langle\psi_2|\langle E_2|E_1\rangle + |\psi_2\rangle\langle\psi_1|\langle E_1|E_2\rangle \right\}. \tag{2.73}$$

As usual, the last two terms correspond to interference between the component states $|\psi_1\rangle$ and $|\psi_2\rangle$. Provided the environment has indeed recorded sufficient which-path information (which will certainly be the case for our above example of air molecules scattering off a macroscopic object over a period of one second), the final environmental states $|E_1\rangle$ and $|E_2\rangle$ will be approximately orthogonal, $\langle E_2|E_1\rangle \approx 0$. Then interferences in the reduced density matrix (2.73) will become suppressed,

$$\hat{\rho}_{\text{particle}} \approx \frac{1}{2} \left\{ |\psi_1\rangle\langle\psi_1| + |\psi_2\rangle\langle\psi_2| \right\}. \tag{2.74}$$

Only measurements that include both the system and the environment could possibly reveal the persistent coherence between the components in the superposition state (2.72). However, in practice, it is impossible to include in our observation all the many environmental degrees of freedom that have interacted with the system. Joos and Zeh [7, p. 224] poignantly summarized this state of affairs as "the interference terms still exist, but they are not *there*." That is, the interference terms remain present at the *global* level of the system–environment superposition (2.72) but have become unobservable at the *local* level of the system as described by the reduced density matrix (2.74). Of course, typically some of the environmental degrees of freedom will be part of our observation. For example, we will directly intercept a certain fraction of the light scattered off the system.[18] We can then simply regard these environmental degrees of freedom as part of the system. However, the important point is that there still remains a comparably large number of other environmental degrees of freedom that will not be observed directly.

Because of this very large number of environmental degrees of freedom interacting with the system and our inability to directly manipulate them, the creation of system–environment entanglement described by (2.72) is virtually impossible to undo in practice. Thus the environment-induced loss of

[18] In fact, this is how in practice observers will usually gather information about the system (see Sect. 2.9).

local phase coherence, i.e., of the well-defined phase relations between the components in the superposition necessary for the observation of interference effects, is usually *irreversible for all practical purposes.* (The case of—truly or effectively—reversible decoherence will be discussed in Sect. 2.13.)

This practically irreversible delocalization of phase relations into the composite system–environment state induced by inevitable and ubiquitous environmental monitoring constitutes precisely the process of decoherence. (The environment-induced selection of a preferred basis can be viewed as a consequence of this decoherence process and will be discussed in Sect. 2.8 below.) It leads to *effectively nonunitary dynamics* for the local system that may manifest themselves (for example) in the decay of interference patterns [see (2.74)]. The environment, composed of many subsystems, acts as an amplifying, higher-order measuring device for the state of the system (note that the very definition of measurement hinges on the property of irreversibility).

Environmental monitoring and the resulting decoherence processes therefore provide a solution to the problem of the nonobservability of interference discussed in Sect. 2.5.3. As we have discussed, to observe interference effects between components of a superposition state, these components must have not been measured, i.e., which-state information must not be available. But all physical systems encountered in nature are open quantum systems that interact strongly with their surroundings. These surroundings continuously acquire information about the system, leading to a constant "leakage" of coherence from the system into the environment. In our example of the scattering of environmental particles, the larger the system, the more particles will bounce off the surface per unit time and carry away which-path information. Accordingly, mesoscopic and macroscopic objects described by superpositions of wave packets that are well-separated in position space are usually extremely prone to decoherence, which explains why such superpositions and the corresponding spatial interference patterns are so exceedingly difficult to observe in nature.

We reiterate our warning, spelled out in Sect. 2.4.6 (see p. 48), against a misinterpretation of reduced density matrices as describing a proper mixture of states. Although formally identical to a mixed-state density matrix, the reduced density matrix (2.74) describes an only improper mixture. Therefore we must not conclude from (2.74) that the system actually *is* in either of the two states $|\psi_1\rangle$ or $|\psi_2\rangle$. After all, both of these components remain fully and equally present in the composite quantum state (2.72). The purely unitary interaction with the environment by itself simply cannot single out either one of the components. This issue will be analyzed in more detail in Sect. 8.1.

Our above account of decoherence relied on a rather intuitive argument about how the relative environmental states $|E_1\rangle$ and $|E_2\rangle$ become distinguishable such that the environment can act as a which-path detector for the system. A large part of research on decoherence is devoted to a realistic modeling of system–environment interactions in various physical situations of interest. This yields precise dynamical descriptions for the time evolution of the environmental states and the resulting influence on the dynamics of the reduced density matrix of the system. Typically, these models show that the different environmental relative states $|E_i(t)\rangle$ (which are quantum-correlated with the distinct system states $|s_i\rangle$ monitored by the environment) become distinguishable on very short timescales as a result of the system–environment interaction.

Specifically, as we shall see, for many system–environment models the overlap of the different relative environmental states $|E_i(t)\rangle$ is found to follow an exponential decay, i.e.,

$$\langle E_i(t)|E_j(t)\rangle \propto \mathrm{e}^{-t/\tau_\mathrm{d}} \tag{2.75}$$

for $i \neq j$. Here and in the following we shall take $t = 0$ to correspond to the time at which the interaction is "switched on" (for times $t < 0$ the system and environment are usually assumed to be completely uncorrelated). The quantity τ_d denotes the characteristic decoherence timescale, which can be evaluated numerically for particular choices of the parameters in each model.

Through (2.73), these decay characteristics (such as the functional form, relevant timescales, etc.) for the overlap of the environmental states $|E_i(t)\rangle$ then translate directly into the characteristic time dependences and timescales for the suppression of coherences between the superposition-state components $|s_i\rangle$ of the system that are one-to-one correlated with the $|E_i(t)\rangle$ [see (2.72)]. In order to quantitatively study the decoherence dynamics for a given system–environment model, we must therefore evaluate the explicit form of the $|E_i(t)\rangle$. We will study a first example in Sect. 2.10 below. More realistic models will be treated later in the book.

In particular, Chap. 3 will be devoted to an explicit modeling of spatial decoherence induced by the scattering of environmental particles such as air molecules and photons. To anticipate, we will find that the overlap of the relative environmental states decays as

$$\langle E_x(t)|E_{x'}(t)\rangle \propto \mathrm{e}^{-\Lambda|x-x'|^2 t}. \tag{2.76}$$

Here $|E_x(t)\rangle$ denotes the relative state of the environment that is one-to-one correlated with the state of the system that describes localization of the system's center of mass at position x. The constant Λ is determined by the particular physical properties of the scattering process. We see that the overlap (2.76) decreases exponentially with time. Expression (2.76) is therefore an example of the frequently encountered exponential-decay behavior (2.75). Environmental particles continuously scatter off the object, steadily increasing the amount of which-path information (namely, of information about the position of the system) that becomes encoded in the environment.

We also observe from (2.76) that the characteristic decoherence timescale τ_d [see (2.75)] is inversely proportional to the square of the coherent separation $|x - x'|$ between different (center-of-mass) positions x and x' of the system. This result can be understood through the following intuitive argument (we will make the explanation much more rigorous in Chap. 3). If the positions x and x' are close together, the corresponding final relative states of the scattered environmental particles can be expected to be very similar, making it difficult for the environment to resolve the spatial difference between these two positions. Accordingly, a noticeable decay of the overlap $\langle E_x(t)|E_{x'}(t)\rangle$ will require the accumulation of a large number of scattering events, each of which encodes only highly incomplete which-path information. Conversely, the more the center-of-mass positions x and x' become separated, the more easily the environment will be able to distinguish between these two positions. In the limit where $|x - x'|$ becomes sufficiently large such that each individual scattering event completely resolves the difference between the two center-of-mass positions, the rate of the exponential decay of the overlap $\langle E_x(t)|E_{x'}(t)\rangle$ becomes independent of the separation $|x - x'|$ and is simply proportional to the total scattering rate Γ_{tot}. In this limit the expression (2.76) is replaced by

$$\langle E_x(t)|E_{x'}(t)\rangle \propto \mathrm{e}^{-\Gamma_{\text{tot}} t}. \tag{2.77}$$

We will derive and discuss the expressions (2.76) and (2.77) in Chap. 3. Other important decoherence models and their dynamics will be discussed in detail in Chap. 5.

2.8 Environment-Induced Superselection

In the previous section, we showed how the system–environment interaction leading to decoherence may be expressed in form of the standard von Neumann measurement scheme [see (2.72)]. This, however, implies that the preferred-basis problem discussed in the context of von Neumann measurements (see Sect. 2.5.2) applies also here. That is, we may now ask: What singles out the particular components (i.e., the states $|\psi_1\rangle$ and $|\psi_2\rangle$ in the above example) as those between which interference is (locally) suppressed through the interaction with the environment?

To answer this question, recall that in our example the interaction between the system and environment is described by the evolution (2.71),

$$|\psi_1\rangle\,|E_0\rangle \longrightarrow |\psi_1\rangle\,|E_1\rangle\,, \tag{2.78a}$$
$$|\psi_2\rangle\,|E_0\rangle \longrightarrow |\psi_2\rangle\,|E_2\rangle\,, \tag{2.78b}$$

with $\langle E_1|E_2\rangle \longrightarrow 0$. We motivated this form of the interaction by referring to the fact that the states $|\psi_1\rangle$ and $|\psi_2\rangle$ correspond to sufficiently distinct physical states of the system (such as different spatial positions and orientations of a body), and that this difference is resolved by the environment continuously monitoring the system.

Let us now introduce the conjugate states $|\psi_\pm\rangle$ of the system, defined as even and odd superpositions of $|\psi_1\rangle$ and $|\psi_2\rangle$,

$$|\psi_\pm\rangle \equiv \frac{1}{\sqrt{2}} \left(|\psi_1\rangle \pm |\psi_2\rangle \right). \tag{2.79}$$

Looking back at (2.70), we immediately see that $\psi_+(x)$ corresponds to the previously considered initial (superposition) state of the system. Since $|\psi_1\rangle$ and $|\psi_2\rangle$ represent states localized in space, $|\psi_+\rangle$ and $|\psi_-\rangle$ correspond to nonclassical delocalized states. Would the interaction (2.78) with the environment also lead to the suppression of interference between the components $|\psi_+\rangle$ and $|\psi_-\rangle$?

Recall that, in order for this suppression to happen, the interaction between the system and the environment must be such as to distinguish between the states $|\psi_+\rangle$ and $|\psi_-\rangle$. That is, the environment has to be able to encode which-state information about $|\psi_+\rangle$ and $|\psi_-\rangle$. It is now easy to see that the dynamics (2.78) do not accomplish this goal. The form of the system–environment interaction implies the evolution

$$|\psi_\pm\rangle |E_0\rangle \longrightarrow \frac{1}{\sqrt{2}} \left(|\psi_1\rangle |E_1\rangle \pm |\psi_2\rangle |E_2\rangle \right). \tag{2.80}$$

The system states $|\psi_1\rangle$ and $|\psi_2\rangle$ are by assumption distinct, $\langle\psi_1|\psi_2\rangle = 0$. Furthermore, following our above discussion, the final environmental states $|E_1\rangle$ and $|E_2\rangle$ resolve the difference between these states, i.e., $\langle E_1|E_2\rangle \longrightarrow 0$. Hence the final composite state on the right-hand side of (2.80) is of the nonseparable Bell type (2.7). A measurement performed on the environment in the $\{|E_1\rangle, |E_2\rangle\}$ basis will reveal no information whatsoever about which of the two states $|\psi_+\rangle$ and $|\psi_-\rangle$ has been prepared initially. This is so because we will always obtain the outcomes $|E_1\rangle$ and $|E_2\rangle$ with equal probabilities, regardless of whether the system started out in the state $|\psi_+\rangle$ or the state $|\psi_-\rangle$. Moreover, we can rewrite the right-hand side of (2.80) in infinitely equivalent many ways (see the discussion in Sect. 2.5.2) and still retain the Bell-state structure (2.80),

$$\frac{1}{\sqrt{2}} \left(|\psi_1\rangle |E_1\rangle \pm |\psi_2\rangle |E_2\rangle \right) = \frac{1}{\sqrt{2}} \left(|\psi_1'\rangle |E_1'\rangle \pm |\psi_2'\rangle |E_2'\rangle \right) = \dots, \tag{2.81}$$

with $\langle\psi_1'|\psi_2'\rangle = 0$ and $\langle E_1'|E_2'\rangle \longrightarrow 0$. Accordingly, there exists no measurement in any basis $\{|E_1'\rangle, |E_2'\rangle\}$ that could be performed on the environment and would allow us to distinguish between the two states $|\psi_+\rangle$ and $|\psi_-\rangle$.

Hence the particular interaction (2.78) between the system and the environment is completely insensitive to the difference between the states $|\psi_+\rangle$ and $|\psi_-\rangle$. No information is encoded in the environment as a result of this interaction that would allow us to distinguish these two states by looking at the environment. Thus the environment cannot bring about any suppression of interference between $|\psi_+\rangle$ and $|\psi_-\rangle$.

We can also understand this point more directly by considering a superposition of the states $|\psi_+\rangle$ and $|\psi_-\rangle$,

$$|\psi\rangle = \frac{1}{\sqrt{2}} \left(|\psi_+\rangle + |\psi_-\rangle \right). \tag{2.82}$$

But from the definition (2.79) it is immediately obvious that this superposition is simply equal to the original localized state $|\psi_1\rangle$. Thus it follows from the dynamics (2.78) that the state (2.82) will not get entangled with the environment at all,

$$|\psi\rangle |E_0\rangle \xrightarrow{(2.78)} |\psi\rangle |E_1\rangle. \tag{2.83}$$

Because of the absence of any environmental entanglement in (2.83), there will be no local damping of interference between the components $|\psi_+\rangle$ and $|\psi_-\rangle$.

We can generalize these observations to the main result of this section. Some states of the system are more prone to decoherence than others, and the sensitivity of a particular state is determined by the structure of the system–environment interaction. The *preferred states* of the system emerge dynamically as those states that are the least sensitive, or the most *robust*, to the interaction with the environment, in the sense that they become least entangled with the environment in the course of the evolution and are thus most immune to decoherence. This is commonly referred to as the *stability criterion* for the selection of preferred states, introduced and discussed by Zurek in two seminal papers in the early 1980s [8, 9].

The resulting environmental dynamical selection of preferred states was first studied and formalized in detail by Zurek [8, 9] under the heading of *environment-induced superselection* (in the literature the abbreviation "einselection" is also frequently used). The term "superselection" is chosen in reference to so-called superselection rules frequently used in areas such as elementary-particle physics [78–83] (see also Chap. 6 of [17]). Such (kinematical) superselection rules were historically *postulated* to *a priori* exclude certain never-observed physical states, such as superpositions of a proton and a neutron (which were thus considered forbidden by virtue of a "charge superselection rule").

Similarly, we may say that the interaction with the environment "superselects" the observable states of the system: Some states are robust in spite of the environmental interaction, while other states are rapidly decohered and become therefore unobservable in practice. However, in contrast with the postulated "exact" superselection rules, environment-induced superselection represents *effective* superselection rules that dynamically emerge from the (structure of the) system–environment interaction. Environment-induced superselection may therefore be regarded as a more powerful concept than that of postulated superselection rules, since it *explains* [4,8,9,83–85] why we do not observe certain states, instead of simply precluding the existence of such states from the outset.[19] Already early on, Zeh [4] had emphasized the importance of dynamical stability conditions and suggested that the openness of quantum systems may explain the "superselection" of preferred quantities (such as charge, handedness of sugar molecules, parity of ammonia molecules, etc.).

For the simple example of the system–environment interaction given by (2.78), we immediately see that a system that starts out in either one of the states $|\psi_1\rangle$ and $|\psi_2\rangle$ will not get entangled with the environment: The final composite system–environment state will at all times remain in the separable product forms $|\psi_1\rangle\,|E_1\rangle$ and $|\psi_2\rangle\,|E_2\rangle$, respectively. Thus the environment can be regarded as carrying out a quantum nondemolition measurement on the system, i.e., a measurement that does not disturb the state of the system (see also Sect. 2.5.1). In the case of the scattering of environmental particles discussed above, $|\psi_1\rangle$ and $|\psi_2\rangle$ represent spatially well-localized states, and accordingly the environment-superselected *preferred observable* is the position of the system. Conversely, if the system is described by the superposition (2.70) of $|\psi_1\rangle$ and $|\psi_2\rangle$, it will become maximally entangled with the environment, leading to decoherence in the $\{|\psi_1\rangle\,,|\psi_2\rangle\}$ basis. The system observable corresponding to the measurement of a superposition of positions would therefore be "difficult" to measure, in the sense that the spatial interference terms in the reduced matrix corresponding to a measurement of this observable decay rapidly due to the interaction with the environment. Broadly speaking, the fragility of such spatial superpositions thus means that they are hard to observe in practice. As originally shown by Zurek [8, 9, 12] and Joos and Zeh [7], this provides us with a powerful explanation for why most physical systems are usually found in well-defined positions rather than in delocalized superposition states (see Sect. 2.8.4).

Zurek also coined the term *pointer states* [8] for the preferred states selected by the stability criterion. This terminology is motivated by the idea that, because of their robustness, the environment-superselected preferred states correspond to the physical quantities that are most easily "read off" at the level of the system, analogous to the reading-off of a pointer on the dial of a measurement apparatus. Since robustness is a hallmark of states in classical physics (see Sect. 2.1), pointer states can be viewed as a "stand-in" for such classical states (one thus says that pointer states represent *quasiclassical* states). In the following, we shall use the terms "environment-superselected states," "preferred states," and "(preferred) pointer states" interchangeably for the states selected by the interaction with the environment in the sense of the stability criterion.

In some cases the environment-superselected states form a basis for the Hilbert space of the system, i.e., decoherence leads to the emergence of a

[19] However, it should be noted that the "exact" superselection rules can often be well motivated from symmetry and other arguments and are therefore not simply introduced via arbitrary *ad hoc* postulates.

pointer basis for the system. This situation is encountered, for example, in the context of systems effectively described by a discrete two-dimensional state space. On the other hand, there are important cases of interest in which the environment-superselected states form an overcomplete set of states. That is, the pointer states are not necessarily mutually orthogonal and therefore do not represent proper basis states for the Hilbert space of the system. An important example that will be discussed in more detail in Sect. 5.2 is the environment-induced superselection of a continuum of coherent states in phase space, i.e., of minimum-uncertainty Gaussian wave packets well localized in both position and momentum [6,86]. Such coherent states have a finite overlap and therefore do not represent a proper basis.

Let us now reconsider the von Neumann measurement scheme (2.52) (see Sect. 2.5.1) and the associated preferred-basis problem (Sect. 2.5.2) by including a subsequent interaction with the environment. Following our above considerations, this interaction will select preferred pointer states for the system–apparatus combination, which are those states that get least entangled with the environment. This will be the case for the states $|s_i\rangle\,|a_i\rangle$ on the right-hand side of (2.52) if the interaction with the environment has the form of a nondemolition measurement of $|s_i\rangle\,|a_i\rangle$,[20]

$$|s_i\rangle\,|a_i\rangle\,|E_0\rangle \longrightarrow |s_i\rangle\,|a_i\rangle\,|E_i\rangle \qquad \text{for all } i. \tag{2.84}$$

For simplicity, we may assume that the environment is in direct interaction dominantly with the (usually macroscopic) apparatus but only insignificantly with the (often microscopic) system. Then (2.84) means that the states $|a_i\rangle$—which describe the different pointer positions of the apparatus—are the robust preferred states superselected by the environment. (This observation is another direct motivation for Zurek's pointer-state terminology.) In turn, the form of the system–apparatus dynamics then implies the selection of the states of the system which can be measured by the apparatus interacting with the environment, i.e., which can be reliably recorded through the formation of dynamically stable system–apparatus quantum correlations [8,9].

We may also look at this issue from a slightly different angle by realizing that (2.84) corresponds to the requirement that the interaction with the environment does not disturb the established quantum correlations between the state $|s_i\rangle$ of the system and the corresponding apparatus state $|a_i\rangle$. This criterion was first suggested and explored by Zurek [8]. It can be viewed as a generalization of the concept of "faithful measurement" to the realistic

[20]Note that while for two subsystems, say, $\mathcal{S}$ and $\mathcal{A}$, there always exists a diagonal decomposition of the final state of the form $\sum_n c_n\,|s_n\rangle\,|a_n\rangle$ (see Sect. 2.15.1), for three subsystems (for example, $\mathcal{S}$, $\mathcal{A}$, and $\mathcal{E}$) a decomposition of the form $\sum_n c_n\,|s_n\rangle\,|a_n\rangle\,|e_n\rangle$ is not always possible. This implies that the total Hamiltonian that induces a time evolution of the kind (2.84) must be of a special form. For a comment regarding these restrictions on the form of the evolution operator and the possibility of a resulting disagreement with experimental evidence, see [64].

case in which the environment is included. Faithful measurement in the usual sense concerns the requirement that the measuring apparatus act as a reliable "mirror" of the states of the system by forming only correlations of the form $|s_i\rangle\,|a_i\rangle$ but not $|s_i\rangle\,|a_j\rangle$ with $i \neq j$. But since any realistic measurement process must include the inevitable coupling of the apparatus to its environment, the measurement could hardly be considered faithful as a whole if the interaction with the environment disturbed the correlations between the system and the apparatus.[21] In fact, this ability of the apparatus to serve as a robust and faithful "indicator" of the state of the system amounts to the very *definition* of a measurement device. Environment-induced superselection applied to the von Neumann scheme therefore explains why measurement apparatuses seem to be *designed* to measure only certain physical quantities but not others.

Let us now return to the case of a system interacting with an environment without the explicit presence of an apparatus. Thus far, we have considered the simplified ideal situation in which the effective system–environment dynamics were directly written in the von Neumann form, as in (2.78). This allowed us to immediately infer the preferred states of the system. In general, however, we are simply given a total Hamiltonian $\hat{H}$, which describes the intrinsic dynamics of the system and environment as well as the interaction between the system and environment. Typically, this total Hamiltonian is decomposed into the relevant three parts,

$$\hat{H} = \hat{H}_{\mathcal{S}} + \hat{H}_{\mathcal{E}} + \hat{H}_{\text{int}}, \tag{2.85}$$

where $\hat{H}_{\mathcal{S}}$ and $\hat{H}_{\mathcal{E}}$ denote the self-Hamiltonians of the system and environment representing the intrinsic dynamics, and $\hat{H}_{\text{int}}$ is the system–environment interaction Hamiltonian.

We may now ask the important question: Given an arbitrary total Hamiltonian defining the system–environment model, how are we to determine the pointer states for the system? Of course, our general selection criterion remains the same: The preferred set of pointer states corresponds to the set of states of the system that are most robust under the influence of the environment. However, how are we to find out which states fulfill this requirement? This problem will be discussed in the following.

2.8.1 Pointer States in the Quantum-Measurement Limit

Let us first consider an important special case that not only applies to many situations relevant to decoherence, but also provides us with a simple and intuitive criterion for determining the preferred set of pointer states of the system. Suppose the energy scales associated with the system–environment interaction Hamiltonian $\hat{H}_{\text{int}}$ in (2.85) are much larger than the energy scales

[21] For fundamental limitations on the precision of von Neumann measurements of operators that do not commute with a globally conserved quantity, see the Wigner–Araki–Yanase theorem [87, 88].

of the self-Hamiltonians $\hat{H}_{\mathcal{S}}$ and $\hat{H}_{\mathcal{E}}$. That is, we consider the *quantum-measurement limit* in which the interaction between the system and the environment is so strong as to completely dominate the evolution of the system. In other words, the intrinsic dynamics of the system and the environment are negligible in comparison with the evolution induced by the interaction. Then the total Hamiltonian (2.85) can be approximated by the interaction Hamiltonian $\hat{H}_{\text{int}}$ alone,

$$\hat{H} \approx \hat{H}_{\text{int}}. \tag{2.86}$$

Let us now find the set of robust states of the system that do not get entangled under the evolution generated by this Hamiltonian. To do so, we look for system states $|s_i\rangle$ such that the composite system–environment state, when starting from a product state $|s_i\rangle |E_0\rangle$ at time $t = 0$, remains in the product form $|s_i\rangle |E_i(t)\rangle$ at all subsequent times $t > 0$ under the action of the interaction Hamiltonian $\hat{H}_{\text{int}}$. That is, we demand that[22]

$$\mathrm{e}^{-\mathrm{i}\hat{H}_{\text{int}}t} |s_i\rangle |E_0\rangle = \lambda_i |s_i\rangle \mathrm{e}^{-\mathrm{i}\hat{H}_{\text{int}}t} |E_0\rangle \equiv |s_i\rangle |E_i(t)\rangle . \tag{2.87}$$

This immediately yields our main result: In the quantum-measurement limit (2.86) the pointer states $|s_i\rangle$ of the system will simply be given by the eigenstates of the part of the interaction Hamiltonian $\hat{H}_{\text{int}}$ pertaining to the Hilbert space of the system. This important result was first obtained by Zurek in his famous paper of 1981 [8]. It is quite intuitive: After all, the eigenstates $|s_i\rangle$ will be stationary under the action of $\hat{H} = \hat{H}_{\text{int}}$.

We may also equivalently express this result in terms of *pointer observables*, which are simply linear combinations of the pointer projectors $|s_i\rangle\langle s_i|$,

$$\hat{O}_{\mathcal{S}} = \sum_i o_i |s_i\rangle\langle s_i|. \tag{2.88}$$

Since the $|s_i\rangle$ are eigenstates of $\hat{H}_{\text{int}}$, it follows that each term $|s_i\rangle\langle s_i|$ commutes with $\hat{H}_{\text{int}}$, and therefore also $\hat{O}_{\mathcal{S}}$ itself commutes with $\hat{H}_{\text{int}}$,

$$\left[\hat{O}_{\mathcal{S}}, \hat{H}_{\text{int}}\right] = 0. \tag{2.89}$$

This important condition is often referred to as the *commutativity criterion* in the literature. It was first introduced and discussed by Zurek [8, 9] and has since become one of the cornerstones of the formal body of the decoherence program.

Thus, we have found two alternative but equivalent ways of determining the environment-superselected quantities in the quantum-measurement limit (2.86). The pointer *states* are given by those states of the system that are eigenstates of the interaction Hamiltonian, and the corresponding pointer *observables* of the system are those observables commuting with the interaction

[22]We assume here that $\hat{H}_{\text{int}}$ is not explicitly time-dependent.

Hamiltonian. Of course, pointer states and pointer observables are a largely synonymous concept, since the latter are simply linear combinations of the pointer-state projectors [see (2.88)].

In many cases of interest, the structure of the interaction Hamiltonian is such as to immediately reveal the corresponding system eigenstates and thus the preferred set of pointer states. For instances, the interaction Hamiltonian is often given in the product form

$$\hat{H}_{\text{int}} = \hat{S} \otimes \hat{E}, \tag{2.90}$$

with $\hat{S}$ and $\hat{E}$ denoting operators acting on the Hilbert spaces of the system and environment, respectively. Thus the environment-superselected observables will be those observables that commute with $\hat{S}$. If $\hat{S}$ is Hermitian, the product structure $\hat{S} \otimes \hat{E}$ of the interaction Hamiltonian (2.90) has a very intuitive interpretation: $\hat{S}$ represents the physical quantity that is directly monitored by the environment. A frequently encountered example for $\hat{S}$ is the position operator $\hat{x}$, such that the interaction Hamiltonian is of the form

$$\hat{H}_{\text{int}} = \hat{x} \otimes \hat{E}. \tag{2.91}$$

This interaction Hamiltonian describes the continuous monitoring of the position of the system by the environment. That is, the environment is performing an effective nondemolition measurement in the position basis of the system. Thus we can immediately anticipate that spatial superpositions will become decohered. If we denote the eigenstates of $\hat{x}$ by $|x\rangle$, the explicit system–environment evolution generated by the interaction Hamiltonian (2.91) is

$$\mathrm{e}^{-\mathrm{i}\hat{H}_{\text{int}}t} |x\rangle |E_0\rangle = |x\rangle \, \mathrm{e}^{-\mathrm{i}x\hat{E}t} |E_0\rangle \equiv |x\rangle |E_x(t)\rangle \, . \tag{2.92}$$

Here we have used the subscript "x" in $|E_x(t)\rangle$ to denote the fact that the state of the environment now contains information about the position of the system. As discussed in Sect. 2.7, the time dependence of the overlap between the states $|E_x(t)\rangle$ for different values of x will then yield the resulting decoherence dynamics and timescales [see (2.76) and (2.77)].

We note that one can show that it is always possible to write an arbitrary interaction Hamiltonian $\hat{H}_{\text{int}}$ in form of a diagonal decomposition of (unitary but not necessarily Hermitian) system and environment operators $\hat{S}_\alpha$ and $\hat{E}_\alpha$, respectively,

$$\hat{H}_{\text{int}} = \sum_\alpha \hat{S}_\alpha \otimes \hat{E}_\alpha. \tag{2.93}$$

This interaction Hamiltonian is an obvious generalization of the single-term Hamiltonian (2.90). If the $\hat{S}_\alpha$ are Hermitian, (2.93) describes the simultaneous environmental monitoring of different observables $\hat{S}_\alpha$ on the system. A sufficient condition for $\{|s_i\rangle\}$ to form a set of pointer states of the system is then given by the requirement that the $|s_i\rangle$ be simultaneous eigenstates of the operators $\hat{S}_\alpha$,

$$\hat{S}_\alpha |s_i\rangle = \lambda_i^{(\alpha)} |s_i\rangle \qquad \text{for all } \alpha \text{ and } i, \tag{2.94}$$

since in this case

$$\begin{aligned} e^{-i\hat{H}_{\text{int}}t} |s_i\rangle |E_0\rangle &= e^{-i(\sum_\alpha \hat{S}_\alpha \otimes \hat{E}_\alpha)t} |s_i\rangle |E_0\rangle \\ &= |s_i\rangle e^{-i\left(\sum_\alpha \lambda_i^{(\alpha)} \hat{E}_\alpha\right)t} |E_0\rangle \\ &\equiv |s_i\rangle |E_i(t)\rangle . \end{aligned} \tag{2.95}$$

That is, the system described by the initial state $|s_i\rangle$ does not get entangled with the environment, and thus $|s_i\rangle$ represents an environment-superselected preferred state. Fortunately, in decoherence models of practical interest, it is rare to encounter interaction Hamiltonians (2.93) involving more than two different observables $\hat{S}_\alpha$. In fact, the interaction Hamiltonians of the models discussed in Chap. 5 will all be of the simple single-term form (2.90).

The pointer-state condition (2.94) can be strengthened to the concept of *pointer subspaces*, first introduced by Zurek in 1982 [9] and recently further explored (especially in the context of quantum computing) by several authors under the heading of *decoherence-free subspaces* (DFS) [89–98]. Pointer subspaces, or DFS, are subspaces of the Hilbert space of the system in which *every* state in the subspace is immune to decoherence. (Recall that in general superpositions of pointer states will not be pointer states themselves.)

One important condition for this to happen is that the preferred states $|s_i\rangle$ defined by the requirement (2.94) form an orthonormal basis $\{|s_i\rangle\}$ of the subspace, and that the eigenvalues $\lambda_i^{(\alpha)}$ in (2.94) are independent of the index i, i.e., that all $|s_i\rangle$ are simultaneous *degenerate* eigenstates of each $\hat{S}_\alpha$,

$$\hat{S}_\alpha |s_i\rangle = \lambda^{(\alpha)} |s_i\rangle \qquad \text{for all } \alpha \text{ and } i. \tag{2.96}$$

Then any state $|\psi\rangle$ of the subspace can be expressed as a superposition $|\psi\rangle = \sum_i c_i |s_i\rangle$ of the states $|s_i\rangle$, and the superposition will evolve as

$$\begin{aligned} e^{-i\hat{H}_{\text{int}}t} |\psi\rangle |E_0\rangle &= e^{-i\hat{H}_{\text{int}}t} \left(\sum_i c_i |s_i\rangle \right) |E_0\rangle \\ &= \left(\sum_i c_i |s_i\rangle \right) e^{-i(\sum_\alpha \lambda^{(\alpha)} \hat{E}_\alpha)t} |E_0\rangle \\ &\equiv |\psi\rangle |E_\psi(t)\rangle . \end{aligned} \tag{2.97}$$

Thus the state $|\psi\rangle$ does not become entangled with the environment. Since $|\psi\rangle$ was completely arbitrary, all states in the subspace spanned by the set of orthonormal states $|s_i\rangle$ obeying (2.96) will be immune to decoherence.

We can interpret the condition (2.96) also as implying that there is no term in the interaction Hamiltonian that would act jointly on both the system and the environment in a nontrivial manner [see (2.97)]. Thus, if the system starts out in a state within the subspace spanned by basis vectors fulfilling this condition and if it has no correlations with the environment, the system–environment combination will automatically remain in a separable state at all subsequent times (provided that $\hat{H}_{\mathcal{S}}$ does not take the state out of the subspace; see the comment below). The system does not get entangled with its environment, and thus no decoherence occurs. Also note that, since (2.96) implies that the action of a given interaction operator $\hat{S}_\alpha$ is the same for all basis states $|s_i\rangle$ of the DFS, the existence of a DFS corresponds to a symmetry in the structure of the system–environment interaction, i.e., to a *dynamical symmetry*.

In the more general case where the self-Hamiltonian $\hat{H}_{\mathcal{S}}$ of the system also plays a role we would additionally need to ensure that the subspace remains decoherence-free over time under the action of $\hat{H}_{\mathcal{S}}$. That is, we will need to demand that none of the basis states $|s_i\rangle$ of the DFS will drift out of the subspace under the evolution generated by $\hat{H}_{\mathcal{S}}$. Otherwise an initially decoherence-free state would in general become prone to decoherence. More formally, we can express this condition by saying that $\hat{H}_{\mathcal{S}}$ must not project any of the basis states $|s_i\rangle$ into a Hilbert subspace outside of the subspace spanned by the basis $\{|s_i\rangle\}$. A third condition for a DFS to exist is given by the requirement that the system $\mathcal{S}$ and the environment $\mathcal{E}$ must be completely uncorrelated at some initial time $t = 0$. It can be rigorously proved that the three conditions listed above are both sufficient and necessary for the existence of a DFS.[23]

Pointer subspaces, or DFS, have attracted much interest over the past decade because of their relevance to quantum computing. There the basic idea is to encode the fragile quantum information stored in the quantum computer in such subspaces so as to naturally protect it from decoherence. We will describe this approach in more detail in Sect. 7.5 (see also the review article by Lidar and Whaley [101]). The ideas behind pointer subspaces have also been used to propose methods for taming decoherence in other areas of interest, for example, in the context of superposition states of macroscopically distinguishable states in Bose–Einstein condensates [102] (see Sect. 6.4.1). Finally, we shall note here that the concept of DFS has recently been extended and generalized to the formalism of *noiseless subsystems* (or "noiseless quantum codes"), first developed by Knill, Laflamme, and Viola [98] (see also [100]). A brief review of this topic can be found in Sect. 7 of [101].

[23] One of the earliest proofs (by Zanardi and Rasetti [92]) employed a standard Hamiltonian approach. Lidar, Chuang, and Whaley [90], and independently Zanardi [93], used an approach based on the Born–Markov master-equation formalism described in Chap. 4. This was later generalized to the non-Markovian setting by Bacon, Lidar, and Whaley [99]. A general proof using the operator-sum formalism (see Sect. 2.15.4) was given by Lidar, Bacon, and Whaley [94]. The conditions can also be arrived at by employing the language of quantum error correction (see Sect. 7.4) or the so-called "stabilizer" formalism (see, e.g., [100]). For a comparison of these different formulations, see Sect. 5.6 of [101].

2.8.2 Pointer States in the Quantum Limit of Decoherence

Let us now consider the situation in which the modes of the environment are "slow" in comparison with the evolution of the system. That is, we assume that the highest frequencies (i.e., energies) available in the environment are smaller than the separation between the energy eigenstates of the system. In this situation, the environment will be able to monitor only quantities which are constants of motion. In the case of nondegeneracy, this quantity will be the energy of the system, thus leading to the environment-induced superselection of energy eigenstates for the system (i.e., eigenstates of the self-Hamiltonian of the system). This process was first investigated in detail by Paz and Zurek [103], who called it the *quantum limit of decoherence.*

In introductory textbooks on quantum mechanics, such energy eigenstates (for closed systems) are usually attributed a special role since they are stationary under the action of the Hamiltonian. Indeed, much of the subsequent problems considered in these textbooks are then concerned with calculating the energy eigenstates for different Hamiltonians describing various (microscopic) physical systems such as the hydrogen atom. However, the superposition principle would predict that arbitrary superpositions of such energy eigenstates should nonetheless be perfectly legitimate.

It is therefore very important to emphasize that the selection of energy eigenstates in the limit where the evolution is dominated by the self-Hamiltonian of the system is *not* equivalent to a situation in which the presence of the environment could simply be neglected altogether (which would correspond to the aforementioned closed-system case commonly treated in textbooks). Quite to the contrary, the environment plays the crucial role of continuously monitoring the energy of the system, which leads to decoherence in the energy eigenbasis of the system and hence to the local suppression of superpositions of energy eigenstates. Thus, although the form of the pointer states is dominated by the eigenstates of the self-Hamiltonian of the system, it is (once again) the presence of the environment that imposes an effective superselection rule on the space of observable superpositions.

2.8.3 General Methods for Determining the Pointer States

In most realistic cases of interest, the commutativity criterion (2.89) can usually only be fulfilled approximately [84, 86]. For example, as mentioned in Sect. 2.5.4, individual eigenstates of the position operator are not proper quantum states of physical objects. Although, for example, the interaction Hamiltonian (2.91) together with the commutativity criterion (2.89) would imply the selection of such eigenstates as the preferred set of pointer states of the system, in practice the preferred states will be narrow position-space wave packets, i.e., they will represent only approximate eigenstates of position, with a finite spread in position space. In other situations it may simply be impossible to find any observable for which the commutativity criterion (2.89) would be fulfilled exactly.

Thus, although the diagonalization of the interaction Hamiltonian or the application of the commutativity criterion (2.89) are conceptually intuitive methods for determining the preferred states and observables in simple cases for which the quantum-measurement limit holds, they are often not adequate and practicable in more complex situations. The goal of finding exact pointer states that are *completely* immune to the immersion into the environment is therefore often weakened to the aim of determining the states of the system that are *most* robust to the environmental interaction. Approximate pointer states are then our best representation of quasiclassical states that are minimally affected by the inevitable interaction with the environment. Also, in many realistic physical situations of interest, both the self-Hamiltonian of the system and the system–environment interaction Hamiltonian both contribute in roughly equal strengths (for example, this is the case in the model for quantum Brownian motion discussed in Sect. 5.2). Then neither the quantum-measurement limit of negligible intrinsic dynamics nor the quantum limit of decoherence corresponding to a "slow" environment applies.

Therefore more general methods are needed for determining the preferred states selected by the interaction with the environment. Such methods were developed by Zurek [84, 104] and Zurek, Habib, and Paz [86] under the name of the *predictability-sieve* strategy. The basic idea consists of using a suitable measure for the amount of decoherence introduced into the system, such as the purity or the von Neumann entropy of the reduced density matrix (see Sect. 2.4.3). One then computes the time dependence of the purity or von Neumann entropy (or whichever measure has been chosen) for a large set of initial states of the system evolving under the total system–environment Hamiltonian. The states most immune to decoherence will be those which lead to the smallest decrease in purity or the smallest increase in von Neumann entropy. Application of this method leads to a ranking of the possible preferred states with respect to their "classicality," i.e., their robustness with respect to the interaction with the environment.

We thus picture ourselves "sifting" through the Hilbert space, sorting the states according to their robustness to environmental interactions. The most stable states will also be the most *predictable* (thus motivating the terminology "predictability sieve" introduced by Zurek). This is so because the fact that the loss of purity or the increase in von Neumann entropy is minimized for these states also means that the loss of "information" about the state of the system is minimized (recall that a pure state, which has zero von Neumann entropy, corresponds to a maximum amount of knowledge about the state of the system).

In general, the preferred states selected by the predictability sieve may differ depending on which measure is used to rank the robustness of the states. We already mentioned two measures, namely, purity and von Neumann entropy, but other meaningful measures of decoherence and robustness exist and more are likely to be suggested in the future. For particular models it has been explicitly shown that the states picked out by the predictability sieve are rather insensitive to the particular choice of the measure. For example, in the model for quantum Brownian motion (see Sect. 5.2), different measures lead to the same minimum-uncertainty wave packets [6, 17, 84, 105, 106]. It is reasonable to anticipate that, at least in the macroscopic limit, the resulting stable pointer states obtained from different criteria will turn out to be very similar (see also the discussion by Zurek in [16]).

2.8.4 Selection of Quasiclassical Properties

Let us summarize our results for the environment-induced selection of preferred states and discuss the implications for the general preferred-basis problem outlined in Sect. 2.5.2 and for our observation of only particular physical quantities in the world around us.

System–environment interaction Hamiltonians frequently describe a scattering process of surrounding particles (photons, air molecules, etc.) interacting with the system under study. Since the force laws describing such processes typically depend on some power of distance (such as $\propto r^{-2}$ in Newton's or Coulomb's force law), the interaction Hamiltonian will usually commute with the position operator. According to the commutativity requirement (2.89), the pointer states will therefore be approximate eigenstates of position. The fact that position is typically the determinate property of our experience can thus be explained by referring to the dependence of most interactions on distance. This origin of the special role of position in the quantum-to-classical transition was clearly pointed out and analyzed for the first time by Zurek [8, 9, 13]. Subsequently, the scattering model of Joos and Zeh [7] showed directly how surrounding photons and air molecules continuously measure the spatial structure of small objects such as dust particles, leading to rapid decoherence into an (improper) mixture of narrow position-space wave packets (see Chap. 3).

Similar results sometimes hold even for microscopic systems (usually found in energy eigenstates; see below) when they occur in distinct spatial structures that couple strongly to the surrounding medium. For instance, chiral molecules such as sugar are always observed to be in chirality eigenstates (left-handed and right-handed), which are superpositions of different energy eigenstates [107, 108] (see also Sect. 3.2.4 of [17]). Once again, this can be explained by the fact that the distinct spatial structure of these molecules is continuously monitored by the environment through scattering processes. This environmental monitoring gives rise to a much stronger coupling to the "outside world" than could typically be achieved by a measuring device that was intended to measure, say, parity or energy. Furthermore, any attempt to prepare such molecules in energy eigenstates will lead to immediate decoherence into the environmentally stable chirality eigenstates, thus selecting position as the preferred quantity.

On the other hand, it is well known that microscopic systems such as atoms are typically found in energy eigenstates, even if the interaction Hamiltonian depends on a different observable than energy. This is easily explained by noting that in such cases the evolution of the system is dominated by the self-Hamiltonian, while the environment is comparably slow (in the sense that its largest frequencies are smaller than the spacing of the energy eigenstates of the system). As discussed in Sect. 2.8.2, this quantum limit of decoherence [103] results in the environment-induced selection of energy eigenstates, with interference between different energy eigenstates being continuously suppressed due to the environmental monitoring of the energy of the system.

Another interesting example is the environment-induced superselection of charge. In nature, only eigenstates of the charge operator are observed, but never superpositions of different charges. The corresponding superselection rules were first only postulated [78, 79] but were subsequently explained by Giulini, Kiefer, and Zeh [85] (see also [83]) by referring to the interaction of the charge with its own Coulomb (far) field. This field plays the role of the environment, leading to immediate decoherence of charge superpositions into (improper) mixtures of charge eigenstates.

To summarize, we have distinguished three different cases for the type of preferred pointer states emerging from interactions with the environment:

1. *The quantum-measurement limit.* When the evolution of the system is dominated by $\hat{H}_{\text{int}}$, i.e., by the interaction with the environment, the preferred states will be eigenstates of $\hat{H}_{\text{int}}$ (and thus often eigenstates of position).
2. *The quantum limit of decoherence.* When the environment is slow and the self-Hamiltonian $\hat{H}_{\mathcal{S}}$ dominates the evolution of the system, a case frequently encountered in the microscopic domain, the preferred states will be energy eigenstates, i.e., eigenstates of $\hat{H}_{\mathcal{S}}$ [103].
3. *The intermediary regime.* When the evolution of the system is governed by $\hat{H}_{\text{int}}$ and $\hat{H}_{\mathcal{S}}$ in roughly equal strengths, the resulting preferred states will represent a compromise between the first two cases. For instance, in the frequently studied model of quantum Brownian motion (see Sect. 5.2) the interaction Hamiltonian $\hat{H}_{\text{int}}$ describes monitoring of the position of the system. However, through the intrinsic dynamics induced by $\hat{H}_{\mathcal{S}}$ this monitoring also leads to indirect decoherence in momentum. This combined influence of $\hat{H}_{\text{int}}$ and $\hat{H}_{\mathcal{S}}$ results in the emergence of preferred states localized in phase space, i.e., in both position and momentum (see Sect. 5.2.5 and [16, 17, 86, 106, 109]).

In fact, these three regimes can in turn be understood as limiting cases of the general rule that the preferred pointer states are the states of the system that are most robust under the evolution generated by the *total* Hamiltonian. Depending on which term dominates in this total Hamiltonian (in terms of the relative energy scales), the pointer basis is then formed either by eigenstates of the interaction Hamiltonian, by eigenstates of the self-Hamiltonian, or by states arising as a compromise between the interaction Hamiltonian and self-Hamiltonian.

The clear merit of the approach of environment-induced superselection to the preferred-basis problem lies in the fact that the preferred basis is not chosen in an *ad hoc* manner so as to simply make our measurement records determinate or to match our experience of which physical quantities are usually perceived as determinate (for example, position). Instead the selection is motivated on physical, observer-free grounds, namely, through the structure of the system–environment interaction Hamiltonian. The vast space of possible quantum-mechanical superpositions is reduced so much because the laws governing physical interactions depend on only a few physical quantities (position, momentum, charge, and the like), and the fact that precisely these are the properties that appear determinate to us is explained by the dependence of the preferred basis on the form of the interaction. The appearance of classicality is therefore grounded in the structure of the physical laws governing the system–environment interactions.

2.9 Redundant Encoding of Information in the Environment and "Quantum Darwinism"

Thus far we have focused on what can be observed at the level of the system. The environment has simply played the role of a "sink" that carries away which-path (or, more generally, which-state) information about the system. We have assumed that we do not further observe the environment or interact with it. At the same time, since the interaction between the system and the environment constantly encodes information about the system in the environment, the environment constitutes a huge resource for the indirect acquisition of information about the system. In fact, it is important to realize that in most (if not all) cases observers gather information about the state of a system through indirect observations, namely, by intercepting fragments of environmental degrees of freedom that have interacted with the system in the past and thus contain information about the state of the system. Probably the most common example for such indirect acquisition of information is the visual registration of photons that have scattered off the object of interest. We see things not by directly interacting with them, but through the light that has "measured" the spatial structure of the object.

Why is this realization of the importance of indirect observation of systems through the interception of environmental degrees of freedom so important? To answer this question, recall that in Sect. 2.1.1 we pointed out that a characteristic feature of classical physics is the fact that the state of a system can be found out and agreed upon by many independent observers (who all can be initially completely ignorant about the state) without disturbing this state. In this sense, classical states preexist objectively, resulting in our notion of "classical reality." By contrast, measurements on a closed quantum system will in general alter the state of the system.

It is therefore impossible to regard quantum states of a closed system as existing in the way that classical states do. This raises the question of how classical reality emerges from within the quantum substrate, i.e., how observables are "objectified" in the above sense. The environment-induced superselection of preferred states discussed in the previous Sect. 2.8 has certainly made a significant contribution toward answering this question by explaining why only a certain subset of the possible states in the Hilbert space of the system are actually observable. Nonetheless, the problem sketched in the previous paragraph remains, as any direct measurement performed on the system would, in general, still alter the state of the system.

As clearly recognized and systematically investigated for the first time by the Los Alamos group of Zurek, Ollivier, Poulin, and Blume-Kohout [16,66,76,84,104,110–113], it is here that the function of the environment as a which-state monitor plays another important role. Since information about the system is encoded in the environment, we can acquire this information without having to directly interact (and thereby disturb) the system itself. Thus the role of the environment is now broadened, namely, from the selection of preferred states for the system of interest and the delocalization of local phase coherence between these states to the transmission of information about the state of the system. The key question, first spelled out by the Los Alamos group, is then the following. How, and which kind of, information is both *redundantly* and *robustly* stored in a large number of distinct fragments of the environment in such a way that multiple observers can retrieve this information without disturbing the state of the system, thereby achieving effective classicality of the state?[24]

The research of the Los Alamos group into answering this question is carried out under the headings of the *"environment as a witness"* program (the recognition of the role of the environment as a communication channel) and *quantum Darwinism* (the study of what information about the system can be stably stored and proliferated by the environment) [66, 110–113]. To explicitly quantify the degree of completeness and redundancy of information imprinted on the environment, the measure of (classical [110,111] or quantum [16, 112, 113]) mutual information has usually been used. Roughly speaking, this quantity represents the amount of information (expressed in terms of Shannon [114–116] or von Neumann entropies) about the system that can be acquired by measuring (a fragment of) the environment.[25]

---

[24]The importance of redundancy in quantum measurement was first emphasized by Zurek [76] already in the early 1980s (see also [9]).

[25]Note that the amount of information contained in each fragment is always somewhat less than the maximum information provided by the system itself (as given by the von Neumann entropy of the system) [113].

The measure of classical mutual information is based on the choice of particular observables of the system $\mathcal{S}$ and the environment $\mathcal{E}$ and quantifies how well one can predict the outcome of a measurement of a given observable of $\mathcal{S}$ by measuring some observable on a fraction of $\mathcal{E}$ [110, 111]. The quantum mutual information $\mathcal{I}_{\mathcal{S}:\mathcal{E}}$, used in more recent studies by Blume-Kohout and Zurek [112, 113], can be viewed as a generalization of classical mutual information and is defined as $\mathcal{I}_{\mathcal{S}:\mathcal{E}} \equiv S(\hat{\rho}_{\mathcal{S}}) + S(\hat{\rho}_{\mathcal{E}}) - S(\hat{\rho})$, where $\hat{\rho}_{\mathcal{S}}$, $\hat{\rho}_{\mathcal{E}}$, and $\hat{\rho}$ are the density matrices of $\mathcal{S}$, $\mathcal{E}$, and the composite system $\mathcal{SE}$, respectively, and $S(\hat{\rho})$ is the von Neumann entropy (2.32). Thus $\mathcal{I}_{\mathcal{S}:\mathcal{E}}$ measures the amount of entropy produced by destroying all (quantum) correlations between $\mathcal{S}$ and $\mathcal{E}$, i.e., it quantifies the degree of correlations between $\mathcal{S}$ and $\mathcal{E}$. Results derived from the two measures—classical mutual information based on the Shannon entropies of the subsystems, and quantum mutual information defined in terms of the von Neumann entropies—have been found to nearly coincide [16, 110–113]. This robustness with respect to the particular choice of measure is due to the fact that the difference between the two measures (the so-called *quantum discord* [117]) disappears when decoherence is sufficiently effective to select a well-defined pointer basis, as shown by Ollivier and Zurek [117]. Therefore the quantum discord for the information about the system decohered by all of the environment (except for the fragment that is being intercepted) must be small.

Ollivier, Poulin, and Zurek [110, 111] and Blume-Kohout and Zurek [112, 113] found that the observable of the system that can be imprinted most completely and redundantly in many distinct subsets of the environment coincides with the pointer observable selected by the system–environment interaction as discussed in Sect. 2.8. Conversely, most other states do not seem to be redundantly storable. Thus it suffices to probe a comparably very small fraction of the environment to infer a large amount of the maximum information about the pointer state of the system. On the other hand, if the observer tried to measure other observables on the same fragment, she would learn virtually nothing, as information about the corresponding observables of the system is not redundantly stored. Thus the environment-superselected states of the system play a twofold role: They are the states that are least perturbed by the interaction with the environment, and they also the states that are most easily found out, without disturbing the system, by probing environmental degrees of freedom. Since the same information about the pointer observable is stored independently in many fragments of the environment, multiple observers can measure this observable on different fragments and will automatically agree on the findings. In this sense, one can ascribe *effectively objective existence* to the environment-superselected states (see also Sect. 8.2.3).

The research of the Los Alamos group into the objectification of observables along the lines outlined in this section is only in its beginnings. Important aspects, such as the explicit dynamical evolution of the objectification process [111] and the role of the assumptions and definitions in the current treatments of the "objectification through redundancy" idea, are currently still under investigation, as are studies involving more detailed and realistic system–environment models. However, it should already have become clear that the approach of departing from the closed-system view and of describing observations as the interception of information that is redundantly and robustly stored in the environment represents a very promising candidate for a purely quantum-mechanical account of the emergence of classicality from the quantum domain.

2.10 A Simple Model for Decoherence

Let us now illustrate the basic features and dynamics of environment-induced decoherence and superselection in the context of a model first introduced and studied by Zurek in 1982 [9]. The model is solved quite easily and will therefore nicely introduce the reader to the more complicated decoherence models discussed in Chaps. 3 and 5. (Later, in Sect. 5.4.1, we will discuss a more complex version of the model.) Despite its simplicity, the properties of the model continue to capture the interest of researchers to this date [118]. Moreover, the model seems to be realistic in certain circumstances. For example, recently it has been shown that the model is capable of explaining the behavior of the so-called Loschmidt echo observed in experiments on nuclear magnetic resonance (NMR) [118–120].

The model consists of a central quantum two-level system $\mathcal{S}$ linearly coupled to an environment $\mathcal{E}$ composed of a collection of N other quantum two-level systems. The obvious examples for "true" quantum two-level systems are spin-$\frac{1}{2}$ particles (where the two possible states correspond to "spin up" and "spin down" along some axis) and photons with two different orientations of polarization for a given orientation of an axis. Interestingly, as we shall explain in more detail in Sect. 5.1, it turns out that numerous other physical systems can be represented by quantum two-level systems. Therefore the model of a spin interacting with a collection of other spins is relevant to many more physical situations of interest than one might initially suspect.

Furthermore, since such (true or effective) two-state systems are the basis of all proposed implementations of quantum computers (see Chap. 7), they are often called "qubits," short for "quantum bits." Recall that a bit in classical computers is the smallest unit of information storage, represented by the two possible values "0" ("off") and "1" ("on"). Similarly, a qubit is represented by a quantum two-level system with basis states $|0\rangle$ and $|1\rangle$, which can be thought of corresponding to the classical bit states "0" and "1." However, the key difference between a qubit and a classical bit is grounded in the fact that a quantum two-level system can be prepared in arbitrary superpositions $\alpha\,|0\rangle + \beta\,|1\rangle$ of its basis states $|0\rangle$ and $|1\rangle$. The interest in quantum computing and in the generation of mesoscopic and macroscopic "Schrödinger cats" has made the experimental realization of such qubit systems a major area of research (see Chap. 6).

Let us now return to our simple model for decoherence. In the following, let us denote the two basis states of $\mathcal{S}$ by $|0\rangle$ and $|1\rangle$, while we shall use $|\uparrow\rangle_i$ and $|\downarrow\rangle_i$, $i = 1 \dots N$, to represent the basis states of the N two-level systems of the environment. We describe the total system–environment combination by a 2^{N+1}-dimensional tensor- product Hilbert space

$$\mathcal{H} = \mathcal{H}_{\mathcal{S}} \otimes \mathcal{H}_{\mathcal{E}_1} \otimes \mathcal{H}_{\mathcal{E}_1} \otimes \cdots \otimes \mathcal{H}_{\mathcal{E}_N}, \tag{2.98}$$

where $\mathcal{H}_{\mathcal{S}}$ denotes the Hilbert space of the system and $\mathcal{H}_{\mathcal{E}_i}$ represents the Hilbert space of the ith environmental spin.

We will now assume that the interaction Hamiltonian $\hat{H}_{\text{int}}$ governing the interaction between the system and the environment completely dominates the evolution, i.e., we shall neglect any intrinsic dynamics of the system and the environment. Both system and environment are entirely static, and the only dynamical process is the formation of correlations between the two partners. This corresponds to the quantum-measurement limit described in Sect. 2.8.1 above.

The interaction Hamiltonian $\hat{H}_{\text{int}}$ (and thus the total system–environment Hamiltonian $\hat{H}$) is chosen to be of the form

$$\begin{aligned} \hat{H} = \hat{H}_{\text{int}} &= \frac{1}{2}\left(|0\rangle\langle 0| - |1\rangle\langle 1|\right) \otimes \left(\sum_{i=1}^{N} g_i \left[(|\uparrow\rangle\langle\uparrow|)_i - (|\downarrow\rangle\langle\downarrow|)_i\right] \bigotimes_{i' \neq i} \hat{I}_{i'} \right) \\ &\equiv \frac{1}{2}\hat{\sigma}_z \otimes \left(\sum_{i=1}^{N} g_i \hat{\sigma}_z^{(i)} \bigotimes_{i' \neq i} \hat{I}_{i'} \right). \end{aligned} \tag{2.99}$$

Here $\hat{I}_i = (|\uparrow\rangle\langle\uparrow|)_i + (|\downarrow\rangle\langle\downarrow|)_i$ denotes the identity operator for the ith environmental two-level system, and $\hat{\sigma}_z$ and $\hat{\sigma}_z^{(i)}$ are the Pauli z-spin operators of the system and the ith environmental spin, respectively. From now on in this book, we shall simplify our notation by omitting the explicit inclusion of the identity operators in interaction Hamiltonians and simply write (2.99) as

$$\hat{H} = \hat{H}_{\text{int}} = \frac{1}{2}\hat{\sigma}_z \otimes \sum_{i=1}^{N} g_i \hat{\sigma}_z^{(i)} \equiv \frac{1}{2}\hat{\sigma}_z \otimes \hat{E}. \tag{2.100}$$

From (2.100) we observe that the central spin couples linearly through its z-spin component to each of the environmental spins, with a coupling strength given by the constants g_i, $i = 1 \dots N$. Comparison with (2.90) shows that the interaction Hamiltonian (2.100) is already in diagonal form, which allows us to immediately infer the pointer states of the system selected by the interaction with the environment. We see that the environment monitors the observable $\hat{\sigma}_z$, i.e., the z-spin component, of the system. Thus, without any calculations, we can conclude that we expect the eigenbasis $\{|0\rangle, |1\rangle\}$ of $\hat{\sigma}_z$ to emerge as the dynamically selected preferred basis of the system. Superposition states of $|0\rangle$ and $|1\rangle$ will be prone to decoherence, whereas $|0\rangle$ and $|1\rangle$ will represent the dynamically robust states of the system.

We also observe that, since the Hamiltonian (2.100) commutes with the Pauli operator $\hat{\sigma}_z$ of the system, the populations of the system are conserved quantities under the action of this Hamiltonian. There is no exchange of energy between the system and the environment, and therefore the interaction with the environment can influence only the degree of coherence present in the system. This model is therefore an example of *decoherence without dissipation*, showing that decoherence is a pure quantum effect without any classical counterpart (see also Sect. 2.11).

Additionally, the Hamiltonian (2.100) is diagonal in the basis states $\{|\uparrow\rangle_i, |\downarrow\rangle_i\}$ of the environmental spins. Thus the energy eigenstates $|n\rangle$ of the environment part $\hat{E}$ of this Hamiltonian are given by products of these individual basis states, i.e., by states of the form

$$|n\rangle = |\uparrow\rangle_1 |\downarrow\rangle_2 \cdots |\uparrow\rangle_N, \qquad \text{etc.}, \tag{2.101}$$

where $0 \leq n \leq 2^N - 1$. The energy ϵ_n associated with such an eigenstate $|n\rangle$ is

$$\epsilon_n = \sum_{i=1}^{N} (-1)^{n_i} g_i, \tag{2.102}$$

where $n_i = 1$ if the ith environmental spin is in the "down" state $|\downarrow\rangle_i$, and $n_i = 0$ otherwise.

The eigenstates of the total Hamiltonian $\hat{H} = \hat{H}_{\text{int}}$, see (2.100), are therefore simply of the form $|0\rangle|n\rangle$ and $|1\rangle|n\rangle$. Since these states form a basis of the composite system $\mathcal{SE}$, any arbitrary pure state $|\Psi\rangle$ of $\mathcal{SE}$ can be expanded as

$$|\Psi\rangle = \sum_{n=0}^{2^N-1} (c_n |0\rangle |n\rangle + d_n |1\rangle |n\rangle). \tag{2.103}$$

Let us now assume that at $t = 0$, before the interaction between $\mathcal{S}$ and $\mathcal{E}$ is turned on, the system and the environment are completely uncorrelated, i.e., that the initial state $|\Psi(0)\rangle$ of $\mathcal{SE}$ factorizes into a product state of the form

$$|\Psi(0)\rangle = (a|0\rangle + b|1\rangle) \sum_{n=0}^{2^N-1} c_n |n\rangle. \tag{2.104}$$

Then the time evolution of the state of $\mathcal{SE}$ generated by the action of the Hamiltonian (2.100) is given by

$$|\Psi(t)\rangle = \mathrm{e}^{-\mathrm{i}\hat{H}t} |\Psi(0)\rangle = a|0\rangle |\mathcal{E}_0(t)\rangle + b|1\rangle |\mathcal{E}_1(t)\rangle, \tag{2.105}$$

where

$$|\mathcal{E}_0(t)\rangle = |\mathcal{E}_1(-t)\rangle = \sum_{n=0}^{2^N-1} c_n e^{-i\epsilon_n t/2} |n\rangle . \tag{2.106}$$

The interpretation of the time-evolved state (2.105) is simple and follows our general discussion of the role of the environment as a "which-state" detector (see Sect. 2.6.3). The basis states $|0\rangle$ and $|1\rangle$ of the system have become correlated with the corresponding relative states $|\mathcal{E}_0(t)\rangle$ and $|\mathcal{E}_1(t)\rangle$ of the environment. The more distinguishable the states $|\mathcal{E}_0(t)\rangle$ and $|\mathcal{E}_1(t)\rangle$ are, i.e., the smaller the overlap $\langle\mathcal{E}_1(t)|\mathcal{E}_0(t)\rangle$ is, the more information becomes encoded in the environment that would allow one to distinguish between the states $|0\rangle$ and $|1\rangle$ of the system by looking at the environment. Accordingly, we expect that interference between these states $|0\rangle$ and $|1\rangle$ will become progressively damped as the overlap $\langle\mathcal{E}_1(t)|\mathcal{E}_0(t)\rangle$ decreases.

This motivates us to interpret the overlap $\langle\mathcal{E}_1(t)|\mathcal{E}_0(t)\rangle$ as a *decoherence factor* $r(t)$, which is given by [see (2.106)]

$$r(t) \equiv \langle\mathcal{E}_1(t)|\mathcal{E}_0(t)\rangle = \sum_{n=0}^{2^N-1} |c_n|^2 e^{-i\epsilon_n t}, \tag{2.107}$$

with $|c_n|^2 \leq 1$ and $\sum_{n=0}^{2^N-1} |c_n|^2 = 1$. If $r(t) \longrightarrow 0$, then the off-diagonal terms $|0\rangle\langle 1|$ and $|1\rangle\langle 0|$ of the reduced density matrix $\hat{\rho}_\mathcal{S}(t)$ of the system expressed in the $\{|0\rangle, |1\rangle\}$ basis disappear, i.e.,

$$\begin{aligned} \hat{\rho}_\mathcal{S}(t) &= \mathrm{Tr}_\mathcal{E}\, \hat{\rho}(t) \equiv \mathrm{Tr}_\mathcal{E}\, |\Psi(t)\rangle\langle\Psi(t)| \\ &= |a|^2|0\rangle\langle 0| + |b|^2|1\rangle\langle 1| + ab^* r(t)|0\rangle\langle 1| + a^*br^*(t)|1\rangle\langle 0| \\ &\longrightarrow\ |a|^2|0\rangle\langle 0| + |b|^2|1\rangle\langle 1|. \end{aligned} \tag{2.108}$$

This describes the local damping of interferences between the states $|0\rangle$ and $|1\rangle$. The expression (2.108) also explicitly confirms that the diagonal elements of $\hat{\rho}_\mathcal{S}(t)$ are independent of time and that thus the two-level populations are unchanged by the interaction, as already pointed out above.

The interesting question, of course, concerns the time evolution of the decoherence factor $r(t)$ and the dependence of this factor on the number of environmental spins. From (2.107) we see that $r(t)$ corresponds to the addition of 2^N vectors of lengths $|c_n|^2$ rotating in the complex plane with different frequencies proportional to ϵ_n. For any fixed time t, this amounts to a two-dimensional random-walk problem and was investigated in detail by Zurek [9] and Cucchietti, Paz, and Zurek [118]. Each of the 2^N steps of the random walk has length $|c_n|^2$ and follows the direction given by the phase $\epsilon_n t$. Since $\sum_{n=0}^{2^N-1} |c_n|^2 = 1$, the average step length $\langle|c|^2\rangle$ will be equal to 2^{-N}. From standard random-walk theory it then follows that the average squared length of the complex vector $r(t)$ will scale as

$$\langle|r(t)|^2\rangle \propto \langle|c|^2\rangle = 2^{-N}. \tag{2.109}$$

This shows that the degree of suppression of coherences in the $\{|0\rangle, |1\rangle\}$ scales *exponentially* with the size N of the environment. Furthermore, Zurek [9] and Cucchietti, Paz, and Zurek [118] demonstrated that, for sufficiently large N and a large class of distributions of the couplings g_i, $r(t)$ follows an approximately Gaussian decay with time,

$$r(t) \approx \mathrm{e}^{-\Gamma^2 t^2}. \tag{2.110}$$

The precise value of the decay constant Γ^2 is determined by the initial state of the environment and the distribution of the couplings g_i. An example for the time evolution of the decoherence factor (2.107) is shown in Fig. 2.6. We observe rapid decay of $|r(t)|$ for both $N = 20$ and $N = 100$.

However, it is important to realize that, as long as the number of environmental spins (and thus the number of degrees of freedom in the environment) is finite, there always exists a characteristic recurrence time τ_{rec} for which the decoherence factor (2.107) will return to its initial value of one. This is simply a consequence of the fact that (2.107) is a sum of functions that are periodic in time. Any such sum of periodic functions must in turn be periodic, too. The value of the recurrence time depends on the initial state of the environment and the distribution of the couplings g_i. For example, in the limiting case of a highly "nonrandom" initial state of the environment with no initial correlations between the environmental spins,

$$|\Psi(0)\rangle = (a\,|0\rangle + b\,|1\rangle) \prod_{i=1}^{N} \frac{1}{\sqrt{2}} \left(|\uparrow\rangle_i + |\downarrow\rangle_i\right), \tag{2.111}$$

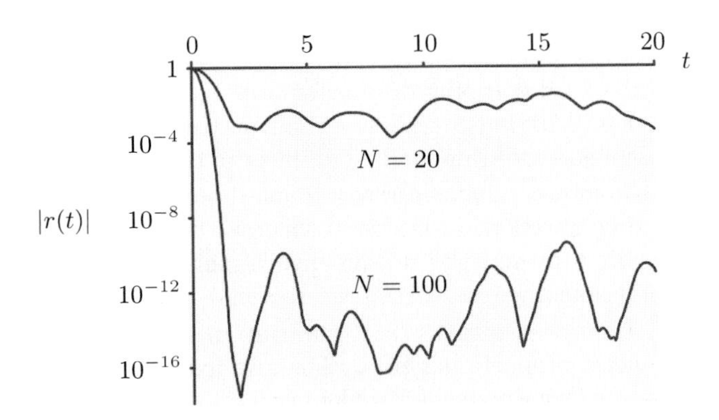

Fig. 2.6. Evolution of the absolute value of the decoherence factor $r(t)$, see (2.107), describing the time dependence of interference terms in a central static spin system interacting with a collection of N static environmental spins for two different values $N = 20$ and $N = 100$. The initial state of each environmental spin was assumed to be completely random, with the environmental spins completely uncorrelated with each other and with the system. The couplings were chosen randomly from the interval $[0, 1]$.

and $g_i = g$ for all i, (2.107) yields

$$r(t) = \cos^N(gt). \tag{2.112}$$

This decoherence factor is periodic with frequency $g/2\pi$. Thus only for times $t < \pi/g$ the absolute value $|r(t)|$ of the decoherence factor will exhibit decay, and a full revival of coherence will occur at times equal to integer multiples of $\tau_{\text{rec}} = \pi/g$. Note also that, in the extreme situation in which the environment happens to start out in one of the eigenstates $|n\rangle$ of $\hat{E}$, $r(t)$ will not decay at all, and thus no decoherence will occur.

In realistic cases, such highly ordered initial states and symmetrical system–environment couplings are unlikely to be relevant. Here the characteristic recurrence time τ_{rec} is typically extremely long and of the Poincaré type with $\tau_{\text{rec}} \propto N!$. For macroscopic environments of realistic but finite sizes, τ_{rec} can exceed the lifetime of the universe, as first pointed out by Zurek [9]. Thus the loss of coherence from the system is typically irreversible for all practical purposes not only because of our practical inability to control and observe the environment, but also because the timescale for the recurrence of coherence is astronomically large in virtually all physically realistic situations.

2.11 Decoherence Versus Dissipation

Dissipation, the loss of energy from the system, is a classical effect. If we let a system interact with another system (which we may call the environment), energy exchange will often take place between the two partners, and the systems will then ultimately approach thermal equilibrium. The characteristic timescale on which this happens is typically referred to as the *relaxation timescale* for the system. If a particular system–environment interaction leads to dissipation in the system, then the strength of the system–environment interaction is a measure of the relaxation time. As the interaction strength decreases, the relaxation times becomes longer, and vice versa.

However, if dissipation is absent or negligible, we must not conclude that there is no (or only negligible) interaction between the system and its environment. Even if a particular type of interaction does not lead to dissipation, it will in general still result in decoherence, which is a pure quantum effect. That is, the environment may in general obtain which-path (or, more generally, which-state) information without absorbing any energy from the system. The model discussed in the previous Sect. 2.10 has provided us with a first explicit example of such decoherence without dissipation. Thus decoherence *may*, but does not have to, be accompanied by dissipation, whereas the presence of dissipation also implies the occurrence of decoherence.

If dissipation and decoherence are both present, then they are usually quite easily distinguished because of their very different timescales. As numerous theoretical and experimental studies have demonstrated, the decoherence timescale is typically many orders of magnitude shorter than the relaxation timescale. In Sect. 5.2.4, we will show that a rule-of-thumb estimate for the ratio of the relaxation timescale τ_{r} to the decoherence timescale τ_{d} for a massive object described by a superposition of two different positions a distance Δx apart is

$$\frac{\tau_{\mathrm{r}}}{\tau_{\mathrm{d}}} \sim \left(\frac{\Delta x}{\lambda_{\mathrm{dB}}} \right)^2 . \tag{2.113}$$

Here, λ_{dB} is the thermal de Broglie wavelength of the object,

$$\lambda_{\mathrm{dB}} = \frac{\hbar}{\sqrt{2mk_{\mathrm{B}}T}}, \tag{2.114}$$

where m is the mass of the object, T its temperature, and k_{B} is the Boltzmann constant. Equation (2.113) was derived by Zurek in 1984 [12], providing a first general estimate for the decoherence timescale (then still called "decorrelation timescale").

In his paper (and also in his subsequent article in *Physics Today* [13]), Zurek gave the following numerical example to illustrate the typically enormous differences between the decoherence and relaxation timescales at the macroscopic level. Zurek considered a macroscopic object of mass $m = 1$ g at room temperature ($T = 300$ K). For this object, λ_{dB} is tiny, namely, $\lambda_{\mathrm{dB}} \approx 10^{-23}$ m. Clearly, this number is microscopic by all accounts. Suppose now the particle is described by a coherent superposition of two mesoscopically or even macroscopically distinct locations. Then, the ratio $(\Delta x/\lambda_{\mathrm{dB}})^2$ in (2.113) will be very large, and it follows that the decoherence timescale τ_{d} for this superposition will be overwhelmingly shorter than the dissipation (relaxation) timescale τ_{r}. Zurek [12,13] used the example of the object's being described by a superposition of two spatial locations a macroscopic distance $\Delta x = 1$ cm apart. Then the ratio (2.113) is on the order of 10^{40}. That is, the typical decoherence time for an object described by such a superposition is some 40 orders of magnitude shorter than the relaxation time required to reach thermal equilibrium!

For macroscopic objects, the dissipative influence of the environment is therefore usually completely negligible with respect to the dynamics of the system on any timescale relevant to the decoherence induced by this environment. For example, photons scattering off a bowling ball will hardly affect the ball's motion in any way, while they will lead to virtually instantaneous decoherence of a superposition state involving macroscopically distinguishable positions of the ball (see also Fig. 1.3).

Thus this crucial difference between relaxation and decoherence timescales explains why we observe macroscopic objects to follow perfectly Newtonian trajectories—effectively "created" through the action of decoherence (see Sect. 5.2.5)—with often hardly any manifestation of dissipation, such as a slowing-down of the object. To use an example going back to Joos [17, p. 79], the planet Jupiter has been revolving around the sun on a Newtonian trajectory for billions of years, while its motional state has remained virtually unaffected by any dissipative loss (friction) due to the sunlight scattered off the planet's surface. If relaxation and decoherence timescales were indeed similar, this fact would obviously be very difficult to explain.

2.12 Decoherence Versus Classical Noise

In this book, we consistently reserve the term "decoherence" to describe the consequences of (usually in practice irreversible) quantum entanglement with some environment, in agreement with the historically established meaning and the vast body of literature on environmental decoherence. As described in detail in Sects. 2.6 and 2.7, the basic mechanism of decoherence can be formalized as follows. Suppose we have a set of pointer states $\{|s_n\rangle\}$ of the system $\mathcal{S}$ defined by their property of getting least getting entangled in the course of the interaction with the environment $\mathcal{E}$,

$$|s_n\rangle\,|E_0\rangle \longrightarrow |s_n\rangle\,|E_n\rangle\,. \tag{2.115}$$

Then an arbitrary pure-state superposition $\sum_n c_n\,|s_n\rangle$ of these pointer states $|s_n\rangle$ results in the familiar entangled state

$$\left(\sum_n c_n\,|s_n\rangle\right)|E_0\rangle \longrightarrow \sum_n c_n\,|s_n\rangle\,|E_n\rangle\,. \tag{2.116}$$

As described in Sect. 2.7, in the formalism of reduced density matrices this leads to a decay of the off-diagonal terms in the $\{|s_n\rangle\}$ basis,

$$\hat{\rho}_{\mathcal{S}} = \sum_{nn'} c_n\,c_{n'}^*|s_n\rangle\langle s_{n'}|\langle E_{n'}|E_n\rangle \longrightarrow \sum_n |c_n|^2\,|s_n\rangle\langle s_n|, \tag{2.117}$$

since the relative environmental states $|E_n\rangle$ become rapidly mutually orthogonal (distinguishable). Evidently, the interaction (2.116) does not result in any dynamical change of any of the component states of $\mathcal{S}$; its only effect is pure, nonlocal entanglement. Thus decoherence should be understood as a distinctly quantum-mechanical effect with no classical analog. This is the formalism and the physical interpretation of decoherence used in this book.

However, the reader may find descriptions of several other processes in the literature that are sometimes referred to as "decoherence" or that are claimed to be "equivalent" to decoherence in some (formal, physical, etc.) sense. Most of these associations are based on the observation that different processes may all lead to the disappearance of off-diagonal elements (in some basis) in the density matrix of the system. However, it is important to emphasize that the density-matrix description is only a formal tool that, somewhat misleadingly and deceptively, hides the crucial physical and conceptual differences between the processes whose effects may have a similar representation in the density-matrix formalism. While in some cases, if one takes a practical point of view (say, as a quantum engineer or experimentalist), it may occasionally indeed be justified to associate processes different from pure environmental entanglement with the term "decoherence," the fundamental interpretation of such processes is different from "true" decoherence in the sense used in this book. This is not merely a philosophical issue: There always exists some experimental procedure that would, at least in principle, be able to distinguish between the different physical processes underlying formally similar density-matrix descriptions.

The most common and important example of "fake decoherence" (to use a term coined by Joos [17]) is to interpret the result of an ensemble average over different noisy realizations of a system as the description of a decoherence process. This represents an example where formal similarities arising in the density-matrix formalism lead to the (incorrect) classification as a physical decoherence process. We may trace back the source of this confusion to the use of density matrices to describe two physically very different settings. On the one hand, the density-matrix formalism is often used to describe true *physical ensembles*, i.e., a collection of N physical systems in which each individual system is described by a pure state (see also the discussion in Sect. 2.4.5). For example, suppose that each system is represented by a two-dimensional Hilbert space spanned by the basis states $|0\rangle$ and $|1\rangle$, and that the pure state of the ith system is given by

$$|\psi_i\rangle = \frac{1}{\sqrt{2}} \left(|0\rangle + |1\rangle \, \mathrm{e}^{\mathrm{i}\phi_i} \right), \tag{2.118}$$

where $\mathrm{e}^{\mathrm{i}\phi_i}$ is some system-specific relative phase factor. These phase factors may come about in two different ways. First, each system may have been prepared in a slightly different initial state from the beginning. Alternatively, all systems may start out in the same state but may be subject to slightly different Hamiltonians. For example, the potential in the Hamiltonians $\hat{H}_i$ of each system may contain random fluctuations, $V_i(t) = V_0(t) + \delta V_i(t)$ [121]. In either case, the differences in the relative-phase factors for the different systems can thus be thought of as being due to the presence of *classical noise* (applied during the preparation of the system and/or during the subsequent evolution).

If we now take the average over the ensemble $\{|\psi_i\rangle\}$ of the pure states of all N systems, we obtain the ensemble density matrix

$$\hat{\rho} = \frac{1}{N} \sum_{i=1}^{N} |\psi_i\rangle\langle\psi_i|$$
$$= \frac{1}{2}|0\rangle\langle 0| + \frac{1}{2}|1\rangle\langle 1| + \left(\frac{1}{2N} \sum_{i=1}^{N} \mathrm{e}^{\mathrm{i}\phi_i} \right) |1\rangle\langle 0| + \left(\frac{1}{2N} \sum_{i=1}^{N} \mathrm{e}^{-\mathrm{i}\phi_i} \right) |0\rangle\langle 1|. \tag{2.119}$$

In the limit of large N, the sums over the random phase factors $\mathrm{e}^{\pm\mathrm{i}\phi_i}$ average out to zero, and therefore

$$\hat{\rho} \longrightarrow \frac{1}{2}|0\rangle\langle 0| + \frac{1}{2}|1\rangle\langle 1|. \tag{2.120}$$

Evidently, this "ensemble dephasing" leads to a density matrix that has the same diagonal form as the reduced density matrix of an individual system subject to decoherence in the $\{|0\rangle, |1\rangle\}$ basis. However, in the case of (2.119) and (2.120) the disappearance of the interference terms $|0\rangle\langle 1|$ and $|1\rangle\langle 0|$ is simply the result of a mathematical averaging procedure over many members in a physical ensemble of system (or, put differently, over the different instances of particular noise processes). It does not correspond to decoherence in our sense, which describes the delocalization of phase coherence for individual systems. Another example of "fake decoherence," this time resulting from an averaging over *dynamical* phases in the density-matrix formalism, has been critically analyzed in [122].

In fact, as we have pointed out in Sect. 2.4.5, the correct description of an ensemble of N physical systems, each of which is described by a pure state $|\psi_i\rangle$ of the form (2.118), is given by a pure-state (tensor) product of all states $|\psi_i\rangle$,

$$|\Psi\rangle = \prod_{i=1}^{N} |\psi_i\rangle \,. \tag{2.121}$$

This state "lives" in a 2^N-dimensional Hilbert space, and thus the corresponding density matrix has size $2^N \times 2^N$, rather than 2×2 as in (2.119). Without environmental interactions, each individual system evolves completely unitarily, and therefore the phase relations ϕ_i remain well-defined at all times. Only formally these phases *appear* to be "washed out" when the density-matrix formalism is used to represent ensemble averages over many physical systems, as in (2.119), but they are not physically delocalized from the systems and remain therefore, at least in principle, experimentally accessible to the observer of each of these systems. If we knew the time evolution of the noise process, we could always apply a suitable unitary countertransformation to the system that would completely reverse the effect of this noise on the system in each individual run of the experiment. This reversal of ensemble dephasing has been implemented in practice, for example, in NMR experiments. Here a collection of spins eventually dephases, since noise effects cause each spin to precess at a slightly different frequency. However, by application of a suitably chosen control pulse the spin ensemble can be "refocused" (this is the so-called spin-echo technique).

Noise—the addition of random fluctuations to the Hamiltonian of the system—does not create any system–environment entanglement and can be completely undone (at least in principle) by local operations. By contrast, decoherence means that the system becomes entangled with environmental degrees of freedom such that phase relations are for all practical purposes irreversibly lost from the system and can no longer be accessed by local measurements performed on the system. In turn, information about the system becomes imprinted in the environment. Thus, while in the case of classical noise the system is perturbed by the environment, (2.115) shows that decoherence describes a situation in which *the system perturbs the environment* (see also Sect. IV.C of [16]). The nonlocal nature of quantum states then implies that this "distortion" of the environment by the system in turn influences what can be observed at the level of the system (as formally described by the reduced density matrix). However, as discussed in Sect. 2.11, this influence takes place on typically much shorter timescales than the perturbation of the system due to classical noise imparted by the environment.

Stochastic fluctuations (i.e., classical-noise processes) have often also been used to *simulate* the influence of the environment on the system. Decoherence is then modeled through stochastic "kicks" applied to the system (see, e.g., [121, 123–125]). Of course, the same issues as discussed above apply here as well. The application of classical noise to a system corresponds to the unitary evolution of a closed system governed by a randomly fluctuating Hamiltonian, while decoherence is represented by an open quantum system becoming deterministically entangled with another quantum system (i.e., the environment). In general, the quantum states of the system resulting from these two situations will have different observable properties [121]. Only in certain cases the simulation of decoherence by classical noise will lead to formally similar results at the level of a density-matrix description. However, the fundamental conceptual differences between noise and decoherence remain even in such cases. In Sect. 7.3, we will further discuss these differences in the context of quantum computing, in particular with respect to the experimental and theoretical simulation of decoherence by noise.

2.13 Virtual Decoherence and Quantum "Erasure"

As discussed in Sect. 2.7, the effective irreversibility of the delocalization of local phase relations—induced by our inability to control the large number of degrees of freedom interacting with the system—is a hallmark of decoherence. We may distinguish such "real" decoherence from a (truly or effectively) reversible delocalization of phases, which we shall refer to as *virtual decoherence*.

The possibility of a *true* reversal of phase delocalization (i.e., the *relocalization* of the superposition at the level of the system) is based on the fact that the global system–environment evolution is completely unitary. Application of an appropriate unitary countertransformation to the system–environment combination then disentangles the system from the environment and dynamically recovers the original separable system–environment product state. This complete reversal of decoherence—corresponding, in effect, to a reversal of the arrow of time—describes an actual *recoherence* process involving the *local* recombination of wave-function components. This is analogous to the reversible Stern–Gerlach experiment [126, 127], where the two atomic beams separated by the action of the magnetic field are subsequently recombined (refocused). In practice, such reversible decoherence could be experimentally studied through the coupling of the system to a second, fully controlled system acting as an "artificial environment." We will discuss a concrete proposal for the experimental realization of this idea in Sect. 6.1.4.

The *influence* of decoherence on the system can also be *effectively* reversed by reconstructing the original pre-decoherence superposition state of the system. An example of such an effective reconstruction of the original state has become known under the heading of *quantum erasure* [75, 128–130]. A similar idea underlies the concept of error correction in quantum computers (see Sect. 7.4). The basic scheme of quantum "erasure" (the use of the quotation marks will be motivated below) is typically described in the literature as follows (see, e.g., [130]). We again consider the double-slit experiment, with which-path detectors measuring through which slit the particle passes. The composite system–detector evolution is then of the form (2.65), i.e.,

$$\frac{1}{\sqrt{2}}\left(|\psi_1\rangle + |\psi_2\rangle\right)|\text{“ready”}\rangle \longrightarrow \frac{1}{\sqrt{2}}\left(|\psi_1\rangle\,|1\rangle + |\psi_2\rangle\,|2\rangle\right). \tag{2.122}$$

We now couple the which-path detector to an explicit read-out device. The measurement interaction is assumed to be of the form

$$|\pm\rangle\,|\Phi_0\rangle \longrightarrow |\pm\rangle\,|\Phi_\pm\rangle\,, \tag{2.123}$$

where $|\Phi_0\rangle$ is the initial state of the read-out device, and $|\pm\rangle$ are the conjugate states of the which-path detector,

$$|\pm\rangle \equiv \frac{1}{\sqrt{2}}\left(|1\rangle \pm |2\rangle\right). \tag{2.124}$$

The resulting evolution of the total system composed of the particle, the which-path detector, and the read-out device is then

$$
\begin{aligned}
\frac{1}{\sqrt{2}}\left(|\psi_1\rangle + |\psi_2\rangle\right)|\text{“ready”}\rangle\,|\Phi_0\rangle \longrightarrow\; & \frac{1}{\sqrt{2}}\left(|\psi_1\rangle\,|1\rangle + |\psi_2\rangle\,|2\rangle\right)|\Phi_0\rangle \\
=\; & \frac{1}{\sqrt{2}}\left(|\psi_+\rangle\,|+\rangle + |\psi_-\rangle\,|-\rangle\right)|\Phi_0\rangle \\
\longrightarrow\; & \frac{1}{\sqrt{2}}\left(|\psi_+\rangle\,|+\rangle\,|\Phi_+\rangle + |\psi_-\rangle\,|-\rangle\,|\Phi_-\rangle\right),
\end{aligned}
\tag{2.125}
$$

where $|\psi_\pm\rangle$ are the conjugate states defined in (2.79).

The evolution (2.125) shows that the read-out device becomes quantum-correlated with states of the which-path detector (namely, $|\pm\rangle$) that contain no information about the path of the particle.[26] In the standard collapse picture of quantum mechanics, the measurement carried out by the read-out device then projects the final state on the right-hand side of (2.125) onto either one of the components $|\psi_+\rangle\,|+\rangle\,|\Phi_+\rangle$ and $|\psi_-\rangle\,|-\rangle\,|\Phi_-\rangle$. In other words, the measurement "forces" the system into a state that does not allow us to infer the path of the particle.

Hence interference between the path components $|\psi_1\rangle$ and $|\psi_2\rangle$ (as embodied in the superpositions $|\psi_\pm\rangle$) is restored in a *nonlocal* manner, in contrast with an actual (local) recoherence process. It is in this sense that the initial superposition $|\psi_+\rangle$ (or the phase-reversed superposition $|\psi_-\rangle$) of the system is reconstructed. The erasure can therefore also be delayed until *after* the registration of the particle on the screen. Such "delayed-choice" experiments (see, e.g., [131]) go back in spirit to a thought experiment first proposed by Wheeler [41]. Quantum "erasure" has been demonstrated in several experiments [131–135], including a double-slit experiment with photons [135].

It should be emphasized, however, that the term "quantum erasure" is quite misleading. To see this, let us distinguish two cases. First, we may assume that the interaction (2.122) between the particle and the which-path detector is irreversible (describing a measurement in the true sense—recall that irreversibility may be regarded as a defining property of measurements). In the open-systems picture, this irreversibility would arise from inevitable environmental interactions and the resulting delocalization of phase relations between the component states $|\psi_1\rangle\,|1\rangle$ and $|\psi_2\rangle\,|2\rangle$. Then the records of the which-path measurement will be indelibly imprinted in the inaccessible environment, and therefore we cannot actually erase this information.[27]

---

[26] This is so because, if we performed a measurement on the detector described by either one of the states $|+\rangle$ or $|-\rangle$ to find out which of the two possible paths the detector has registered, we would obtain the outcomes $|1\rangle$ and $|2\rangle$ with equal probabilities.

[27] Of course, the particular source of the irreversibility of the which-path measurement is not important for this argument. For example, we may also simply assume that the irreversibility is due to a wave-function collapse.

Alternatively, we may regard (2.122) as describing a reversible (and thus only *virtual*) which-path "measurement."[28] In classical physics, "erasure" would correspond to an irreversible destruction of information (e.g., via a transformation into heat) [136]. The analogous process in the quantum setting would be represented by an irreversible change of the (virtual) "results" $|1\rangle$ and $|2\rangle$ of the reversible which-path "measurement" into new states $|d_1\rangle$ and $|d_2\rangle$ that depend in an *uncontrolled* manner on the previously registered path of the particle and thus no longer encode which-path information. Hence, from the point of view of which-path information, $|d_1\rangle$ and $|d_2\rangle$ are indistinguishable and may thus be jointly denoted by $|d_0\rangle$. This *reset* [136,137] could be accomplished by coupling the detector to an (uncontrolled) "sink" for information (realized, e.g., as a thermal bath), thereby inducing an effectively irreversible evolution of the form

$$\begin{aligned} \frac{1}{\sqrt{2}}\left(|\psi_1\rangle\,|1\rangle + |\psi_2\rangle\,|2\rangle\right)|\chi_0\rangle &\longrightarrow \frac{1}{\sqrt{2}}\,|\psi_1\rangle\,|d_1\rangle\,|\chi_1\rangle + |\psi_2\rangle\,|d_2\rangle\,|\chi_2\rangle \\ &\equiv \frac{1}{\sqrt{2}}\left(|\psi_1\rangle\,|\chi_1\rangle + |\psi_2\rangle\,|\chi_2\rangle\right)|d_0\rangle\,, \qquad (2.126) \end{aligned}$$

where $|\chi_0\rangle$ is the initial state of the sink. The difference between the original which-path detector states $|1\rangle$ and $|2\rangle$ must have been deterministically transferred to the sink, thereby increasing the entropy of the detector and compensating for the gain of information from the initial reversible which-path "measurement." Therefore the states $|\chi_1\rangle$ and $|\chi_2\rangle$ of the sink are distinguishable, and thus a true erasure of which-path information as implemented by (2.126) would evidently introduce additional (and this time inevitably irreversible) spatial decoherence of the particle. Evidently, the erasure process (2.126) is very different from the evolution (2.125). In fact, for (2.125) to hold, an irreversible destruction (i.e., an actual *erasure*) of which-path information in the above sense (2.126) must *not* take place.

2.14 Resolution into Subsystems

Note that decoherence derives from the presupposition of the existence and the possibility of a division of the world into "the system" and "the environment." In the decoherence program, the term "environment" is usually

[28] Here the use of quotation marks is intended to make clear the fact that it is difficult to regard a reversible interaction as a proper measurement. Note also that the transformation of the entangled particle–detector state from the first to the second line of (2.125) leaves open the question of what was "measured" by the which-path detector in the first place. From the first line of (2.125), we would deduce a "measurement" in the basis $\{|\psi_1\rangle\,, |\psi_2\rangle\}$ of the system, while the second line would imply a "measurement" in the conjugate basis $\{|\psi_+\rangle\,, |\psi_-\rangle\}$. (This, of course, is simply the preferred-basis problem discussed in Sect. 2.5.2.)

understood as the "remainder" of the system, in the sense that the degrees of freedom of the environment are typically not (cannot be, do not need to be) controlled and are not directly relevant to the observation under consideration, but that nonetheless the environment includes all those degrees of freedom which contribute significantly to the evolution of the state of the system. This is essentially the definition originally given in Zurek's first paper on decoherence and pointer states [8, p. 1520].

This system–environment dualism is generally associated with quantum entanglement, which always describes a correlation between parts of the universe. As long as the universe is not resolved into individual subsystems, there is no measurement problem: The state vector $|\Psi\rangle$ of the entire universe[29] evolves deterministically according to the Schrödinger equation, which poses no interpretive difficulty. The measurement problem arises only once we decompose the total Hilbert space $\mathcal{H}$ of the universe into a product of two spaces $\mathcal{H}_1 \otimes \mathcal{H}_2$ and would like to assign an individual state to one of the two subsystems. Zurek [16, p. 718] puts it like this:

> In the absence of systems, the problem of interpretation seems to disappear. There is simply no need for "collapse" in a universe with no systems. Our experience of the classical reality does not apply to the universe as a whole, seen from the outside, but to the systems within it.

Moreover, terms like "observation," "correlation," and "interaction" will naturally make little sense without a division into systems. Zeh has suggested that the locality of the observer defines an observation in the sense that any observation arises from the ignorance of a part of the universe, and that this locality also defines the "facts" that can occur in a quantum system. Landsman [139, pp. 45–46] argues similarly:

> The essence of a "measurement," "fact" or "event" in quantum mechanics lies in the non-observation, or irrelevance, of a certain part of the system in question. (...) A world without parts declared or forced to be irrelevant is a world without facts.

However, the assumption of a decomposition of the universe into subsystems (as necessary as it appears to be for the emergence of the measurement problem and for the formulation of the decoherence program) is definitely nontrivial. By definition, the universe as a whole is a closed system, and therefore there are no "unobserved degrees of freedom" of an external environment. We thus cannot apply the theory of decoherence to the universe in its entirety in order to determine the "global" quasiclassical observables. Also, there exists no general criterion for how the total Hilbert space is to be divided into subsystems, while at the same time much of what is called a property of the system will depend on its correlation with other systems. This

[29] If we dare to postulate this total state—see counterarguments by Auletta [138].

problem becomes particularly acute if one would like decoherence not only to motivate explanations for the subjective perception of classicality (see, e.g., Sects. 8.2 and 9.4), but moreover to allow for the definition of quasiclassical "macrofacts." Zurek [104, p. 1820] was one of the first to clearly point out this conceptual difficulty:

> In particular, one issue which has been often taken for granted is looming big, as a foundation of the whole decoherence program. It is the question of what are the "systems" which play such a crucial role in all the discussions of the emergent classicality. (...) [A] compelling explanation of what are the systems—how to define them given, say, the overall Hamiltonian in some suitably large Hilbert space—would be undoubtedly most useful.

A frequently proposed idea is to abandon the notion of an "absolute" resolution and instead to postulate the intrinsic relativity of the distinct state spaces and properties that emerge through the correlation between these relatively defined spaces (see, for example, the proposals, unrelated to decoherence, in [140–143]). This relative view of systems and correlations has counterintuitive, in the sense of nonclassical, implications. However, as in the case of quantum entanglement, these implications need not be taken as paradoxes that demand further resolution. Accepting some properties of nature as counterintuitive is indeed a satisfactory path to take in order to arrive at a description of nature that is as complete and objective as is allowed by the range of our experience (which is based on inherently local observations).

2.15 Formal Tools and Their Interpretation

Let us now briefly outline a few formal procedures that are often used and referred to in the context of decoherence. In Sect. 2.15.1, we will introduce the Schmidt decomposition theorem, which tells us that a bipartite state can always be written in a diagonal form reminiscent of the final state at the conclusion of a von Neumann measurement. Both meaningful and spurious connections between the Schmidt decomposition and decoherence will be discussed. In Sect. 2.15.2, we will introduce the Wigner representation of the density matrix. This representation is often a useful tool for visualizing decoherence in phase space. In Sect. 2.15.3, we will then show that it is possible to view any given density matrix as a traced-over pure-state density matrix of a larger system. Finally, in Sect. 2.15.4, we shall introduce the so-called operator-sum formalism. While application of this formalism to more complex decoherence models is usually not feasible, it provides a very general framework for describing the reduced dynamics of a system interacting with an environment.

2.15.1 The Schmidt Decomposition

Consider two systems $\mathcal{A}$ and $\mathcal{B}$ endowed with Hilbert spaces $\mathcal{H}_\mathcal{A}$ and $\mathcal{H}_\mathcal{B}$, respectively. Then the *Schmidt decomposition theorem* [144] tells us that an arbitrary pure state $|\Psi\rangle$ of the composite system $\mathcal{AB}$ can always be written in the diagonal (or "Schmidt") form

$$|\Psi\rangle = \sum_i \lambda_i \, |a_i\rangle \, |b_i\rangle \, . \tag{2.127}$$

Here the *Schmidt states* $|a_i\rangle$ and $|b_i\rangle$ form orthonormal bases (the so-called *Schmidt bases*) of $\mathcal{H}_\mathcal{A}$ and $\mathcal{H}_\mathcal{B}$, respectively, and the expansion coefficients λ_i are complex numbers fulfilling $\sum_i |\lambda_i|^2 = 1$. However, by absorbing the phase factors into the Schmidt basis states, these coefficients can be chosen to be real and nonnegative numbers $\sqrt{p_i}$ obeying $\sum_i p_i = 1$, and thus the Schmidt decomposition can be written as

$$|\Psi\rangle = \sum_i \sqrt{p_i} \, |a_i\rangle \, |b_i\rangle \, . \tag{2.128}$$

Furthermore, it can be shown that this decomposition is unique if and only if the coefficients $\sqrt{p_i}$ are all different from one another.

The issue of the uniqueness of the Schmidt decomposition has already been touched upon in our discussion of the preferred-basis problem in Sect. 2.5.2. There we showed that the basis ambiguity in the final composite system–apparatus state (which arises as the result of a von Neumann measurement interaction and is in the diagonal Schmidt form) leads to the problem that the measured observable cannot be uniquely inferred from this state.

The Schmidt theorem also has immediate implications for the reduced density matrices $\hat{\rho}_\mathcal{A}$ and $\hat{\rho}_\mathcal{B}$ of $\mathcal{A}$ and $\mathcal{B}$, respectively. Using (2.128), $\hat{\rho}_\mathcal{A}$ is given by

$$\begin{aligned} \hat{\rho}_\mathcal{A} &= \mathrm{Tr}_\mathcal{B} \, |\Psi\rangle\langle\Psi| \\ &= \sum_k \langle b_k| \left[\sum_{ij} \sqrt{p_i p_j} |a_i\rangle\langle a_j| \otimes |b_i\rangle\langle b_j| \right] |b_k\rangle \\ &= \sum_i p_i |a_i\rangle\langle a_i|, \end{aligned} \tag{2.129}$$

and similarly

$$\begin{aligned} \hat{\rho}_\mathcal{B} &= \mathrm{Tr}_\mathcal{A} \, |\Psi\rangle\langle\Psi| \\ &= \sum_i p_i |b_i\rangle\langle b_i|. \end{aligned} \tag{2.130}$$

Thus $\hat{\rho}_{\mathcal{A}}$ and $\hat{\rho}_{\mathcal{B}}$ are diagonal in the bases $\{|a_i\rangle\}$ and $\{|b_i\rangle\}$, respectively, and have the *same* spectrum $\{p_i\}$. Applied to the decoherence-relevant case of a system interacting with an environment, the Schmidt theorem means that we can always find an orthonormal basis in which the reduced density matrix of our system of interest becomes exactly diagonal.

Thus the Schmidt decomposition of a density matrix refers simply to its formal diagonalization and the fact that this diagonalization can always be done. While this decomposition plays an important role in many areas of quantum physics, its relevance to a description of decoherence and to interpretive problems in quantum mechanics has sometimes been misrepresented. This might be due to the fact that, since the effect of decoherence is described in terms of a vanishing of off-diagonal elements of the (reduced) density matrix in a specific environment-selected basis, decoherence may formally appear to correspond to a density-matrix diagonalization, which is precisely the subject of the Schmidt decomposition.

In fact, the Schmidt basis, obtained by diagonalizing the density matrix of the system at each instant of time, has been frequently studied with respect to its ability to yield a preferred basis (see, for example, [5,6,145,146]). This has led some to consider the Schmidt basis states as describing "instantaneous pointer states" [145]. However, the Schmidt decomposition is a purely mathematical procedure that has no *a priori* physical or interpretive meaning. The Schmidt theorem applied to reduced density matrices simply states that every such density matrix can be diagonalized in *some* basis. However, this basis will not necessarily represent quasiclassical properties (see, e.g., [147, 148]). This is in stark contrast with the environment-induced *approximate* diagonalization of the reduced density matrix in the pointer basis selected by the structure of the system–environment interaction. The important distinction between pointer states and the states that diagonalize the density matrix was first clearly noted by Zurek [8], laying the foundation for subsequent developments, such as the predictability sieve discussed in Sect. 2.8.3.

Other approaches have refrained from computing instantaneous Schmidt states and have instead allowed for a characteristic decoherence time to pass during which the reduced density matrix becomes approximately diagonal in the basis selected by the structure of the system–environment interaction (this process can, for example, be described by an appropriate master equation; see Chap. 4). Schmidt states are then calculated by diagonalizing this decohered density matrix. Since decoherence usually leads to rapid diagonality of the reduced density matrix in the pointer basis to a very good approximation, the resulting Schmidt states are typically very similar to the environment-superselected states, *except* in cases where the latter states are very nearly degenerate.

This particular situation of near-degeneracy is readily illustrated [146]. Consider a system described by a two-dimensional Hilbert space. Suppose the action of decoherence has resulted in the system's being represented by a reduced density matrix (expressed in the environment-superselected basis) of the form

$$\rho = \begin{pmatrix} 1/2+\delta & \omega^* \\ \omega & 1/2-\delta \end{pmatrix}, \tag{2.131}$$

with $|\omega| \ll 1$ (strong decoherence) and $\delta \ll 1$ (near-degeneracy). If decoherence led to exact diagonality, $\omega = 0$, the eigenvectors of ρ would be, for any fixed value of δ, proportional to $(0,1)$ and $(1,0)$ (corresponding to the "ideal" pointer states). However, for $\omega > 0$ (approximate diagonality) and $\delta \longrightarrow 0$ (degeneracy), the eigenvectors—and thus the Schmidt states—become proportional to $(\pm|\omega|/\omega, 1)$. This implies that, in the case of degeneracy, the Schmidt decomposition of the reduced density matrix may yield states that are very different from the environment-superselected states, even if the decohered (rather than the instantaneous) reduced density matrix is diagonalized.

In summary, it is important to emphasize that the resilience to environmental entanglement is the relevant criterion for obtaining the preferred quasiclassical pointer states. These states cannot, in general, be arrived at by simply diagonalizing the instantaneous density matrix of the system. However, the eigenstates of the decohered reduced density matrix will, in many situations, approximate the quasiclassical stable pointer states well, especially when these latter states are sufficiently nondegenerate.

2.15.2 The Wigner Representation

The *Wigner representation* (or *Wigner function*) is often used as an alternative to the density matrix for systems described by a continuous degree of freedom. Typically, this degree of freedom is the position x of the system, and we shall here focus on this case. Given the (pure-state or mixed-state) position-space density matrix $\rho(x,x') \equiv \langle x|\,\hat{\rho}\,|x'\rangle$ of the system, the Wigner function is defined as [149, 150]

$$W(x,p) \equiv \frac{1}{2\pi}\int_{-\infty}^{+\infty} \mathrm{d}y\, \mathrm{e}^{\mathrm{i}py}\rho(x+y/2, x-y/2), \tag{2.132}$$

where p is the momentum of the particle (or, more generally, the variable conjugate to x).

The Wigner function bears some similarities to a (classical) probability distribution in phase space. For example, $W(x,p)$ is a real-valued function of x and p, and the probability distributions $P(x) \equiv \rho(x,x)$ and $P(p) \equiv \widetilde{\rho}(p,p) \equiv \langle p|\,\hat{\rho}\,|p\rangle$ for x and p can be recovered as the marginals of $W(x,p)$,

$$P(x) = \rho(x,x) = \int \mathrm{d}p\, W(x,p), \tag{2.133}$$

$$P(p) = \widetilde{\rho}(p,p) = \int \mathrm{d}x\, W(x,p), \tag{2.134}$$

with $\int \mathrm{d}x \int \mathrm{d}p\, W(x,p) = \mathrm{Tr}(\hat{\rho}) = 1$.

However, there are important caveats that show that one must not take too far this tempting identification between the Wigner function and a probability distribution in phase space. Most importantly, the Wigner function will in general take on *negative* values in some regions, which demonstrates that it cannot represent a proper probability distribution. (In fact, one can show that the Wigner function of a pure state has no negative values if and only if the state is Gaussian [151]). Of course, this finding should not be all that surprising. After all, the uncertainty relation of quantum mechanics forbids the simultaneous determination of position and momentum with arbitrary precision. Therefore, quite simply, an actual quantum phase-space distribution cannot possibly exist.

Despite these cautionary remarks, in many cases the Wigner function allows us to nicely visualize position–momentum correlations and quantum interferences in phase space. The use of the Wigner function to represent "double slit–like" superpositions and to study the dynamics of phase-space decoherence goes back to Zurek [13]. In this book, we will employ the Wigner function in our study of quantum Brownian motion in Sect. 5.2.

To get a feel for what the Wigner function looks like, let us consider a superposition of two Gaussian wave packets separated by a distance Δx in position space (see Fig. 2.7). The corresponding Wigner function is shown in Fig. 2.8. We observe two main peaks together with an oscillatory pattern. The main peaks, often also called the *direct peaks*, are located in the classically expected phase-space regions (i.e., separated by Δx in position and identical with respect to their extension in the momentum direction). We may therefore view these peaks as akin to describing the "classical" phase-space distribution described by the two superimposed peaks. We clearly see the significant spread in the momentum direction, which is a consequence of the fairly tight localization in position space: According to the uncertainty principle, we cannot have perfect localization in both position and momentum,

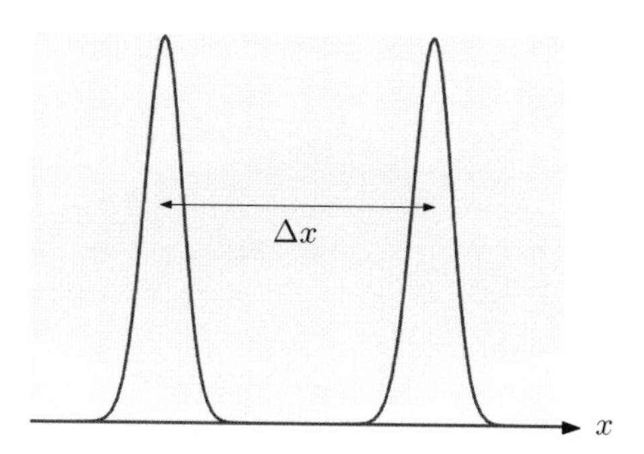

Fig. 2.7. Illustration of a superposition of two Gaussian wave packets separated by Δx in space.

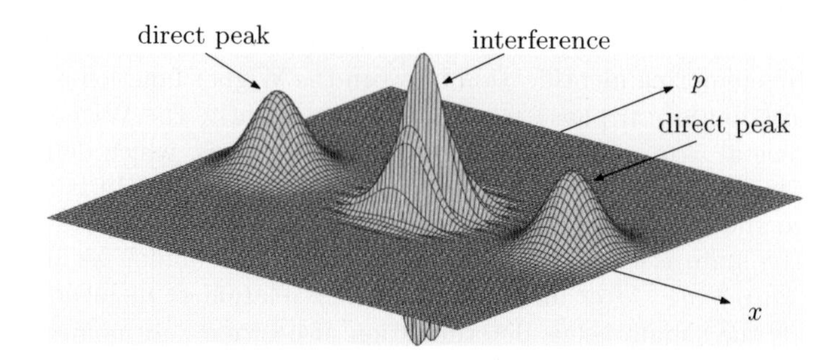

Fig. 2.8. Wigner representation of a superposition of two Gaussian wave packets separated in position space (see Fig. 2.7).

and the better the localization in position, the broader will be the spread in the momentum direction.

The oscillatory pattern visible in Fig. 2.8, with its ridges parallel to the line joining the two main peaks, is the telltale sign for the presence of quantum interference between the two wave packets. When the interaction with an environment monitoring the position of the system leads to a suppression of spatial coherence, these oscillations become damped (see Fig. 2.9). This provides us with an intuitive method for visualizing the action of decoherence in phase space.

One can show that, for the superposition state of two spatially separated Gaussian wave packets considered here, the frequency f of the oscillatory pattern in the Wigner function is directly related to the separation Δx between the wave packets in position space via

$$f = \Delta x/\hbar. \tag{2.135}$$

Thus, broadly speaking, the oscillations in the Wigner function will get more rapid as the superposition state becomes more nonclassical. We also clearly

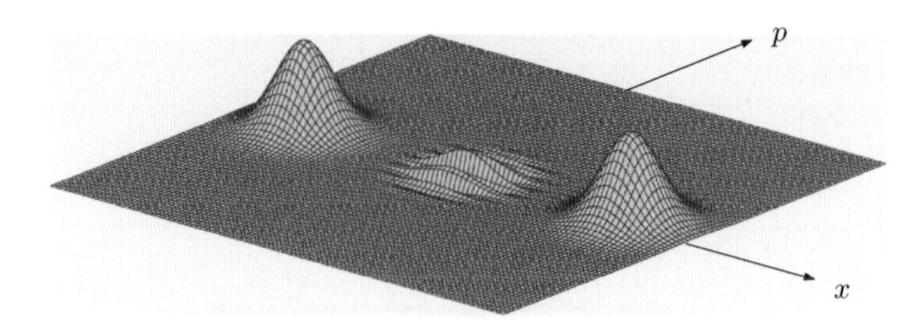

Fig. 2.9. The interaction with the environment and the resulting suppression of interference manifests itself through the damping of the oscillatory pattern in the Wigner representation (compare Fig. 2.8).

see from Fig. 2.8 that the Wigner function takes on negative values, in particular in the region of the oscillatory pattern. As discussed above, this is a direct visual indicator that we cannot interpret the Wigner function as a classical phase-space probability distribution.

2.15.3 "Purifying" the Environment

When discussing the trace operation in Sect. 2.4.6, we have shown how to obtain the reduced state of a system from the pure state of a larger composite system—in our case, the system of interest together with its environment. It should not come as a surprise that we can also "invert" this procedure to *purify* any given nonpure state (in a Hilbert space of finite dimension). By this statement we mean the ability of regarding an arbitrary nonpure state as the reduced state of the pure state of a larger system.

The proof of this statement is absolutely straightforward. Suppose our system of interest $\mathcal{S}$ is described by some (in general nonpure) density matrix $\hat{\rho}_{\mathcal{S}}$ in a Hilbert space $\mathcal{H}_{\mathcal{S}}$. From Sect. 2.4 [see especially (2.20)] we know that any density matrix can be written in the form $\hat{\rho}_{\mathcal{S}} = \sum_i p_i |\psi_i\rangle\langle\psi_i|$, where the p_i are nonnegative real numbers obeying $\sum_i p_i = 1$. Let us now introduce another (fictitious) system, which we may call the "environment" $\mathcal{E}$ (although, of course, no particular physical role or representation needs to be attached to this second system). We take the Hilbert space $\mathcal{H}_{\mathcal{E}}$ of $\mathcal{E}$ to have the same dimensionality as $\mathcal{H}_{\mathcal{S}}$ and choose an orthonormal basis $\{|\phi_i\rangle\}$ of $\mathcal{H}_{\mathcal{E}}$. Let us define the pure state

$$|\Psi\rangle \equiv \sum_i \sqrt{p_i}\, |\psi_i\rangle\, |\phi_i\rangle \tag{2.136}$$

of the composite $\mathcal{SE}$ system. The corresponding pure-state density matrix is then

$$\hat{\rho} = |\Psi\rangle\langle\Psi| = \sum_{ij} \sqrt{p_i p_j} |\psi_i\rangle\langle\psi_j| \otimes |\phi_i\rangle\langle\phi_j| . \tag{2.137}$$

If we now take the trace over our fictitious "environment" $\mathcal{E}$, we simply recover the original density matrix $\hat{\rho}_{\mathcal{S}}$ of $\mathcal{S}$,

$$\mathrm{Tr}_{\mathcal{E}}\hat{\rho} = \sum_k \langle\phi_k| \left[\sum_{ij} \sqrt{p_i p_j} |\psi_i\rangle\langle\psi_j| \otimes |\phi_i\rangle\langle\phi_j| \right] |\phi_k\rangle \tag{2.138}$$

$$= \sum_i p_i |\psi_i\rangle\langle\psi_i| = \hat{\rho}_{\mathcal{S}} . \tag{2.139}$$

This "purification theorem" has an important implication. In discussions of decoherence one can always assume, without loss of generality, that the environment is in a pure state before its interaction with the system of interest. This is so because in cases where the environment is *not* in a pure state to begin with, we can always purify it through the introduction of an additional (fictitious) environment.

2.15.4 The Operator-Sum Formalism

Let us introduce a completely general formalism for representing the influence of the environment on the reduced density matrix of the system, without any reference to the particular form of the Hamiltonian. The derivation of this so-called *operator-sum formalism* (often also referred to as the *Kraus-operator formalism* [152]) is very simple, and goes as follows.

To begin, suppose that the system $\mathcal{S}$ and its environment $\mathcal{E}$ are initially completely uncorrelated, such that the joint density matrix is

$$\hat{\rho}(0) = \hat{\rho}_{\mathcal{S}}(0) \otimes \hat{\rho}_{\mathcal{E}}(0). \tag{2.140}$$

We can then write down the diagonal decomposition of the initial density matrix $\hat{\rho}_{\mathcal{E}}(0)$ of the environment,

$$\hat{\rho}_{\mathcal{E}}(0) = \sum_i p_i |E_i\rangle\langle E_i|. \tag{2.141}$$

Here the coefficients p_i obey $\sum_i p_i = 1$, and the environmental states $|E_i\rangle$ form a set of orthonormal basis states of the Hilbert space $\mathcal{H}_{\mathcal{E}}$ of the environment.

Given the usual unitary time evolution operator $\hat{U}(t) = \mathrm{e}^{-\mathrm{i}\hat{H}t}$, where $\hat{H}$ is the total Hamiltonian (here assumed to be time-independent) of the system–environment combination, the evolution of the reduced density matrix $\hat{\rho}_{\mathcal{S}}(t)$ of the system is given by

$$\hat{\rho}_{\mathcal{S}}(t) = \mathrm{Tr}_{\mathcal{E}} \left\{ \hat{U}(t) \left[\hat{\rho}_{\mathcal{S}}(0) \otimes \left(\sum_i p_i |E_i\rangle\langle E_i| \right) \right] \hat{U}^\dagger(t) \right\}. \tag{2.142}$$

If we explicitly carry out the trace operation in the basis $\{|E_i\rangle\}$, (2.142) becomes

$$\begin{aligned} \hat{\rho}_{\mathcal{S}}(t) &= \sum_j \langle E_j| \hat{U}(t) \left[\hat{\rho}_{\mathcal{S}}(0) \otimes \left(\sum_i p_i |E_i\rangle\langle E_i| \right) \right] \hat{U}^\dagger(t) \, |E_j\rangle \\ &= \sum_{ij} \sqrt{p_i} \, \langle E_j| \hat{U}(t) \, |E_i\rangle \otimes \hat{\rho}_{\mathcal{S}}(0) \otimes \sqrt{p_i} \, \langle E_i| \hat{U}^\dagger(t) \, |E_j\rangle \\ &\equiv \sum_{ij} \hat{E}_{ij} \otimes \hat{\rho}_{\mathcal{S}}(0) \otimes \hat{E}_{ij}^\dagger. \end{aligned} \tag{2.143}$$

In the last line we have defined the *Kraus operators* [152]

$$\hat{E}_{ij} \equiv \sqrt{p_i} \, \langle E_j| \hat{U}(t) \, |E_i\rangle \tag{2.144}$$

acting on the Hilbert space $\mathcal{H}_{\mathcal{S}}$ of the system.

The Kraus operators contain all the available information about the initial state of the environment and about the dynamics of the system–environment combination. Thus they entirely encapsulate the effect of the environment on the reduced density matrix of the system. The Kraus operators depend on the particular choice of the basis $\{|E_i\rangle\}$ used in performing the trace, but are otherwise uniquely determined by the initial state of the environment $\mathcal{E}$ and the Hamiltonian of the joint $\mathcal{SE}$ system.

Since the evolution of the composite $\mathcal{SE}$ system is unitary, the Kraus operators satisfy the *completeness* constraint

$$\sum_{ij} \hat{E}_{ij}\hat{E}_{ij}^{\dagger} = \hat{I}_{\mathcal{S}}. \tag{2.145}$$

This can be readily seen from the fact that

$$\begin{aligned}
\sum_{ij} \hat{E}_{ij}\hat{E}_{ij}^{\dagger} &= \sum_{ij} p_i \langle E_j|\,\hat{U}(t)\,|E_i\rangle \langle E_i|\,\hat{U}^{\dagger}(t)\,|E_j\rangle \\
&= \sum_i p_i \langle E_i|\,\hat{U}^{\dagger}(t) \left(\sum_j |E_j\rangle\langle E_j| \right) \hat{U}(t)\,|E_i\rangle \\
&= \sum_i p_i \langle E_i|\,\hat{U}^{\dagger}(t)\hat{U}(t)\,|E_i\rangle \\
&= \left(\sum_i p_i \right) \hat{I}_{\mathcal{S}} \\
&= \hat{I}_{\mathcal{S}}.
\end{aligned} \tag{2.146}$$

Equation (2.145) can therefore be viewed as an indicator for the unitary evolution of $\mathcal{SE}$. If this relation was *not* fulfilled, we would have to infer the existence of another environment $\mathcal{E}'$ interacting with $\mathcal{SE}$, which would lead to an effectively nonunitary evolution of $\mathcal{SE}$.

The operator-sum (or Kraus-operator) approach neatly represents the effect of the environment as a sequence of (in general nonunitary) transformations of the reduced density matrix generated by the operators $\hat{E}_{ij}$, see (2.144). This approach provides a compact and transparent framework for the formal representation of the evolution of the reduced density matrix. However, its practical usefulness for explicitly calculating the decoherence dynamics in concrete situations of interest is somewhat limited, for several reasons.

First, the task of computing the Kraus operators (2.144) corresponds to diagonalizing the full Hamiltonian $\hat{H}$. Except for a very few simple decoherence models (such as the model described in Sect. 2.10), this diagonalization is usually impossible to carry out in practical applications. Second, the Kraus operators contain all (i.e., both unitary and nonunitary) contributions to the evolution of the reduced density matrix. When discussing decoherence, however, we are usually interested only in the nonunitary terms (since these are the contributions responsible for decoherence). This desired clear distinction into nonunitary and unitary terms is not afforded by the above operator-sum formalism. Third, the dynamics described by the Kraus operators are often unnecessarily detailed, as they represent the *exact* evolution of the reduced density matrix, including, for example, all possible back-action effects from the system on the environment. In many situations of practical interest, such effects can be neglected, and one is then able to derive simplified ("master") equations for the approximate evolution of the reduced density matrix of the system. Such equations will be discussed in detail in Chaps. 4 and 5.

2.16 Summary

Let us summarize some of the main points and results of this chapter.

- Quantum states differ fundamentally from states in classical physics. They do not simply describe a catalog of values of physical quantities that could be arbitrarily enlarged through additional measurements, since in general a measurement fundamentally alters the quantum state.
- Coherent superpositions are at the heart of quantum mechanics and cannot be interpreted as classical ensembles of states. Superpositions describe individual physical states. In the laboratory, their existence and quantum nature are typically verified by means of interference experiments.
- Quantum entanglement describes nonlocal quantum correlations between systems and is at the heart of decoherence. Entanglement goes far beyond the classical concept of probability relations between systems.
- The concept of density matrices plays an important role in the formal description of decoherence. We distinguished three different types of density matrices:
  - *Pure-state density matrices* are the direct density-matrix equivalent of pure states and represent a completely known (physical) state of the system.
  - *Mixed-state density matrices* describe a classical probability distribution (i.e., an ignorance-interpretable ensemble) of pure states.
  - *Reduced density matrices* exhaustively describe the statistics of all possible measurements that can be performed on a subsystem of a larger system. In the context of decoherence, they are used to describe the local measurement statistics for a system entangled with an environment. Reduced density matrices are obtained by a (nonunitary) trace operation over all degrees of freedom other than those of the system of interest. They represent improper ensembles and therefore must not be interpreted as mixed-state density matrices.
- The measurement problem, and the more general problem of the quantum-to-classical transition, is composed of three main issues:
  - The preferred-basis problem (what determines the preferred physical quantities of our experience?).

  - The problem of the nonobservability of interference (why is it so hard to observe interference effects?).
  - The problem of outcomes (why do measurements seem to have outcomes at all, and what selects the particular observed outcome?).

As we have indicated in this chapter and will discuss further in other places of this book (see, e.g., Chap. 8), it is reasonable to conclude that decoherence is capable of solving the first two problems, whereas the third problem is intrinsically linked to matters of interpretation that are mostly outside of the scope of decoherence.

- Bohr's famous "complementarity principle" can be understood as a consequence of entanglement. A second system $\mathcal{S}'$ entangled with the system of interest $\mathcal{S}$ can encode information about $\mathcal{S}$ in the following sense. The less the overlap between the relative states of $\mathcal{S}'$ that are quantum-correlated with the components in a superposition state of $\mathcal{S}$, the better $\mathcal{S}'$ is able to distinguish between these components and the more difficult it is to observe interference between the components through measurements performed on $\mathcal{S}$ only.
- Decoherence is based on the idea that physical systems are immersed into environments that play the role of the second system $\mathcal{S}'$: Through entanglement between the system and its environment, the environment continuously monitors certain states of the system. Which states are monitored in a particular situation depends on the form of the specific system–environment interaction.
- This environmental monitoring implies that information about certain states of the system becomes encoded in the environment. If the system is described by a superposition of these states, we would (if only in principle) be able to distinguish these component states by measuring the environment, which forces out the decay of interference between the components at the local level of the system. To observationally confirm the existence of the superposition, we would need to perform measurements on the composite system–environment system, which is impossible for all practical purposes in most physically realistic situations. Thus coherence is *practically irreversibly delocalized* into the larger system–environment combination through uncontrolled environmental entanglement and thus becomes effectively unavailable to the observer who has only access to the system.
- This decoherence process has three interrelated consequences:
  - The local damping of coherence between the states monitored by the environment. This provides a solution to one component of the measurement problem, namely, the problem of the nonobservability of interference.
  - The environment-induced superselection of preferred states (pointer states) for the system. These are the states most immune to entanglement with the environment and thus to decoherence. Because of their

robustness, they can be regarded as effectively classical. We discussed how to determine these pointer states, given the total Hamiltonian describing the global system–environment dynamics. Environment-induced superselection provides a very general explanation for the emergence of the preferred physical quantities in nature (such as position or charge, instead of superpositions thereof) and thus an effective solution to the preferred-basis problem.
  - The robust and redundant encoding of information about the environment-superselected states in the environment, allowing observers to gain information about the system by intercepting fragments of the environment. This approach mirrors closely how information is usually gathered by observers and minimizes the disturbance of the system imparted by the observer.
- We emphasized that decoherence is a pure quantum-mechanical effect without classical analog. In particular, decoherence does not need to be accompanied by dissipation. If it is, then decoherence typically happens on timescales vastly shorter than the timescales for dissipation (especially on macroscopic scales). Also, classical noise processes have to be clearly distinguished from decoherence, both from conceptual and observational points of view.

3 Decoherence Is Everywhere: Localization Due to Environmental Scattering

It is fair to say that localization induced by the scattering of environmental particles really lies at the heart of the decoherence program and the quantum-classical transition, for several reasons.

Together with the emission of thermal radiation (see Sect. 6.2.5), environmental scattering is the dominant and ubiquitous process for decoherence in the macroscopic domain. Air molecules, light (optical photons), background radioactivity, cosmic muons, solar neutrinos, and even the 3 K cosmic background radiation present everywhere in the universe continuously monitor the position of the quantum system of interest. This results in system–environment entanglement that delocalizes local phase relations between spatially separated wave-function components, leading to decoherence in position space (i.e., to localization). While sophisticated experimental setups may be able to shield the system from some of these environmental particles (such as air molecules and light), it is prohibitively difficult, if not impossible, to exclude other influences.

The quantum nature of decoherence is again important in understanding why environmental scattering by essentially *any* particle is relevant. As already pointed out in Chap. 1 and more generally discussed in Sect. 2.11, in the classical picture of scattering the scatterer will only be influenced by the scattered particle if the mass of the latter particle is sufficiently large in comparison with the mass of the scatterer. Simply put, in determining the Newtonian dynamics of a billiard ball we do not need to concern ourselves with the influence of air molecules, photons, etc. By contrast, in the quantum picture any interaction between the scattering system and the scattered particle may lead to entanglement and thus to decoherence, irrespective of the mass ratio. Typically, the more macroscopic the object, the larger its scattering cross section, and thus the stronger and faster it is decohered by environmental scattering. This behavior can be understood from the fact that, as discussed in Sect. 2.12, decoherence corresponds to a situation in which the system perturbs the environment (a process which in turn influences the reduced dynamics of the system). It is thus intuitively clear that the degree of this perturbation—and thus the amount of decoherence introduced into the system—will increase with the size of the system.

Environmental scattering and the resulting process of spatial localization also play a key role for historical reasons. The attempt to directly identify narrow wave packets in position space with particles and their trajectories goes back to the early years of quantum mechanics. Based on ideas inspired by the theory of relativity and the photoelectric effect, the French physicist Louis de Broglie suggested in the early 1920s that every particle of mass m and velocity v has a quantum "matter wave" of wavelength $\lambda = h/mv$ associated with it [34, 153–155]. Schrödinger was heavily influenced by de Broglie's Ph.D. thesis of 1924 [156], as evidenced in a letter to Einstein dated November 16, 1925:

> I have been intensely concerned these days with Louis de Broglie's ingenious theory. It is extraordinarily exciting, but still has some very grave difficulties.

Only weeks later, Schrödinger identified de Broglie's matter wave with a wave function ψ and formulated his famous wave equation that specified the time evolution of this wave function, thereby giving birth to wave mechanics.

However, in developing his theory, Schrödinger became immediately aware of the problem of how to describe particles by spatially extended waves. It is well known that, for a free particle, unitary time evolution tends to coherently spread out any spatially localized wave packet (Fig. 3.1). For example, if at time $t = 0$ a free particle is described by a wave packet of the form

$$\psi(x, t = 0) = \left(\frac{1}{\sqrt{\pi}\sigma} \right)^{1/2} \exp\left[-\frac{x^2}{2\sigma^2} \right], \tag{3.1}$$

the position probability density $|\psi(x,t)|^2$ at a later time $t > 0$ is given by

$$\begin{aligned} |\psi(x,t)|^2 &= \frac{1}{\sqrt{\pi}\sigma \left[1 + \hbar^2 t^2/(m^2\sigma^4)\right]^{1/2}} \exp\left[-\frac{x^2}{\sigma^2 \left[1 + \hbar^2 t^2/(m^2\sigma^4)\right]} \right] \\ &\equiv \frac{1}{\sqrt{\pi}\sigma(t)} \exp\left[-\frac{x^2}{\sigma^2(t)} \right]. \end{aligned} \tag{3.2}$$

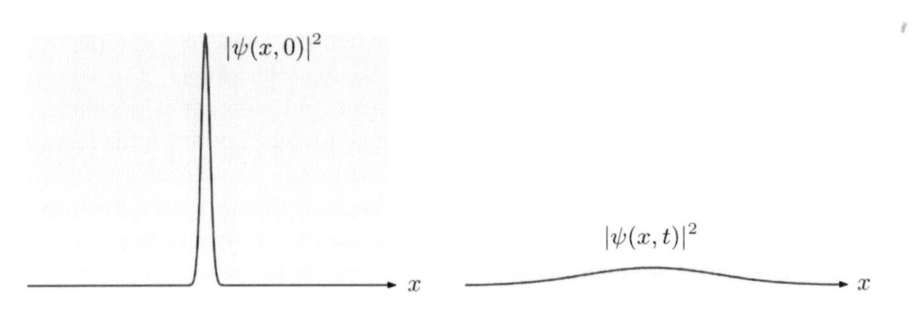

Fig. 3.1. Coherent spread of the probability density $|\psi(x,t)|^2$ for a free-particle Gaussian wave packet under unitary time evolution.

This means that the width σ of the wave packet grows as

$$\sigma(t) = \sigma \left[1 + \hbar^2 t^2/(m^2\sigma^4)\right]^{1/2}. \tag{3.3}$$

For microscopic particles, this spreading occurs on very short timescales. For example, if the Gaussian wave packet describes a particle with the mass of an electron ($m \approx 10^{-30}$ kg) and has an initial width of $\sigma = 1$ Å, unitary time evolution spreads the wave packet to a width on the order of $10^{16}\sigma = 10^6$ m (or 1,000 km!) within a second. This effect of coherent spreading for free-particle wave functions posed a serious difficulty to approaches that tried to relate narrow wave packets to particles. How could we ever directly identify any objects—that evidently remain in spatially well-defined regions—with quantum-mechanical wave packets, if these packets immediately and coherently disperse over macroscopic distances?

Since the attempt to establish a one-to-one correspondence between wave packets and particles had failed for the case of isolated free particles, Schrödinger turned his attention to situations in which particles are subject to a potential. In particular, in a famous 1926 paper entitled "The continuous transition from micromechanics to macromechanics"[1] [157], he used the example of the quantum harmonic oscillator to show that, by forming a particular linear combination of the position-space solutions of the Schrödinger equation, each of which represents a spatially spread-out wave, the resulting wave packet is not only narrow at $t = 0$, but also remains narrow at all subsequent times $t > 0$, with its peak oscillating back and forth just like a classical point mass.

Schrödinger, of course, had discovered the coherent states of the quantum harmonic oscillator, i.e., minimum-uncertainty wave packets in phase space that follow classical trajectories. This success led Schrödinger to the upbeat prediction that [157, p. 666]

> in much the same way, one will be able to construct wave packets which follow elliptical Kepler orbits with large quantum numbers and which are the wave-mechanical image of the electron in the hydrogen atom; it is just that one might encounter more mathematical difficulties than in the particularly simple example discussed here.[2]

---

[1]The original German title reads: *"Der stetige Übergang von der Mikro- zur Makromechanik."*

[2]The original German text reads: *"Es läßt sich mit Bestimmtheit voraussehen, daß man auf ganz ähnliche Weise auch die Wellengruppen konstruieren kann, welche auf hochquantigen Keplerellipsen umlaufen und das undulationsmechanische Bild des Wasserstoffelektrons sind; nur sind da die rechentechnischen Schwierigkeiten größer als in dem hier behandelten, ganz besonders einfachen Schulbeispiel."*

However, this—as it turns out, too optimistic[3]—assessment did not touch upon the core problem identified earlier, namely, of how to describe free particles in terms of wave packets. Several decades after the publication of Schrödinger's paper, an understanding of the importance of the openness of quantum systems and of the resulting decoherence effects [4–6, 8, 9, 12] brought us a major step closer to resolving the difficulty of how to reconcile the predictions of the Schrödinger equation for free particles with the fact that every macroscopic objects appears localized. The seminal paper of Joos and Zeh [7] showed, through detailed models and numerical results, how the ubiquitous scattering of photons, air molecules, and other environmental particles effectively suppresses the coherent spreading on extremely short timescales. This decoherence process leads to a reduced density matrix for the system that represents an (improper) ensemble of position-space wave packets whose widths rapidly decrease toward the thermal de Broglie wave length (2.114).

In the following, we will study a model for scattering-induced decoherence that goes back to the original model of Joos and Zeh [7]. It is now understood that the derivation of the equation of motion for the reduced density matrix as presented in [7] contains a subtle flaw, which results in a decoherence rate that is larger by a factor of 2π compared to the correct result [159–161]. Of course, at the time of the publication of their paper (1985), Joos and Zeh were merely interested in simple estimates of decoherence timescales, since measurements of such timescales seemed out of reach for any experiment. However, over the past few years a rapid progress in experimental techniques has enabled researchers to make very precise measurements of decoherence timescales even for mesoscopic and macroscopic objects (see, e.g., the interference experiments with C_{70} molecules described in Sect. 6.2). These measurements are sensitive to differences on the order of the aforementioned factor of 2π, and it has therefore become important to employ a theory of environmental scattering that gives the quantitatively correct results.

Furthermore, the original work by Joos and Zeh was based on an assumption about the relative wavelengths of the object and the scattered environmental particles. A more general approach that relaxed this assumption but did not address the aforementioned flaw in the derivation (thus still leading

[3]The case of Kepler orbits does not give rise to the exact coherent states as in the idealized example of a quantum harmonic oscillator. Another interesting example is the case of Hyperion, a moon of Saturn that has been observed to exhibit chaotic tumbling. Chaotic dynamics imply an exponential sensitivity of the (classical) trajectory on the initial phase-space parameters, which cannot be specified with arbitrary precision due to the Heisenberg uncertainty principle. Thus an initially narrow phase-space wave packet describing Hyperion would very rapidly spread over large spatial regions. In fact, Zurek [158] estimated that within ≈ 20 years the quantum state would be a highly nonlocal coherent superposition of macroscopically distinguishable orientations of the satellite's major axes. Thus even on planetary scales the problem of coherent spreading may arise.

to the additional incorrect factor of 2π) was subsequently given by Gallis and Fleming [162]. The first derivation remedying this flaw and arriving at the correct result for the decoherence rate was presented by Diósi [159]. Several years later, Hornberger and Sipe [160] revisited the problem of environmental scattering and obtained the correct result using a sophisticated approach quite different from that of Diósi. While their reevaluation is very careful and thorough, it is also rather complicated. Recently, Adler [161] showed that the correct result can be obtained in a significantly easier manner by means of a little trick. Other derivations have been discussed by, e.g., Halliwell [163] and Hornberger [164, 165]. Our following presentation will mainly adhere to the derivations of Hornberger and Sipe [160] and Adler [161], with the goal of obtaining the correct result while keeping the calculations as simple as possible.

This chapter is organized as follows. Section 3.1 will introduce our scattering model. In Sect. 3.2, we will compute the explicit form of the decoherence factor that describes the decay of spatial coherences. In Sect. 3.3, we will treat two important limiting cases, namely, the situation of environmental wavelengths that are short (Sect. 3.3.1) and long (Sect. 3.3.2) in comparison with the coherent separation between the two center-of-mass positions in a spatial superposition state of the system. Section 3.4 will apply our theory to the scattering of photons (Sect. 3.4.1) and air molecules (Sect. 3.4.2). Finally, in Sect. 3.5, we will discuss the explicit dynamics resulting from such scattering-induced decoherence by studying the time evolution of the reduced density matrix of the system.

3.1 The Scattering Model

We consider an object (the system $\mathcal{S}$) that scatters a collection of environmental particles (the environment $\mathcal{E}$), see Fig. 3.2. We shall make the usual assumption that $\mathcal{S}$ and $\mathcal{E}$ are initially completely uncorrelated, i.e., that the (pure-state) density matrix $\hat{\rho}$ for the composite $\mathcal{SE}$ arrangement factorizes at $t = 0$,

$$\hat{\rho}(0) = \hat{\rho}_{\mathcal{S}}(0) \otimes \hat{\rho}_{\mathcal{E}}(0). \tag{3.4}$$

Let us first consider the case of a single scattering event and study its influence on the density matrix $\hat{\rho}_{\mathcal{S}}$ of the system (Fig. 3.3). In the following, we will use very basic quantum-mechanical scattering theory. Reviews of this subject matter can be found in most textbooks on quantum mechanics (e.g., in Chap. 7 of [58]), but the reader should be able to follow the derivation without having to consult other literature.

Let us denote the initial state of the system with center-of-mass location $\boldsymbol{x}$ by a position eigenstate $|\boldsymbol{x}\rangle$ (with eigenvalue $\boldsymbol{x}$), and let us write $|\chi_i\rangle$ for the initial state of the incoming environmental particle. Then the effect of the scattering event can be formally expressed by the action of the scattering operator $\hat{S}$ (the so-called "S-matrix") on the initial state,

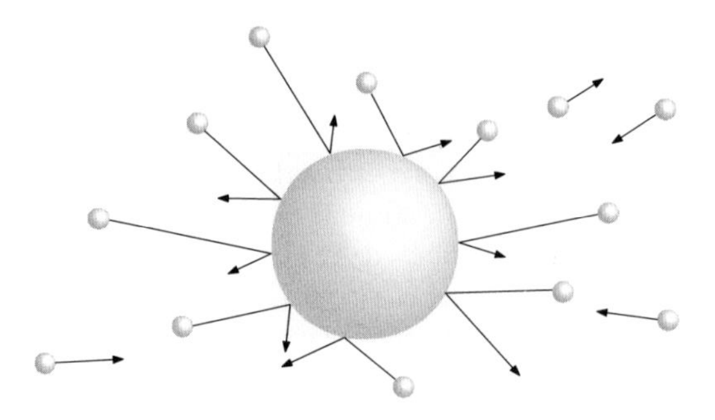

Fig. 3.2. Environmental scattering. A collection of particles, such as photons or air molecules, scatters off the object of interest.

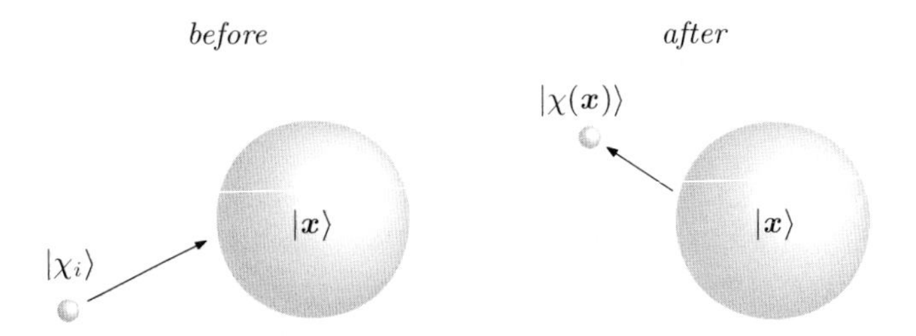

Fig. 3.3. A single scattering event. The incoming environmental particle, described by the initial state $|\chi_i\rangle$, elastically scatters off a stationary object of interest, whose center-of-mass state is represented by a position eigenstate $|\boldsymbol{x}\rangle$. The range of the scattering potential is sufficiently short, such that the states of the object and the incoming particle are initially uncorrelated. The object is assumed to be much more massive than the scattered environmental particle, such that the object is not disturbed by the scattering event. Thus the interaction yields the evolution $|\boldsymbol{x}\rangle|\chi_i\rangle \longrightarrow |\boldsymbol{x}\rangle|\chi(\boldsymbol{x})\rangle$, where $|\chi(\boldsymbol{x})\rangle$ is the final state of the scattered particle [see also (3.12)].

$$|\boldsymbol{x}\rangle|\chi_i\rangle \longrightarrow \hat{S}|\boldsymbol{x}\rangle|\chi_i\rangle , \tag{3.5}$$

where we have used the separability assumption (3.4) for the initial state. At the risk of stating the obvious, we note that $|\chi_i\rangle$ is of course independent of the location $\boldsymbol{x}$ of the scattering object. The state $|\boldsymbol{x}\rangle$ can be thought of the state $|\boldsymbol{x}=\boldsymbol{0}\rangle$ (corresponding to the scattering center being located at the origin) translated by the action of the momentum operator $\hat{\boldsymbol{p}}$ of the system,[4]

$$|\boldsymbol{x}\rangle = \mathrm{e}^{-\mathrm{i}\hat{\boldsymbol{p}}\cdot\boldsymbol{x}/\hbar}|\boldsymbol{x}=\boldsymbol{0}\rangle . \tag{3.6}$$

We can hence rewrite (3.5) as

[4]Since our model will be used to calculate explicit numerical values of decoherence timescales, we shall retain $\hbar$ throughout the following derivation.

$$\mathrm{e}^{-\mathrm{i}\hat{\boldsymbol{p}}\cdot\boldsymbol{x}/\hbar}\,|\boldsymbol{0}\rangle\,|\chi_i\rangle \longrightarrow \hat{S}\mathrm{e}^{-\mathrm{i}\hat{\boldsymbol{p}}\cdot\boldsymbol{x}/\hbar}\,|\boldsymbol{0}\rangle\,|\chi_i\rangle = \hat{S}\mathrm{e}^{-\mathrm{i}(\hat{\boldsymbol{p}}+\hat{\boldsymbol{q}})\cdot\boldsymbol{x}/\hbar}\,|\boldsymbol{0}\rangle\,\mathrm{e}^{\mathrm{i}\hat{\boldsymbol{q}}\cdot\boldsymbol{x}/\hbar}\,|\chi_i\rangle\,. \quad (3.7)$$

Here, $\hat{\boldsymbol{q}}$ is the momentum operator for the scattered particle, and thus $\hat{\boldsymbol{P}} \equiv \hat{\boldsymbol{p}} + \hat{\boldsymbol{q}}$ is the momentum operator for the composite $\mathcal{SE}$ system.

If we now make the assumption that the scattering interaction is invariant under spatial translations of the joint system $\mathcal{SE}$ (i.e., under translations generated by the operator $\hat{\boldsymbol{P}}$),

$$\left[\hat{S}, \hat{\boldsymbol{P}}\right] = 0, \quad (3.8)$$

we can pull the first exponential on the right-hand side of (3.7) out to the front to obtain

$$\hat{S}\,|\boldsymbol{x}\rangle\,|\chi_i\rangle = \mathrm{e}^{-\mathrm{i}(\hat{\boldsymbol{p}}+\hat{\boldsymbol{q}})\cdot\boldsymbol{x}/\hbar}\hat{S}\,|\boldsymbol{0}\rangle\,\mathrm{e}^{\mathrm{i}\hat{\boldsymbol{q}}\cdot\boldsymbol{x}/\hbar}\,|\chi_i\rangle\,. \quad (3.9)$$

To proceed further, a key assumption is now made, namely, that the interaction between the object and the scattered particle does not lead to any recoil of the object. This means that the particle remains essentially undisturbed by the scattering event, and that therefore the only consequence of the scattering process is the creation of quantum correlations (entanglement) between the system and the environment.

This is a good assumption in situations where the scattering system is much more massive than the scattered particle such that recoil effects on the system can be neglected. This can be expected to hold for the case of microscopic or mesoscopic particles scattered off macroscopic objects. However, we should note that timescales for spatial decoherence quoted in the literature have sometimes been computed using this assumption even for scattering processes in which the ratio of the mass of the scattering object to the mass of the scattered environmental particle is either close to one or even less than one. An example is the case of small molecules and free electrons decohered by the scattering of air molecules [166]. Clearly, in this regime the no-recoil assumption cannot be expected to be physically realistic, and one should therefore resort to more general models, such as that of Diósi [159] (which does take recoil into account).

For the following treatment, we should adopt the above assumption of a recoil-free scattering process by taking the scattering object to be much more massive than the environmental particle. Then the action of the S-matrix on the composite system–environment state does not affect the center-of-mass state $|\boldsymbol{x}\rangle$ of the system, i.e., the center of mass remains at position $\boldsymbol{x}$. We can therefore evaluate (3.7) further,

$$\begin{aligned}\hat{S}\,|\boldsymbol{x}\rangle\,|\chi_i\rangle &= \mathrm{e}^{-\mathrm{i}(\hat{\boldsymbol{p}}+\hat{\boldsymbol{q}})\cdot\boldsymbol{x}/\hbar}\,|\boldsymbol{0}\rangle\,\hat{S}_0\mathrm{e}^{\mathrm{i}\hat{\boldsymbol{q}}\cdot\boldsymbol{x}/\hbar}\,|\chi_i\rangle \\ &= \mathrm{e}^{-\mathrm{i}\hat{\boldsymbol{p}}\cdot\boldsymbol{x}/\hbar}\,|\boldsymbol{0}\rangle\,\mathrm{e}^{-\mathrm{i}\hat{\boldsymbol{q}}\cdot\boldsymbol{x}/\hbar}\hat{S}_0\mathrm{e}^{\mathrm{i}\hat{\boldsymbol{q}}\cdot\boldsymbol{x}/\hbar}\,|\chi_i\rangle \\ &= |\boldsymbol{x}\rangle\,\mathrm{e}^{-\mathrm{i}\hat{\boldsymbol{q}}\cdot\boldsymbol{x}/\hbar}\hat{S}_0\mathrm{e}^{\mathrm{i}\hat{\boldsymbol{q}}\cdot\boldsymbol{x}/\hbar}\,|\chi_i\rangle\,. \end{aligned} \quad (3.10)$$

Here, we have added the subscript "0" to the scattering operator $\hat{S}$ to indicate that this operator refers specifically to a scattering process in which the scattering center is located at the origin. The last line of (3.10) can be used to define the translated scattering operator

$$\hat{S}_{\boldsymbol{x}} \equiv \mathrm{e}^{-\mathrm{i}\hat{\boldsymbol{q}}\cdot\boldsymbol{x}/\hbar}\hat{S}_0\mathrm{e}^{\mathrm{i}\hat{\boldsymbol{q}}\cdot\boldsymbol{x}/\hbar}, \tag{3.11}$$

which describes scattering with the scattering center located at the position $\boldsymbol{x}$. To summarize, the scattering interaction is then given by

$$|\boldsymbol{x}\rangle\,|\chi_i\rangle \longrightarrow \hat{S}\,|\boldsymbol{x}\rangle\,|\chi_i\rangle = |\boldsymbol{x}\rangle\,\hat{S}_{\boldsymbol{x}}\,|\chi_i\rangle \equiv |\boldsymbol{x}\rangle\,|\chi(\boldsymbol{x})\rangle\,, \tag{3.12}$$

where we have introduced the abbreviation $|\chi(\boldsymbol{x})\rangle$ for the final state of the environmental particle scattered at $\boldsymbol{x}$. We see from (3.12) that the scattering process can be thought of as a measurement-like interaction that establishes correlations between the state $|\boldsymbol{x}\rangle$, describing the center-of-mass position of the system, and the final state $|\chi(\boldsymbol{x})\rangle$ of the particle, which encodes which-path information about the location $\boldsymbol{x}$ of the scattering center.

From (3.12) it follows that the scattering process transforms the initial composite density matrix

$$\hat{\rho}(0) = \hat{\rho}_{\mathcal{S}}(0) \otimes \hat{\rho}_{\mathcal{E}}(0) = \int \mathrm{d}\boldsymbol{x} \int \mathrm{d}\boldsymbol{x}'\, \rho_{\mathcal{S}}(\boldsymbol{x}, \boldsymbol{x}', 0)|\boldsymbol{x}\rangle\langle\boldsymbol{x}'| \otimes |\chi_i\rangle\langle\chi_i| \tag{3.13}$$

into the density matrix

$$\hat{\rho} = \int \mathrm{d}\boldsymbol{x} \int \mathrm{d}\boldsymbol{x}'\, \rho_{\mathcal{S}}(\boldsymbol{x}, \boldsymbol{x}', 0)|\boldsymbol{x}\rangle\langle\boldsymbol{x}'| \otimes |\chi(\boldsymbol{x})\rangle\langle\chi(\boldsymbol{x}')|. \tag{3.14}$$

Using (2.50), the corresponding final reduced density matrix $\hat{\rho}_{\mathcal{S}}$ of the system is then given by

$$\hat{\rho}_{\mathcal{S}} = \mathrm{Tr}_{\mathcal{E}}\hat{\rho} = \int \mathrm{d}\boldsymbol{x} \int \mathrm{d}\boldsymbol{x}'\, \rho_{\mathcal{S}}(\boldsymbol{x}, \boldsymbol{x}', 0)|\boldsymbol{x}\rangle\langle\boldsymbol{x}'|\langle\chi(\boldsymbol{x}')|\chi(\boldsymbol{x})\rangle. \tag{3.15}$$

Expressed in the position basis, the scattering-induced evolution of the reduced density matrix can therefore be summarized as

$$\rho_{\mathcal{S}}(\boldsymbol{x}, \boldsymbol{x}', 0) \longrightarrow \rho_{\mathcal{S}}(\boldsymbol{x}, \boldsymbol{x}', 0)\langle\chi(\boldsymbol{x}')|\chi(\boldsymbol{x})\rangle. \tag{3.16}$$

We see that the local suppression of spatial coherence due to the scattering event is quantified by the overlap $\langle\chi(\boldsymbol{x}')|\chi(\boldsymbol{x})\rangle$ of the relative states of the scattered particle (see also our general discussion in Sects. 2.6.3 and 2.7). We shall now calculate the time dependence of this overlap.

3.2 Calculating the Decoherence Factor

To proceed, let us formally write the scattering operator $\hat{S}_0$ in terms of another operator $\hat{T}$ (the "T-matrix") as

$$\hat{S}_0 = \hat{I} + \mathrm{i}\hat{T}. \tag{3.17}$$

The T-matrix is well-known from standard quantum-mechanical scattering theory (see, e.g., [58]). It is used here since its elements in the momentum eigenbasis $\{|\boldsymbol{q}\rangle\}$ of the scattered particle are conveniently defined in terms of the scattering amplitude $f(\boldsymbol{q}, \boldsymbol{q}')$ as

$$\langle \boldsymbol{q} | \hat{T} | \boldsymbol{q}' \rangle = \frac{\mathrm{i}}{2\pi\hbar q} \delta(q - q') f(\boldsymbol{q}, \boldsymbol{q}') = \frac{\mathrm{i}}{2\pi\hbar m} \delta(E - E') f(\boldsymbol{q}, \boldsymbol{q}'). \tag{3.18}$$

Here m is the mass of the environmental particle, and $q = |\boldsymbol{q}|$. The term $\delta(E - E')$ ensures energy conservation, as mandated by the assumption of recoil-free elastic scattering. We note that $|f(\boldsymbol{q}, \boldsymbol{q}')|^2$ is an experimentally accessible quantity, namely, the differential cross section $\frac{\mathrm{d}\sigma}{\mathrm{d}\Omega}$ for the scattering process, which is defined as

$$\frac{\mathrm{d}\sigma}{\mathrm{d}\Omega} \equiv \frac{\text{scattered flux}}{\text{incident flux}}. \tag{3.19}$$

Since the scattering operator $\hat{S}_0$ is unitary,

$$\hat{S}_0 \hat{S}_0^\dagger = \hat{I}, \tag{3.20}$$

we obtain the following relation from the definition (3.17) of the T-matrix,

$$\hat{T}\hat{T}^\dagger + \mathrm{i}(\hat{T} - \hat{T}^\dagger) = 0. \tag{3.21}$$

This result will come in handy below.

Using the expression (3.12) for the environmental states, the overlap $\langle \chi(\boldsymbol{x}') | \chi(\boldsymbol{x}) \rangle$ is given by

$$\langle \chi(\boldsymbol{x}') | \chi(\boldsymbol{x}) \rangle = \langle \chi_i | \hat{S}_{\boldsymbol{x}'}^\dagger \hat{S}_{\boldsymbol{x}} | \chi_i \rangle. \tag{3.22}$$

Evidently, this expression is simply equal to the expectation value of the product $\hat{S}_{\boldsymbol{x}'}^\dagger \hat{S}_{\boldsymbol{x}}$ of the translated scattering operators (3.11) in the (pure) state $|\chi_i\rangle$ of the incident environmental particle (the generalization to mixed states is straightforward). Using the trace rule (2.17), we can therefore equivalently rewrite (3.22) as

$$\langle \chi(\boldsymbol{x}') | \chi(\boldsymbol{x}) \rangle = \mathrm{Tr}_{\mathcal{E}} \left\{ \hat{\rho}_{\mathcal{E}}(0) \hat{S}_{\boldsymbol{x}'}^\dagger \hat{S}_{\boldsymbol{x}} \right\}, \tag{3.23}$$

with $\hat{\rho}_{\mathcal{E}}(0) = |\chi_i\rangle\langle\chi_i|$. This is the expression of interest that we shall now evaluate. Let us begin by assuming the environmental particle is restricted to a box-normalization volume V. Following [160], we define momentum eigenkets $|\widetilde{\boldsymbol{q}}\rangle$ which are normalized over the volume V,

$$|\widetilde{\boldsymbol{q}}\rangle \equiv \left[\frac{(2\pi\hbar)^3}{V} \right]^{1/2} |\boldsymbol{q}\rangle. \tag{3.24}$$

The kets $|\widetilde{\boldsymbol{q}}\rangle$ constitute an orthonormal basis over the volume V in the sense that

$$\sum_{q \in \boldsymbol{Q}_V} |\widetilde{\boldsymbol{q}}\rangle\langle\widetilde{\boldsymbol{q}}| = \hat{I}_V. \tag{3.25}$$

Here $\boldsymbol{Q}_V$ is the set of all momenta corresponding to the space of wave functions that fulfill periodic boundary conditions in the volume V, and $\hat{I}_V$ denotes the identity operator in this space.

We can then expand the initial density matrix $\hat{\rho}_{\mathcal{E}}(0)$ of the environmental particle in terms of an ensemble of the normalized momentum eigenstates $|\widetilde{\boldsymbol{q}}\rangle$,

$$\hat{\rho}_{\mathcal{E}}(0) = \frac{(2\pi\hbar)^3}{V} \sum_{q \in \boldsymbol{Q}_V} \mu(\boldsymbol{q}) |\widetilde{\boldsymbol{q}}\rangle\langle\widetilde{\boldsymbol{q}}|, \tag{3.26}$$

where $\mu(\boldsymbol{q})$ is the momentum-space density. The great advantage of using this expansion is that the operator $\hat{T}$, see (3.17) and (3.18), is diagonal in the full-space momentum basis $\{|\boldsymbol{q}\rangle\}$ and therefore also in the properly normalized V-volume basis $\{|\widetilde{\boldsymbol{q}}\rangle\}$.

This feature enables us to evaluate the overlap $\langle\chi(\boldsymbol{x}')|\chi(\boldsymbol{x})\rangle$, see (3.23), in a fairly straightforward manner. Using the expansion (3.26) for the state of the environment in (3.23), we first get

$$\begin{aligned}
\langle\chi(\boldsymbol{x}')|\chi(\boldsymbol{x})\rangle &= \frac{(2\pi\hbar)^3}{V} \sum_{q \in \boldsymbol{Q}_V} \mu(\boldsymbol{q}) \, \langle\widetilde{\boldsymbol{q}}| \, \hat{S}^\dagger_{\boldsymbol{x}'} \hat{S}_{\boldsymbol{x}} \, |\widetilde{\boldsymbol{q}}\rangle \\
&= \frac{(2\pi\hbar)^3}{V} \sum_{q \in \boldsymbol{Q}_V} \mu(\boldsymbol{q}) \, \langle\widetilde{\boldsymbol{q}}| \, \mathrm{e}^{-\mathrm{i}\hat{\boldsymbol{q}}\cdot\boldsymbol{x}'/\hbar} \hat{S}^\dagger_0 \mathrm{e}^{-\mathrm{i}\hat{\boldsymbol{q}}\cdot(\boldsymbol{x}-\boldsymbol{x}')/\hbar} \hat{S}_0 \mathrm{e}^{\mathrm{i}\hat{\boldsymbol{q}}\cdot\boldsymbol{x}/\hbar} \, |\widetilde{\boldsymbol{q}}\rangle \\
&= \frac{(2\pi\hbar)^3}{V} \sum_{q \in \boldsymbol{Q}_V} \mu(\boldsymbol{q}) \mathrm{e}^{\mathrm{i}\boldsymbol{q}\cdot(\boldsymbol{x}-\boldsymbol{x}')/\hbar} \, \langle\widetilde{\boldsymbol{q}}| \, \hat{S}^\dagger_0 \mathrm{e}^{-\mathrm{i}\hat{\boldsymbol{q}}\cdot(\boldsymbol{x}-\boldsymbol{x}')/\hbar} \hat{S}_0 \, |\widetilde{\boldsymbol{q}}\rangle \, .
\end{aligned} \tag{3.27}$$

Now using that $\hat{S}_0 = \hat{I} + \mathrm{i}\hat{T}$ [see (3.17)] and noting the relation (3.21), this expression becomes

$$\begin{aligned}
&\langle\chi(\boldsymbol{x}')|\chi(\boldsymbol{x})\rangle \\
&= \frac{(2\pi\hbar)^3}{V} \sum_{q \in \boldsymbol{Q}_V} \mu(\boldsymbol{q}) \mathrm{e}^{\mathrm{i}\boldsymbol{q}\cdot(\boldsymbol{x}-\boldsymbol{x}')/\hbar} \, \langle\widetilde{\boldsymbol{q}}| \left(\hat{I} - \mathrm{i}\hat{T}^\dagger\right) \mathrm{e}^{-\mathrm{i}\hat{\boldsymbol{q}}\cdot(\boldsymbol{x}-\boldsymbol{x}')/\hbar} \left(\hat{I} + \mathrm{i}\hat{T}\right) |\widetilde{\boldsymbol{q}}\rangle \\
&= \frac{(2\pi\hbar)^3}{V} \sum_{q \in \boldsymbol{Q}_V} \mu(\boldsymbol{q}) \left[1 - \langle\widetilde{\boldsymbol{q}}| \, \hat{T}\hat{T}^\dagger \, |\widetilde{\boldsymbol{q}}\rangle + \mathrm{e}^{\mathrm{i}\boldsymbol{q}\cdot(\boldsymbol{x}-\boldsymbol{x}')/\hbar} \, \langle\widetilde{\boldsymbol{q}}| \, \hat{T}^\dagger \mathrm{e}^{-\mathrm{i}\hat{\boldsymbol{q}}\cdot(\boldsymbol{x}-\boldsymbol{x}')/\hbar} \hat{T} \, |\widetilde{\boldsymbol{q}}\rangle\right] .
\end{aligned} \tag{3.28}$$

Here, any exponentials associated with the second term within the square brackets in the last line have dropped out, because the operator $\hat{T}$ commutes with the momentum operator [see (3.18)].

Let us evaluate the second and third term in the sum in the last line of (3.28) by inserting a complete set of momentum eigenstates $|\widetilde{\boldsymbol{q}}\rangle$, see (3.25), and then using (3.18). This yields

$$\langle\chi(\boldsymbol{x}')|\chi(\boldsymbol{x})\rangle = \frac{(2\pi\hbar)^3}{V}\sum_{\boldsymbol{q}\in\boldsymbol{Q}_V}\mu(\boldsymbol{q})\left[1-\sum_{\boldsymbol{q}'\in\boldsymbol{Q}_V}\left(1-\mathrm{e}^{\mathrm{i}(\boldsymbol{q}-\boldsymbol{q}')\cdot(\boldsymbol{x}-\boldsymbol{x}')/\hbar}\right)\left|\langle\widetilde{\boldsymbol{q}}|\,\hat{T}\,|\widetilde{\boldsymbol{q}}'\rangle\right|^2\right]. \tag{3.29}$$

We now go to the continuum limit of momentum states $|\widetilde{\boldsymbol{q}}\rangle$ by replacing the sum by an integral according to the usual transformation relation

$$\frac{(2\pi\hbar)^3}{V}\sum_{\boldsymbol{q}\in\boldsymbol{Q}_V}\longrightarrow\int\mathrm{d}\boldsymbol{q}. \tag{3.30}$$

Using that $\int\mathrm{d}\boldsymbol{q}\,\mu(\boldsymbol{q})=1$ and employing definition (3.24), expression (3.29) then reads

$$\langle\chi(\boldsymbol{x}')|\chi(\boldsymbol{x})\rangle = 1-\int\mathrm{d}\boldsymbol{q}\,\mu(\boldsymbol{q})\frac{(2\pi\hbar)^3}{V}\int\mathrm{d}\boldsymbol{q}'\left(1-\mathrm{e}^{\mathrm{i}(\boldsymbol{q}-\boldsymbol{q}')\cdot(\boldsymbol{x}-\boldsymbol{x}')/\hbar}\right)\left|\langle\boldsymbol{q}|\,\hat{T}\,|\boldsymbol{q}'\rangle\right|^2. \tag{3.31}$$

Following the approach of Adler [161], let us now introduce a time parameter T which denotes the elapsed time in the scattering process. Then, from (3.16) and (3.31), we may write the evolution of the reduced density matrix as

$$\begin{aligned}\rho_S(\boldsymbol{x},\boldsymbol{x}',T)-\rho_S(\boldsymbol{x},\boldsymbol{x}',0) &= -\rho_S(\boldsymbol{x},\boldsymbol{x}',0)\\ &\times\int\mathrm{d}\boldsymbol{q}\,\mu(\boldsymbol{q})\frac{(2\pi\hbar)^3}{V}\int\mathrm{d}\boldsymbol{q}'\left(1-\mathrm{e}^{\mathrm{i}(\boldsymbol{q}-\boldsymbol{q}')\cdot(\boldsymbol{x}-\boldsymbol{x}')/\hbar}\right)\left|\langle\boldsymbol{q}|\,\hat{T}\,|\boldsymbol{q}'\rangle\right|^2.\end{aligned} \tag{3.32}$$

Note that in writing (3.32), we have assumed that the characteristic timescale for the free evolution of the object is much longer than the average time between scattering events, such that the only change of the density matrix of the object during the interval T arises from the scattering events.

Let us now evaluate the squared matrix element $\left|\langle\boldsymbol{q}|\,\hat{T}\,|\boldsymbol{q}'\rangle\right|^2$ appearing in this equation. Using the form

$$\langle\boldsymbol{q}|\,\hat{T}\,|\boldsymbol{q}'\rangle=\frac{\mathrm{i}}{2\pi\hbar m}\delta(E-E')f(\boldsymbol{q},\boldsymbol{q}') \tag{3.33}$$

of (3.18), we thus need to calculate

$$\left|\langle\boldsymbol{q}|\,\hat{T}\,|\boldsymbol{q}'\rangle\right|^2=\frac{1}{(2\pi\hbar m)^2}\,\delta^2(E-E')\,|f(\boldsymbol{q},\boldsymbol{q}')|^2\,. \tag{3.34}$$

Squared energy delta functions $\delta^2(E - E')$ commonly appear in the derivation of Fermi's Golden Rule for the transition probability in time-dependent perturbation theory (see, e.g., Chap. 18 of [167]). In fact, the matrix element $\left| \langle \boldsymbol{q} | \hat{T} | \boldsymbol{q}' \rangle \right|^2$ is evidently simply the probability of making the scattering-induced transition $|\boldsymbol{q}\rangle \longrightarrow |\boldsymbol{q}'\rangle$.

Then, following the method suggested by Adler [161] (which is akin to the usual approach employed in the derivation of Fermi's Golden Rule), we can handle the squared energy delta function by using the Fourier-integral representation of the energy delta function,

$$\delta(E - E') = \lim_{T \to \infty} \frac{1}{2\pi\hbar} \int_{-T/2}^{T/2} \mathrm{d}t \, \mathrm{e}^{\mathrm{i}(E-E')t/\hbar}. \tag{3.35}$$

This allows us to write

$$\begin{aligned} \delta^2(E - E') &= \delta(E - E') \lim_{T \to \infty} \frac{1}{2\pi\hbar} \int_{-T/2}^{T/2} \mathrm{d}t \, \mathrm{e}^{\mathrm{i}(E-E')t/\hbar} \\ &= \delta(E - E') \lim_{T \to \infty} \frac{1}{2\pi\hbar} \int_{-T/2}^{T/2} \mathrm{d}t \\ &= \delta(E - E') \lim_{T \to \infty} \frac{T}{2\pi\hbar}, \end{aligned} \tag{3.36}$$

where the second line follows from the fact that the first delta function in front of the integral vanishes unless $E = E'$.

In the general derivation of Fermi's Golden Rule, the parameter T can be interpreted as the time interval during which the interaction is "turned on." Equation (3.36) then shows that the transition probability grows approximately linearly with T for sufficiently large T. Therefore, if we let T denote a time interval much longer than the typical time required for a single scattering event to take place, we can write (3.36) as

$$\delta^2(E - E') = \delta(E - E') \frac{T}{2\pi\hbar} = \delta(q - q') \frac{m}{q} \frac{T}{2\pi\hbar}. \tag{3.37}$$

We shall also assume that T is significantly shorter than the characteristic decoherence time of the central particle induced by the scattering of a large number of particles. The chosen regime for T is reasonable, since it will in general require many collisions between the object and the environmental particles to induce an appreciable degree of spatial decoherence.

We now use (3.37) in (3.34) and insert the resulting expression into (3.32). The delta function $\delta(q - q')$ in (3.37) enforces momentum conservation, $|\boldsymbol{q}'| = |\boldsymbol{q}|$. Writing $\mathrm{d}\boldsymbol{q}' \equiv q^2 \mathrm{d}\hat{n}'$, where $\mathrm{d}\hat{n}'$ is a solid-angle differential in momentum space, we readily obtain

$$\rho_S(\boldsymbol{x}, \boldsymbol{x}', T) - \rho_S(\boldsymbol{x}, \boldsymbol{x}', 0) = -\rho_S(\boldsymbol{x}, \boldsymbol{x}', 0) \times \frac{T}{V} \int \mathrm{d}\boldsymbol{q}\, \mu(\boldsymbol{q}) v(q) \int \mathrm{d}\hat{n}' \left(1 - \mathrm{e}^{\mathrm{i}(\boldsymbol{q} - q\hat{\boldsymbol{n}}') \cdot (\boldsymbol{x} - \boldsymbol{x}')/\hbar}\right) |f(\boldsymbol{q}, q\hat{\boldsymbol{n}}')|^2 . \quad (3.38)$$

Here $v(q)$ denotes the speed of particles with momentum q. For scattering of massive particles we have $v(q) = q/m$, while for scattering of photons and other massless particles $v(q) = c$, with c representing the speed of light.

Equation (3.38) describes the effect of a single scattering event. A collection of N independent scattering events may then be represented by simply multiplying the integral on the right-hand side of (3.38) by N. Then N/V is the total number density of environmental particles. For simplicity, let us assume that the incoming particles are isotropically distributed in space, i.e., that every initial direction $\boldsymbol{q}/|\boldsymbol{q}|$ is equally likely. We may then write the momentum probability distribution $\mu(\boldsymbol{q})$ as

$$\mu(\boldsymbol{q}) \equiv \frac{1}{4\pi} \left(\frac{N}{V}\right)^{-1} \varrho(q)\, \mathrm{d}q\, \mathrm{d}\hat{n}, \quad (3.39)$$

where the prefactor is chosen such that $\int \mathrm{d}q\, \varrho(q) = N/V$ (recall that $\int \mathrm{d}\boldsymbol{q}\, \mu(\boldsymbol{q}) = 1$), i.e., $\varrho(q)$ denotes the number density of incoming particles with magnitude of momentum equal to q.

Dividing (3.38) by T and taking the differential limit of small T, we obtain our final result for the time evolution of the reduced density matrix,

$$\frac{\partial \rho_S(\boldsymbol{x}, \boldsymbol{x}', t)}{\partial t} = -F(\boldsymbol{x} - \boldsymbol{x}') \rho_S(\boldsymbol{x}, \boldsymbol{x}', t), \quad (3.40)$$

with the decoherence factor $F(\boldsymbol{x} - \boldsymbol{x}')$ given by

$$F(\boldsymbol{x} - \boldsymbol{x}') = \int \mathrm{d}q\, \varrho(q) v(q) \int \frac{\mathrm{d}\hat{n}\, \mathrm{d}\hat{n}'}{4\pi} \left(1 - \mathrm{e}^{\mathrm{i}q(\hat{\boldsymbol{n}} - \hat{\boldsymbol{n}}') \cdot (\boldsymbol{x} - \boldsymbol{x}')/\hbar}\right) |f(q\hat{\boldsymbol{n}}, q\hat{\boldsymbol{n}}')|^2 . \quad (3.41)$$

From (3.40) we see that $F(\boldsymbol{x} - \boldsymbol{x}')$ plays the role of a *localization rate*, i.e., it denotes the characteristic decoherence rate at which spatial coherences between two positions $\boldsymbol{x}$ and $\boldsymbol{x}'$ become locally suppressed.

Equations (3.40) and (3.41) represent our main result. We have found the expression that quantifies the influence of a collection of incoming particles scattering off our object of interest on the interference (off-diagonal) terms in the position density matrix of the object. As we have not yet specified the form of the scattering cross section $|f(q\hat{\boldsymbol{n}}, q\hat{\boldsymbol{n}}')|^2$, our expression is considered fairly general.

Let us summarize the assumptions employed in our derivation of (3.40) and (3.41):

1. There are no initial correlations between the system and the environment [see (3.4)]. This assumption is made in virtually all decoherence models (see also Chap. 5).

2. The scattering interaction is invariant under translations of the composite object–environment system [see (3.8)].
3. The center-of-mass state of the object is not disturbed by the scattering event.
4. The rate of scattering is much faster than the characteristic rate of change of the state of the system induced by the system's self-Hamiltonian.
5. The distribution of the different directions of the incoming particles is isotropic [see (3.39)].

Most of these assumptions are fairly innocuous and can be rather well justified in the context of the physical situation that is to be modeled here, namely, the scattering of a large number of microscopic or mesoscopic particles per unit time off a macroscopic body. The potentially most limiting assumption is that of no recoil (assumption 3 above). As already pointed out before, this assumption renders our localization model unrealistic for situations in which the scattering object and the scattered environmental particles have similar masses.

3.3 Full Versus Partial Which-Path Resolution

Suppose now our object of interest is described by a coherent superposition of two well-localized wave packets a distance $\Delta x = |\boldsymbol{x} - \boldsymbol{x}'|$ apart. Based on our results (3.40) and (3.41), how fast will such a superposition become decohered as a consequence of environmental scattering? To answer this question, we shall now discuss two limiting cases: The *short-wavelength limit* in which each scattered particle completely resolves the separation Δx, and the *long-wavelength limit* in which many such scattering events are required to encode a significant amount of which-path information in the environment and to thereby resolve this separation.

3.3.1 The Short-Wavelength Limit

First, let us consider the situation in which the typical wavelength λ_0 of the scattered particle is much shorter than the coherent separation $\Delta x = |\boldsymbol{x} - \boldsymbol{x}'|$, that is, $\lambda_0 \ll \Delta x$. Then we expect that the particle will be able to well resolve this separation and thus carry away a maximum of which-path information, inducing a maximum amount of decoherence in the system per scattering event.

To check whether this prediction is indeed correct, let us have a look at the expression (3.41) for $F(\boldsymbol{x} - \boldsymbol{x}')$. According to de Broglie relation, the wavelength λ_0 corresponds to a momentum $q_0 = 2\pi\hbar/\lambda_0$, and thus $\lambda_0 \ll \Delta x$ implies that $q_0 \Delta x/\hbar \gg 1$. Therefore the exponential in the integral on the right-hand side of (3.41) will oscillate very rapidly and thus in average not contribute to the integral when compared to the constant term of one. In this limit

$$F(\boldsymbol{x}-\boldsymbol{x}') = \int \mathrm{d}q\, \varrho(q) v(q) \int \frac{\mathrm{d}\hat{n}\,\mathrm{d}\hat{n}'}{4\pi} \left|f(q\hat{\boldsymbol{n}}, q\hat{\boldsymbol{n}}')\right|^2 . \tag{3.42}$$

The integral

$$\int \mathrm{d}\hat{n}' \left|f(q\hat{\boldsymbol{n}}, q\hat{\boldsymbol{n}}')\right|^2 \tag{3.43}$$

is simply equal to the total cross section for momentum $\boldsymbol{q} = q\hat{\boldsymbol{n}}$. The second volume integral over $\mathrm{d}\hat{n}$ then averages over all possible directions $\hat{\boldsymbol{n}}$ of $\boldsymbol{q}$, and thus we obtain

$$\int \frac{\mathrm{d}\hat{n}\,\mathrm{d}\hat{n}'}{4\pi} \left|f(q\hat{\boldsymbol{n}}, q\hat{\boldsymbol{n}}')\right|^2 = \sigma_{\mathrm{tot}}(q), \tag{3.44}$$

where $\sigma_{\mathrm{tot}}(q)$ denotes the total cross section for momentum q (irrespective of the direction). Therefore (3.42) becomes

$$F(\boldsymbol{x}-\boldsymbol{x}') = \int \mathrm{d}q\, \varrho(q) v(q) \sigma_{\mathrm{tot}}(q). \tag{3.45}$$

We can simplify this expression even further by recognizing that the integral is equal to the total scattering rate $\varGamma_{\mathrm{tot}}$,

$$F(\boldsymbol{x}-\boldsymbol{x}') = \varGamma_{\mathrm{tot}}. \tag{3.46}$$

Note that this result means that there is an upper limit to the decoherence rate when going to larger and larger separations Δx. This, of course, constitutes a completely reasonable finding: For any given separation Δx there is a wavelength of the environmental particle that allows for complete resolution of this separation and therefore for a maximum amount of spatial decoherence. Increasing this separation further therefore cannot lead to even stronger decoherence.

Inserting (3.46) into (3.40), we thus see that the off-diagonal elements $\rho_S(\boldsymbol{x}, \boldsymbol{x}')$, $\boldsymbol{x} \neq \boldsymbol{x}'$, of the reduced density matrix of the system change as

$$\frac{\partial \rho_S(\boldsymbol{x}, \boldsymbol{x}', t)}{\partial t} = -\varGamma_{\mathrm{tot}} \rho_S(\boldsymbol{x}, \boldsymbol{x}', t). \tag{3.47}$$

For short time intervals over which the internal dynamics of the system are negligible, the time evolution of the interference terms is therefore given by

$$\rho_S(\boldsymbol{x}, \boldsymbol{x}', t) = \rho_S(\boldsymbol{x}, \boldsymbol{x}', 0) \mathrm{e}^{-\varGamma_{\mathrm{tot}} t}. \tag{3.48}$$

This shows that in the limit of maximum decoherence, spatial interference terms become exponentially suppressed at a rate set by the total scattering rate $\varGamma_{\mathrm{tot}}$.

3.3.2 The Long-Wavelength Limit

Let us now consider the opposite limit in which the typical wavelength λ_0 of the incoming particle is much larger than the coherent separation $\Delta x = |\boldsymbol{x} - \boldsymbol{x}'|$. This implies that an individual scattered particle will not be able to resolve the separation Δx, and it will thus carry away an only insufficient amount of which-path information. Therefore we anticipate that it will take a large number of scattering events to induce a significant degree of spatial localization of the object.[5]

Let us investigate the influence of our assumption $\lambda_0 \gg \Delta x$ (or, equivalently, $q_0 \Delta x/\hbar \ll 1$ for a typical value q_0 of momentum) on the form of the expression for $F(\boldsymbol{x} - \boldsymbol{x}')$, see (3.41). We expand the exponential in (3.41) up to first order in the argument $q(\hat{\boldsymbol{n}} - \hat{\boldsymbol{n}}')(\boldsymbol{x} - \boldsymbol{x}')$, i.e.,

$$1 - \mathrm{e}^{\mathrm{i}q(\hat{\boldsymbol{n}} - \hat{\boldsymbol{n}}')(\boldsymbol{x} - \boldsymbol{x}')/\hbar} \approx -\frac{\mathrm{i}}{\hbar} q(\hat{\boldsymbol{n}} - \hat{\boldsymbol{n}}') \cdot (\boldsymbol{x} - \boldsymbol{x}') + \frac{1}{2\hbar^2} q^2 \left[(\hat{\boldsymbol{n}} - \hat{\boldsymbol{n}}') \cdot (\boldsymbol{x} - \boldsymbol{x}')\right]^2 . \tag{3.49}$$

Now note that the first term in this expansion does not contribute to the integral in the expression for $F(\boldsymbol{x} - \boldsymbol{x}')$, see (3.41). This is readily seen from the fact that, since $f(q\hat{\boldsymbol{n}}, q\hat{\boldsymbol{n}}') = f^*(q\hat{\boldsymbol{n}}', q\hat{\boldsymbol{n}})$ [see (3.18)] and thus $|f(q\hat{\boldsymbol{n}}, q\hat{\boldsymbol{n}}')|^2 = |f(q\hat{\boldsymbol{n}}', q\hat{\boldsymbol{n}})|^2$, we encounter the situation of a product of a function that is odd in $(\hat{\boldsymbol{n}}, \hat{\boldsymbol{n}}')$, namely, $(\hat{\boldsymbol{n}} - \hat{\boldsymbol{n}}') \cdot (\boldsymbol{x} - \boldsymbol{x}')$, with a function that is even in $(\hat{\boldsymbol{n}}, \hat{\boldsymbol{n}}')$, i.e., $|f(q\hat{\boldsymbol{n}}, q\hat{\boldsymbol{n}}')|^2$. Integrated over all directions $\hat{\boldsymbol{n}}$ and $\hat{\boldsymbol{n}}'$ of $\boldsymbol{q}$ and $\boldsymbol{q}'$, the contribution to the integral due to this product term thus averages out to zero.

The expression for $F(\boldsymbol{x} - \boldsymbol{x}')$ in the limit $q_0 \Delta x/\hbar \ll 1$ then becomes

$$F(\boldsymbol{x} - \boldsymbol{x}') = \int \mathrm{d}q\, \varrho(q) v(q) q^2 \int \frac{\mathrm{d}\hat{n}\, \mathrm{d}\hat{n}'}{8\pi\hbar^2} \left[(\hat{\boldsymbol{n}} - \hat{\boldsymbol{n}}') \cdot (\boldsymbol{x} - \boldsymbol{x}')\right]^2 |f(q\hat{\boldsymbol{n}}, q\hat{\boldsymbol{n}}')|^2 . \tag{3.50}$$

We can further simplify this equation by assuming that the particular orientation of scattering center (and thus of the coordinate system) does not influence the scattering process. First, this allows us to average the term $\left[(\hat{\boldsymbol{n}} - \hat{\boldsymbol{n}}') \cdot (\boldsymbol{x} - \boldsymbol{x}')\right]^2$ over all possible directions $(\boldsymbol{x} - \boldsymbol{x}')$. This average is given by

$$\begin{aligned}
(\boldsymbol{x} - \boldsymbol{x}')^2 \frac{1}{3} \sum_{i=x,y,z} \left[(\hat{\boldsymbol{n}} - \hat{\boldsymbol{n}}') \cdot \hat{\boldsymbol{i}}\right]^2 &= \frac{1}{3} (\boldsymbol{x} - \boldsymbol{x}')^2 (\hat{\boldsymbol{n}} - \hat{\boldsymbol{n}}')^2 \\
&= \frac{2}{3} (\boldsymbol{x} - \boldsymbol{x}')^2 (1 - \hat{\boldsymbol{n}} \cdot \hat{\boldsymbol{n}}') \\
&= \frac{2}{3} (\boldsymbol{x} - \boldsymbol{x}')^2 (1 - \cos\Theta) ,
\end{aligned} \tag{3.51}$$

[5] We note that the original derivation of Joos and Zeh [7] assumed the limit $\lambda_0 \gg \Delta x$ from the outset (see the comment following equation (3.45) of [7]) and thus was intended to lead to cautious lower-bound estimates of localization rates.

where Θ is the angle between the incoming and the outgoing trajectory of the scattered particle. Thus

$$F(\boldsymbol{x}-\boldsymbol{x}') = (\boldsymbol{x}-\boldsymbol{x}')^2 \int \mathrm{d}q\, \varrho(q) v(q) q^2 \frac{2}{3\hbar^2} \int \frac{\mathrm{d}\hat{\boldsymbol{n}}\, \mathrm{d}\hat{\boldsymbol{n}}'}{8\pi} (1-\cos\Theta) \left| f(q\hat{\boldsymbol{n}}, q\hat{\boldsymbol{n}}') \right|^2 . \tag{3.52}$$

Second, we may take the cross section $|f(q\hat{\boldsymbol{n}}, q\hat{\boldsymbol{n}}')|^2$ to be isotropic, i.e., to be dependent only on the magnitude q of the momentum and the scattering angle Θ. Then we may further simplify (3.52) by carrying out some of the angular integrations, which yields

$$F(\boldsymbol{x}-\boldsymbol{x}') = (\boldsymbol{x}-\boldsymbol{x}')^2 \int \mathrm{d}q\, \varrho(q) v(q) q^2 \frac{2\pi}{3\hbar^2} \int \mathrm{d}\cos\Theta\ (1-\cos\Theta) \left| f(q, \cos\Theta) \right|^2 . \tag{3.53}$$

Apart from the angular weighting term $(1-\cos\Theta)$, the second integral is similar to the expression for the total cross section $\sigma_{\mathrm{tot}}(q)$ of the scattering process, see (3.44). Let us therefore interpret

$$\sigma_{\mathrm{eff}}(q) \equiv \frac{2\pi}{3} \int \mathrm{d}\cos\Theta\ (1-\cos\Theta) \left| f(q, \cos\Theta) \right|^2 \tag{3.54}$$

as the effective cross section for the scattering interaction, which is consequently on the order of the total cross section $\sigma_{\mathrm{tot}}(q)$.

Let us now explore the resulting time dependence of interference terms $\rho_S(\boldsymbol{x}, \boldsymbol{x}')$, $\boldsymbol{x} \neq \boldsymbol{x}'$, of the reduced density matrix of the object. Referring back to (3.40) and using (3.53) and (3.54), the change of the reduced density matrix due to scattering is given by

$$\frac{\partial \rho_S(\boldsymbol{x}, \boldsymbol{x}', t)}{\partial t} = -\Lambda (\boldsymbol{x}-\boldsymbol{x}')^2 \rho_S(\boldsymbol{x}, \boldsymbol{x}', t), \tag{3.55}$$

where we have introduced the *scattering constant*

$$\Lambda \equiv \int \mathrm{d}q\, \varrho(q) v(q) \frac{q^2}{\hbar^2} \sigma_{\mathrm{eff}}(q), \tag{3.56}$$

which encapsulates the physical details of the interaction.

Over timescales that are short in comparison with the internal dynamics of the system, (3.55) shows that the scattering events lead to a suppression of off-diagonal terms that increases exponentially with time and with the squared separation $(\Delta x)^2 = (\boldsymbol{x}-\boldsymbol{x}')^2$,

$$\rho_S(\boldsymbol{x}, \boldsymbol{x}', t) = \rho_S(\boldsymbol{x}, \boldsymbol{x}', 0) \mathrm{e}^{-\Lambda (\Delta x)^2 t} . \tag{3.57}$$

We see that the scattering constant Λ quantifies the rate at which spatial coherences over a given distance Δx are suppressed. Equation (3.57) therefore motivates the introduction of a decoherence timescale $\tau_{\Delta x}$ given by

$$\tau_{\Delta x} = \frac{1}{\Lambda(\Delta x)^2}, \tag{3.58}$$

which is the characteristic time required to damp spatial coherences over a distance Δx by a factor of e. The inverse quantity $\tau_{\Delta x}^{-1}$ therefore plays the role of a decoherence rate.

By contrast with the result (3.48) obtained for the opposite limiting case of environmental wavelengths that were assumed to be very short in comparison with the separation Δx, we see from (3.58) that the damping is now explicitly, and strongly, dependent on Δx. This difference is rather easily understood from our earlier discussion. If the wavelength of the particle is much smaller than Δx, each scattering event leads to a complete resolution of this separation, and thus the localization rate will be independent of Δx. On the other hand, if the particle's wavelength is much larger than Δx, it will require a large number of particles to encode an appreciable amount of which-path information in the environment, and this amount can be anticipated to increase, for a given number of scattering events, as Δx becomes larger.

Of course, the decoherence rate cannot exceed the value obtained in the short-wavelength limit, since this case already corresponds to a maximum of which-path information acquired by the environment. Therefore the long-wavelength decoherence timescale given by (3.58) cannot become shorter without bounds for increasingly large separations Δx, as one might be tempted to conclude from (3.58). At some point, Δx would become larger than the typical wavelength of the environment. The long-wavelength assumption would then be violated, and we would need to instead consider the short-wavelength case in which the decoherence rate indeed saturates to a value independent of the separation Δx [see (3.46)].

3.4 Decoherence Due to Scattering of Thermal Photons and Air Molecules

Let us now explore the application of our decoherence model to concrete physical situations of interest. In the following, we shall consider environments composed of thermal photons and air molecules, which represent ubiquitous sources of decoherence in nature.

3.4.1 Photon Scattering

Let us first consider the case of a photon-gas environment at temperature T interacting with a small object. Following the original example of Joos and Zeh [7], we model this object as a dielectric sphere of radius a, and we assume that the dielectric constant ε is independent of the frequency of the scattered photon. In the present case of photon scattering, the long-wavelength limit of environmental scattering (see Sect. 3.3.2) is usually appropriate. At any rate, considering this limit (rather than the short-wavelength limit of Sect. 3.3.1) ensures that we obtain lower bounds on the decoherence rate.

The relevant differential cross section is then given by the Rayleigh law [168] (averaged over the different polarizations of the scattered radiation),

$$|f(q\hat{\boldsymbol{n}}, q\hat{\boldsymbol{n}}')|^2 = \left(\frac{q}{\hbar}\right)^4 a^6 \left(\frac{\varepsilon - 1}{\varepsilon + 2}\right)^2 \frac{1}{2}\left(1 + \cos^2\Theta\right). \tag{3.59}$$

We see that this differential cross section depends on the magnitude q of the momentum of the environmental particle and on the scattering angle Θ between the directions $\hat{\boldsymbol{n}}$ and $\hat{\boldsymbol{n}}'$ of the particle before and after the scattering.

Let us now calculate the effective cross section $\sigma_{\text{eff}}(q)$ for the Rayleigh cross section (3.59). From (3.54) we obtain

$$\begin{aligned}\sigma_{\text{eff}}(q) &= \left(\frac{q}{\hbar}\right)^4 a^6 \left(\frac{\varepsilon - 1}{\varepsilon + 2}\right)^2 \frac{\pi}{3}\int \mathrm{d}\cos\Theta\,(1 - \cos\Theta)\left(1 + \cos^2\Theta\right) \\ &= \frac{8\pi}{9}\left(\frac{q}{\hbar}\right)^4 a^6 \left(\frac{\varepsilon - 1}{\varepsilon + 2}\right)^2 .\end{aligned} \tag{3.60}$$

Next, to compute the scattering constant Λ, see (3.56), we need to obtain the momentum density $\varrho(q)$. Assuming black-body radiation, the average occupation number of photons of energy cq at temperature T is given by the Planck distribution

$$\langle n(\boldsymbol{q})\rangle_T = \frac{2}{\mathrm{e}^{cq/k_{\mathrm{B}}T} - 1}, \tag{3.61}$$

where c denotes the speed of light, and the factor of 2 is due to the fact that there are two possible polarization directions of the photon. To obtain the number density $\varrho(\boldsymbol{q})$, we must multiply $\langle n(\boldsymbol{q})\rangle_T$ by the number of states per unit volume with momenta between $\boldsymbol{q}$ and $\boldsymbol{q} + \mathrm{d}\boldsymbol{q}$, which is given by (assuming an isotropic distribution of momenta)

$$\frac{1}{(2\pi\hbar)^3} d^3 q = \frac{1}{2\pi^2\hbar^3} q^2 \mathrm{d}q. \tag{3.62}$$

This yields the number density

$$\varrho(q) = \frac{1}{\pi^2\hbar^3}\left(\frac{q^2}{\mathrm{e}^{cq/k_{\mathrm{B}}T} - 1}\right). \tag{3.63}$$

Let us now compute the resulting scattering constant Λ. Using (3.56), we obtain

$$\begin{aligned}\Lambda &= \int \mathrm{d}q\, \varrho(q) v(q) q^2 \sigma_{\text{eff}}(q) \\ &= \int \mathrm{d}q\, \frac{1}{\pi^2\hbar^3}\left(\frac{q^2}{\mathrm{e}^{cq/k_{\mathrm{B}}T} - 1}\right) c\frac{q^2}{\hbar^2}\frac{8\pi}{9}\left(\frac{q}{\hbar}\right)^4 a^6 \left(\frac{\varepsilon - 1}{\varepsilon + 2}\right)^2 \\ &= \frac{8}{9\pi\hbar^9} a^6 c \left(\frac{\varepsilon - 1}{\varepsilon + 2}\right)^2 \int \mathrm{d}q\, \frac{q^8}{\mathrm{e}^{cq/k_{\mathrm{B}}T} - 1}.\end{aligned} \tag{3.64}$$

The integral can be computed using the definition of the Riemann ζ-function for integer arguments n,

$$\zeta(n) = \frac{1}{(n-1)!} \int_0^\infty \mathrm{d}x \, \frac{x^{n-1}}{\mathrm{e}^x - 1}. \tag{3.65}$$

Evaluating this expression for $x = cq/k_\mathrm{B}T$ and $n = 9$, we finally obtain

$$\Lambda = 8! \frac{8}{9\pi} a^6 c \left(\frac{\varepsilon - 1}{\varepsilon + 2} \right)^2 \left(\frac{k_\mathrm{B}T}{\hbar c} \right)^9 \zeta(9) \propto a^6 T^9, \tag{3.66}$$

where $\zeta(9) \approx 1.002$. We emphasize the extremely strong dependence of Λ on the size a of the object and the temperature T of the photon gas. For example, increasing T by a factor of two makes the decoherence rate more than 500 times larger.

Evaluating the constants appearing in (3.66), and assuming $(\varepsilon - 1)/(\varepsilon + 2) \approx 1$, yields the scattering rate

$$\Lambda \approx 10^{20} \frac{1}{\mathrm{cm}^2 \; \mathrm{s}} \left(\frac{a}{\mathrm{cm}} \right)^6 \left(\frac{T}{\mathrm{K}} \right)^9. \tag{3.67}$$

Let us use this expression to compute numerical estimates for Λ for a few cases of interest. For photons at room temperature ($T = 300$ K) scattering off a dust grain of size $a = 10^{-3}$ cm, we obtain $\Lambda \approx 10^{24}$ cm^{-2} s^{-1}. Using (3.58), this means that spatial interferences over distances on the order of as little as $\Delta x = |\boldsymbol{x} - \boldsymbol{x}'| \approx 10^{-12}$ cm will become effectively suppressed within the time span of a second. Similarly, interferences over separations on the order of the size of the object ($\Delta x \approx 10^{-3}$ cm) will become significantly damped within a mere 10^{-18} seconds. For the same object ($a = 10^{-3}$ cm) immersed into cosmic microwave background radiation (which consists of photons at the characteristic temperature of 3 K), the scattering constant is reduced by a factor of 10^{18}, i.e., $\Lambda \approx 10^6$ cm^{-2} s^{-1}. Even this rate still corresponds to a very efficient decoherence process, as coherences over distances on the order of 10^{-3} cm will be locally suppressed within the time span of a second.

If we reduce the size of the object down to, say, $a = 10^{-6}$ cm, corresponding to a very small dust particle or a large molecule, we obtain $\Lambda \approx 10^6$ cm^{-2} s^{-1} for a photon gas at room temperature and $\Lambda \approx 10^{-12}$ cm^{-2} s^{-1} for cosmic microwave background radiation. We see that in the latter case, decoherence is relatively slow, as the reader might have intuitively expected. Our estimate means that, if the 3 K background radiation was the only source of environmental scattering, interferences could here in principle be maintained for several seconds over separations up to 10^6 cm before becoming noticeably degraded.

All values for Λ calculated here are summarized in Table 3.1. In Table 3.2, we have also listed the corresponding decoherence timescales (3.58) for coherent separations Δx equal to the size a of the object. Needless to say, the

Table 3.1. Estimates for the scattering constant Λ, in units of $\mathrm{cm}^{-2}\ \mathrm{s}^{-1}$, for a dust grain of size $a = 10^{-3}$ cm and for a large molecule of size $a = 10^{-6}$ cm immersed into radiation and gaseous environments. The quantity $\Lambda(\Delta x)^2$ corresponds to the characteristic rate at which spatial coherences over a distance Δx become suppressed as a consequence of decoherence. The values for photon environments shown in the first two rows were computed from (3.67), whereas the values for scattering of air molecules (at room temperature) given in the last two rows were obtained from (3.73) and (3.74).

Environment	Dust grain	Large molecule
Cosmic background radiation	10^6	10^{-12}
Photons at room temperature	10^{24}	10^6
Best laboratory vacuum	10^{20}	10^{14}
Air at normal pressure	10^{37}	10^{31}

Table 3.2. Estimates of decoherence timescales $\tau_{\Delta x} = \Lambda^{-1}(\Delta x)^{-2}$ (in seconds) for the suppression of spatial interferences over a distance Δx equal to the size a of the object ($\Delta x = a = 10^{-3}$ cm for a dust grain and $\Delta x = a = 10^{-6}$ cm for a large molecule). The timescales were computed using the values for Λ listed in Table 3.1.

Environment	Dust grain	Large molecule
Cosmic background radiation	1	10^{24}
Photons at room temperature	10^{-18}	10^6
Best laboratory vacuum	10^{-14}	10^{-2}
Air at normal pressure	10^{-31}	10^{-19}

shortness of these timescales is truly astonishing and indicates the extreme speed and efficiency of decoherence. Our estimates demonstrate that spatial interference effects are extremely difficult to observe for "ordinary" objects (such as dust grains) immersed into similarly "ordinary" environments (such as thermal photons).

Even coherence properties of electrons and other microscopic charged particles are significantly influenced by the scattering of thermal radiation [7, 166]. For the scattering of radiation by charged particles, the relevant effective cross section is given by the Thompson cross section,

$$\frac{\mathrm{d}\sigma}{\mathrm{d}\Omega}(\Theta) = \left(\frac{Q^2}{mc^2}\right)^2 \frac{1}{2}\left(1 + \cos^2\Theta\right). \tag{3.68}$$

The so-called Compton radius Q^2/mc^2 of the charged particle has units of length and plays here a role analogous to that of the spatial size a of the object in the case of Rayleigh scattering.

The scattering constant Λ corresponding to the scattering cross section (3.68) can be computed in a similar way as above for Rayleigh scattering. One then finds a temperature dependence of $\Lambda \propto T^5$. Numerical estimates for the scattering constant for the case of a free electron range from $\Lambda \approx 10^0$ cm^{-2} s^{-1} for an environment of thermal (300 K) photons down to $\Lambda \approx 10^{-15}$ cm^{-2} s^{-1} for an environment of solar neutrinos [166].

Although these values are much smaller than those obtained for mesoscopic and macroscopic objects above, they still show that the coherent spreading of wave packets induced by the free Schrödinger evolution is efficiently suppressed at the level of the system. Recall that in the beginning of this chapter we had shown that the wave function of an electron with an initial spatial width of 1 Å would very quickly become completely delocalized under purely unitary time evolution, reaching a width of about $\Delta x = 10^8$ cm within a second [see (3.3)]. Our numerical estimates demonstrate that localization of such a spread-out wave packet would occur within less than a second even if only the environment of solar neutrinos was taken into account. Thermal photons, on the other hand, would lead to an extremely short decoherence time of about 10^{-16} s for the electron described by this delocalized wave packet. However, it is very important to keep in mind that for microscopic particles such as electrons the assumption of a recoil-free interaction (on which our above derivation of the expression for the localization rate is based) is usually not physically reasonable. Therefore we should not take these numerical estimates too literally.

3.4.2 Scattering of Air Molecules

Let us now consider the scattering of air molecules. Needless to say, this type of environment constitutes a very important source of decoherence in the everyday world around us. Air molecules are matter particles whose thermal de Broglie wavelength (2.114) is typically very short. For example, an O_2 molecule at room temperature has a thermal de Broglie wavelength of about $\lambda_{\text{dB}} \approx 2 \times 10^{-11}$ m, or about 20 Å. At $T = 3$ K, the wavelength is ten times longer, i.e., $\lambda_{\text{dB}} \approx 2 \times 10^{-10}$ m, or about 200 Å. If we again take "dust" particles with typical extensions between $a = 10^{-6}$ cm and $a = 10^{-3}$ cm as our objects of interest, then it is clear that we are here in the regime in which the relevant wavelength of the scattering environmental particle is much shorter than the size of the object.

In this case, we may take the cross section $|f(q\hat{\boldsymbol{n}}, q\hat{\boldsymbol{n}}')|^2$ to be a constant $|f|^2$ such that the cross section integrated over all orientations is simply equal to the geometric cross section πa^2 of the object,

$$\int \mathrm{d}\hat{n}\, |f|^2 = 4\pi\, |f|^2 \stackrel{!}{=} \pi a^2, \tag{3.69}$$

which implies $|f|^2 = a^2/4$.

Since our goal is to determine lower-bound estimates of the decoherence rate, let us again employ the long-wavelength limit of Sect. 3.3.2. That is, we shall assume that the coherent separation Δx between the two position-space components of the spatial superposition describing the object is much shorter than the wavelengths of the scattering air molecules.

Then (3.54) gives $\sigma_{\mathrm{eff}} = \pi a^2/3$. If we insert this value into our general expression (3.56) for the scattering constant Λ in the long-wavelength limit, we obtain

$$\begin{aligned} \Lambda &= \int \mathrm{d}q\, \varrho(q) v(q) \frac{q^2}{\hbar^2} \sigma_{\mathrm{eff}}(q) \\ &= \frac{\pi a^2}{3\hbar^2} \int \mathrm{d}q\, \varrho(q) v(q) q^2 \\ &\equiv \frac{\pi a^2}{3\hbar^2} \left\langle v(q) q^2 \right\rangle_{\varrho}, \end{aligned} \tag{3.70}$$

where $\langle v(q)q^2\rangle_\varrho$ denotes the average value of the quantity $v(q)q^2$ for a given distribution $\varrho(q)$.

Since we deal with an environment of gas particles, we take the number density $\varrho(q)$ of air molecules with momentum q to be distributed according to the Maxwell–Boltzmann distribution,

$$\varrho(q) = \frac{N}{V} 4\pi q^2 \left(\frac{1}{2\pi m k_{\mathrm{B}} T} \right)^{3/2} \exp\left[-\frac{q^2}{2 m k_{\mathrm{B}} T} \right], \tag{3.71}$$

where N/V is the total density of the air. The average $\left\langle v(q)q^2 \right\rangle_\varrho$ appearing in (3.70), with $v(q) = q/m$, is then easily computed by carrying out the Gaussian integral,

$$\left\langle v(q) q^2 \right\rangle_{\varrho} = \int \mathrm{d}q\, \varrho(q) v(q) q^2 = 4 \frac{N}{V} (m/\pi)^{1/2} (2 k_{\mathrm{B}} T)^{3/2}. \tag{3.72}$$

Thus the scattering constant (3.70) is

$$\Lambda = \frac{8}{3\hbar^2} \frac{N}{V} (2\pi m)^{1/2} a^2 (k_{\mathrm{B}} T)^{3/2}. \tag{3.73}$$

We see that here the dependence of the scattering rate on the size a of the object and the temperature T is much less dramatic than in the case of photon scattering, where we had found $\Lambda \propto T^9$ [see (3.66)]. The a^2 dependence is a direct consequence of the use of the geometric cross section, while Rayleigh scattering is characterized by a much stronger a^6 scaling [see (3.59)].

Using (3.73), we can estimate some typical values for Λ. Air at normal pressure contains about 3×10^{19} particles per cm^3 and has a mass of about 29 kg per kmol (1 kmol contains about 6×10^{26} particles). Thus the mass m of an individual air molecule is approximately $m = 0.5 \times 10^{-25}$ kg. Evaluating (3.73) with these values leads to

$$\Lambda(\text{normal pressure}) \approx 10^{39} \frac{1}{\mathrm{cm}^2\ \mathrm{s}} \left(\frac{a}{\mathrm{cm}} \right)^2 \left(\frac{T}{\mathrm{K}} \right)^{3/2}. \tag{3.74}$$

Let us again focus on the two objects previously considered, namely, a dust grain of size $a = 10^{-3}$ cm and a large molecule of size $a = 10^{-6}$ cm (see Table 3.1).

At room temperature ($T = 300$ K), we then obtain from (3.73) the values $\Lambda \approx 10^{37}$ cm^{-2} s^{-1} for a dust grain ($a = 10^{-3}$ cm) and $\Lambda \approx 10^{31}$ cm^{-2} s^{-1} for a large molecule ($a = 10^{-6}$ cm). On the other hand, good laboratory vacuums achieve particle densities on the order of a few hundred particles per cm^3 [169, p. 267]. Since the scattering constant is simply proportional to the number density of the environmental particles, we thus obtain values for Λ that are smaller by a factor of about 10^{17} than those for air at normal pressure. This yields $\Lambda \approx 10^{20}$ cm^{-2} s^{-1} and $\Lambda \approx 10^{14}$ cm^{-2} s^{-1} for the dust grain and large molecule, respectively. Our values for Λ are summarized in Table 3.1.

Let us again illustrate the effectiveness of decoherence with a few examples. Suppose the large molecule is prepared in a superposition of two distinct positions a distance $\Delta x = 10$ Å apart. Using (3.58), the scattering of air molecules at normal pressure will then lead to decoherence of such a superposition on a timescale $\tau_{\Delta x} \approx 10^{-17}$ s. Even in the best available laboratory vacuums, this decoherence time will still be on the order of a mere second. The shortness of the decoherence times is of course even more impressive if we consider a larger object such as the aforementioned dust grain. Even for a microscopic separation $\Delta x = 10$ Å (which is much less than the size of the object), (3.58) and (3.74) predict a decoherence time at normal air pressure of $\tau_{\Delta x} \approx 10^{-23}$ s. A few other values of characteristic decoherence timescales for the case in which the coherent separation Δx is on the order of the size of the object are listed in Table 3.2.

These decoherence times are extremely short and demonstrate impressively the effectiveness of scattering-induced decoherence in locally suppressing spatial coherences. Such estimates show why it is so difficult to observe spatial interference patterns for mesoscopic and macroscopic particles not only in the everyday world around us, but even under rather sophisticated laboratory conditions.

Once again, our numerical estimates should be interpreted with a grain of salt. For example, for the case of a molecule immersed into an environment of air molecules, the ratio between the mass of the object and the masses of the scattered environmental particles may not be sufficiently large for the no-recoil assumption underlying our model to be appropriate.

3.4.3 Comparison with Experiments

There are several experiments that have explicitly measured rates of spatial decoherence of atoms and molecules due to scattering of background particles such as photons or background-gas molecules. A very impressive experiment that directly demonstrates the gradual decohering influence of a scattering environment on the spatial coherence of C_{60} and C_{70} molecules will be described in detail in Sect. 6.2.

Here we shall therefore only briefly note a couple of other experiments, which have used atom interferometers of the Mach–Zehnder type in order to observe interference patterns for matter particles [170]. In such an interferometer, the path of each atom is coherently split within the interferometer and subsequently recombined. This results in an spatial interference pattern in form of an oscillatory dependence of the atomic flux on position.

Kokorowski et al. [171] used a Mach–Zehnder interferometer to observe the loss of coherence of sodium atoms due to scattering of controlled numbers of photons emitted from a laser beam. They measured the loss of contrast of the interference pattern as a function of the mean number of photons that were spontaneously scattered by atoms within the interferometer for different path separations (these separations correspond to the quantity Δx used in our model). The experiment verified the theoretically predicted exponential decay of this contrast, and thus of spatial coherence, with the square of the separation Δx and with time [see (3.57)]. In a similar experiment, Uys, Perreault, and Cronin [172] studied spatial decoherence of sodium atoms due to scattering of gas molecules within the interferometer and found, again, good agreement with theoretical predictions.

3.5 Illustrating the Dynamics of Decoherence

In the previous Sect. 3.4, we focused on estimating typical decoherence timescales of spatial superpositions for various objects immersed into different types of environments. Let us now illustrate the explicit dynamics of the decoherence process. We shall concentrate here on the case of a particle moving in one spatial dimension.

Recall that, in the long-wavelength limit, the interference terms of the reduced density matrix of the object decay due to the influence of the scattering environment as [see (3.55)]

$$\frac{\partial \rho_S(x, x', t)}{\partial t} = -\Lambda (x - x')^2 \rho_S(x, x', t). \tag{3.75}$$

Let us now include the term describing the unitary evolution under the "free-particle" Hamiltonian[6] (setting $\hbar \equiv 1$ in the remainder of this chapter)

$$H(x, p) = \frac{p^2}{2m} \equiv -\frac{1}{2m}\frac{\partial^2}{\partial x^2}. \tag{3.76}$$

This term is given by the Liouville–von Neumann equation,

[6] Of course, our particle is not truly free, since it is subject to collisions with the environment.

$$\frac{\mathrm{d}\hat{\rho}(t)}{\mathrm{d}t} = -\mathrm{i}\left[\hat{H}(t), \hat{\rho}(t)\right]. \tag{3.77}$$

Expressed in the position representation, we then obtain the full equation of motion for the particle,

$$\frac{\partial \rho_S(x,x',t)}{\partial t} = -\frac{\mathrm{i}}{2m}\left(\frac{\partial^2}{\partial x'^2} - \frac{\partial^2}{\partial x^2}\right)\rho_S(x,x',t) - \Lambda(x-x')^2\rho_S(x,x',t). \tag{3.78}$$

In Chap. 4, we will recognize this equation as an example of a master equation for decoherence, i.e., of an equation of motion that describes the evolution of the reduced density matrix of an object interacting with an environment.

It is now fairly straightforward to construct solutions of the equation of motion (3.78) [7]. We will closely follow Joos et al. [17] in our subsequent discussion. Let us use the Gaussian ansatz

$$\begin{aligned}\rho_S(x,x',t) = \exp\big\{-A(t)(x-x')^2 - \mathrm{i}B(t)(x-x')(x+x') \\ -C(t)(x+x')^2 - D(t)\big\},\end{aligned} \tag{3.79}$$

where $A(t)$, $B(t)$, $C(t)$, and $D(t)$ are time-dependent coefficients which depend on the state at $t=0$. If $\rho_S(x,x',t)$ is Hermitian, then these coefficients are real-valued functions of time.

Each of the coefficients $A(t)$, $B(t)$, $C(t)$, and $D(t)$ appearing in (3.79) has a particular interpretation (see Fig. 3.4). The width of the Gaussian in the "off-diagonal" $x=-x'$ direction quantifies the range of spatial coherence. From (3.79) we see that this width is inversely proportional to $\sqrt{A(t)}$, which motivates the introduction of a characteristic *coherence length*

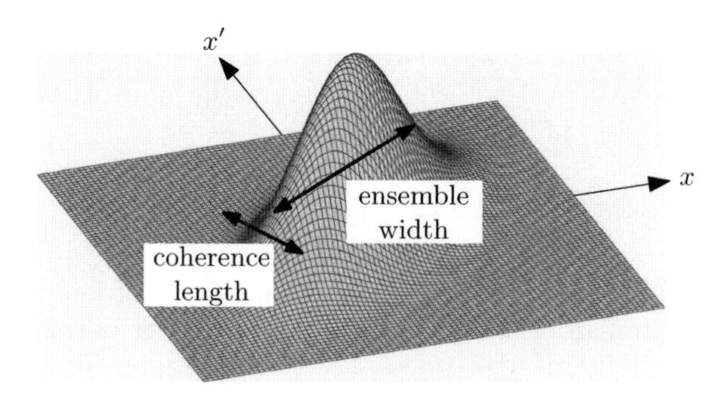

Fig. 3.4. Illustration of some of the key quantities relevant to the description of the Gaussian density matrix (3.79). The extension of the density matrix in the off-diagonal $x=-x'$ direction is quantified by the coherence length $\ell(t) \equiv 1/\sqrt{8A(t)}$, which measures the characteristic distance over which the system can exhibit spatial interference effects. The ensemble width $\Delta X(t) \equiv 1/\sqrt{8C(t)}$ represents the size of the probability distribution $P(x,t) \equiv \rho_S(x,x,t)$ for different positions x.

$$\ell(t) \equiv \frac{1}{\sqrt{8A(t)}}. \tag{3.80}$$

On the other hand, the width of $\rho_S(x, x', t)$ in the diagonal $x = x'$ direction corresponds to the size of the probability distribution $P(x, t) \equiv \rho_S(x, x, t)$, which describes the probability of finding the system at x upon a measurement of its position. This width is inversely proportional to $\sqrt{C(t)}$ [see (3.79)]. Accordingly, we shall introduce the *ensemble width* $\Delta X(t)$ as

$$\Delta X(t) = \frac{1}{\sqrt{8C(t)}}, \tag{3.81}$$

which quantifies the total size of the position-space ensemble.

The coefficient $B(t)$, together with $A(t)$ and $C(t)$, is related to the spread $\Delta P(t)$ of the momentum distribution, which is given by

$$\Delta P(t) = \sqrt{2}\left[A(t) + \frac{B^2(t)}{4C(t)}\right]^{1/2}. \tag{3.82}$$

We can use (3.81) and (3.82) to write down the corresponding uncertainty relation,

$$\Delta X(t)\Delta P(t) = \frac{1}{2}\left[\frac{A(t)}{C(t)} + \frac{B^2(t)}{4C(t)}\right]^{1/2}. \tag{3.83}$$

Finally, the term $D(t)$ ensures that the trace of the density matrix is a conserved quantity at all times.

Inserting our Gaussian ansatz (3.79) into the equation of motion (3.78) yields a set of coupled differential equations for the coefficients $A(t)$, $B(t)$, $C(t)$, and $D(t)$ that can be solved explicitly for arbitrary initial conditions (see, e.g., Appendix A2 of [17]). Here, let us illustrate the decoherence dynamics of the reduced density matrix by choosing a particular initial state, namely, a pure-state Gaussian wave packet of width b centered at $x = 0$,

$$\psi(x, 0) = \left[\frac{1}{2\pi b^2}\right]^{1/4} \exp\left(\frac{-x^2}{4b^2}\right). \tag{3.84}$$

This yields the initial density matrix

$$\rho_S(x, x', 0) = \left[\frac{1}{2\pi b^2}\right]^{1/2} \exp\left(-\frac{x^2 + x'^2}{4b^2}\right), \tag{3.85}$$

which corresponds to the initial values $A(0) = C(0) = 1/8b^2$ and $B(0) = D(0) = 0$ for the coefficients appearing in (3.79).

The density matrix (3.85) is plotted in Fig. 3.5. We see that the Gaussian is symmetric with respect to the x and x' axes, i.e., the coherence length and ensemble width are of equal magnitudes. For this initial state (3.85), the

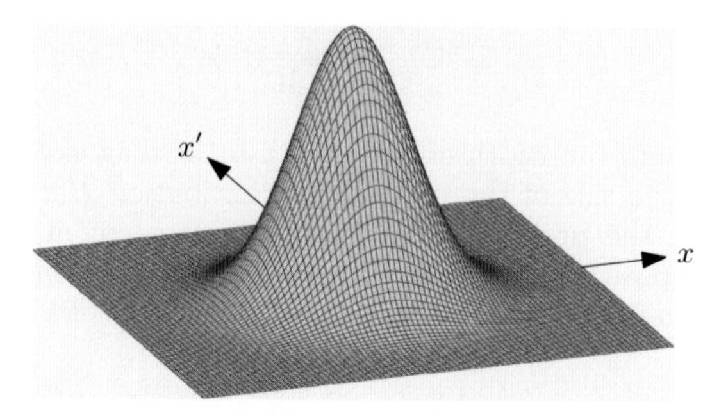

Fig. 3.5. Illustration of the initial density matrix $\rho_S(x, x', 0)$ [see (3.85)]. The density matrix has the shape of a symmetric Gaussian in the (x, x') plane, with a coherence length $\ell(0)$ (the extension along the "off-diagonal" $x = -x'$) equal to the ensemble width $\Delta X(0)$ (the width along the diagonal $x = x'$).

resulting time dependence of the coefficients $A(t)$, $B(t)$, $C(t)$, and $D(t)$ is easily computed. Let us explicitly state the expressions for $A(t)$ and $C(t)$, which read (see [7] or Appendix A2 of [17] for the derivation)

$$A(t) = \frac{\Lambda t^3 + 3m^2b^2/2 + 2\Lambda^2 b^2 t^4 + 12\Lambda m^2 b^4 t}{3t^2 + 8\Lambda b^2 t^3 + 12m^2 b^4} \tag{3.86}$$

and

$$C(t) = \frac{1}{2}\left[\frac{t^2}{m^2b^2} + \frac{8}{3}\frac{\Lambda t^3}{m^3} + 4b^2\right]^{-1}. \tag{3.87}$$

Before studying the resulting time evolution of related key quantities such as the coherence length, ensemble width, and purity of the reduced density matrix, let us briefly discuss how we expect the initial Gaussian (3.85) to evolve over time if decoherence due to environmental scattering is taken into account. We anticipate that the interaction with the environment will make it increasingly hard to observe spatial coherences over a given distance $\Delta x = |x - x'|$. That is, we expect the coherence length $\ell(t)$ [see (3.80)] to decrease over time, "squeezing" the Gaussian (3.85) in the direction perpendicular to the "classical" diagonal $x = x'$. In the limit of complete decoherence, the Gaussian would then approach the shape of the quasiclassical distribution, i.e., that of an infinitely narrow ridge along the diagonal $x = x'$. At the same time, we anticipate the squeezing in the $x = -x'$ direction to be accompanied by a spreading in the $x = x'$ direction.

Let us now confirm these intuitions by studying the time evolution of some of the relevant quantities, using the explicit expressions (3.86) and (3.87) for the coefficients $A(t)$ and $C(t)$. Probably the quantity of most interest to decoherence is the coherence length $\ell(t)$ defined in (3.80). Using (3.86), the time dependence of this quantity is given by

$$\ell(t) = \frac{1}{\sqrt{8A(t)}} = \frac{1}{2}\left[\frac{3t^2 + 8\Lambda b^2 t^3 + 12m^2 b^4}{2\Lambda t^3 + 3m^2 b^2 + 4\Lambda^2 b^2 t^4 + 24\Lambda m^2 b^4 t}\right]^{1/2}. \tag{3.88}$$

As expected, at $t = 0$ the coherence length is simply equal to the width b of the wave packet. In the absence of environmental interactions ($\Lambda = 0$), $\ell(t)$ becomes

$$\ell_{\text{free}}(t) = \frac{1}{2}\left[\frac{t^2}{m^2 b^2} + 4b^2\right]^{1/2} = \Delta X_{\text{free}}(t). \tag{3.89}$$

This expression simply describes the free coherent spread of the wave packet discussed at the beginning of this chapter. It is analogous to (3.3) with the obvious identification $\sigma \equiv \sqrt{2}b$ (recall also that we set $\hbar \equiv 1$ in this section).

Let us study the time evolution of $\ell(t)$ [see (3.88)] in more detail. In the following, we shall measure time in units of $\tau_b \equiv 1/\Lambda b^2$, which is the characteristic localization timescale for spatial coherences over a distance $\Delta x = b$ (the initial width of the wave packet). A plot of $\ell(t)$ is shown in Fig. 3.6.

Let us first investigate the time evolution of $\ell(t)$ for times t much shorter than the localization timescale τ_b. Expanding (3.88) around $\Lambda b^2 t = 0$ up to first order in $\Lambda b^2 t$ yields

$$\ell(t) \approx b\left(1 - 4\Lambda b^2 t\right). \tag{3.90}$$

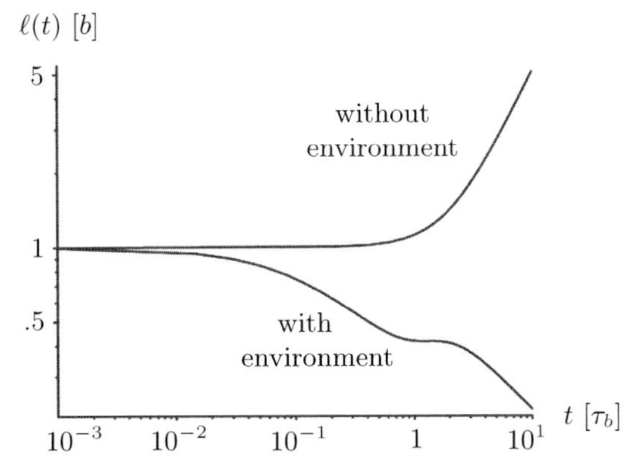

Fig. 3.6. Time dependence of the coherence length $\ell(t)$ given by (3.88) for a Gaussian wave packet, as studied by Joos et al. [17]. The time t is measured in units of the characteristic localization timescale $\tau_b = 1/\Lambda b^2$, and $\ell(t)$ is displayed in units of $\ell(0) = b$, the initial width of the wave packet. The upper curve shows the behavior without environmental interaction ($\Lambda = 0$). The lower curve displays the effects of the environmental scattering, showing that the distance over which the system is able to maintain spatial coherences decreases rapidly with time.

We thus see that, due to the influence of environmental scattering, for short times the coherence length decreases linearly with time, and that the rate of decrease is proportional to the scattering constant Λ.

On the other hand, we may consider the limit in which t is much larger than the localization timescale τ_b. In view of the fact that the scattering rate Λ may take on very large values for macroscopic objects (see Table 3.1), this does not mean that t itself needs to be large for this limit to apply. For example, even with a rather low value of $\Lambda \approx 10^{20}$ cm^{-2} s^{-1} for a dust grain of size $a = 10^{-3}$ cm and an initial coherent spread of 10% of this size, i.e., $b = 10^{-4}$ cm, we have $\tau_b \approx 10^{-12}$ s. Thus, for a macroscopic object, the limit $t \gg \tau_b$ will typically be reached within a tiny fraction of a second. In this limit, (3.88) approaches the asymptotic expression

$$\ell(t) \longrightarrow \frac{1}{\sqrt{2\Lambda t}}. \tag{3.91}$$

Thus in this case the coherence length follows a $t^{-1/2}$ dependence and is independent of the width b of the initial state (3.84).

Let us now consider the time evolution of the total size $\Delta X(t)$ of the ensemble. From (3.81) and (3.87), we obtain

$$\Delta X(t) = \frac{1}{\sqrt{8C(t)}} = \frac{1}{2}\left[\frac{t^2}{m^2b^2} + \frac{8}{3}\frac{\Lambda t^3}{m^3} + 4b^2\right]^{1/2}. \tag{3.92}$$

In the absence of any environmental interactions ($\Lambda = 0$), i.e., for a purely unitary time evolution of the free particle described by the initial wave function (3.84), $\Delta X(t)$ coincides with the expression for the coherence length $\ell(t)$ evaluated for $\Lambda = 0$ [see (3.89)],

$$\Delta X_{\text{free}}(t) = \frac{1}{2}\left[\frac{t^2}{m^2b^2} + 4b^2\right]^{1/2}. \tag{3.93}$$

Thus, without the environment, the spreading proceeds at exactly the same rate in both the $x = x'$ direction (describing the ensemble of positions) and the $x = -x'$ direction (representing spatial coherence). This, of course, is the expected behavior for a free particle: If the environment is absent, there exists no physical mechanism that would single out one of these two directions over the other direction. However, once the environment is included, an asymmetry is introduced. Now the scattering events act as a form of measurement device for the position of the system. Broadly speaking, the scattered particles will try to resolve differences $|x - x'|$ between two positions x and x' while being insensitive to the sum $(x + x')$ of two positions. This asymmetry of the environmental monitoring process with respect to the coordinates $(x + x')$ and $(x - x')$ is encapsulated in the form of the equation of motion for $\rho_S(x, x', t)$ [see the right-hand side of (3.78)] and leads to the different time evolutions for $\Delta X(t)$ and $\ell(t)$, as seen by comparing the expressions (3.92) and (3.88).

With the environmental interaction present, the additional term $\frac{8}{3}\frac{\Lambda t^3}{m^3}$ appearing in (3.92) leads to an even *faster* increase of the width of the ensemble with time (see Fig. 3.7).[7] In our model, the fact that the presence of the environment accelerates this "classical" spreading is due to the fact that the scattering events lead to an increase in the mean energy of the system. This energy increase is a consequence of our assumption that these scattering events do not result in a recoil of the system (see p. 121).

We can thus visualize the time evolution of the density matrix as a competition between two opposite influences, as illustrated in Fig. 3.8. On the one hand, the overall spatial extension $\Delta X(t)$ of the position-space ensemble tends to spread out more and more with time, induced by both the unitary free-particle evolution and the increase in the mean energy due to the scattering events. On the other hand, these scattering events also play the role of a continuous monitoring of the position of the system, leading to a decrease of the coherence length $\ell(t)$ and thus to the suppression of spatial coherences over increasingly short distances.

[7] At a first glance, this might seem like a surprising result. After all, in the introduction to this chapter we suggested that decoherence *counteracts* the spreading of the wave packet. However, there is no contradiction here. The crucial point to bear in mind here is the difference between *coherent* and *ensemble* spreading. Coherent spreading means that the system can be found in a *superposition* of increasingly separated positions, e.g., by performing an interference experiment. By contrast, ensemble spreading simply corresponds to an increased range of possible positions in which the system can be found upon a measurement of its position.

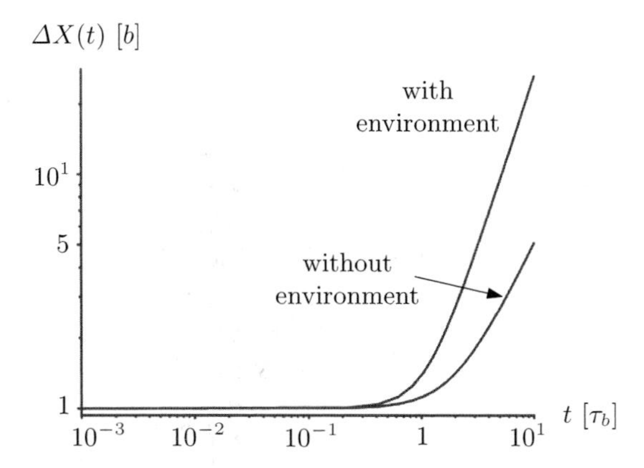

Fig. 3.7. Time evolution of the ensemble width $\Delta X(t)$ given by (3.92) for a Gaussian wave packet, as studied by Joos et al. [17]. The time t is measured in units of the characteristic localization timescale $\tau_b = 1/\Lambda b^2$, and $\Delta X(t)$ is plotted in units of the initial width $\Delta X(0) = b$ of the wave packet.

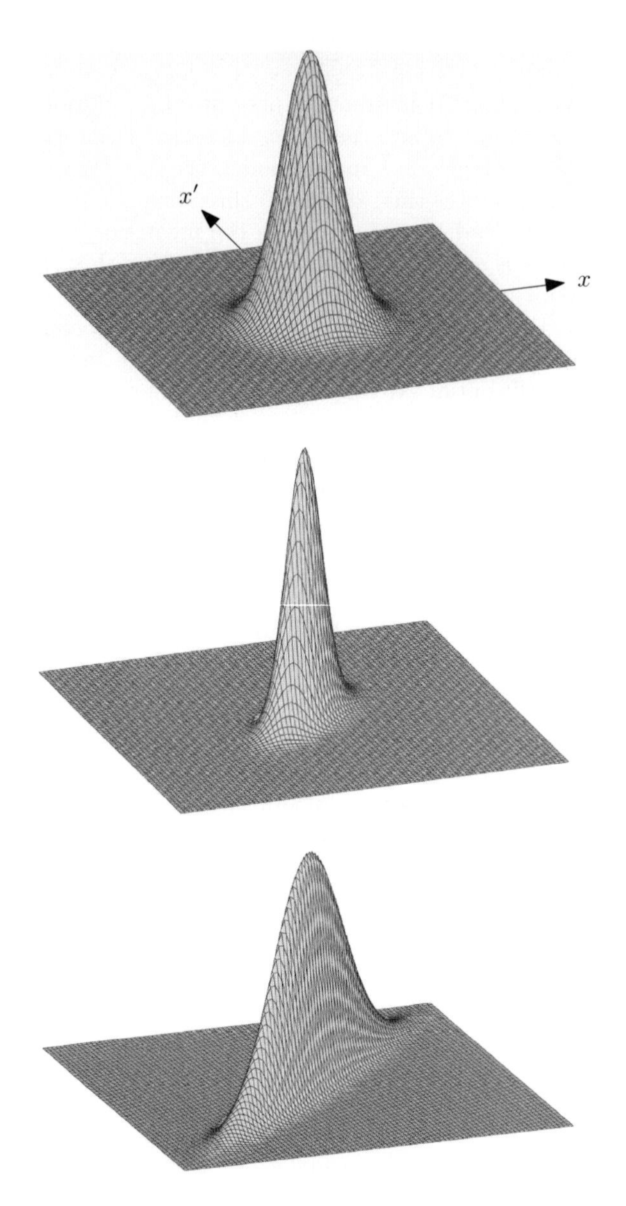

Fig. 3.8. Time evolution of the Gaussian wave packet (3.79), with the initial state given by (3.85), under the influence of a scattering environment. The coefficients $A(t)$ and $C(t)$ appearing in (3.79) are determined by (3.86) and (3.87), respectively. The width of the Gaussian in the off-diagonal direction $x = -x'$ (representing spatial coherences) becomes progressively reduced, with the density matrix approaching a quasiclassical probability distribution of positions concentrated along the diagonal $x = x'$.

To quantify the interplay between the reduction of the coherence length $\ell(t)$ and the increase of the ensemble size $\Delta X(t)$, one often considers a combined dimensionless quantity in form of the ratio (see, e.g., [17, 173]),

$$\delta(t) \equiv \frac{\ell(t)}{\Delta X(t)} = \sqrt{\frac{C(t)}{A(t)}}. \tag{3.94}$$

A plot of $\delta(t)$ is shown in Fig. 3.9. Without the environment present, $\delta(t)$ remains constant, i.e., the coherent spreading proceeds at the same rate in the $x = x'$ and $x = -x'$ directions [see (3.89) and the subsequent discussion]. With the environment included, the coherence length $\ell(t)$ decreases while the overall width $\Delta X(t)$ increases. However, $\Delta X(t)$ increases at a slower rate than the rate of decrease of $\ell(t)$, leading to a net decrease of $\delta(t)$.

Another reason for our introduction of the quantity $\delta(t)$ is that it allows us to easily establish an immediate connection to another measure of decoherence, namely, the purity $\varsigma = \mathrm{Tr}\,(\hat{\rho}^2)$ of a density matrix $\hat{\rho}$ as introduced in Sect. 2.4.3 [see (2.30)]. For the reduced density matrix $\rho_S(t)$ of our model, the purity is found to be

$$\varsigma(t) = \sqrt{\frac{C(t)}{A(t)}}. \tag{3.95}$$

But evidently this coincides with the expression (3.94) for $\delta(t)$, i.e., the quantity $\delta(t)$ also describes the purity of the reduced density matrix in our scattering model. From Fig. 3.9 we see that, as expected, the environmental scattering leads to a decrease of the purity of the reduced density matrix, reflecting the gradual action of decoherence.

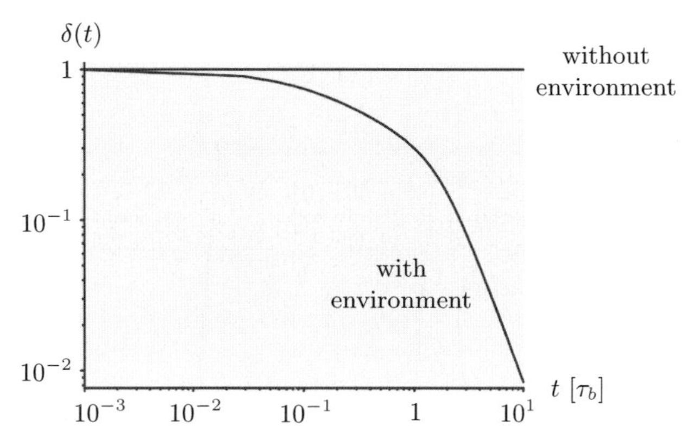

Fig. 3.9. Time dependence of the ratio $\delta(t) \equiv \ell(t)/\Delta X(t)$ [see (3.94)] with and without environmental scattering. The quantity $\delta(t)$ also describes the time evolution of the purity of the local density matrix in our model. The time t is again measured in units of the characteristic localization timescale $\tau_b = 1/\Lambda b^2$.

Finally, let us illustrate the time evolution of a superposition of two equal-weight Gaussian wave packets of the form (3.84) centered around $x = \pm x_0$, i.e.,

$$\begin{aligned}\Psi(x,0) &= \psi_{x_0}(x,0) + \psi_{-x_0}(x,0) \\ &= \left[\frac{1}{2\pi b^2}\right]^{1/4} \left\{ \exp\left(\frac{-(x-x_0)^2}{4b^2}\right) + \exp\left(\frac{-(x+x_0)^2}{4b^2}\right)\right\}. \end{aligned} \tag{3.96}$$

This wave function is illustrated in Fig. 3.10.

If $x_0 \gg b$, such a quantum state would correspond to the nonclassical situation of a particle being described by a superposition of two well-defined, distinguishable positions separated by a distance $2x_0$. The corresponding density matrix is

$$\begin{aligned}\rho_S(x,x',0) &= \Psi^*(x',0)\Psi(x,0) \\ &= \psi^*_{x_0}(x',0)\psi_{x_0}(x,0) + \psi^*_{-x_0}(x',0)\psi_{-x_0}(x,0) \\ &\quad + \psi^*_{x_0}(x',0)\psi_{-x_0}(x,0) + \psi^*_{-x_0}(x',0)\psi_{x_0}(x,0). \end{aligned} \tag{3.97}$$

This density matrix is shown as the top plot of Fig. 3.11. The two peaks along the off-diagonal $x = -x'$ correspond to the spatial interference terms $\psi^*_{x_0}(x',0)\psi_{-x_0}(x,0)$ and $\psi^*_{-x_0}(x',0)\psi_{x_0}(x,0)$ between the two wave packets $\psi_{x_0}(x,0)$ and $\psi_{-x_0}(x,0)$ introduced in (3.96).

The subsequent time evolution of the density matrix (3.97) as determined by the equation of motion (3.78) is illustrated in Fig. 3.11. We observe that, as expected, interference terms corresponding to spatial coherence between the two regions around $x = \pm x_0$ are gradually suppressed due to the presence of the environment. In the limit of complete decoherence, we obtain an (albeit improper!) ensemble of the two states describing localization of the system around $+x_0$ and $-x_0$, respectively. That is, measurements that per-

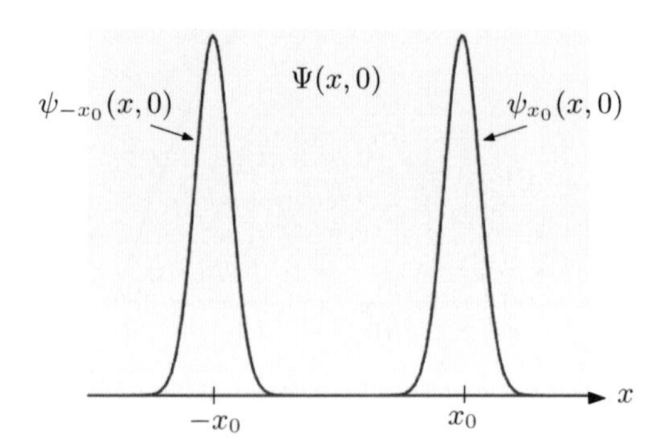

Fig. 3.10. Illustration of the superposition (3.96) of two Gaussian wave packets centered around $+x_0$ and $-x_0$, respectively.

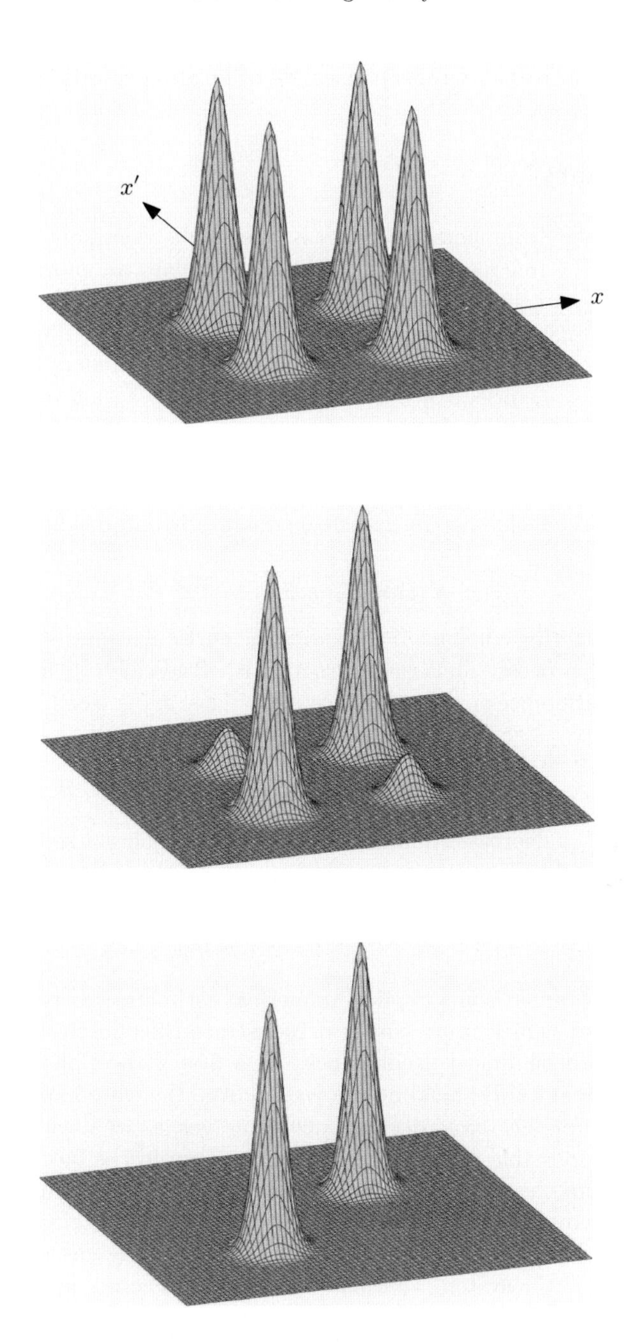

Fig. 3.11. Time evolution of the density matrix $\rho_S(x, x', t)$ for the superposition (3.96) of two Gaussian wave packets. Interference terms along the off-diagonal $x = -x'$ become progressively damped by the scattering of environmental particles.

tain to the system alone will be unable to distinguish this ensemble from the corresponding classical (proper) ensemble of localized positions.

3.6 Summary

Scattering of environmental particles is one of the dominant sources of decoherence in the macroscopic domain and thus plays an enormously important role in the emergence of classicality in the everyday world around us. Scattering-induced decoherence is also ideally suited for illustrating basic conceptual aspects of decoherence, for instance, the understanding of this decoherence as a consequence of the environment's acting as a continuous monitoring device for the position of the system.

We found that for the case in which each scattering event encodes only incomplete which-path information in the environment (the "long-wavelength limit"), spatial interference terms are exponentially damped according to [see (3.57)]

$$\rho_S(\boldsymbol{x}, \boldsymbol{x}', t) = \rho_S(\boldsymbol{x}, \boldsymbol{x}', 0)\mathrm{e}^{-\Lambda(\boldsymbol{x}-\boldsymbol{x}')^2 t}, \tag{3.98}$$

where the scattering constant Λ embodies the particular physical properties of the scattering process. Thus the characteristic timescale $\tau_{\Delta x}$ for decoherence of spatial interferences over a distance $\Delta x = |\boldsymbol{x} - \boldsymbol{x}'|$ is [see (3.58)]

$$\tau_{\Delta x} = \frac{1}{\Lambda(\Delta x)^2}. \tag{3.99}$$

We see that $\tau_{\Delta x}$ decreases quadratically with the coherent separation Δx. If this separation becomes larger than the typical wavelength of the environmental particles, it is completely resolved by each scattering event. Accordingly, the decoherence rate saturates and is simply given by the total scattering rate [see (3.48)].

We also obtained some explicit numerical estimates for typical decoherence timescales, which impressively demonstrate the effectiveness and ubiquity of scattering-induced decoherence. At a first glance, one may have assumed that, by suitably shielding the system from the environment, one might be able to completely avoid decoherence. However, as we have seen, such full shielding is impossible to achieve in practice. Even in the best available laboratory vacuum the density of air molecules remains large enough to lead to strong and virtually immediate decay of macroscopic spatial interferences. If we were able to complete prevent all these surrounding gas molecules from interacting with the system, thermal photons would continue to induce rapid decoherence. And even if we managed to reduce such effects, decoherence due to an inescapable environment of solar neutrinos and microwave background radiation would still occur. These arguments also show why position is the ubiquitous preferred basis on macroscopic scales (and often even on microscopic scales, e.g., in the case of chiral molecules mentioned in Sect. 2.8.4).

The recognition that quantum systems are practically impossible to shield from the influence of a scattering environment and that these scattering processes have a crucial influence on the dynamics and observable quantum properties of the systems has supplied a missing key ingredient in Schrödinger's attempt for a true "wave mechanics" in which wave packets are directly identified with the spatially extended objects around us. The reduced density matrix describing an object in contact with a scattering environment can be used to define an improper ensemble of wave packets whose width is given by the progressively decreasing coherence length. Each such wave packet may then be identified (at least in the operational sense of local measurements) with the appearance of a well-localized particle.

As we will discuss in more detail in Sect. 6.2, recent experiments of the double-slit type involving massive molecules have allowed us to directly confirm the qualitative and quantitative predictions of models for scattering-induced decoherence. As such experiments push toward the generation of interference patterns and the observation of the gradual influence of decoherence for larger and larger objects, we will be able to directly test these models in the laboratory on increasingly macroscopic scales.

4 Master-Equation Formulations of Decoherence

In this chapter, we will introduce the description of decoherence dynamics in terms of so-called master equations. Such master equations directly and conveniently yield the time evolution of the reduced density matrix $\hat{\rho}_{\mathcal{S}}(t)$ for the open quantum system $\mathcal{S}$ interacting with an environment $\mathcal{E}$. They relieve us from the need of having to first determine the dynamics of the total system–environment combination and to then trace out the degrees of freedom of the environment. The master-equation approach is motivated by two issues. First, we are usually not interested in the dynamics of the environment or of the global system–environment combination. All we really care about is the *influence* of the environment on our system of interest. Second, it is often impossible to analytically determine the time evolution of the density matrix. In such cases, one can use approximation schemes that lead to master equations for the *approximate* evolution of the reduced density matrix.

In contrast with the dynamics of the density matrix of a closed system, the evolution equation for the reduced density matrix will of course, in general, be nonunitary, since the interaction with the environment will typically change the amount of coherence present in the system. Put into more formal terms, since the trace operation used to obtain the reduced density matrix is a nonunitary operation, the master equation must in general also be nonunitary.

This chapter is organized as follows. In Sect. 4.1, we will introduce the basic ideas and the general formalism of master equations. In Sect. 4.2, we will then discuss in some detail a very important type of master equation, namely, the so-called Born–Markov master equation. This master equation plays a key role in studies of decoherence. The derivation of this equation, given in Sect. 4.2.2, is inevitably somewhat technical, so some readers may prefer to just glance over this section or skip it altogether. In Sect. 4.3, we will discuss so-called Lindblad master equations, a special and more simplified form of the general Born–Markov master equation that is valid under an additional assumption. As we shall see, Lindblad master equations allow for a very intuitive representation of decoherence processes. Finally, in Sect. 4.4, we will briefly discuss the case of non-Markovian dynamics.

We note that master equations also play an important role in the theory of so-called quantum dynamical semigroups. Since our focus here is on the modeling of decoherence, we shall not further discuss this mathematical connection between master equations and dynamical semigroups. Instead we refer interested readers to Chap. 7 of [17] and to the book by Breuer and Petruccione [18] for further information on this topic.

4.1 General Formalism

In the "ordinary" formalism of decoherence, the reduced density matrix $\hat{\rho}_{\mathcal{S}}(t)$ is computed via

$$\hat{\rho}_{\mathcal{S}}(t) = \mathrm{Tr}_{\mathcal{E}} \left\{ \hat{\rho}_{\mathcal{SE}}(t) \right\} \equiv \mathrm{Tr}_{\mathcal{E}} \left\{ \hat{U}(t) \hat{\rho}_{\mathcal{SE}}(0) \hat{U}^{\dagger}(t) \right\}, \tag{4.1}$$

where $\hat{U}(t)$ denotes the time-evolution operator for the composite system $\mathcal{SE}$. As is evident from (4.1), this approach inevitably requires that we first determine the dynamics $\hat{\rho}_{\mathcal{SE}}(t)$ of the total system $\mathcal{SE}$ before we can arrive at the reduced description through the trace operation. As we have indicated in the introduction to this chapter, this task is very difficult, if not impossible, to carry out in practice for most reasonably complex model systems.

By contrast, in the master-equation formalism, we instead calculate $\hat{\rho}_{\mathcal{S}}(t)$ directly from an expression of the form

$$\hat{\rho}_{\mathcal{S}}(t) = \hat{V}(t) \hat{\rho}_{\mathcal{S}}(0), \tag{4.2}$$

where the operator $\hat{V}(t)$ is the so-called *dynamical map* that generates the evolution of $\hat{\rho}_{\mathcal{S}}(t)$. Since $\hat{V}(t)$ represents an operator that in turn acts on another operator, it is commonly referred to as a "superoperator." Equation (4.2) is called a *master equation* for $\hat{\rho}_{\mathcal{S}}(t)$, and it represents the most general form that such a master equation may take.

Obviously, if the master equation is exact, then (4.1) and (4.2) must be equivalent by definition, i.e., we must have the identity

$$\hat{V}(t) \hat{\rho}_{\mathcal{S}}(0) \equiv \mathrm{Tr}_{\mathcal{E}} \left\{ \hat{U}(t) \hat{\rho}_{\mathcal{SE}}(0) \hat{U}^{\dagger}(t) \right\}, \tag{4.3}$$

and the master equation would amount to nothing else but a trivial rewriting of (4.1).

Therefore, the power of master equations is only unlocked once we impose certain assumptions about the system–environment states and dynamics. Such assumptions then allow us to determine the approximate time evolution of $\hat{\rho}_{\mathcal{S}}(t)$ even when it is impossible to calculate the exact global dynamics $\hat{\rho}_{\mathcal{SE}}(t)$. In fact, here we shall restrict our attention to master equations (obtained from certain approximations) that can be written as first-order differential equations that are local in time, i.e., that can be expressed in the form

$$\frac{\mathrm{d}}{\mathrm{d}t}\hat{\rho}_{\mathcal{S}}(t) = \hat{\mathcal{L}}\left[\hat{\rho}_{\mathcal{S}}(t)\right] = \underbrace{-\mathrm{i}[\hat{H}'_{\mathcal{S}}, \hat{\rho}_{\mathcal{S}}(t)]}_{\text{unitary evolution}} + \underbrace{\hat{\mathcal{D}}[\hat{\rho}_{\mathcal{S}}(t)]}_{\text{decoherence}} . \tag{4.4}$$

This equation is local in time in the sense that the change of $\hat{\rho}_{\mathcal{S}}$ at time t depends only on $\hat{\rho}_{\mathcal{S}}$ evaluated at t but not at any other times $t' \neq t$. The superoperator $\hat{\mathcal{L}}$ appearing in (4.4) acts on $\hat{\rho}_{\mathcal{S}}(t)$ and typically depends on the initial state of the environment and the different terms in the Hamiltonian. To convey the physical intuition behind this superoperator, $\hat{\mathcal{L}}$ has been decomposed into two parts:

1. A *unitary* part that is given by the usual Liouville–von Neumann commutator with the Hamiltonian $\hat{H}'_{\mathcal{S}}$. It is important to note that this Hamiltonian is in general *not* identical to the unperturbed free Hamiltonian $\hat{H}_{\mathcal{S}}$ of $\mathcal{S}$ that would generate the evolution of $\mathcal{S}$ in absence of the environment, because the presence of the environment often perturbs the free Hamiltonian, leading to a renormalization of the energy levels of the system. We emphasize that this effect (often called the *Lamb-shift* contribution) has nothing do with the nonunitary evolution induced by the environment but only alters the unitary part of the reduced dynamics.
2. A *nonunitary* part $\hat{\mathcal{D}}[\hat{\rho}_{\mathcal{S}}(t)]$ that represents decoherence (and possibly also dissipation) due to the environment. This term will be of the most interest to us in the following.

We note that, if the evolution of the system is completely unitary, we have $\hat{\mathcal{D}}[\hat{\rho}_{\mathcal{S}}(t)] = 0$, and thus (4.4) simply becomes

$$\frac{\mathrm{d}}{\mathrm{d}t}\hat{\rho}_{\mathcal{S}}(t) = -\mathrm{i}\left[\hat{H}'_{\mathcal{S}}, \hat{\rho}_{\mathcal{S}}(t)\right] . \tag{4.5}$$

This equation differs from the standard Liouville–von Neumann equation for closed systems only in the use of the environment-shifted system Hamiltonian $\hat{H}'_{\mathcal{S}}$ instead of the unperturbed Hamiltonian $\hat{H}_{\mathcal{S}}$.

In the next section, we shall derive an explicit general expression for $\hat{\mathcal{L}}$ (i.e., for $\hat{H}'_{\mathcal{S}}$ and $\hat{\mathcal{D}}[\hat{\rho}_{\mathcal{S}}(t)]$) under a certain set of assumptions. Applied to a given model, the resulting master equation then allows us to compute the time evolution of $\hat{\rho}_{\mathcal{S}}(t)$ for all times t.

4.2 The Born–Markov Master Equation

The Born–Markov master equation plays an enormously important role in the study of open quantum systems and decoherence. It allows one to treat many decoherence problems in a mathematically simple, and often closed, form. Comparisons between the predictions of models based on this equation and experimental data have shown that the Born and Markov assumptions on which the master equation is based are reasonable in many cases. However, we should emphasize already at this stage that there exist various important physical systems (for example, low-temperature solid-state systems) which do not obey Markovian dynamics and which therefore often cannot be adequately modeled using the Born–Markov master equation. We will come back to this issue in Sect. 4.4 below.

4.2.1 Structure of the Born–Markov Master Equation

The Born–Markov master equation is based on two core approximations that can be broadly stated as follows (the formal representation and physical justification of these approximations is discussed in more detail in Sect. 4.2.2 below):

1. *The Born approximation.* The system–environment coupling is sufficiently weak and the environment is reasonably large such that changes of the density operator of the environment are negligible and the system–environment state remains in an approximate product state at all times, i.e.,
$$\hat{\rho}(t) \approx \hat{\rho}_{\mathcal{S}}(t) \otimes \hat{\rho}_{\mathcal{E}}, \tag{4.6}$$
with $\hat{\rho}_{\mathcal{E}}$ approximately constant at all times.
2. *The Markov approximation.* "Memory effects" of the environment are negligible, in the sense that any self-correlations within the environment created by the coupling to the system decay rapidly compared to the characteristic timescale over which the state of the system varies noticeably.

Assume now these assumptions hold. Suppose further that the system $\mathcal{S}$ has self-Hamiltonian $\hat{H}_{\mathcal{S}}$ and that its coupling to the environment is described by the interaction Hamiltonian $\hat{H}_{\text{int}}$, which we may write in the diagonal form [see (2.93)]
$$\hat{H}_{\text{int}} = \sum_{\alpha} \hat{S}_{\alpha} \otimes \hat{E}_{\alpha}. \tag{4.7}$$

Here the system and environment operators $\hat{S}_{\alpha}$ and $\hat{E}_{\alpha}$ are unitary but not necessarily Hermitian. As explained in Sect. 2.8, the intuitive physical interpretation of the diagonal decomposition (4.7) is that the operators $\hat{S}_{\alpha}$ correspond to the physical quantities of the system that are continuously "monitored" by the environment (provided the $\hat{S}_{\alpha}$ are Hermitian).

Then the evolution of the reduced density operator $\hat{\rho}_{\mathcal{S}}(t)$ is given by the *Born–Markov master equation*

$$\frac{\mathrm{d}}{\mathrm{d}t}\hat{\rho}_{\mathcal{S}}(t) = -\mathrm{i}\left[\hat{H}_{\mathcal{S}}, \hat{\rho}_{\mathcal{S}}(t)\right] - \sum_{\alpha}\left\{\left[\hat{S}_{\alpha}, \hat{B}_{\alpha}\hat{\rho}_{\mathcal{S}}(t)\right] + \left[\hat{\rho}_{\mathcal{S}}(t)\hat{C}_{\alpha}, \hat{S}_{\alpha}\right]\right\}. \tag{4.8}$$

The system operators $\hat{B}_{\alpha}$ and $\hat{C}_{\alpha}$ appearing in this equation are defined as

$$\hat{B}_\alpha \equiv \int_0^\infty \mathrm{d}\tau \sum_\beta \mathcal{C}_{\alpha\beta}(\tau)\hat{S}_\beta^{(I)}(-\tau), \tag{4.9a}$$

$$\hat{C}_\alpha \equiv \int_0^\infty \mathrm{d}\tau \sum_\beta \mathcal{C}_{\beta\alpha}(-\tau)\hat{S}_\beta^{(I)}(-\tau). \tag{4.9b}$$

Here $\hat{S}_\alpha^{(I)}(-\tau)$ denotes the system operator $\hat{S}_\alpha$ in the so-called interaction picture. The interaction-picture formalism will be used heavily in the derivation of the Born–Markov master equation below and is reviewed in the Appendix. From now on, we shall simplify our notation by omitting the superscript "(I)" for indicating operators in the interaction picture. Instead we shall introduce the convention that all operators bearing explicit time arguments are meant to be understood as interaction-picture operators. For the density operator, however, we will uphold the superscript notation to make clear the distinction from the Schrödinger-picture density operator, which also carries a time argument (since in the Schrödinger picture quantum states are explicitly time-dependent).

The quantity $\mathcal{C}_{\alpha\beta}(\tau)$ appearing in (4.9) is given by

$$\mathcal{C}_{\alpha\beta}(\tau) \equiv \left\langle \hat{E}_\alpha(\tau)\hat{E}_\beta \right\rangle_{\hat{\rho}_\mathcal{E}}, \tag{4.10}$$

where the average is taken over the initial state $\hat{\rho}_\mathcal{E}$ of the environment (recall that the Born approximation (4.6) means that $\hat{\rho}_\mathcal{E}$ remains approximately constant at all times). The $\mathcal{C}_{\alpha\beta}(\tau)$ will be referred to as the *environment self-correlation functions* in the following. The reason for this terminology is easy to understand. The operators $\hat{E}_\alpha$ (provided they are Hermitian) can be thought of as observables "measured" on the environment by the interaction between the system and the environment. The environment self-correlation functions then tell us to what extent the result of such a "measurement" of a particular $\hat{E}_\alpha$ is correlated with the result of a "measurement" of the same observable carried out a time τ later. Broadly speaking, these functions thus quantify to what degree the environment retains information over time about its interaction with the system. In fact, the Markov approximation corresponds to the assumption of a rapid decay of these environment self-correlation functions relative to the timescale set by the evolution of the system.

The form (4.8) of the Born–Markov master equation may look fairly daunting and difficult to apply to concrete cases. To explicitly determine the quantities $\hat{B}_\alpha$ and $\hat{C}_\alpha$ [see (4.9)] for a given model, we would need to calculate (i) the interaction-picture operators $\hat{S}_\alpha(\tau)$ and $\hat{E}_\alpha(\tau)$ (whose time evolution is given by the system and environment Hamiltonians $\hat{H}_\mathcal{S}$ and $\hat{H}_\mathcal{E}$, respectively; see Appendix), and (ii) the environment self-correlation functions $\mathcal{C}_{\alpha\beta}(\tau)$, see (4.10).

However, in many situations of interest, the master equation (4.8) simplifies considerably. For example, in most cases we have to consider only a single system observable $\hat{S}$ that is monitored by the environment. Then the sum in the diagonal decomposition (4.7) of the interaction Hamiltonian reduces to a single term,

$$\hat{H}_{\text{int}} = \hat{S} \otimes \hat{E}. \tag{4.11}$$

Also, the time dependence of the interaction-picture operators $\hat{S}_\alpha(\tau)$ and $\hat{E}_\alpha(\tau)$ is often very simple (sometimes even trivial), facilitating the calculation of the quantities $\hat{B}_\alpha$ and $\hat{C}_\alpha$ [see (4.9)]. We shall see how this works in the next Chap. 5, where we will apply the Born–Markov master equation (4.8) to some concrete examples.

4.2.2 Derivation of the Born–Markov Master Equation

In this section, we shall derive the Born–Markov master equation (4.8) from first principles. Readers are encouraged to work through the main steps of the derivation, in particular in order to get a better understanding of the assumptions that enter into this derivation.

Interaction-Picture Description

First, let us begin by decomposing the total system–environment Hamiltonian in the usual form as

$$\hat{H} = \hat{H}_{\mathcal{S}} + \hat{H}_{\mathcal{E}} + \hat{H}_{\text{int}}. \tag{4.12}$$

Since we would like to apply perturbation theory, it is convenient to switch to the interaction picture. Then

$$\hat{H}_0 \equiv \hat{H}_{\mathcal{S}} + \hat{H}_{\mathcal{E}} \tag{4.13}$$

denotes the total free Hamiltonian, and $\hat{H}_{\text{int}}$ represents the interaction that is to be treated perturbatively.

We now transform the interaction Hamiltonian $\hat{H}_{\text{int}}$ and the total system–environment density operator $\hat{\rho}(t)$ to the interaction picture. These transformations are given by [see (A.4) and (A.5)]

$$\hat{H}_{\text{int}}(t) = \mathrm{e}^{\mathrm{i}\hat{H}_0 t} \hat{H}_{\text{int}} \mathrm{e}^{-\mathrm{i}\hat{H}_0 t}, \tag{4.14}$$

$$\begin{aligned} \hat{\rho}^{(I)}(t) &= \mathrm{e}^{\mathrm{i}\hat{H}_0 t} \hat{\rho}(t) \mathrm{e}^{-\mathrm{i}\hat{H}_0 t} \\ &= \mathrm{e}^{\mathrm{i}\hat{H}_0 t} \mathrm{e}^{-\mathrm{i}\hat{H} t} \hat{\rho} \mathrm{e}^{\mathrm{i}\hat{H} t} \mathrm{e}^{-\mathrm{i}\hat{H}_0 t}. \end{aligned} \tag{4.15}$$

The time evolution of $\hat{\rho}^{(I)}(t)$ is determined by the interaction-picture Liouville–von Neumann equation [see (A.9)],

$$\frac{\mathrm{d}}{\mathrm{d}t} \hat{\rho}^{(I)}(t) = -\mathrm{i} \left[\hat{H}_{\text{int}}(t), \hat{\rho}^{(I)}(t) \right]. \tag{4.16}$$

Thus the evolution of the interaction-picture density operator is generated only by the interaction-picture coupling part $\hat{H}_{\text{int}}(t)$ of the full Hamiltonian.

Iterative Solution by Integration

If we now formally integrate (4.16), we get

$$\hat{\rho}^{(I)}(t) = \hat{\rho}(0) - \mathrm{i} \int_0^t \mathrm{d}t' \left[\hat{H}_{\mathrm{int}}(t'), \hat{\rho}^{(I)}(t') \right]. \tag{4.17}$$

Let us insert this expression for $\hat{\rho}^{(I)}(t)$ back into the right-hand side of (4.16). This yields

$$\begin{aligned} \frac{\mathrm{d}}{\mathrm{d}t}\hat{\rho}^{(I)}(t) &= -\mathrm{i} \left[\hat{H}_{\mathrm{int}}(t), \left\{ \hat{\rho}(0) - \mathrm{i} \int_0^t \mathrm{d}t' \left[\hat{H}_{\mathrm{int}}(t'), \hat{\rho}^{(I)}(t') \right] \right\} \right] \\ &= -\mathrm{i} \left[\hat{H}_{\mathrm{int}}(t), \hat{\rho}(0) \right] - \int_0^t \mathrm{d}t' \left[\hat{H}_{\mathrm{int}}(t), \left[\hat{H}_{\mathrm{int}}(t'), \hat{\rho}^{(I)}(t') \right] \right]. \end{aligned} \tag{4.18}$$

Since we have [see (A.12) for the derivation]

$$\hat{\rho}_{\mathcal{S}}^{(I)}(t) = \mathrm{Tr}_{\mathcal{E}} \left\{ \hat{\rho}^{(I)}(t) \right\}, \tag{4.19}$$

we can transform (4.18) into an equation for the reduced density operator $\hat{\rho}_{\mathcal{S}}^{(I)}(t)$ by simply taking the trace over the environment,

$$\frac{\mathrm{d}}{\mathrm{d}t}\hat{\rho}_{\mathcal{S}}^{(I)}(t) = -\mathrm{i}\,\mathrm{Tr}_{\mathcal{E}} \left[\hat{H}_{\mathrm{int}}(t), \hat{\rho}(0) \right] - \int_0^t \mathrm{d}t'\, \mathrm{Tr}_{\mathcal{E}} \left[\hat{H}_{\mathrm{int}}(t), \left[\hat{H}_{\mathrm{int}}(t'), \hat{\rho}^{(I)}(t') \right] \right]. \tag{4.20}$$

In fact, without loss of generality, one can assume that

$$\mathrm{Tr}_{\mathcal{E}} \left[\hat{H}_{\mathrm{int}}(t), \hat{\rho}(0) \right] = 0. \tag{4.21}$$

It is easy to show that this can always be achieved by a formal redefinition of the Hamiltonians $\hat{H}_0$ and $\hat{H}_{\mathrm{int}}$. Then (4.20) simplifies to

$$\frac{\mathrm{d}}{\mathrm{d}t}\hat{\rho}_{\mathcal{S}}^{(I)}(t) = - \int_0^t \mathrm{d}t'\, \mathrm{Tr}_{\mathcal{E}} \left[\hat{H}_{\mathrm{int}}(t), \left[\hat{H}_{\mathrm{int}}(t'), \hat{\rho}^{(I)}(t') \right] \right]. \tag{4.22}$$

All calculations leading up to this equation have been exact and no approximations have been made. Let us now pause for a moment and ask ourselves the following question. What features are still absent from (4.22) that would be desirable for a master equation for $\hat{\rho}_{\mathcal{S}}^{(I)}(t)$? Note that the right-hand side of (4.22) still depends on the total system–environment density operator $\hat{\rho}^{(I)}$ evaluated at all times between 0 and t. Thus (4.22) is not yet a time-local differential equation of the form (4.4) that would determine the change of $\hat{\rho}_{\mathcal{S}}^{(I)}$ at time t solely in terms of $\hat{\rho}_{\mathcal{S}}^{(I)}(t)$ and a set of known quantities. Therefore two desirable properties emerge:

(P1) We would like to express our master equation entirely in terms of the *reduced* density operator $\hat{\rho}_{\mathcal{S}}^{(I)}(t)$ and the initial state of the environment, i.e., we would like to eliminate any terms pertaining to a time-dependent state of the environment. A master equation that fulfills this property is called an *integro-differential equation*, since the differential change of the quantity of interest at a certain time t depends on an integral of a function of this quantity over all times $t' \leq t$.

(P2) We would like to eliminate any dependences of the change of $\hat{\rho}_{\mathcal{S}}^{(I)}$ at time t on $\hat{\rho}_{\mathcal{S}}^{(I)}$ evaluated at times $t' < t$.

The trick to accomplish these goals lies in the application of the Born and Markov approximations mentioned above. Imposing the Born approximation will eliminate the dependence on the full density operator $\hat{\rho}^{(I)}$ on the right-hand side of (4.22) and thus take care of (P1). The Markov approximation will then transform the resulting integro-differential master equation into a time-local differential equation, thereby achieving the goal stated in (P2). As we will discuss below, both approximations can be motivated from assumptions of weak system–environment couplings and of an environment that is large compared to the size of the system.

Before proceeding, let us note that in the following we will assume the absence of initial correlations between the system and the environment, i.e.,

$$\hat{\rho}(0) = \hat{\rho}^{(I)}(0) = \hat{\rho}_{\mathcal{S}}(0) \otimes \hat{\rho}_{\mathcal{E}}(0). \tag{4.23}$$

This is a reasonable assumption for the case of weak system–environment coupling considered here.

Imposing the Born Approximation

As already mentioned in Sect. 4.2.1 [see (4.6)], the Born approximation assumes that the density operator $\hat{\rho}$ of the system–environment combination remains at all times in an approximate product form and that temporal changes in the density matrix $\hat{\rho}_{\mathcal{E}}$ of the environment can be neglected,

$$\hat{\rho}(t) \approx \hat{\rho}_{\mathcal{S}}(t) \otimes \hat{\rho}_{\mathcal{E}} \quad \forall t \geq 0, \tag{4.24}$$

with $\hat{\rho}_{\mathcal{E}} = \hat{\rho}_{\mathcal{E}}(0)$. The equivalent condition then also holds in the interaction picture,

$$\hat{\rho}^{(I)}(t) \approx \hat{\rho}_{\mathcal{S}}^{(I)}(t) \otimes \hat{\rho}_{\mathcal{E}} \quad \forall t \geq 0. \tag{4.25}$$

This is easily seen from the fact that $\hat{\rho}^{(I)}(t) = \mathrm{e}^{\mathrm{i}\hat{H}_0 t}\hat{\rho}(t)\mathrm{e}^{-\mathrm{i}\hat{H}_0 t}$ [see (4.15)]. Since $\hat{H}_0 = \hat{H}_{\mathcal{S}} + \hat{H}_{\mathcal{E}}$ acts separately on the system and the environment, it cannot establish any entanglement between the two partners.

The physical intuition behind the Born approximation is that the interaction between the system and the environment is sufficiently *weak* (the so-called "weak-coupling limit"), and that the environment is *large* in comparison with the size of the system, such that (i) the density operator for the environment does not change significantly as a consequence of the interaction with the system, and that (ii) the system and the environment remain in a separable state at all times.

It turns out that the Born approximation is a reasonable assumption in many cases of physical interest. Usually, the system is coupled to a very large environment which, viewed as a whole, undergoes only negligibly small changes in the course of the system–environment interaction compared to the change of the state of the system. That is, the back-action of the system on the environment can in such cases be ignored. Note that (4.25) allows for arbitrarily large changes of the density operator of the system, and it does not exclude the occurrence of environmental excitations in the environment induced by the interaction with the system.

Using the Born approximation (4.25), the master equation (4.22) can now be written as

$$\frac{\mathrm{d}}{\mathrm{d}t}\hat{\rho}_{\mathcal{S}}^{(I)}(t) = -\int_0^t \mathrm{d}t' \, \mathrm{Tr}_{\mathcal{E}} \left[\hat{H}_{\mathrm{int}}(t), \left[\hat{H}_{\mathrm{int}}(t'), \hat{\rho}_{\mathcal{S}}^{(I)}(t') \otimes \hat{\rho}_{\mathcal{E}} \right] \right]. \tag{4.26}$$

Evidently (by virtue of the Born approximation), we have fulfilled our desired property (P1) stated above. Both sides of our master equation are now expressed entirely in terms of the reduced density operator $\hat{\rho}_{\mathcal{S}}^{(I)}(t)$, with the state of the environment separated out and reduced to the initial state $\hat{\rho}_{\mathcal{E}}$.

Imposing the Markov Approximation

The expression (4.26) is still an integro-differential equation, i.e., computing the change of $\hat{\rho}_{\mathcal{S}}^{(I)}$ at time t requires knowledge of $\hat{\rho}_{\mathcal{S}}^{(I)}$ at all previous times $t' < t$. However, as we shall now show, by imposing a suitable approximation (namely, the Markov approximation) we can transform (4.26) into a time-local master equation of the form (4.4).

Let us start by writing the interaction Hamiltonian in the diagonal form $\hat{H}_{\mathrm{int}} = \sum_\alpha \hat{S}_\alpha \otimes \hat{E}_\alpha$ [see (4.7)] and transforming it to the interaction picture,

$$\begin{aligned}
\hat{H}_{\mathrm{int}}(t) &= \mathrm{e}^{\mathrm{i}\hat{H}_0 t} \hat{H}_{\mathrm{int}} \mathrm{e}^{-\mathrm{i}\hat{H}_0 t} \\
&= \sum_\alpha \left(\mathrm{e}^{\mathrm{i}\hat{H}_{\mathcal{S}} t} \hat{S}_\alpha \mathrm{e}^{-\mathrm{i}\hat{H}_{\mathcal{S}} t} \right) \otimes \left(\mathrm{e}^{\mathrm{i}\hat{H}_{\mathcal{E}} t} \hat{E}_\alpha \mathrm{e}^{-\mathrm{i}\hat{H}_{\mathcal{E}} t} \right) \\
&= \sum_\alpha \hat{S}_\alpha(t) \otimes \hat{E}_\alpha(t).
\end{aligned} \tag{4.27}$$

Then (4.26) can be expressed as

$$\frac{\mathrm{d}}{\mathrm{d}t}\hat{\rho}_{\mathcal{S}}^{(I)}(t)$$
$$= -\int_0^t \mathrm{d}t' \sum_{\alpha\beta} \mathrm{Tr}_{\mathcal{E}} \left[\hat{S}_\alpha(t) \otimes \hat{E}_\alpha(t), \left[\hat{S}_\beta(t') \otimes \hat{E}_\beta(t'), \hat{\rho}_{\mathcal{S}}^{(I)}(t') \otimes \hat{\rho}_{\mathcal{E}}\right]\right]. \tag{4.28}$$

Let us now define

$$\mathcal{C}_{\alpha\beta}(t,t') \equiv \mathrm{Tr}_{\mathcal{E}} \left\{\hat{E}_\alpha(t)\hat{E}_\beta(t')\hat{\rho}_{\mathcal{E}}\right\} = \left\langle \hat{E}_\alpha(t)\hat{E}_\beta(t')\right\rangle_{\hat{\rho}_{\mathcal{E}}}. \tag{4.29}$$

In the following, we shall suppose that the environment is in a stationary state (i.e., in equilibrium),

$$\left[\hat{H}_{\mathcal{E}}, \hat{\rho}_{\mathcal{E}}\right] = 0. \tag{4.30}$$

This implies that we can write

$$\mathcal{C}_{\alpha\beta}(t,t') = \mathrm{Tr}_{\mathcal{E}} \left\{\hat{E}_\alpha(t-t')\hat{E}_\beta\hat{\rho}_{\mathcal{E}}\right\} \equiv \mathcal{C}_{\alpha\beta}(t-t'), \tag{4.31}$$

which are precisely the environment self-correlation functions (4.10) introduced above.

We see that, due to the assumption (4.30), these environment self-correlation functions do not depend on the absolute time t [as in (4.29)] but only on the time *interval* between the results of the two identical "measurements" (represented by a particular operator $\hat{E}_\alpha$) whose degree of correlation is quantified by (4.31).[1]

Inserting (4.31) into (4.28) and writing out the double commutator, we obtain

$$\frac{\mathrm{d}}{\mathrm{d}t}\hat{\rho}_{\mathcal{S}}^{(I)}(t) =$$
$$-\int_0^t \mathrm{d}t' \sum_{\alpha\beta} \left\{\mathcal{C}_{\alpha\beta}(t-t')\left[\hat{S}_\alpha(t)\hat{S}_\beta(t')\hat{\rho}_{\mathcal{S}}^{(I)}(t') - \hat{S}_\beta(t')\hat{\rho}_{\mathcal{S}}^{(I)}(t')\hat{S}_\alpha(t)\right]\right.$$
$$\left. + \mathcal{C}_{\beta\alpha}(t'-t)\left[\hat{\rho}_{\mathcal{S}}^{(I)}(t')\hat{S}_\beta(t')\hat{S}_\alpha(t) - \hat{S}_\alpha(t)\hat{\rho}_{\mathcal{S}}^{(I)}(t')\hat{S}_\beta(t')\right]\right\}. \tag{4.32}$$

In deriving this expression, we have used the basic fact that the trace is invariant under cyclic permutations of the operators, such that

$$\mathrm{Tr}_{\mathcal{E}} \left\{\hat{E}_\alpha\hat{E}_\beta\hat{\rho}_{\mathcal{E}}\right\} = \mathrm{Tr}_{\mathcal{E}} \left\{\hat{\rho}_{\mathcal{E}}\hat{E}_\alpha\hat{E}_\beta\right\} = \mathrm{Tr}_{\mathcal{E}} \left\{\hat{E}_\alpha\hat{\rho}_{\mathcal{E}}\hat{E}_\beta\right\}. \tag{4.33}$$

We have also separated out the operators $\hat{S}_\alpha(t)$ as well as the density operator $\hat{\rho}_{\mathcal{S}}^{(I)}(t')$ from the trace operation. We are allowed to do this because the trace is taken over the environment only.

[1] This simplification is not always appropriate. See, e.g., Sect. 3.4.3 of [18] for an interesting counterexample.

Let us now further simplify (4.33) by imposing the Markov approximation. It turns out that in many physical situation of interest, the environment can be assumed to very quickly "forget" any internal self-correlations that have been established in the course of the interaction with the system. In other words, in such situations the environment does not "keep track of its history"—any dynamically established quantum correlations between parts of the environment are destroyed on a timescale τ_{corr} much shorter than the characteristic timescale τ_S over which the reduced interaction-picture density operator $\hat{\rho}_S^{(I)}(t)$ of the system changes noticeably. This assumption $\tau_{\text{corr}} \ll \tau_S$ constitutes the Markov approximation. It is appropriate if, as in our case, the environment is only weakly coupled to the system, and if the environment is at a sufficiently high temperature.

Readers may be familiar with Markovian processes from classical probability theory (see Sect. 1.4 of [18] for a good introduction to this subject). Broadly speaking, a stochastic process is Markovian if the probability of a particular event is independent of all earlier events. That is, each step in the process is independent of the previous steps. In other words, the system retains no memory of its past—the steps are completely uncorrelated. The analogy with a "Markovian environment" should now be clear.

What are the consequences of the Markov approximation for our goal of obtaining a time-local master equation in differential form? Let us look back at the current structure (4.32) of the master equation. Applied to our derivation, the Markov approximation means that the environment self-correlations functions $\mathcal{C}_{\alpha\beta}(t-t')$ [see (4.31)] are sharply peaked around $(t-t')=0$ and decay on a timescale much shorter than the timescale set by τ_S, which quantifies the rate of change of the interaction-picture density operator $\hat{\rho}_S^{(I)}(t)$ of the system. The implications of this property are twofold.

First, since the reduced interaction-picture density operator $\hat{\rho}_S^{(I)}(t)$ of the system changes only insignificantly during the typical time interval τ_{corr} over which the environment self-correlations functions $\mathcal{C}_{\alpha\beta}(t-t')$ vanish, we can replace the retarded-time density operator $\hat{\rho}_S^{(I)}(t')$ by the current-time density operator $\hat{\rho}_S^{(I)}(t)$ in the integrand on the right-hand side of (4.32). The resulting master equation is the so-called *Redfield equation* [174, 175], which fulfills our desired property (P2) of being time-local. In fact, the Markov assumption allows us to simplify the master equation even further. Namely, we realize that, for $t \gg \tau_{\text{corr}}$, we can safely extend the lower limit of the integration on the right-hand side of (4.32) to $-\infty$, since the self-correlation functions $\mathcal{C}_{\alpha\beta}(t-t')$ (and thus the integrand) vanish for $t' \ll t$.

With these two modifications (which are both based on the Markov approximation of negligibly short environmental self-correlation times) and the substitution $t' \longrightarrow \tau \equiv t-t'$, (4.32) becomes

$$
\begin{aligned}
&\frac{\mathrm{d}}{\mathrm{d}t}\hat{\rho}_{S}^{(I)}(t) = \\
&\quad - \int_0^{\infty} \mathrm{d}\tau \sum_{\alpha\beta} \Big\{ \mathcal{C}_{\alpha\beta}(\tau) \Big[\hat{S}_\alpha(t)\hat{S}_\beta(t-\tau)\hat{\rho}_{S}^{(I)}(t) - \hat{S}_\beta(t-\tau)\hat{\rho}_{S}^{(I)}(t)\hat{S}_\alpha(t) \Big] \\
&\qquad + \mathcal{C}_{\beta\alpha}(-\tau) \Big[\hat{\rho}_{S}^{(I)}(t)\hat{S}_\beta(t-\tau)\hat{S}_\alpha(t) - \hat{S}_\alpha(t)\hat{\rho}_{S}^{(I)}(t)\hat{S}_\beta(t-\tau) \Big] \Big\} . \qquad (4.34)
\end{aligned}
$$

As desired, the right-hand side of this master equation does no longer depend on the full time-dependent density operator $\hat{\rho}^{(I)}(t)$, but only on the *reduced* density operator $\hat{\rho}_{S}^{(I)}(t)$ of the system, together with the *initial* state of the environment [which enters through the self-correlation functions $\mathcal{C}_{\alpha\beta}(\tau)$, see (4.31)]. Also, the master equation is local in time. Only completely known quantities, namely, the operators $\hat{S}_\alpha(t)$ and $\hat{E}_\alpha(t)$ (which are determined *a priori* by the form of the Hamiltonian), enter into the master equation with time arguments other than t.

Equation (4.34) establishes our core result, and our derivation is essentially complete. The last remaining step is simply concerned with transforming the equation back to the Schrödinger picture and with making some formal rearrangements.

Transformation Back to the Schrödinger Picture

To transform a density-operator evolution equation written in the interaction picture back to the Schrödinger picture, we can use the following general recipe. From (A.12) we have

$$
\hat{\rho}_{S}^{(I)}(t) = \mathrm{e}^{\mathrm{i}\hat{H}_S t}\hat{\rho}_S(t)\mathrm{e}^{-\mathrm{i}\hat{H}_S t}, \qquad (4.35)
$$

and thus it immediately follows that

$$
\frac{\mathrm{d}}{\mathrm{d}t}\hat{\rho}_{S}^{(I)}(t) = \mathrm{i}\left[\hat{H}_S, \hat{\rho}_{S}^{(I)}(t)\right] + \mathrm{e}^{\mathrm{i}\hat{H}_S t}\left(\frac{\mathrm{d}}{\mathrm{d}t}\hat{\rho}_S(t)\right)\mathrm{e}^{-\mathrm{i}\hat{H}_S t}. \qquad (4.36)
$$

Again using (4.35), this may be rewritten as

$$
\frac{\mathrm{d}}{\mathrm{d}t}\hat{\rho}_S(t) = -\mathrm{i}\left[\hat{H}_S, \hat{\rho}_S(t)\right] + \mathrm{e}^{-\mathrm{i}\hat{H}_S t}\left(\frac{\mathrm{d}}{\mathrm{d}t}\hat{\rho}_{S}^{(I)}(t)\right)\mathrm{e}^{\mathrm{i}\hat{H}_S t}. \qquad (4.37)
$$

Let us apply this procedure to our interaction-picture Born–Markov master equation (4.34). We insert the expression (4.34) for $\frac{\mathrm{d}}{\mathrm{d}t}\hat{\rho}_{S}^{(I)}(t)$ into the right-hand side of (4.37). We pull the time evolution operators $\mathrm{e}^{\pm\mathrm{i}\hat{H}_S t}$ into the square brackets in (4.34) and then use (4.14) and (4.35). For example, for the first term inside the brackets on the right-hand side of (4.34), this will lead to

$$
\begin{aligned}
&\mathrm{e}^{-\mathrm{i}\hat{H}_S t}\hat{S}_\alpha(t)\hat{S}_\beta(t-\tau)\hat{\rho}_S^{(I)}(t)\mathrm{e}^{\mathrm{i}\hat{H}_S t} \\
&\quad= \left(\mathrm{e}^{-\mathrm{i}\hat{H}_S t}\hat{S}_\alpha(t)\mathrm{e}^{\mathrm{i}\hat{H}_S t}\right)\left(\mathrm{e}^{-\mathrm{i}\hat{H}_S t}\hat{S}_\beta(t-\tau)\mathrm{e}^{\mathrm{i}\hat{H}_S t}\right)\left(\mathrm{e}^{-\mathrm{i}\hat{H}_S t}\hat{\rho}_S^{(I)}(t)\mathrm{e}^{\mathrm{i}\hat{H}_S t}\right) \\
&\quad= \hat{S}_\alpha\left(\mathrm{e}^{-\mathrm{i}\hat{H}_S \tau}\hat{S}_\beta\mathrm{e}^{\mathrm{i}\hat{H}_S \tau}\right)\hat{\rho}_S \\
&\quad= \hat{S}_\alpha\hat{S}_\beta(-\tau)\hat{\rho}_S.
\end{aligned} \tag{4.38}
$$

In this way, (4.34) can now be written as (reintroducing the commutator notation)

$$
\begin{aligned}
\frac{\mathrm{d}}{\mathrm{d}t}\hat{\rho}_S(t) = -\mathrm{i}\left[\hat{H}_S,\hat{\rho}_S(t)\right] - \int_0^\infty \mathrm{d}\tau \sum_{\alpha\beta}\Big\{&\mathcal{C}_{\alpha\beta}(\tau)\left[\hat{S}_\alpha,\hat{S}_\beta(-\tau)\hat{\rho}_S(t)\right] \\
&+\mathcal{C}_{\beta\alpha}(-\tau)\left[\hat{\rho}_S(t)\hat{S}_\beta(-\tau),\hat{S}_\alpha\right]\Big\}.
\end{aligned} \tag{4.39}
$$

We can simplify our notation by realizing that the integral on the right-hand side of (4.39) extends over the entire range $\tau = 0\ldots\infty$, and that therefore the overall time dependence of the environment self-correlation functions $\mathcal{C}_{\alpha\beta}(\tau)$ and of the interaction-picture operators $\hat{S}_\beta(-\tau)$ is effectively integrated out. Using the time-independent quantities $\hat{B}_\alpha$ and $\hat{C}_\alpha$ introduced in (4.9),

$$
\hat{B}_\alpha = \int_0^\infty \mathrm{d}\tau \sum_\beta \mathcal{C}_{\alpha\beta}(\tau)\hat{S}_\beta(-\tau), \tag{4.40a}
$$

$$
\hat{C}_\alpha = \int_0^\infty \mathrm{d}\tau \sum_\beta \mathcal{C}_{\beta\alpha}(-\tau)\hat{S}_\beta(-\tau), \tag{4.40b}
$$

the Born–Markov master equation (4.39) then takes the final form (4.8) given in Sect. 4.2, i.e.,

$$
\frac{\mathrm{d}}{\mathrm{d}t}\hat{\rho}_S(t) = -\mathrm{i}\left[\hat{H}_S,\hat{\rho}_S(t)\right] - \sum_\alpha\left\{\left[\hat{S}_\alpha,\hat{B}_\alpha\hat{\rho}_S(t)\right] + \left[\hat{\rho}_S(t)\hat{C}_\alpha,\hat{S}_\alpha\right]\right\}. \tag{4.41}
$$

This completes our derivation. In Chap. 5, we shall illustrate the application of this master equation in the context of several important decoherence models.

4.3 Master Equations in the Lindblad Form

Master equations in the so-called *Lindblad* form refer to a particular (albeit quite general) class of Markovian master equations. They arise from the requirement that the master equation ought to ensure the *positivity* of the reduced density matrix at all times, i.e., that

$$
\langle\psi|\,\hat{\rho}_S(t)\,|\psi\rangle \geq 0 \tag{4.42}
$$

for any pure state $|\psi\rangle$ of the system $\mathcal{S}$ and for all t. This condition is physically reasonable, since we would like to interpret the elements $\langle\psi|\,\hat{\rho}_{\mathcal{S}}(t)\,|\psi\rangle$ of the reduced density matrix as occupation probabilities. If the time evolution of $\hat{\rho}_{\mathcal{S}}(t)$ is exact, the positivity condition will of course be automatically fulfilled. However, for reduced density matrices evolved by means of approximate master equations, (4.42) will not necessarily hold.

It was first shown by Gorini, Kossakowski, and Sudarshan [176] and Lindblad [177] that the most general master equation that ensures the positivity (4.42) of the reduced density matrix $\hat{\rho}_{\mathcal{S}}(t)$ is of the form

$$\frac{\mathrm{d}}{\mathrm{d}t}\hat{\rho}_{\mathcal{S}}(t) = -\mathrm{i}\left[\hat{H}_{\mathcal{S}}, \hat{\rho}_{\mathcal{S}}(t)\right] + \frac{1}{2}\sum_{\alpha\beta}\gamma_{\alpha\beta}\left\{\left[\hat{S}_{\alpha}, \hat{\rho}_{\mathcal{S}}(t)\hat{S}_{\beta}^{\dagger}\right] + \left[\hat{S}_{\alpha}\hat{\rho}_{\mathcal{S}}(t), \hat{S}_{\beta}^{\dagger}\right]\right\}. \tag{4.43}$$

Here, the operators $\hat{S}_{\alpha}$ are the system operators appearing in the diagonal decomposition $\hat{H}_{\mathrm{int}} = \sum_{\alpha}\hat{S}_{\alpha}\otimes\hat{E}_{\alpha}$ of the interaction Hamiltonian [see (4.7)]. The time-independent coefficients $\gamma_{\alpha\beta}$ encapsulate all information about the physical parameters of the decoherence (and possibly dissipation) processes and define a coefficient matrix $\Gamma \equiv (\gamma_{\alpha\beta})$.

Evidently, the Lindblad master equation (4.43) is local in time. Moreover, it is also Markovian. Equations of the form (4.43) can be derived in many different ways from various assumptions and are therefore not necessarily tied to the Born–Markov master equation (4.8). Master equations of the Lindblad form, derived from phenomenological models, were used by many authors already in the 1960s (see, e.g., [178]). However, their rigorous mathematical derivation within the theory of generators and quantum dynamical semigroups was established only in the second half of the 1970s [176,177,179–181].

If we compare (4.43) to our general expression (4.8) for the Born–Markov master equation, we observe a certain formal similarity between the two equations. Yet, evidently, they are not the same. In particular, the operators $\hat{B}_{\alpha}$ and $\hat{C}_{\alpha}$ appearing in (4.8) are in general nontrivial time integrals over the interaction operators $\hat{S}_{\alpha}$ and the environment self-correlation functions [see (4.9)]. Still, the reader may suspect that it could be possible to transform (4.8) into the Lindblad form (4.43). However, it turns out that this cannot always be done: The Lindblad master equation is a special case of the general Born–Markov master equation. It follows that density matrices evolved by the latter equation do not necessarily fulfill the condition (4.42) of positivity [182].

It is, however, possible to bring (4.8) into the Lindblad form (4.43) if another assumption (in addition to the Born–Markov approximations) is imposed. This assumption is often called the *rotating-wave approximation*, which is commonly made in quantum optics. This assumption is justified whenever the typical timescale $\tau_{\mathcal{S}}$ for the evolution of the system is short in comparison with the relaxation timescale of the system. For details on how the rotating-wave approximation is used to transform the Born–Markov master equation into the Lindblad form, we refer the reader to Sect. 3.3.1 of [18].

Conversely, the inability of writing a master equation in the Lindblad form does *not* necessarily imply that the positivity condition (4.42) is violated. A good example is the model for quantum Brownian motion, which we shall discuss in Sect. 5.2. For this model, it is possible to derive the exact, non-Markovian master equation (see Sect. 5.2.7). This equation cannot be brought into the Lindblad form without imposing several approximations and assumptions [183]. Yet, it is clear that, since the master equation is exact, it should automatically enforce the positivity of the density matrix at all times. As it turns out, this is indeed the case. The explanation for this initially puzzling behavior can be found through a careful analysis of the exact dynamics, which shows that in the exact master equation positivity arises through rather subtle time dependences of the coefficients appearing in the master equation [184, 185].

We can simplify the Lindblad master equation (4.43) further by diagonalizing the matrix Γ defined by the coefficients $\gamma_{\alpha\beta}$ appearing in this equation. This diagonalization is always possible because, as one can show, this matrix is positive (i.e., all its eigenvalues κ_μ are ≥ 0). Then (4.43) can be rewritten in the diagonal form [177, 179]

$$\frac{\mathrm{d}}{\mathrm{d}t}\hat{\rho}_S(t) = -\mathrm{i}\left[\hat{H}'_S, \hat{\rho}_S(t)\right] - \frac{1}{2}\sum_\mu \kappa_\mu \left\{\hat{L}^\dagger_\mu \hat{L}_\mu \hat{\rho}_S(t) + \hat{\rho}_S \hat{L}^\dagger_\mu \hat{L}_\mu - 2\hat{L}_\mu \hat{\rho}_S(t)\hat{L}^\dagger_\mu\right\}. \tag{4.44}$$

Here $\hat{H}'_S$ denotes the renormalized ("Lamb-shifted") Hamiltonian of the system. The so-called Lindblad operators (or "Lindblad generators") $\hat{L}_\mu$ are simply appropriate linear combinations of the original operators $\hat{S}_\alpha$, with coefficients determined from the diagonalization of the matrix Γ.

Note that, because the operators $\hat{S}_\alpha$ are not necessarily Hermitian, the Lindblad operators $\hat{L}_\mu$ do not always correspond to physical observables, but when they do, we can rewrite (4.44) in the more compact double-commutator form

$$\frac{\mathrm{d}}{\mathrm{d}t}\hat{\rho}_S(t) = -\mathrm{i}\left[\hat{H}'_S, \hat{\rho}_S(t)\right] - \frac{1}{2}\sum_\mu \kappa_\mu \left[\hat{L}_\mu, \left[\hat{L}_\mu, \hat{\rho}_S(t)\right]\right]. \tag{4.45}$$

In fact, our equation of motion (3.78) for environmental scattering,

$$\frac{\partial \rho_S(x,x',t)}{\partial t} = -\frac{\mathrm{i}}{2m}\left(\frac{\partial^2}{\partial x'^2} - \frac{\partial^2}{\partial x^2}\right)\rho_S(x,x',t) - \Lambda(x-x')^2\rho_S(x,x',t), \tag{4.46}$$

is of the Lindblad form (4.45). To see this, let us evaluate (4.45) with the single Lindblad operator $\hat{L} = \hat{x}$ and the "free"-particle Hamiltonian $\hat{H}'_S = \hat{H}_S = \hat{p}^2/2m$,

$$\frac{\mathrm{d}}{\mathrm{d}t}\hat{\rho}_S(t) = -\frac{\mathrm{i}}{2m}\left[\hat{p}^2, \hat{\rho}_S(t)\right] - \frac{1}{2}\kappa\left[\hat{x}, \left[\hat{x}, \hat{\rho}_S(t)\right]\right]. \tag{4.47}$$

Expressing this master equation in the position representation then yields

$$\frac{\partial \rho_S(x,x',t)}{\partial t} = -\frac{\mathrm{i}}{2m}\left(\frac{\partial^2}{\partial x'^2} - \frac{\partial^2}{\partial x^2}\right)\rho_S(x,x',t) - \frac{1}{2}\kappa\left(x-x'\right)^2\rho_S(x,x',t), \tag{4.48}$$

which is evidently identical to (4.46) if we identify the Lindblad coefficient $\kappa/2$ with the scattering constant Λ, see (3.56).

Lindblad master equations provide us with an intuitive and rather simple way of representing the environmental monitoring of an open quantum system. Most of the real physics behind this monitoring process is hidden in the coefficients κ_μ appearing in (4.44). In fact, if the Lindblad operators are chosen to be dimensionless, the coefficients κ_μ can be viewed as directly representing decoherence rates, since they have units of inverse time. Accordingly, usually the most important and difficult task of estimating decoherence rates using the Lindblad equation consists precisely of determining these coefficients κ_μ.

From (4.45) we immediately see that the decoherence term vanishes if $\hat{\rho}_S(t)$ commutes with each of the Lindblad operators $\hat{L}_\mu$ for all times t,

$$\left[\hat{L}_\mu, \hat{\rho}_S(t)\right] = 0 \qquad \forall \mu, t. \tag{4.49}$$

In this case, $\hat{\rho}_S(t)$ evolves purely unitarily. Since the $\hat{L}_\mu$ are simply linear combinations of the $\hat{S}_\alpha$, (4.49) (typically[2]) implies that

$$\left[\hat{S}_\alpha, \hat{\rho}_S(t)\right] = 0 \qquad \forall \alpha, t. \tag{4.50}$$

The connection to the concept of preferred pointer states (see Sect. 2.8) should now be rather clear. Equation (4.50) implies that simultaneous eigenstates of all $\hat{S}_\alpha$ will be immune to decoherence, which is precisely the pointer-state criterion (2.94).

Finally, let us briefly mention so-called *quantum-jump* and *quantum-trajectory* methods. Here, the evolution of the reduced density matrix is *conditioned* on an explicitly observed sequence of measurement results in the environment. This allows for the (formal) description of a single realization of the system evolving stochastically, conditioned on a particular measurement record. The dynamics are then described by a master equation of the Lindblad type (4.45) for the reduced density matrix $\hat{\rho}_S^C$ conditioned on the measurement records of the Lindblad operators $\hat{L}_\mu$,

$$\mathrm{d}\hat{\rho}_S^C = -\mathrm{i}\left[\hat{H}_S, \hat{\rho}_S^C\right]\mathrm{d}t - \frac{1}{2}\sum_\mu \kappa_\mu\left[\hat{L}_\mu, \left[\hat{L}_\mu, \hat{\rho}_S^C\right]\right]\mathrm{d}t + \sum_\mu \sqrt{\kappa_\mu}\,\mathcal{W}[\hat{L}_\mu]\hat{\rho}_S^C\,\mathrm{d}W_\mu. \tag{4.51}$$

[2] One might of course devise some rather pathological linear combinations of the $\hat{S}_\alpha$ for a particular $\hat{L}_\mu$ such that the commutator (4.49) happens to vanish while the individual commutators (4.50) are nonzero.

Here, $\mathcal{W}[\hat{L}]\hat{\rho} \equiv \hat{L}\hat{\rho}+\hat{\rho}\hat{L}^\dagger - \hat{\rho}\,\mathrm{Tr}\left\{\hat{L}\hat{\rho}+\hat{\rho}\hat{L}^\dagger\right\}$, and the $\mathrm{d}W_\mu$ denote the so-called Wiener increments. Equation (4.51) corresponds to a diffusive "unraveling" of the Lindblad equation (4.45) into individual quantum trajectories that can then be expressed by means of a "stochastic Schrödinger equation." We shall not go into the details here—there exists a large volume of literature on this topic (see, e.g., [186–198]).

4.4 Non-Markovian Dynamics

As we have emphasized before, the Born–Markov master equation is only strictly applicable if the system–environment coupling is weak and memory effects of the environment can be neglected. In many situations of physical interest, however, this assumption is not fulfilled because of low temperatures of the environment and strong interactions of the system with its environment. A concrete example of such non-Markovian dynamics is that of a superconducting qubit strongly coupled to a low-temperature environment of other two-level systems, a situation in which making the Born–Markov approximations is no longer justified [199, 200].

Intuitively, one might expect that pronounced memory effects in the environment will cause strong dependences of the evolution of the reduced density operator on the past history of the full system–environment combination and thus make it impossible to describe the reduced dynamics by a differential equation that is local in time. In some situations this is indeed the case, and one will then, in general, have to resort to solving integro-differential equations such as (4.22), which depend on retarded-time kernels and integrations over the history of the system. Such equations are encountered, for example, in the context of the so-called Nakajima–Zwanzig projection-operator approach [201–203] (see also Chap. 7 of [17] for details). Such techniques lead to equations that are not local in time and are difficult to solve analytically, and they are therefore rarely used in their exact forms.

Surprisingly, however, in many other cases one can show that even non-Markovian dynamics can still be described by a differential equation of the form

$$\frac{\mathrm{d}}{\mathrm{d}t}\hat{\rho}_S(t) = \hat{\mathcal{K}}(t)\hat{\rho}_S(t). \tag{4.52}$$

In contrast with the Born–Markov master equation (4.8), the superoperator $\hat{\mathcal{K}}(t)$ depends here explicitly on t, but there are no dependences on previous times $t' < t$. In particular, the change in $\hat{\rho}_S$ is entirely determined by $\hat{\rho}_S$ evaluated at the same time t only, and no integration over the history of the system in the equation of motion is required. Therefore the master equation (4.52) is still local in time.

Time-local non-Markovian master equations of the form (4.52) can be arrived at in various ways. For example, in the case of quantum Brownian motion (see Sect. 5.2), one can derive a master equation that is non-Markovian (but still assumes the weak-coupling limit) in much the same way as the equation deduced from the Born–Markov master equation (4.8). The two master equations differ only in the range of the time integral appearing in the expressions (4.9) for the operators $\hat{B}_\alpha$ and $\hat{C}_\alpha$. In the Markovian case, the upper limit of the integral is taken to infinity, leading to time-independent expressions for $\hat{B}_\alpha$ and $\hat{C}_\alpha$. Instead, one can let the integral extend from zero to t only, resulting in explicitly time-dependent operators $\hat{B}_\alpha(t)$ and $\hat{C}_\alpha(t)$. With this formally simple modification, a "pre-Markovian" master equation for quantum Brownian motion emerges that exhibits more fine-grained dynamics, especially in the initial regime close to $t = 0$. (This equation is of the Redfield type mentioned in Sect. 4.2.2.)

In fact, for the case of quantum Brownian motion it is possible to derive the *exact* non-Markovian master equation (see Sect. 5.2.7), which has the same structure as the Markovian master equation save for more complicated expressions for the environment self-correlation functions (4.10). The existence of an exact solution of the quantum Brownian motion model also enables us to directly compare the resulting dynamics to the dynamics obtained from the approximate Born–Markov master equation. In this way, we can investigate to what extent the Born–Markov assumptions may constitute a reasonable set of approximations for a given physical situation described by the model for quantum Brownian motion. The issue of Markovian vs. non-Markovian dynamics in quantum Brownian motion will be discussed in more detail in Sects. 5.2.2 and 5.2.7 (see also [183]). In general, it is often possible to arrive at non-Markovian but time-local master equations of the type (4.52) by using the so-called time-convolutionless projection operator technique [204–207]. We shall not go into the details of this approach here.

We conclude that the Born–Markov master equation (4.8), especially when written in the simple Lindblad forms (4.44) or (4.45), allows for a fairly straightforward calculation of the approximate decoherence dynamics and for an intuitive connection between the formalism (the Lindblad operators) and its physical interpretation (the continuous monitoring of system observables by the environment). However, in each physical situation one needs to carefully evaluate to what extent the assumptions underlying the derivation of this master equation, such as weak system–environment couplings and negligible environmental memory effects, are actually appropriate. In the next Chap. 5, when discussing concrete models, we will frequently come back to this issue.

5 A World of Spins and Oscillators: Canonical Models for Decoherence

Needless to say, there are countless different physical systems in nature for which decoherence plays an important role. The task of modeling decoherence in each and every of these systems would therefore appear rather daunting. Each time we are confronted with a different physical system, it would seem that we would need to start from scratch in developing an appropriate system–environment model and in determining its dynamics. How could we accomplish this modeling task given the complexity of the many-body systems and interactions typically encountered on mesoscopic and macroscopic scales?

Fortunately, a very convenient simplification comes to our rescue here. It turns out that in many (if not most) situations of practical interest, the central system interacting with an environment can be mapped onto a very small set of *canonical models*. In such models, the central system is represented either by a particle described by continuous phase-space coordinates X and P moving in some potential (e.g., a harmonic-oscillator potential), or by a spin-$\frac{1}{2}$ particle if the state space of the system is discrete and effectively two-dimensional. Similarly, the environment is modeled as a collection of harmonic oscillators or spin particles. Accordingly, decoherence models in which the environment is represented by harmonic oscillators are commonly referred to as *oscillator-environment models*, and models based on an environment of spin particles are called *spin-environment models*. (A quick note on terminology: In the following, the expression "thermal bath," or "bath" for short, will refer to an environment in thermal equilibrium.)

Since there are four different possible ways in which these choices for the representation of the system and the environment can be combined, we thus obtain a total of four main canonical models for decoherence. Canonical models play an enormously important role in studies of decoherence, since they are very general: The details of a given physical system in contact with a specific physical environment are largely encapsulated in the particular set of parameters used to evaluate the model. Thus once the model is solved for an arbitrary choice of parameters, it can rather easily be applied to a whole host of systems in nature by simply choosing the parameters in such a way as to match the physical situation of interest.

Each of the canonical models gives rise to decoherence dynamics that are unique to the model for some parameter regime and may approximate another model for a different parameter regime. For example, in the limit of weak couplings between the system and the environment, spin-environment models can be mapped onto oscillator-environment models [199, 208] (see Sect. 5.4.2 below).

This chapter is organized as follows. In Sect. 5.1, we shall discuss the physical motivations and assumptions that underlie the mapping of physical systems onto canonical models. Then, in Sect. 5.2, we will describe the first of the canonical models that it is very important to the description of decoherence and the quantum-to-classical transition in the everyday world around us. This is the model for quantum Brownian motion. It describes a particle represented by continuous phase-space coordinates and coupled to an environment of noninteracting harmonic oscillators.

In Sect. 5.3, we then go on to replace the central system with an effective two-level system. The resulting model has become known as the so-called *spin–boson model*, since here a central spin system interacts with a bosonic environment of harmonic oscillators. We will analytically solve a simplified version of this model and also derive the Born–Markov master equation for the general model.

In Sect. 5.4, we will turn to spin-environment models. As we shall discuss below, these models become particularly relevant at very low temperatures such as those encountered in superconducting systems. We shall focus on the physically most relevant case of a central two-level system (represented by a spin-$\frac{1}{2}$ particle) interacting with an environment of other spins. We will also describe the mapping of spin environments onto oscillator environments.

The literature on the different decoherence models and their application to concrete physical situations has become so vast that it would be hopeless to even attempt to capture this field in its entirety. Nonetheless, this chapter will provide readers with a good overview of the main decoherence models and their physical relevance and demonstrate some of the standard techniques employed in solving these models. Inevitably, the discussion will occasionally be somewhat technical, so readers may pick and choose the preferred level of detail with which to study the text. Those who would like to focus on only one of the canonical models to begin with and later come back to the other models are encouraged to start with the model for quantum Brownian motion described in Sect. 5.2. This model not only is of crucial importance in studies of decoherence, but will also allow us to discuss many important physical, formal, and conceptual aspects of decoherence. We will see the Born–Markov master equation formalism in action, analyze the physical meaning of the different terms in the resulting master equation, describe the role and properties of spectral densities, illustrate the dynamics of decoherence and the environment-induced superselection of preferred states in phase space, and discuss general limitations of such decoherence models.

5.1 Mapping onto Canonical Models

First, let us describe some of the physical ideas behind the mapping of the central system (Sect. 5.1.1) and the environment (Sect. 5.1.2) onto harmonic oscillators or spin-$\frac{1}{2}$ particles.

5.1.1 Mapping of the Central System

In many cases, the relevant dynamics of the central system can be described by one or two coordinates of interest, even if the full microscopic structure of the dynamics of system is much more complicated. These coordinates may be either continuous or discrete.

In the case of *continuous* coordinates, the typical situation of physical interest is that of a mass point described by phase-space coordinates $\hat{X}$ (position) and $\hat{P}$ (momentum) and moving in some potential $V(\hat{X})$. In studies of decoherence, one often considers a harmonic-oscillator potential $V(\hat{X}) = M\Omega^2\hat{X}^2/2$, representing the motion of a particle, such as an atom or ion, confined to a "trap," for instance, a crystal lattice or magnetic field. The free-particle case $V(\hat{X}) = 0$ is also of obvious relevance to the quantum-to-classical transition.

The selection of position and momentum as the relevant "classical" coordinates is a consequence of environmental interactions, namely, of environment-induced superselection (see Sect. 2.8). In many important cases, such as quantum Brownian motion (Sect. 5.2), the evolution of the central system is governed in roughly equal strengths by the self-Hamiltonian $\hat{H}_{\mathcal{S}}$ of the system and the system–environment interaction Hamiltonian $\hat{H}_{\text{int}}$. Most interactions in nature are dependent on the distance between the two interacting partners, and thus $\hat{H}_{\text{int}}$ will typically be diagonal in position (see also Sect. 2.8.4). On the other hand, the kinetic-energy term $M\hat{P}^2/2$ in $\hat{H}_{\mathcal{S}}$ is diagonal in momentum. As we shall discuss in Sect. 5.2.5, the pointer states of the system then emerge as a compromise between complete localization in position space and complete localization in momentum space, leading to states that are approximately localized in phase space [86, 106, 109].

The physically most relevant situation involving *discrete* coordinates is that of a central system effectively acting as a two-level system. This typically happens if the system has two energy minimums separated by a barrier and if the energy of the system is sufficiently low (i.e., close to the energy minimums and much less than the barrier height). Then the dynamics of the system are effectively those of a particle confined to a double-well potential (Fig. 5.1). In many physical situations, besides the ground state only one or two excited states (if any) are contained in each well, and these excited states are usually not populated. Thus we obtain an effective two-state system, with the two states corresponding to localization in the "left" and "right" wells, respectively. The two levels can then in turn be mapped onto the two quantum states $|0\rangle$ ("spin up") and $|1\rangle$ ("spin down") of a spin-$\frac{1}{2}$

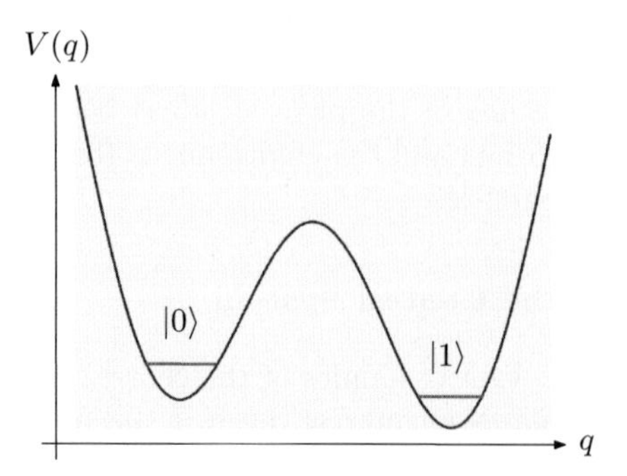

Fig. 5.1. Mapping onto two-state systems. We consider a system described by an extended coordinate q and moving in a double-well potential $V(q)$. In many cases, the system occupies only the lowest-lying state in each well and can therefore be effectively described by two basis states $|0\rangle$ and $|1\rangle$, corresponding to localization in the left and right well, respectively.

particle. Such two-level systems play a particularly important role in quantum computing, where they physically represent the qubits used to encode quantum information (see Chap. 7).

Intrinsic dynamics of the two-level system are created by quantum tunneling between the wells, which results in transitions between the two states $|0\rangle$ and $|1\rangle$. Taking the states $|0\rangle$ and $|1\rangle$ as the eigenstates of the Pauli z-spin operator $\hat{\sigma}_z$, tunneling is therefore mediated by a term in the system's Hamiltonian of the form

$$\hat{H}_{\mathcal{S}}^{\text{tunnel}} = -\frac{1}{2}\Delta_0 \left(|0\rangle\langle 1| + |1\rangle\langle 0|\right) \equiv -\frac{1}{2}\Delta_0 \hat{\sigma}_x, \tag{5.1}$$

where Δ_0 is the so-called tunneling matrix element. The difference in energy (the so-called "asymmetry energy") between the ground states in each of the two wells is often small in comparison with the energy spacing between the states in a given well, such that the low-lying energy levels in each well are approximately lined up. This ensures that a "particle" that is localized in one well—say, the left well corresponding to the state $|0\rangle$—and tunnels through the barrier usually cannot reach an excited level $|1'\rangle \neq |1\rangle$ in the right well [209].

5.1.2 Mapping of the Environment

Let us now turn to the mapping of the environment. As mentioned above, two situations of interest arise here: A description of the environment as a set of harmonic oscillators and as a collection of spin-$\frac{1}{2}$ particles representing quantum two-level systems.

Oscillator Environments

The representation of environments by a large number of harmonic oscillators has a long history going back to quantum electrodynamics and to spin-wave and electron-gas models (see [210] for an overview). Oscillator environments correspond to a quasicontinuum of *delocalized* bosonic field modes, with coherence and energy from the central system becoming effectively irreversibly lost into this extended bosonic environment. By "delocalized modes" we mean to convey the notion that the wave function of each harmonic oscillator (i.e., of each bosonic field mode) is spread out over a large spatial region. This delocalization of the environmental modes is a characteristic property of harmonic-oscillator environments.

Oscillator environments play an enormously important role in the modeling of decoherence processes, mainly because such environments are of great generality. It can be shown that, at sufficiently low energies, a large class of interacting system–environment compounds can effectively be represented by one or two coordinates of the system linearly coupled to an environment of harmonic oscillators. In fact, as was originally demonstrated by Feynman and Vernon [211] (see also the work by Caldeira and Leggett [212]), the interaction with *any* environment can be rigorously mapped onto a system (linearly) coupled to an oscillator environment, provided the interaction is sufficiently weak and second-order perturbation theory can be applied. (We will discuss this mapping in more detail in Sect. 5.4.2 below.)

Spin Environments

Spin environments are typically the appropriate model in the low-temperature setting. Of course, such low temperatures are usually not encountered in the everyday world around us. However, they are routinely attained in the laboratory and are crucial to many experiments. In particular, experiments devoted to the studies of macroscopic quantum coherence and decoherence (see Chap. 6)—most notably, those involving superconducting systems (Sect. 6.3)—require temperatures close to absolute zero in order to operate. Such experiments are also of great interest with respect to the realization of qubit systems that could be used to implement quantum computers (Chap. 7).

It is therefore very important to be able to correctly model decoherence in the low-temperature regime. Experimental evidence shows that in this regime decoherence is typically dominated by interactions with *localized* modes, such as paramagnetic spins, paramagnetic electronic impurities, tunneling charges, defects, and nuclear spins [199, 200, 213] (see also Sects. 6.3 and 6.4.2). The localization of these modes means that, in contrast with oscillator environments, the wave function associated with each of these modes is confined to a small region in space.

Each of the localized modes is described by a finite-dimensional Hilbert space with a finite energy cutoff. Thus we can model these modes as a set of discrete states, which can be mapped onto a spin system. In most cases of interest, there are only two such states of interest, so we can map our environment of localized modes onto an environment of spin-$\frac{1}{2}$ particles. The spin-environment modes may be part either of the physical system of interest itself (thus forming an "internal environment") or of the surrounding substrate. For example, in so-called superconducting quantum interference devices (SQUIDs; see Sect. 6.3 for details) the flux coordinate couples electromagnetically not only to the huge collection of nuclear spins that make up the bulk matter of the SQUID and are located within the penetration depth of the flux, but also to any impurities in the material [214].

There are numerous other examples that show that spin environments are a significant, and often dominant, source of decoherence. When researchers try to create coherent superposition states on macroscopic scales, they typically operate the experiment at very low temperatures in order to "freeze out" the thermal environment and to thus minimize decoherence effects. This thermal environment is precisely the bath of delocalized bosonic modes (i.e., harmonic oscillators) discussed earlier. But lowering the temperature—often down to the vicinity of absolute zero, as in superconducting systems such as SQUIDs—usually does not affect the influence of localized modes such as nuclear spins and impurities that are intrinsically present in the material. These sources of decoherence will then become dominant in this low-temperature regime.

Oscillator Environments versus Spin Environments: Differences and Similarities

Having separately outlined physical motivations behind the use of oscillator and spin environments, let us now synthesize our exposition by discussing some important differences (and similarities) between these two types of environments.

As mentioned above, oscillator environments become universal environments in the weak-coupling limit: The interaction of a system with an arbitrary environment can be mapped onto the linear coupling to an equivalent oscillator environment [211,212]. In this sense one may say that (linearly coupled) oscillator environments presume the validity of the weak-coupling limit. Additionally, the coupling strength of each mode is usually assumed to scale with the number N of oscillators in the environment as $\propto 1/\sqrt{N}$. This scaling ensures that summing up the contributions from all modes yields a value that is independent of N, and that thus the thermodynamic limit $N \longrightarrow \infty$ is well-defined.

This combined assumption of weak and N-dependent coupling implies that, while the central system will in general be strongly affected by the *collective* influence of the oscillators in the environment, each *individual* oscillator in the environment is only negligibly influenced by its coupling to the system. Any effects imparted on the environment by the central system usually decay away very quickly (this corresponds to the Born–Markov approximations introduced in Sect. 4.2). Thus the dynamics of the environment itself are hardly affected by the dynamics of the central system.

As frequently emphasized [199,200,210,215,216], the assumptions of weak and N-dependent coupling are usually *not* physically reasonable in the case of the localized modes represented by spin environments. In many physical situations of interest, the energy scale associated with the strength of the coupling of each environmental spin to the (usually macroscopic) coordinate of the central system dominates over the other relevant energy scales set by the remaining terms in the Hamiltonian. In particular, interactions between the spins in the environment are typically extremely weak. This is due to the fact that the wave functions of the different environmental modes are localized, such that there is only little spatial overlap between these wave functions. Because of their restricted phase space, the environmental spins couple usually also only very weakly to any other external (delocalized) modes. Furthermore, it has been argued [199, 200] that in the spin-environment case it is usually neither necessary nor physically reasonable to use a strongly N-dependent scaling of the couplings as in the case of an oscillator environment (although spin-coupling scalings such as $\propto 1/N$ have sometimes been assumed in the literature; see, e.g., [217]).

Thus the situation of interest in the case of spin environments is the strong-coupling regime described by coupling coefficients that are independent of (or only weakly dependent on) the number of spins in the environment. This leads to a very different behavior of the environment—and thus in turn of the system coupled to the environment—than in the case of an oscillator environment. Typically, the spin-environment modes become very strongly "slaved" to the relevant coordinate describing the central system, and thus the dynamics of the environment are heavily influenced by the dynamics of the central spin [199].

This behavior allows for dynamical changes of the environment even in the zero-temperature limit when the intrinsic fluctuations (i.e., the dynamics induced by the self-Hamiltonian and the interspin interactions) in the environment are frozen out. One consequence of this effect is that, as already mentioned above, the spin environment typically exerts a significant decohering influence on the system even at temperatures close to absolute zero (say, at μK temperatures), while the decohering effect of an oscillator bath would in most cases become in comparison negligibly small at such temperatures. The experimental observation of relatively fast decoherence rates at near-zero temperatures would therefore indicate the interaction with localized modes as the dominant source of decoherence. This enables one to experimentally distinguish the two sources of decoherence, namely, delocalized bosonic field modes and localized spin-type modes.

5.2 Quantum Brownian Motion

The model for quantum Brownian motion consists of a particle moving in one spatial dimension and interacting linearly with an environment of independent harmonic oscillators in thermal equilibrium at a temperature T. Quantum Brownian motion constitutes the probably most important and most studied decoherence model. This model not only is the default choice for estimating decoherence effects especially on macroscopic everyday-world scales, but also plays an extremely important role in general studies of dissipative quantum systems [18, 218]. We shall therefore spend some time investigating the subject of quantum Brownian motion.

We will proceed as follows. First, in Sect. 5.2.1, we shall derive the Born–Markov master equation for the model using the formalism developed in Chap. 4. In Sect. 5.2.2, we shall discuss the master equation for the special case of the central particle's being confined to a harmonic-oscillator potential. We further evaluate the resulting master equation in Sect. 5.2.3 by focusing on a specific form of the so-called *spectral density*, which describes the properties of the environment. In Sect. 5.2.4, we will then consider the special case of a high-temperature environment and derive the corresponding *Caldeira–Leggett master equation*, which plays a central role in studies of decoherence. In Sect. 5.2.5, we will analyze the dynamics of quantum Brownian motion by studying the time evolution of a superposition of two Gaussian wave packets in phase space. We will show how the interaction with the environment leads to the emergence of preferred states that are localized in phase space. The dynamical evolution of these pointer states then corresponds to the "quantum version" of Newtonian trajectories. Some cautionary remarks against an overgeneralization of the results of the Caldeira–Leggett model (as well as the general model for quantum Brownian motion) will be made in Sect. 5.2.6. Finally, in Sect. 5.2.7, we will briefly discuss the exact master equation for quantum Brownian motion.

5.2.1 Derivation of the Born–Markov Master Equation

First, let us define our model. As usual, we decompose the total system–environment Hamiltonian $\hat{H}$ into three parts [see (2.85) and (4.12)],

$$\hat{H} = \hat{H}_{\mathcal{S}} + \hat{H}_{\mathcal{E}} + \hat{H}_{\text{int}}. \tag{5.2}$$

For now, we shall leave the self-Hamiltonian $\hat{H}_{\mathcal{S}}$ of the system unspecified. (In Sect. 5.2.2 below, we will specialize on the case of a harmonic-oscillator potential.) The self-Hamiltonian $\hat{H}_{\mathcal{E}}$ of the environment describes a collection of harmonic oscillators. Since, by assumption, these oscillators do not interact with each other, $\hat{H}_{\mathcal{E}}$ is simply the sum of all single-oscillator Hamiltonians, i.e.,

$$\hat{H}_{\mathcal{E}} = \sum_i \left(\frac{1}{2m_i} \hat{p}_i^2 + \frac{1}{2} m_i \omega_i^2 \hat{q}_i^2 \right). \tag{5.3}$$

In obvious notation, m_i and ω_i denote the mass and natural frequency of the ith environmental oscillator, and $\hat{q}_i$ and $\hat{p}_i$ are the canonical position and momentum operators.

The form of the interaction between the system and the environment is such that the position coordinate $\hat{X}$ of the central particle (the system $\mathcal{S}$) couples *linearly* to the positions $\hat{q}_i$ of the oscillators in the environment, with coupling strengths c_i. The interaction Hamiltonian $\hat{H}_{\text{int}}$ thus reads

$$\hat{H}_{\text{int}} = \hat{X} \otimes \sum_i c_i \hat{q}_i \equiv \hat{X} \otimes \hat{E}. \tag{5.4}$$

The assumption of bilinear coupling (linear in both $\hat{X}$ and $\hat{q}_i$) is an important assumption of our model, which is therefore often explicitly referred to as "linear quantum Brownian motion." Our following discussion will be based on this most commonly studied model.

We see that the system–environment interaction is of the form (2.91), which describes the continuous environmental monitoring of the position coordinate of the system. Thus, without any further calculations, we can already anticipate that the coupling (5.4) will, among other possible consequences such as dissipative effects and decoherence in other bases, lead to decoherence of the system in the position basis.

Our first goal is now to determine the environment self-correlation functions (4.10). Since $\hat{H}_{\text{int}}$, see (5.4), contains a single term, only one such function will need to computed, namely,[1]

$$\mathcal{C}(\tau) = \left\langle \hat{E}(\tau) \hat{E} \right\rangle_{\hat{\rho}_{\mathcal{E}}}, \tag{5.5}$$

where $\hat{\rho}_{\mathcal{E}} = \hat{\rho}_{\mathcal{E}}(0)$ [see (4.6)]. Using (5.4), this expression can be evaluated, which yields

$$\mathcal{C}(\tau) = \sum_{ij} c_i c_j \left\langle \hat{q}_i(\tau) \hat{q}_j \right\rangle_{\hat{\rho}_{\mathcal{E}}} = \sum_i c_i^2 \left\langle \hat{q}_i(\tau) \hat{q}_i \right\rangle_{\hat{\rho}_{\mathcal{E}}}. \tag{5.6}$$

For the sake of notational simplicity, we have here omitted the tensor-product symbol "⊗" in products of operators pertaining to different fragments of the environment (and will continue to do so throughout this chapter). The

[1] As in the previous Chap. 4, throughout this chapter we shall use the convention that all operators (other than the density operator) with explicit time arguments should be understood as interaction-picture operators.

vanishing of the terms $i \neq j$ in the last step in (5.6) is due to the fact that the environmental oscillators do not interact with each other and are therefore completely uncorrelated. Thus, for $i \neq j$,

$$\langle \hat{q}_i(\tau)\hat{q}_j \rangle_{\hat{\rho}_\mathcal{E}} = \langle \hat{q}_i(\tau) \rangle_{\hat{\rho}_\mathcal{E}} \langle \hat{q}_j \rangle_{\hat{\rho}_\mathcal{E}} = 0, \tag{5.7}$$

since the expectation value $\langle q_i \rangle_{\hat{\rho}_\mathcal{E}}$ of the position coordinate of the oscillator is equal to zero.

Thus our task of evaluating $\mathcal{C}(\tau)$ is now reduced to the goal of computing the averages $\langle \hat{q}_i(\tau)\hat{q}_i \rangle_{\hat{\rho}_\mathcal{E}}$ [see (5.6)]. This is easily accomplished. Let us switch to the representation of $\hat{q}_i$ in terms of the bosonic creation and annihilation operators $\hat{a}_i^\dagger$ and $\hat{a}_i$,

$$\hat{q}_i = \sqrt{\frac{1}{2m_i\omega_i}} \left(\hat{a}_i + \hat{a}_i^\dagger \right). \tag{5.8}$$

Then the time evolution of the operator $\hat{q}_i$ in the interaction picture can be written as [see (A.4)]

$$\hat{q}_i(\tau) = \mathrm{e}^{\mathrm{i}\hat{H}_\mathcal{E}\tau} \hat{q}_i \mathrm{e}^{-\mathrm{i}\hat{H}_\mathcal{E}\tau} = \sqrt{\frac{1}{2m_i\omega_i}} \left(\hat{a}_i \mathrm{e}^{-\mathrm{i}\omega_i\tau} + \hat{a}_i^\dagger \mathrm{e}^{\mathrm{i}\omega_i\tau} \right). \tag{5.9}$$

Therefore

$$\begin{aligned} \langle \hat{q}_i(\tau)\hat{q}_i \rangle_{\hat{\rho}_\mathcal{E}} &= \frac{1}{2m_i\omega_i} \left\langle \hat{a}_i\hat{a}_i^\dagger \mathrm{e}^{-\mathrm{i}\omega_i\tau} + \hat{a}_i^\dagger\hat{a}_i \mathrm{e}^{\mathrm{i}\omega_i\tau} \right\rangle_{\hat{\rho}_\mathcal{E}} \\ &= \frac{1}{2m_i\omega_i} \left\{ \left\langle \hat{a}_i\hat{a}_i^\dagger \right\rangle_{\hat{\rho}_\mathcal{E}} \mathrm{e}^{-\mathrm{i}\omega_i\tau} + \left\langle \hat{a}_i^\dagger\hat{a}_i \right\rangle_{\hat{\rho}_\mathcal{E}} \mathrm{e}^{\mathrm{i}\omega_i\tau} \right\}. \end{aligned} \tag{5.10}$$

But the quantity

$$N_i = \left\langle \hat{a}_i^\dagger \hat{a}_i \right\rangle_{\hat{\rho}_\mathcal{E}} \tag{5.11}$$

is simply the mean occupation number of the ith environmental oscillator. By assumption, the environment is in thermal equilibrium, which corresponds to

$$N_i \equiv N_i(T) = \frac{1}{\mathrm{e}^{\omega_i/k_\mathrm{B}T} - 1}. \tag{5.12}$$

Using this expression and the commutation relation $\left[\hat{a}_i, \hat{a}_i^\dagger\right] = 1$, we can rewrite (5.10) as

$$\begin{aligned} \langle \hat{q}_i(\tau)\hat{q}_i \rangle_{\hat{\rho}_\mathcal{E}} &= \frac{1}{2m_i\omega_i} \left\{ [1 + N_i(T)]\, \mathrm{e}^{-\mathrm{i}\omega_i\tau} + N_i(T)\mathrm{e}^{\mathrm{i}\omega_i\tau} \right\} \\ &= \frac{1}{2m_i\omega_i} \left\{ [1 + 2N_i(T)] \cos(\omega_i\tau) - \mathrm{i}\sin(\omega_i\tau) \right\} \\ &= \frac{1}{2m_i\omega_i} \left\{ \coth\left(\frac{\omega_i}{2k_\mathrm{B}T} \right) \cos(\omega_i\tau) - \mathrm{i}\sin(\omega_i\tau) \right\}, \end{aligned} \tag{5.13}$$

where in the last step we have used the fact that [see (5.12)]

$$\begin{aligned} 1 + 2N_i(T) &= 1 + \frac{2}{e^{\omega_i/k_B T} - 1} \\ &= \frac{e^{\omega_i/k_B T} + 1}{e^{\omega_i/k_B T} - 1} = \coth\left(\frac{\omega_i}{2k_B T}\right). \end{aligned} \tag{5.14}$$

Hence the environment self-correlation function (5.6) can now be written as

$$\begin{aligned} \mathcal{C}(\tau) &= \sum_i \frac{c_i^2}{2m_i\omega_i} \left\{ \coth\left(\frac{\omega_i}{2k_B T}\right) \cos(\omega_i \tau) - \mathrm{i} \sin(\omega_i \tau) \right\} \\ &\equiv \nu(\tau) - \mathrm{i}\eta(\tau). \end{aligned} \tag{5.15}$$

Here, the functions

$$\begin{aligned} \nu(\tau) &= \frac{1}{2} \sum_i c_i^2 \langle \{\hat{q}_i(\tau), \hat{q}_i\} \rangle_{\hat{\rho}_\mathcal{E}} \\ &= \sum_i \frac{c_i^2}{2m_i\omega_i} \coth\left(\frac{\omega_i}{2k_B T}\right) \cos(\omega_i \tau) \\ &\equiv \int_0^\infty \mathrm{d}\omega \, J(\omega) \coth\left(\frac{\omega}{2k_B T}\right) \cos(\omega\tau), \end{aligned} \tag{5.16}$$

$$\begin{aligned} \eta(\tau) &= \frac{\mathrm{i}}{2} \sum_i c_i^2 \langle [\hat{q}_i(\tau), \hat{q}_i] \rangle_{\hat{\rho}_\mathcal{E}} \\ &= \sum_i c_i^2 \frac{1}{2m_i\omega_i} \sin(\omega_i \tau) \\ &\equiv \int_0^\infty \mathrm{d}\omega \, J(\omega) \sin(\omega\tau) \end{aligned} \tag{5.17}$$

are commonly referred to in the literature as the *noise kernel* and *dissipation kernel*, respectively. The curly brackets $\{\cdot,\cdot\}$ in the first line of (5.16) denote the anticommutator $\{\hat{A}, \hat{B}\} \equiv \hat{A}\hat{B} + \hat{B}\hat{A}$. The function $J(\omega)$, introduced in the last line of (5.16) and (5.17), is defined as

$$J(\omega) \equiv \sum_i \frac{c_i^2}{2m_i\omega_i} \delta(\omega - \omega_i) \tag{5.18}$$

and is called the *spectral density* of the environment. Spectral densities encapsulate the physical properties of the environment and play an immensely important role in theoretical and experimental studies of decoherence. In modeling the environment, one often goes to a continuum limit for the environment in which the description in terms of individual oscillators with discrete frequencies ω_i and masses m_i is replaced by a spectral density $J(\omega)$ corresponding to a continuous spectrum of environmental frequencies ω. We shall discuss examples for such continuous spectral densities below. For the moment being, we shall stick to the original expression (5.18) for $J(\omega)$. Thus far, the benefit of the definition (5.18) therefore merely amounts to a slightly more compact notation of the noise and dissipation kernels (5.16) and (5.17).

Having successfully determined the environment self-correlation function $\mathcal{C}(\tau)$ [see (5.15)], we have completed the main step in the derivation of the desired Born–Markov master equation for quantum Brownian motion. The rest of the derivation is now straightforward. The operators $\hat{B}$ and $\hat{C}$ [see (4.9a) and (4.9b)] are immediately written down as

$$\hat{B} = \int_0^\infty \mathrm{d}\tau\, \mathcal{C}(\tau)\hat{X}(-\tau), \tag{5.19a}$$

$$\hat{C} = \int_0^\infty \mathrm{d}\tau\, \mathcal{C}(-\tau)\hat{X}(-\tau), \tag{5.19b}$$

where

$$\hat{X}(\tau) = \mathrm{e}^{\mathrm{i}\hat{H}_S\tau}\hat{X}\mathrm{e}^{-\mathrm{i}\hat{H}_S\tau} \tag{5.20}$$

is the position operator of the system $\mathcal{S}$ in the interaction picture. Inserting (5.19a) and (5.19b), with $\mathcal{C}(\tau)$ given by (5.15), into our general expression (4.8) for the Born–Markov master equation then yields

$$\begin{aligned}\frac{\mathrm{d}}{\mathrm{d}t}\hat{\rho}_S(t) &= -\mathrm{i}\left[\hat{H}_S, \hat{\rho}_S(t)\right] \\ &\quad - \int_0^\infty \mathrm{d}\tau\, \left\{\mathcal{C}(\tau)\left[\hat{X}, \hat{X}(-\tau)\hat{\rho}_S(t)\right] + \mathcal{C}(-\tau)\left[\hat{\rho}_S(t)\hat{X}(-\tau), \hat{X}\right]\right\}.\end{aligned} \tag{5.21}$$

Using the decomposition $\mathcal{C}(\tau) = \nu(\tau) - \mathrm{i}\eta(\tau)$ [see (5.15)] involving the noise and the dissipation kernels (5.16) and (5.17) and rearranging terms leads to the final form of the master equation,

$$\begin{aligned}\frac{\mathrm{d}}{\mathrm{d}t}\hat{\rho}_S(t) = -\mathrm{i}\left[\hat{H}_S, \hat{\rho}_S(t)\right] - \int_0^\infty \mathrm{d}\tau\, &\left\{\nu(\tau)\left[\hat{X}, \left[\hat{X}(-\tau), \hat{\rho}_S(t)\right]\right]\right. \\ &\left. -\mathrm{i}\eta(\tau)\left[\hat{X}, \left\{\hat{X}(-\tau), \hat{\rho}_S(t)\right\}\right]\right\}.\end{aligned} \tag{5.22}$$

We see that the decohering (and dissipative) influence of the environment is completely described by integrals over the noise and dissipation kernels (5.16) and (5.17) multiplied by products of (Schrödinger-picture and interaction-picture) position operators of the system.

5.2.2 Harmonic Oscillator as the Central System

To simplify the master equation (5.22) even further and to gain an understanding of the physical meaning of the noise and dissipation terms, let us now specialize to the case of the system's being represented by a harmonic oscillator with the standard self-Hamiltonian

$$\hat{H}_{\mathcal{S}} = \frac{1}{2M}\hat{P}^2 + \frac{1}{2}M\Omega^2\hat{X}^2. \tag{5.23}$$

The explicit form of the interaction-picture position operator $\hat{X}(\tau)$ [see (5.20)] is then easily determined by solving the Heisenberg equations of motions for the operators $\hat{X}$ and $\hat{P}$ of the system, which yields

$$\hat{X}(\tau) = \mathrm{e}^{\mathrm{i}\hat{H}_{\mathcal{S}}\tau}\hat{X}\mathrm{e}^{-\mathrm{i}\hat{H}_{\mathcal{S}}\tau} = \hat{X}\cos(\Omega\tau) + \frac{1}{M\Omega}\hat{P}\sin(\Omega\tau). \tag{5.24}$$

Inserting this expression for $\hat{X}(\tau)$ into (5.22) yields the Born–Markov master equation describing the reduced dynamics of a harmonic oscillator linearly and weakly coupled to a thermal environment of harmonic oscillators,

$$\begin{aligned}\frac{\mathrm{d}}{\mathrm{d}t}\hat{\rho}_{\mathcal{S}}(t) = -\mathrm{i}\left[\hat{H}_{\mathcal{S}} + \frac{1}{2}M\widetilde{\Omega}^2\hat{X}^2, \hat{\rho}_{\mathcal{S}}(t)\right] - \mathrm{i}\gamma\left[\hat{X}, \left\{\hat{P}, \hat{\rho}_{\mathcal{S}}(t)\right\}\right] \\ - D\left[\hat{X}, \left[\hat{X}, \hat{\rho}_{\mathcal{S}}(t)\right]\right] - f\left[\hat{X}, \left[\hat{P}, \hat{\rho}_{\mathcal{S}}(t)\right]\right].\end{aligned} \tag{5.25}$$

Here, we have introduced the coefficients $\widetilde{\Omega}^2$, γ, D, and f defined as

$$\widetilde{\Omega}^2 \equiv -\frac{2}{M}\int_0^\infty \mathrm{d}\tau\,\eta(\tau)\cos(\Omega\tau), \tag{5.26a}$$

$$\gamma \equiv \frac{1}{M\Omega}\int_0^\infty \mathrm{d}\tau\,\eta(\tau)\sin(\Omega\tau), \tag{5.26b}$$

$$D \equiv \int_0^\infty \mathrm{d}\tau\,\nu(\tau)\cos(\Omega\tau), \tag{5.26c}$$

$$f \equiv -\frac{1}{M\Omega}\int_0^\infty \mathrm{d}\tau\,\nu(\tau)\sin(\Omega\tau). \tag{5.26d}$$

Each of these coefficients carries a particular physical interpretation, derived from its role in the master equation (5.25). Before discussing these terms one by one, let us first rewrite (5.25) in two other equivalent ways that are frequently used in the literature.

First of all, it is often convenient to express (5.25) in the position representation. Using the basic relation

$$\langle X|\,\hat{P}\hat{\rho}_{\mathcal{S}}(t)\,|X'\rangle = -\mathrm{i}\frac{\partial}{\partial X}\langle X|\,\hat{\rho}_{\mathcal{S}}(t)\,|X'\rangle \equiv -\mathrm{i}\frac{\partial}{\partial X}\rho_{\mathcal{S}}(X, X', t), \tag{5.27}$$

we obtain the equation of motion for the reduced density matrix $\rho_{\mathcal{S}}(X, X', t)$ in the position basis,

$$\begin{aligned}\frac{\partial}{\partial t}\rho_{\mathcal{S}}(X, X', t) = \Bigg[& -\frac{\mathrm{i}}{2M}\left(\frac{\partial^2}{\partial X'^2} - \frac{\partial^2}{\partial X^2}\right) - \frac{\mathrm{i}}{2}M\left(\Omega^2 + \widetilde{\Omega}^2\right)\left(X^2 - X'^2\right) \\ & + \gamma(X - X')\left(\frac{\partial}{\partial X'} - \frac{\partial}{\partial X}\right) - D(X - X')^2 \\ & + \mathrm{i}f(X - X')\left(\frac{\partial}{\partial X'} + \frac{\partial}{\partial X}\right)\Bigg]\rho_{\mathcal{S}}(X, X', t).\end{aligned} \tag{5.28}$$

We may also express (5.25) as a master equation for the Wigner representation $W(X, P)$, see (2.132) and Sect. 2.15.2, of the position-space reduced density matrix $\rho_S(X, X')$. Without giving the derivation here, we shall simply state the result [219],

$$\frac{\partial}{\partial t} W(X,P,t) = \left[-\frac{P}{M}\frac{\partial}{\partial X} + M\left(\Omega^2 + \widetilde{\Omega}^2\right) X \frac{\partial}{\partial P} + \gamma \frac{\partial}{\partial P} P \right. \\ \left. + D\frac{\partial^2}{\partial P^2} - f \frac{\partial}{\partial X}\frac{\partial}{\partial P} \right] W(X,P,t). \quad (5.29)$$

Readers familiar with statistical mechanics may recognize the formal similarity between (5.29) and the Fokker–Planck equation in classical statistical physics.

Let us proceed by discussing the interpretation of the different terms appearing in the master equation (5.25). The first commutator on the right-hand side of (5.25),

$$-\mathrm{i}\left[\hat{H}_S + \frac{1}{2} M \widetilde{\Omega}^2 \hat{X}^2, \hat{\rho}_S(t) \right], \quad (5.30)$$

describes the reversible unitary dynamics of a harmonic oscillator of physical frequency $(\Omega^2 + \widetilde{\Omega}^2)^{1/2}$. Thus we see that $\widetilde{\Omega}^2$ introduces a temperature-independent shift of the natural frequency Ω of the central oscillator, representing an example of the Lamb-shift contribution discussed on p. 155.

Next, the term

$$-\mathrm{i}\gamma \left[\hat{X}, \left\{ \hat{P}, \hat{\rho}_S(t) \right\} \right] \quad (5.31)$$

in (5.25) describes momentum damping—and thus dissipation—due to the interaction with the environment. To see this, consider the time evolution of the expectation value $\langle \hat{P} \rangle_t$ for the momentum of the central oscillator as determined by the equation of motion (5.25). Using the trace rule $\langle \hat{P} \rangle = \mathrm{Tr}_S\,(\hat{\rho}_S \hat{P})$, we obtain

$$\frac{\mathrm{d}}{\mathrm{d}t}\langle \hat{P} \rangle_t = \frac{\mathrm{d}}{\mathrm{d}t}\mathrm{Tr}_S \left\{ \hat{\rho}_S(t) \hat{P} \right\} = \mathrm{Tr}_S \left\{ \frac{\mathrm{d}\hat{\rho}_S(t)}{\mathrm{d}t} \hat{P} \right\} \\ = -M(\Omega^2 + \widetilde{\Omega}^2)\langle \hat{X} \rangle_t - 2\gamma \langle \hat{P} \rangle_t. \quad (5.32)$$

The first term $-M(\Omega^2 + \widetilde{\Omega}^2)\langle \hat{X} \rangle_t$ is due to the usual unitary evolution in the (Lamb-shifted) harmonic-oscillator potential. Let us, for the moment being, disregard this unitary nondissipative term. Then, from (5.32), we immediately see that, due to the interaction with the environment, the expectation value for the momentum of the central system decreases exponentially at a rate 2γ,

$$\langle \hat{P} \rangle_t \propto \mathrm{e}^{-2\gamma t} \langle \hat{P} \rangle_0. \quad (5.33)$$

Thus we have shown that the term (5.31) in the master equation (5.25) indeed describes momentum damping at a rate proportional to γ. Note that this rate depends only on the spectral density but not on the temperature of the environment [see (5.17) and (5.26b)].

Let us now consider the third term on the right-hand side of (5.25),

$$-D\left[\hat{X},\left[\hat{X},\hat{\rho}_S(t)\right]\right]. \tag{5.34}$$

Note that this term is of the Lindblad double-commutator form (4.45), describing environmental monitoring of the position coordinate $\hat{X}$ of the system and thus decoherence in the position basis. This role can also be easily seen by writing out the double commutator in the position representation,

$$D[\hat{X},[\hat{X},\hat{\rho}_S(t)]] \longrightarrow D(X-X')^2\rho_S(X,X',t). \tag{5.35}$$

The reader may recognize the term on the right-hand side as being of a similar form as the expression (3.55) that described spatial localization in the model for environmental scattering discussed in Chap. 3. The coefficient D multiplied by $(X-X')^2$ therefore plays the role of a decoherence rate. That is,

$$\tau_{|X-X'|} = \frac{1}{D(X-X')^2} \tag{5.36}$$

is the timescale on which spatial interferences over a distance $|X-X'|$ become suppressed by a factor of e [see also (3.58)].

The coefficient D also describes diffusion in momentum, as seen from the time evolution of the expectation value $\langle\hat{P}^2\rangle_t$ of the squared momentum of the system. A simple calculation similar to the one leading to (5.32) yields

$$\frac{\mathrm{d}}{\mathrm{d}t}\langle\hat{P}^2\rangle_t = -M(\Omega^2+\widetilde{\Omega}^2)\langle\hat{X}\hat{P}+\hat{P}\hat{X}\rangle_t - 4\gamma\langle\hat{P}^2\rangle_t + 2D. \tag{5.37}$$

Focusing on the time dependence of $\langle\hat{P}^2\rangle_t$ due to the D term only, we obtain the evolution $\langle\hat{P}^2\rangle_t \propto Dt$, which shows that D indeed describes diffusion in momentum space. D is therefore often referred to as the *normal-diffusion coefficient* in the literature on quantum Brownian motion. Here, the terminology "normal diffusion" refers to the linear time dependence of $\langle\hat{P}^2\rangle_t$.

Recall that such normal diffusion is a signature of classical Brownian motion, where the variance σ_x^2 of the normal distribution describing the possible positions of the particle increases as $\sigma_x^2(t) \propto t$. In fact, the same time dependence holds also in the quantum case. It is easy to show that, in the long-time limit $\gamma t \gg 1$, the master equation (5.25) yields for the time dependence of the position-space dispersion

$$\Delta X^2(t) = \frac{D}{2m^2\gamma^2}t. \tag{5.38}$$

That is, the width $\Delta X(t)$ of the ensemble in position space asymptotically scales as $\Delta X(t) \propto \sqrt{t}$, just as in classical Brownian motion. Indeed, this similarity between the quantum and classical settings originally motivated the choice of the term "quantum Brownian motion." Incidentally, (5.37) also confirms our earlier finding (5.33) that the coefficient γ quantifies dissipation, since $\langle \hat{P}^2/2M \rangle$ is the kinetic energy of the system.

The role of D as a diffusion coefficient can also be seen from the master equation (5.29) in the Wigner representation. Singling out the D term, we obtain the evolution equation

$$\frac{\partial}{\partial t} W(X,P,t) = D \frac{\partial^2}{\partial P^2} W(X,P,t), \tag{5.39}$$

which has the well-known form of a diffusion equation in the variable P for the case of a constant diffusion coefficient D.

Finally, we turn our attention to the last term

$$-f \left[\hat{X}, \left[\hat{P}, \hat{\rho}_S(t) \right] \right] \tag{5.40}$$

in the master equation (5.25). This term, too, describes diffusion and decoherence, just like the normal-diffusion term (5.34). This can best be seen from the Wigner form (5.29) of the master equation. The contribution due to the f term alone is

$$\frac{\partial}{\partial t} W(X,P,t) = -f \frac{\partial}{\partial X} \frac{\partial}{\partial P} W(X,P,t). \tag{5.41}$$

This form is similar to the diffusion equation (5.39) in terms of the appearance of a double derivative on the right-hand side of the equation. However, in contrast with the "normal" diffusion equation, in (5.41) this double derivative is composed of single derivatives with respect to two *different* variables X and P. Motivated by this difference, the quantity f is often referred to as the *anomalous-diffusion coefficient* in the literature. As we will show below, in many physical situations of interest the influence of this anomalous-diffusion term (5.40) on the reduced dynamics of the system is negligibly small in comparison with the normal-diffusion term (5.34). In such cases, we may use a simplified version of the master equation (5.25) without the anomalous-diffusion term (see Sect. 5.2.4).

Note that in our model the normal-diffusion and anomalous-diffusion coefficients depend on both the spectral density $J(\omega)$ and the temperature T of the environment [see (5.26) with (5.16) and (5.17)]. This is in contrast to the frequency shift $\widetilde{\Omega}^2$ and the momentum-damping rate γ, which, as already mentioned above, are independent of the temperature of the environment.

Finally, we shall mention that in the literature a (partially) "pre-Markovian" form of the coefficients (5.26) is often discussed (see, e.g., [15, 16]). In this case, the upper limit of the time integral is not yet taken to infinity but only up to the time t (see also the discussion on p. 163). In other words, an explicitly time-dependent version of the coefficients $\widetilde{\Omega}^2(t)$, $\gamma(t)$, $D(t)$, and $f(t)$ is considered,

$$\widetilde{\Omega}^2(t) = -\frac{2}{M} \int_0^t \mathrm{d}\tau\, \eta(\tau) \cos(\Omega\tau)\,, \tag{5.42a}$$

$$\gamma(t) = \frac{1}{M\Omega} \int_0^t \mathrm{d}\tau\, \eta(\tau) \sin(\Omega\tau)\,, \tag{5.42b}$$

$$D(t) = \int_0^t \mathrm{d}\tau\, \nu(\tau) \cos(\Omega\tau)\,, \tag{5.42c}$$

$$f(t) = -\frac{1}{M\Omega} \int_0^t \mathrm{d}\tau\, \nu(\tau) \sin(\Omega\tau)\,. \tag{5.42d}$$

The master equation (5.25) evaluated with these time-dependent coefficients instead of the "Markovian coefficients" (5.26) is then of the Redfield type [174,175] mentioned in Sect. 4.2.2. That is, the master equation is based on (i) the Born approximation and (ii) the first Markov approximation of replacing the retarded-time density operator $\hat{\rho}_S^{(I)}(t')$ by $\hat{\rho}_S^{(I)}(t)$ in the integrand on the right-hand side of the Born master equation (4.32). However, the limit of the integration in (4.32) is *not* extended to infinity.

The resulting master equation [i.e., (5.25) with coefficients given by (5.42)] allows for a more realistic treatment of quantum Brownian motion in situations where the Markov approximation is not (fully) appropriate, for example, in the case of low-temperature environments [109, 220]. In fact, the coefficients (5.42) represent a middle ground between the Markovian setting [the master equation (5.25) evaluated with the time-independent coefficients (5.26)] on the one hand, and the exact master equation for quantum Brownian motion on the other hand (see Sect. 5.2.7), which is valid for arbitrary system–environment interaction strengths and not just in the weak-coupling limit considered here.

The time dependence of the coefficients (5.42) has been studied in detail in the literature (see, e.g., the discussions by Hu, Paz, and Zhang [219] and by Paz and Zurek [15]). For example, in the case of an ohmic spectral density of the environment (see the next Sect. 5.2.3) it can be shown that the frequency shift $\widetilde{\Omega}^2(t)$ and the momentum-damping coefficient $\gamma(t)$ both start out from a value of zero and asymptotically approach their final values on a timescale set by the order of magnitude of the highest frequencies that are present in the environment to a significant degree. The diffusion coefficients $D(t)$ and $f(t)$ exhibit an initial temperature-independent transient and then settle rather quickly into their (temperature-dependent) asymptotic values.

Since the Markovian coefficients (5.26) correspond to taking the limit $t \longrightarrow \infty$ in the integrals in (5.42), the coefficients $\widetilde{\Omega}^2(t)$, $\gamma(t)$, $D(t)$, and $f(t)$ asymptotically approach the values of these Markovian coefficients. In the following, we shall focus on the Markovian case, i.e., on the master equation (5.25) together with the time-independent expressions (5.26) for the coefficients.

5.2.3 Ohmic Decoherence and Dissipation

Let us now further evaluate the coefficients $\widetilde{\Omega}^2$, γ, D, and f [see (5.26)] appearing in the master equation (5.25) for quantum Brownian motion of a central harmonic oscillator. To do so, we must choose a specific functional form for the spectral density $J(\omega)$ appearing in the expressions for $\nu(\tau)$ and $\eta(\tau)$ [see (5.16) and (5.17)].

As indicated above, one typically replaces the discrete sum in the original expression (5.18) for $J(\omega)$ with an (often phenomenologically motivated) continuous function of the environmental frequencies ω. Usually, the frequency dependence of $J(\omega)$ is taken to follow a power-law dependence of the form $J(\omega) \propto \omega^\alpha$. The most common choice for the exponent α is $\alpha = 1$, such that $J(\omega)$ increases linearly with ω. This type of spectral density bears the (historically motivated) name of an *ohmic* spectral density. Other, less important examples include "subohmic" spectral densities characterized by $\alpha < 1$ and "supraohmic" spectral densities for which $\alpha > 1$.

Let us now focus on the case of an ohmic spectral density $J(\omega) \propto \omega$, which we may write explicitly as

$$J(\omega) = \frac{2M\gamma_0}{\pi}\omega. \tag{5.43}$$

Here, the constant γ_0 describes the effective coupling strength between the system and the environment. In our original discrete expression (5.18) for the spectral density, this interaction strength was represented by the sum over terms containing the individual oscillator couplings c_i. As we shall see shortly, γ_0 plays the role of a frequency-independent damping constant. (In phenomenological models of decoherence in concrete experimental situations, one often derives γ_0 from measured data.)

A power-law frequency dependence of $J(\omega)$ would imply that the distribution of environmental frequencies grows without bound. However, this is physically unreasonable. Typically, the linear dependence of $J(\omega)$ on the frequency ω only holds for small values of ω and becomes attenuated toward higher frequencies. One therefore usually chooses a high-frequency cutoff Λ and modifies (5.43) to include a term that damps $J(\omega)$ for frequencies $\omega > \Lambda$. A common choice is a cutoff term of the so-called *Lorentz–Drude* form,

$$J(\omega) = \frac{2M\gamma_0}{\pi}\omega\frac{\Lambda^2}{\Lambda^2 + \omega^2}. \tag{5.44}$$

A plot of this spectral density is shown in Fig. 5.2. We see that, as intended, for environmental frequencies ω exceeding the cutoff Λ the corresponding density $J(\omega)$ of frequencies decreases.

The explicit choice (5.44) for the spectral density $J(\omega)$ now allows us to compute the coefficients $\widetilde{\Omega}^2$, γ, D, and f [see (5.26)]. It turns out that the momentum-damping coefficient γ and the normal-diffusion coefficient D are

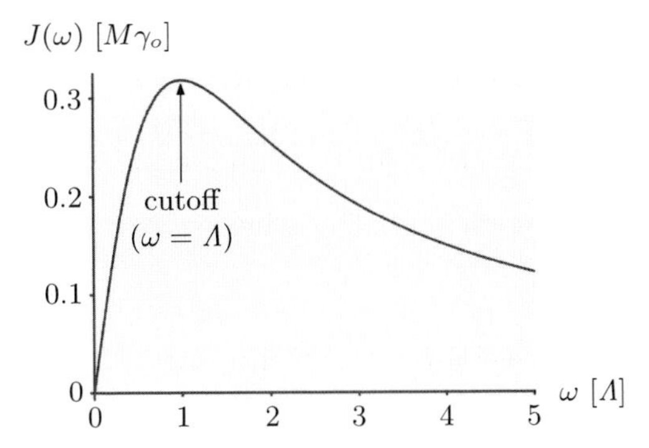

Fig. 5.2. Plot of the spectral density $J(\omega)$ (in units of $M\gamma_0$) as given by (5.44). $J(\omega)$ increases approximately linearly for frequencies ω well below the high-frequency cutoff Λ and decreases subsequently for frequencies $\omega > \Lambda$. The frequency axis is displayed in units of the cutoff Λ.

particularly easy to evaluate. To see this, note that the noise kernel $\nu(\tau)$ [see (5.16)] is simply, apart from a missing prefactor $\sqrt{2/\pi}$, the Fourier cosine transform of the function $J(\omega)\coth(\omega/2k_B T)$. Similarly, the dissipation kernel $\eta(\tau)$ [see (5.17)] is proportional to the Fourier sine transform of $J(\omega)$. Furthermore, a look at the expression for γ [see (5.26b)] shows that this function is proportional to the Fourier sine transform of $\eta(\tau)$.

But, for odd functions $g(-x) = -g(x)$, the double Fourier sine transform returns the original function,

$$\frac{2}{\pi}\int_0^\infty \mathrm{d}k\,\sin(x'k)\int_0^\infty \mathrm{d}x\,g(x)\sin(kx) = g(x'). \tag{5.45}$$

Since our spectral density (5.44) is an odd function of ω, and γ is (apart from a prefactor) the double Fourier sine transform of $J(\omega)$, γ is therefore simply proportional to $J(\omega)$ evaluated at the frequency Ω of the harmonic oscillator of the system [cf. (5.26b) and (5.45)]. Taking into account the prefactor $\frac{2}{\pi}$ arising from the definition of the double Fourier transform (5.45), we thus obtain

$$\gamma = \frac{\pi}{2}\frac{1}{M\Omega}J(\Omega) = \gamma_0\frac{\Lambda^2}{\Lambda^2+\Omega^2}. \tag{5.46}$$

This relation between the momentum-damping coefficient γ appearing in the master equation (5.25) and the effective coupling strength γ_0 introduced in the expression (5.43) for the spectral density justifies our earlier claim that γ_0 can be viewed as a frequency-independent damping constant. If we do not impose any cutoff on the spectral density ($\Lambda \to \infty$), γ and γ_0 are exactly equal.

To calculate the normal-diffusion coefficient D [see (5.26b)], we can follow a similar argument. D is proportional to the double Fourier cosine transform of $J(\omega)\coth(\omega/2k_\mathrm{B}T)$. For even functions $g(-x) = g(x)$, an identity analogous to (5.45) holds for such double Fourier cosine transforms, namely,

$$\frac{2}{\pi}\int_0^\infty \mathrm{d}k\, \cos(x'k) \int_0^\infty \mathrm{d}x\, g(x)\cos(kx) = g(x'). \tag{5.47}$$

Since the function $J(\omega)\coth(\omega/2k_\mathrm{B}T)$ is indeed even in ω, we therefore obtain

$$D = \frac{\pi}{2}J(\Omega)\coth\left(\frac{\Omega}{2k_\mathrm{B}T}\right) = M\gamma_0\Omega\frac{\Lambda^2}{\Lambda^2+\Omega^2}\coth\left(\frac{\Omega}{2k_\mathrm{B}T}\right). \tag{5.48}$$

The temperature dependence of D is illustrated in Fig. 5.3. We observe that D increases with temperature: Decoherence becomes stronger as the temperature of the environment is raised. For high temperatures $k_\mathrm{B}T \gg \Omega$, $\coth(\Omega/2k_\mathrm{B}T) \approx 2k_\mathrm{B}T/\Omega$, and thus D increases approximately linearly with temperature (this high-temperature limit will be studied in more detail in Sect. 5.2.4 below).

The expression for the frequency renormalization $\widetilde{\Omega}^2$ [see (5.26a)] is a Fourier cosine transform of a Fourier sine transform. Although no simple identities of the form (5.45) or (5.47) hold in this case, it is quite straightforward to explicitly evaluate this coefficient. We shall here simply state the result without derivation,

$$\widetilde{\Omega}^2 = -2\gamma_0\frac{\Lambda^3}{\Lambda^2+\Omega^2}. \tag{5.49}$$

Finally, the anomalous-diffusion coefficient f [see (5.26d)] is more difficult to evaluate because of the presence of the term $\coth(\omega/2k_\mathrm{B}T)$. One case that is relatively easy to treat is that of a high-temperature environment such that

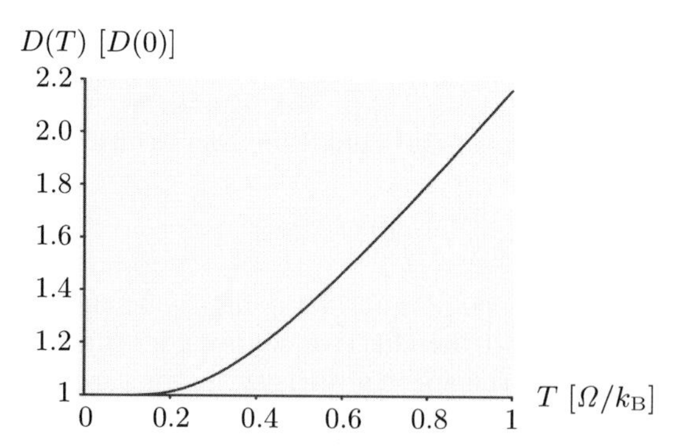

Fig. 5.3. Temperature dependence of the normal-diffusion coefficient D, see (5.48), for the case of the ohmic spectral density (5.44). The vertical axis is shown in units of $D(T=0)$, the zero-temperature value of the normal-diffusion coefficient, and the temperature axis is displayed in units of Ω/k_B.

$k_B T \gg \Omega$ (see the next section). In this limit, the resulting expression for f is smaller than the normal-diffusion coefficient (5.48) by a factor on the order of the cutoff frequency Λ.

5.2.4 The Caldeira–Leggett Master Equation

In the following, we shall have a closer look at the limit of a high-temperature environment. That is, we consider the situation in which the thermal energy $k_B T$ of the environment is much larger than (the energies corresponding to) the natural frequency Ω of the system and the environmental cutoff frequency Λ. Then we can approximate the term $\coth(\Omega/2k_B T)$ appearing in the expression (5.48) for the normal-diffusion coefficient D by $2k_B T/\Omega$, which is the first-order term in the Taylor expansion of $\coth(\Omega/2k_B T)$ around $\Omega/2k_B T = 0$. In this case the coefficient D becomes

$$D \xrightarrow{k_B T \gg \Omega} 2M\gamma_0 k_B T \frac{\Lambda^2}{\Lambda^2 + \Omega^2}. \tag{5.50}$$

We also assume that the cutoff Λ (the upper-limit region of environmental frequencies) is much larger than the characteristic frequency Ω of the system. Then (5.50) takes the form

$$D \xrightarrow{\Lambda \gg \Omega} 2M\gamma_0 k_B T. \tag{5.51}$$

Furthermore, the momentum-damping coefficient γ, see (5.46), approaches the value of the effective coupling strength γ_0,

$$\gamma \xrightarrow{\Lambda \gg \Omega} \gamma_0, \tag{5.52}$$

and the frequency normalization $\widetilde{\Omega}^2$, see (5.49), becomes

$$\widetilde{\Omega}^2 \xrightarrow{\Lambda \gg \Omega} -2\gamma_0 \Lambda. \tag{5.53}$$

Finally, given the assumptions considered here, the anomalous-diffusion coefficient f can be explicitly evaluated to give the result

$$f \xrightarrow[\Lambda \gg \Omega]{k_B T \gg \Omega} \frac{2\gamma_0 k_B T}{\Lambda}. \tag{5.54}$$

To gain further insight into the relevance of the corresponding anomalous-diffusion term $f[\hat{X}, [\hat{P}, \hat{\rho}_S(t)]]$ in the master equation (5.25), let us compare the magnitude of this term to that of the term $D[\hat{X}, [\hat{X}, \hat{\rho}_S(t)]]$. These magnitudes are on the order of, respectively,

$$DX^2 \sim 2M\gamma_0 k_B T X^2 \tag{5.55}$$

and

$$fXP \sim fXM\Omega X \sim \frac{2\gamma_0 k_B T}{\Lambda} M\Omega X^2 = D\frac{\Omega}{\Lambda} X^2. \tag{5.56}$$

Since by assumption $\Lambda \gg \Omega$, we can here safely neglect the anomalous-diffusion term in the master equation (5.25).

Let us now insert our above expressions (5.51), (5.52), and (5.53) for the coefficients D, γ, and $\widetilde{\Omega}^2$ into the Born–Markov master equation (5.25). The result is

$$\frac{\mathrm{d}}{\mathrm{d}t}\hat{\rho}_S(t) = -\mathrm{i}\left[\hat{H}'_S, \hat{\rho}_S(t)\right] - \mathrm{i}\gamma_0 \left[\hat{X}, \left\{\hat{P}, \hat{\rho}_S(t)\right\}\right] - 2M\gamma_0 k_B T \left[\hat{X}, \left[\hat{X}, \hat{\rho}_S(t)\right]\right], \tag{5.57}$$

where the frequency-shifted Hamiltonian $\hat{H}'_S$ of the system is

$$\hat{H}'_S = \hat{H}_S + \frac{1}{2}M\widetilde{\Omega}^2\hat{X}^2 = \frac{1}{2M}\hat{P}^2 + \frac{1}{2}M\left[\Omega^2 - 2\gamma_0\Lambda\right]\hat{X}^2. \tag{5.58}$$

The interpretation of the terms appearing on the right-hand side of (5.57) has already been discussed in the context of the more general form (5.25) of the master equation (see Sect. 5.2.2). The first term is the usual Liouville–von Neumann term with a frequency-shifted Hamiltonian. The second term describes momentum damping (i.e., dissipation) with a characteristic rate proportional to γ_0. The third term represents decoherence in the position basis with consequent momentum diffusion, with a characteristic temperature-dependent localization rate given by the product of the coefficient $M\gamma_0 k_B T$ and the squared separation $(X - X')^2$.

Equation (5.57) is commonly known as the *Caldeira–Leggett master equation* (here written for the case of a central harmonic-oscillator system), which was first derived by Caldeira and Leggett in the early 1980s using a path-integral approach [221]. It is widely used in the modeling of decoherence and dissipation processes. In fact, it has often been quite successfully applied to situations in which the assumptions on which the derivation of the equation is based were not strictly fulfilled (for example, in quantum-optical settings where often $k_B T \lesssim \Lambda$ [222]).

A particularly appealing feature of the Caldeira–Leggett master equation is that it allows for very intuitive and simple estimates of decoherence rates and of the relationship of these rates to the relevant rates for dissipation. To see this, recall [see (2.114)] that the expression for the thermal de Broglie wavelength for a particle of mass M at temperature T is given by (setting, as usual, $\hbar \equiv 1$)

$$\lambda_{\mathrm{dB}} = \frac{1}{\sqrt{2Mk_B T}}. \tag{5.59}$$

Using the position representation, the last term on the right-hand side of (5.57) can therefore be written as

$$-\gamma_0 \left(\frac{X - X'}{\lambda_{\mathrm{dB}}}\right)^2 \rho_S(X, X', t). \tag{5.60}$$

This term describes spatial localization with a decoherence rate $\tau_{|X-X'|}^{-1}$ given by [compare (3.58) and (5.36)]

$$\tau_{|X-X'|}^{-1} = \gamma_0 \left(\frac{X - X'}{\lambda_{\mathrm{dB}}} \right)^2 . \tag{5.61}$$

We see that the ratio of the decoherence rate to the damping rate γ_0 is simply equal to the squared ratio of the spatial separation to the thermal de Broglie wavelength (5.59) of the object. Since the constant γ_0^{-1} quantifies the dissipation (relaxation) timescale of the system, (5.61) is precisely the expression (2.113) given (but not derived) in Sect. 2.11. As mentioned there, the expression (5.61) derived from the model for quantum Brownian motion was first used by Zurek in the early 1980s to introduce the notion of a decoherence timescale [12]. In the same section, we had also pointed out that the thermal de Broglie wavelength is extremely small for macroscopic and even mesoscopic objects. Hence such objects described by superpositions of macroscopically separated center-of-mass positions will typically be decohered on timescales many orders of magnitude shorter than the relaxation timescales.

Over timescales on the order of the decoherence time, we may therefore often neglect the dissipative term in the master equation (5.57). This yields the simplified, "pure-decoherence" master equation

$$\frac{\mathrm{d}}{\mathrm{d}t} \hat{\rho}_{\mathcal{S}}(t) = -\mathrm{i} \left[\hat{H}'_{\mathcal{S}}, \hat{\rho}_{\mathcal{S}}(t) \right] - 2M\gamma_0 k_{\mathrm{B}} T \left[\hat{X}, \left[\hat{X}, \hat{\rho}_{\mathcal{S}}(t) \right] \right] , \tag{5.62}$$

or, expressed in the position basis,

$$\frac{\partial}{\partial t} \rho_{\mathcal{S}}(X, X', t) = \left[-\frac{\mathrm{i}}{2M} \left(\frac{\partial^2}{\partial X'^2} - \frac{\partial^2}{\partial X^2} \right) - \frac{\mathrm{i}}{2} M \left(\Omega^2 - 2\gamma_0 \Lambda \right) \left(X^2 - X'^2 \right) \right.$$
$$\left. - \gamma_0 \left(\frac{X - X'}{\lambda_{\mathrm{dB}}} \right)^2 \right] \rho_{\mathcal{S}}(X, X', t). \tag{5.63}$$

This equation describes the local damping of spatial coherence at a rate given by (5.61). Note that (5.63) has the same structure as the equation of motion (3.78) for the case of environmental scattering. The role of the scattering constant Λ [see (3.56)] in the scattering model is now played by the coefficient $\gamma_0/\lambda_{\mathrm{dB}}^2$. The difference between the two equations (3.78) and (5.63) is simply that (5.63) corresponds to the case of the central system confined to a harmonic-oscillator potential, whereas the scattering model underlying (3.78) considered an unbound system.

However, this difference is far from critical, since it turns out that the form of the Caldeira–Leggett equation (5.57) is in fact not tied to the assumption of a system represented by a harmonic oscillator. It is possible to write down an equation of the form (5.57) for any arbitrary potential of the system. The system does not even need to be described by the canonical position and momentum coordinates $\hat{X}$ and $\hat{P}$. Instead, we may choose to model a discrete two-level system represented by the Pauli spin operators $\hat{\sigma}_x$ and $\hat{\sigma}_z$, as in the case of the spin–boson model (see Sect. 5.3). Then these Pauli spin operators simply take the place of the coordinates $\hat{X}$ and $\hat{P}$ in the master equation, and the distinct physical properties of the system are reflected in the particular choice of the physical quantities entering into the coefficients in the master equation.

5.2.5 Dynamics of Quantum Brownian Motion

Let us now return to the general form of the master equation (5.25) for quantum Brownian motion (with the central system represented by a harmonic oscillator) and illustrate the resulting time evolution of the system generated by this equation. These dynamics have been studied in detail by many authors. We will here mostly follow the analysis given by Paz, Habib, and Zurek [184].

We shall take the initial wave function of the system to be the coherent superposition

$$\Psi(X, t = 0) = \Psi_1(X) + \Psi_2(X), \tag{5.64}$$

where $\Psi_1(X)$ and $\Psi_2(X)$ are Gaussian wave packets given by

$$\Psi_{1,2}(X) = N \exp\left[-\frac{(X \mp X_0)^2}{2\delta^2}\right] \exp\left(\pm \mathrm{i} P_0 X\right). \tag{5.65}$$

Here, N is a normalization constant, and δ is the width of the Gaussian in position space. Thus we consider superpositions of two Gaussians that are initially symmetrically located in phase space, namely, at positions $\pm X_0$ with opposite and equal momenta P_0.

This strategy will allow us to investigate the decoherence dynamics in position and momentum space and thus to address key questions such as: Does decoherence in quantum Brownian motion also affect superpositions of momenta, or only superpositions of positions? How do the characteristic timescales for decoherence in these two bases differ? To answer these questions, we shall consider the decoherence dynamics for two limiting cases of interest [184]. The first situation is that of the superposition (5.64) of two Gaussians separated in position only ($X_0 > \delta$, $P_0 = 0$),

$$\Psi_{\text{pos}}(X, t = 0) = N\left\{\exp\left[-\frac{(X - X_0)^2}{2\delta^2}\right] + \exp\left[-\frac{(X + X_0)^2}{2\delta^2}\right]\right\}. \tag{5.66}$$

The second case corresponds to the superposition (5.64) with $\Psi_1(X)$ and $\Psi_2(X)$ separated in momentum only ($X_0 = 0$, $P_0 > 1/\delta$),

$$\Psi_{\text{mom}}(X, t = 0) = 2N \exp\left[-\frac{X^2}{2\delta^2}\right] \cos\left(P_0 X\right). \tag{5.67}$$

Since we are now working in a phase-space picture described by the coordinates X and P, it will be convenient to switch to the Wigner representation $W(X,P)$, see (2.132), of the position-space density matrix $\rho_S(X,X') = \Psi(X)\Psi^*(X')$. The Wigner function corresponding to the initial state (5.64) is rather easy to compute, since the Wigner representation of a Gaussian wave packet is again a Gaussian localized in the (X,P) plane. The Wigner representation of the two direct terms $\Psi_1(X)\Psi_1^*(X')$ and $\Psi_2(X)\Psi_2^*(X')$ can be shown to be of the form [184]

$$W_{1,2}(X,P,t=0) = N^2\delta^2 \exp\left[-\frac{(X \mp X_0)^2}{\delta^2}\right] \exp\left[-\delta^2(P \mp P_0)^2\right]. \quad (5.68)$$

The contribution from the off-diagonal (interference) terms $\Psi_1(X)\Psi_2^*(X')$ and $\Psi_2(X)\Psi_1^*(X')$ reads in the Wigner representation

$$W_{\text{int}}(X,P,t=0) = 2N^2\delta^2 \exp\left[-\frac{X^2}{\delta^2} - \delta^2 P^2\right] \cos\left[2X_0P + 2P_0X\right]. \quad (5.69)$$

The resulting full Wigner function of the initial state (5.64) is therefore

$$W(X,P,t=0) = W_1(X,P,t=0) + W_2(X,P,t=0) + W_{\text{int}}(X,P,t=0). \quad (5.70)$$

This function is illustrated in the top row of Fig. 5.4 for the two different limiting cases $\Psi_{\text{pos}}(X,t=0)$, see (5.66), and $\Psi_{\text{mom}}(X,t=0)$, see (5.67), representing separation in, respectively, position and momentum only. We clearly see the localization of the two "classical" (direct) peaks along the X and P axes, respectively (see also our general discussion in Sect. 2.15.2). We also observe the characteristic oscillatory pattern between these peaks. Note that the "ridges" and "valleys" of this pattern are parallel to the axis joining the peaks, i.e., parallel to the axis corresponding to the variable in which the separation occurs. As evident from (5.69), the wavenumber of the oscillations is directly determined by the separations X_0 and P_0 [see also (2.135)].

Let us now study the subsequent time evolution of the Wigner function (5.70) as given by the Born–Markov master equation (5.29) expressed in the Wigner representation. Without going into the details of the derivation here, one can show that the resulting time-evolved Wigner function can again be written in the form [184]

$$W(X,P,t) = W_1(X,P,t) + W_2(X,P,t) + W_{\text{int}}(X,P,t), \quad (5.71)$$

with the direct and interference terms given by

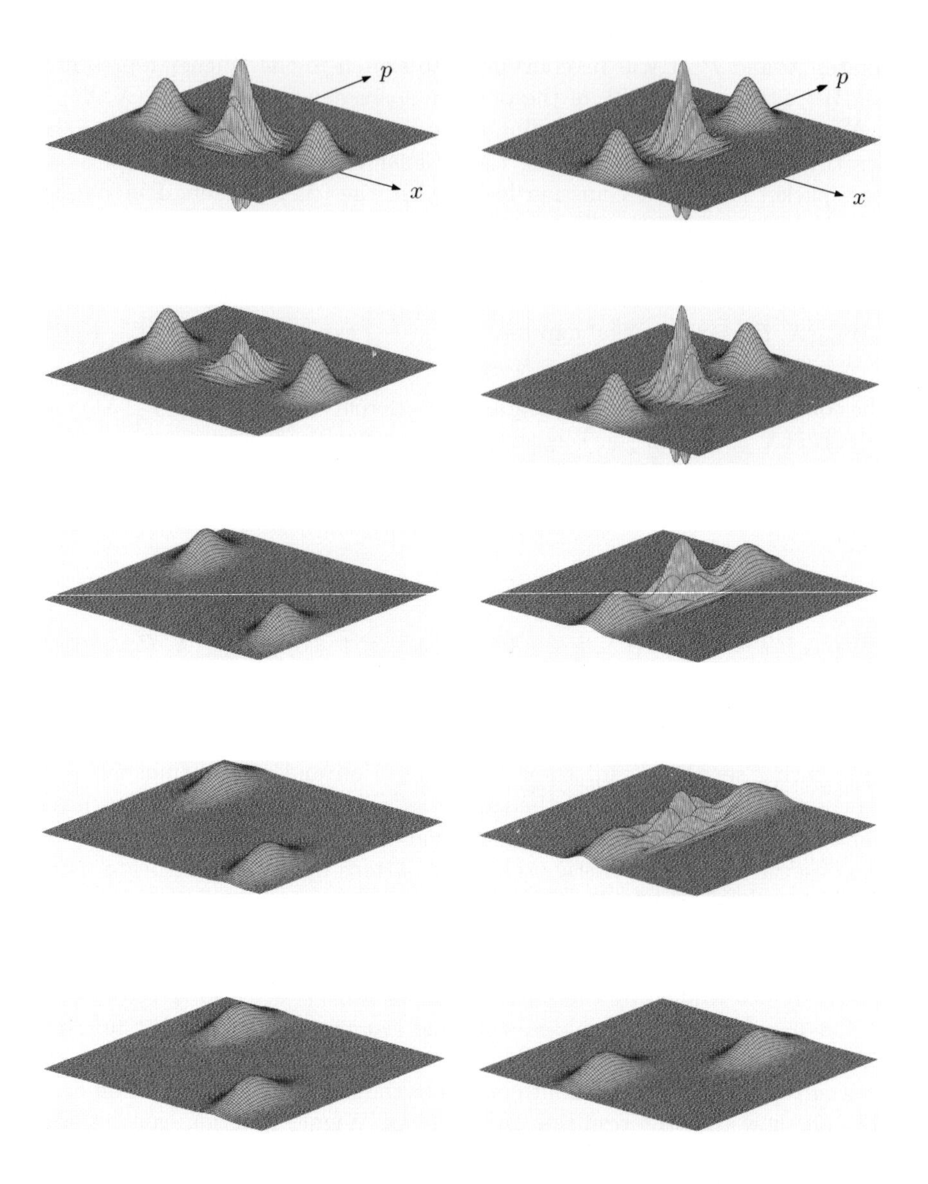

Fig. 5.4. Time evolution of a superposition of two Gaussian wave packets in the Wigner picture, as studied by Paz, Habib, and Zurek [184]. The left column corresponds to a separation of the initial wave packets in position only, whereas the right column represent the dynamics of a superposition of wave packets separated in momentum only. Interference between the two wave packets is represented by oscillations between the direct peaks. The interaction with the environment damps these oscillations, a process which represents decoherence. However, this damping occurs on different timescales for the two initial conditions.

$$W_{1,2}(X,P,t) = N^2\delta^2\frac{\delta_2}{\delta_1}\exp\left[-\frac{(X\mp X_c)^2}{\delta_1^2}\right] \times \exp\left[-\delta_2^2\left(P\mp P_c-\beta(X\mp X_c)\right)^2\right], \quad (5.72a)$$

$$W_{\text{int}}(X,P,t) = 2N^2\delta^2\frac{\delta_2}{\delta_1}\exp(-A_{\text{int}})\exp\left[-\frac{X^2}{\delta_1^2}-\delta_2^2(P-\beta X)^2\right] \times \cos\left[2\kappa_p P + 2(\kappa_x-\beta\kappa_p)X\right]. \quad (5.72b)$$

Here, the coefficients δ_1, δ_2, X_c, P_c, β, A_{int}, κ_x, and κ_p are in general time-dependent, with initial values [compare (5.72) with (5.68) and (5.69)]

$$\delta_1^2 = \delta_2^2 = \delta^2, \quad (5.73a)$$

$$\kappa_x = P_c = P_0, \quad \kappa_p = X_c = X_0, \quad (5.73b)$$

$$A_{\text{int}} = 0. \quad (5.73c)$$

Explicit (however rather complicated) general expressions for these coefficients obtained from the *exact* solution of the quantum Brownian model (i.e., the solution derived without making the Born–Markov approximations used in our above treatment; see Sect. 5.2.7 below) have been given by Paz, Habib, and Zurek [184] for arbitrary choices of the spectral density of the environment. More simple and concrete expressions for these coefficients can be obtained if we focus on the Born–Markov approximation and choose a particular form of the spectral density.

Regardless, the precise form of the coefficients will not be of further interest to us here. Instead, we shall try to get a good qualitative picture of the time evolution of the Wigner function as described by (5.71) and (5.72). This evolution is shown in Fig. 5.4 for the two limiting cases (5.66) and (5.67) of initial states. We see that for both initial conditions, the interaction with the environment leads to a suppression of the oscillations representing quantum interference between the wave packets. However, the timescales on which this decoherence happens are observed to be very different: The superposition (5.66) of two wave packets that are initially separated in position only is decohered much faster than the superposition (5.67) of momenta. In fact, as we shall analyze in more detail below, in the first case decoherence occurs on a timescale that is virtually independent of the relaxation and dynamical timescales of the system. By contrast, in the second case of a superposition of momenta quantum, interference is damped on the timescale set by the intrinsic dynamics of the system, i.e., by the self-Hamiltonian of the system.

This observation is quite easily explained. The interaction Hamiltonian explicitly depends on the position of the system. This means that the environment directly monitors this position, which leads to strong and rapid decoherence in the position basis. The momentum of the system, on the other hand, does not appear in the interaction Hamiltonian and is thus not directly monitored by the environment. Therefore decoherence in momentum can occur only in an indirect manner, i.e., as the result of the monitoring of position and the resulting decoherence in position. In the case in which the wave packets are initially separated in momentum only, the state (5.67) is already well-localized in position. Therefore the environmental monitoring of the position of the system cannot make this state even more localized in position, and the superposition (5.67) is initially insensitive to the monitoring by the environment, and thus to decoherence.

However, on a *dynamical* timescale this superposition evolves into a superposition of positions, since the opposite momenta $\pm P$ associated with the two direct peaks shift these peaks into opposite X-directions (see the right column of Fig. 5.4). Now the state becomes sensitive to the environmental monitoring of position, which in turn leads to the damping of the oscillations. This explains why decoherence in momentum does happen in spite of the absence of a momentum term in the interaction Hamiltonian, and why this decoherence occurs on a dynamical timescale rather than on the much shorter timescale for decoherence in position (i.e., the timescale corresponding to decoherence in the quantity that is directly monitored by the environment). In the following, we shall make these arguments more precise by analyzing the time dependence of the quantities appearing in the direct and interference terms (5.72a) and (5.72b) of the Wigner function (5.71).

First, let us look at the dynamics of the Gaussian peaks described by the direct terms $W_{1,2}(X, P, t)$. These peaks follow "classical" trajectories given by $X(t) = \pm X_c(t)$ and $P(t) = \pm P_c(t)$, which depend on the particular choice of the spectral density but not on the temperature of the environment. The widths of the Gaussian peaks, given by δ_1 (the width in position) and δ_2 (the width in momentum), change over time due to the presence of the environment. We also see that the environment leads to a distortion of the shape of the Gaussians, quantified by the coefficient β. Both the widths $\delta_{1,2}$ and the distortion β depend on the spectral density and the temperature of the environment.

Of most interest for our discussion of decoherence is the interference term $W_{\text{int}}(X, P, t)$, see (5.72b), representing the characteristic oscillations that are a signature of interference between the wave packets $\Psi_1(x)$ and $\Psi_1(x)$. First of all, we see that the (temperature-dependent) wavenumbers κ_x and κ_p of these oscillations along the X and P axes are altered by the interaction with the environment. The quantity most relevant to decoherence is the term $\exp(-A_{\text{int}})$, which (as we shall see) describes the damping of the oscillations and thus decoherence. We can interpret this term as quantifying the visibility of the interference fringes, since $\exp(-A_{\text{int}})$ equals the peak-to-peak ratio between the interference term $W_{\text{int}}(X, P)$ (which takes its maximum value at $X = P = 0$) and the direct terms $W_{1,2}(X, P)$ (which assume their peak values at $X = \pm X_c$ and $P = \pm P_c$),

$$\exp(-A_{\text{int}}) = \frac{1}{2} \frac{W_{\text{int}}(X = 0, P = 0)}{\left[W_1(X = X_c, P = P_c) W_2(X = -X_c, P = -P_c)\right]^{1/2}}. \tag{5.74}$$

In view of this interpretation, the term $\exp(-A_{\text{int}})$ is therefore often called the "fringe-visibility function." One can show that the quantity A_{int} is given by [184]

$$A_{\text{int}} = \frac{X_0^2}{\delta^2} + \delta^2 P_0^2 - \frac{\kappa_p^2}{\delta_2^2} - \delta_1^2 \kappa_x^2, \tag{5.75}$$

where all quantities appearing in this expression should be understood as explicitly time-dependent. At time $t = 0$, we have $\delta_1 = \delta_2 = \delta$, $\kappa_x = P_0$, $\kappa_p = X_0$ [see (5.73)], and thus $A_{\text{int}} = 0$. For subsequent times $t > 0$, A_{int} increases but is always bounded from above by its asymptotic (time-independent) maximum value of

$$A_{\text{int}}^{\text{max}} = \frac{X_0^2}{\delta^2} + \delta^2 P_0^2. \tag{5.76}$$

In fact, one can derive an explicit equation of motion for A_{int} from the master equation for quantum Brownian motion. The result is [184]

$$\frac{\mathrm{d}}{\mathrm{d}t} A_{\text{int}} = 4D\kappa_p^2 - 4f\kappa_p(\kappa_x - \beta\kappa_p), \tag{5.77}$$

where D and f are the usual normal-diffusion and anomalous-diffusion coefficients in the Born–Markov approximations. (If the Redfield equation or the exact master equation [219] are used, these coefficients will be explicitly time-dependent; see also Sect. 5.2.7.)

Equation (5.77) shows that the time evolution of A_{int}, and thus the dynamics of the suppression of the interference terms, depends on several parameters. First of all, we see that the change of A_{int} over time exhibits an explicit direct dependence on the diffusion coefficients D and f, but neither on the momentum-damping (friction) coefficient γ appearing in the master equation (5.25) nor on the free evolution given by the renormalized Hamiltonian $\hat{H}'_{\mathcal{S}}$. However, the time evolution of A_{int} is *indirectly* influenced by γ and $\hat{H}'_{\mathcal{S}}$, since it turns out that the dynamics of κ_x and κ_p appearing in (5.77) depend on γ and $\hat{H}'_{\mathcal{S}}$.

On the other hand, as we will see below, this indirect influence of γ and $\hat{H}'_{\mathcal{S}}$ on A_{int} is typically very weak in comparison with the direct dependence of A_{int} on the diffusion coefficients. This already hints at an important conclusion. In many cases the evolution of A_{int}, and thus the suppression of the oscillatory terms in the Wigner function, occurs on a timescale that is essentially independent of the relaxation timescale set by γ and of the dynamical timescale determined by $\hat{H}'_{\mathcal{S}}$.

Since first term $4D\kappa_p^2$ on the right-hand side of (5.77) is strictly non-negative, the normal-diffusion coefficient D will always correspond to a suppression of the interference terms and thus to decoherence. From (5.72b) we see that κ_p quantifies the wavenumber of the Wigner oscillations along the momentum direction. As evident from (5.77), the sign of the second term is not fixed but rather depends on the interplay between κ_x and $\beta\kappa_p$. Thus the anomalous-diffusion coefficient f may switch its role from aiding the suppression of interference to a counteracting of this suppression.

A crucial feature of the time evolution of A_{int} as given by (5.77) is the direct dependence on the initial conditions, in particular, on the initial separations X_0 and P_0. The dependence enters via the coefficients κ_x and κ_p, which at time $t = 0$ take the values $\kappa_x = P_0$ and $\kappa_p = X_0$ [see (5.73b)]. As we shall see now, this dependence is responsible for the very different timescales for decoherence in position and momentum as mentioned above (see also Fig. 5.4).

First, let us focus on the limiting case (5.66) in which the initial state corresponds to a superposition of two Gaussians separated in position space only. Since $\kappa_p = X_0$ at $t = 0$, we see that the contribution from the normal-diffusion term $4D\kappa_p^2$ in the evolution equation (5.77) for A_{int} immediately plays a role in damping interference.

Let us study the time dependence of A_{int} as given by (5.77) for the simple Caldeira–Leggett case of a high-temperature ohmic environment discussed in Sect. 5.2.4 above. Here, the anomalous-diffusion coefficient f becomes negligible in comparison with the normal-diffusion coefficient D, and D can be approximated by the value $2M\gamma_0 k_{\mathrm{B}}T$ [see (5.51)]. Furthermore, let us make the simplifying assumption that the interference fringes do not move much over time, i.e., that we can neglect the time dependence of κ_p and replace it by its initial value of X_0. With these assumptions, (5.77) is readily solved to give

$$A_{\mathrm{int}} = 8M\gamma_0 k_{\mathrm{B}}TX_0^2 t, \tag{5.78}$$

or, using the thermal de Broglie wavelength $\lambda_{\mathrm{dB}} = 1/\sqrt{2Mk_{\mathrm{B}}T}$,

$$A_{\mathrm{int}} = \gamma_0 \frac{(2X_0)^2}{\lambda_{\mathrm{dB}}^2} t. \tag{5.79}$$

Thus the oscillations in the Wigner function are exponentially damped at a decoherence rate

$$\tau_{\Delta X}^{-1} = \gamma_0 \frac{(\Delta X)^2}{\lambda_{\mathrm{dB}}^2}, \tag{5.80}$$

where ΔX is the coherent separation in position space. We thus recover our previous result (5.61) for the decoherence rate in the context of the Caldeira–Leggett case.

Let us now consider the situation (5.67) in which the superposition of Gaussians is initially only separated in momentum, i.e., $X_0 = 0$. Thus $\kappa_p = 0$ at $t = 0$, and consequently we initially have $\mathrm{d}A_{\mathrm{int}}/\mathrm{d}t = 0$. That is, A_{int} will remain at its initial value of zero until $\kappa_p = 0$ begins to differ from zero, and thus at this initial stage no decoherence occurs. For a weakly damped system, one can then show [184] that κ_p evolves as $\kappa_p(t) \propto X_c(t)$, i.e., its value follows the trajectory of the direct peaks. Thus an increase in κ_p will be tied to a change in X_c. X_c starts out from an initial value of zero and then evolves on a dynamical timescale. Now the diffusion terms in (5.77) can have an effect on $A_{\rm int}$, resulting in positive values of $A_{\rm int}$, which in turn leads to the suppression of the oscillations of the Wigner interference term $W_{\rm int}(X, P, t)$, see (5.72b).

Therefore, in the case of a superposition of Gaussian wave packets separated in momentum only, decoherence is related to two distinct timescales. First, a comparably slow dynamical timescale on which the superposition of positions is generated (i.e., on which X_c becomes sufficiently different from its initial value of zero) such that the environmental monitoring of the position of the system can have an effect. Second, once this superposition of positions is dynamically created, it is rapidly decohered on the decoherence timescale for such position-space superpositions (as given by, e.g., (5.80) for the case of high-temperature ohmic environments). Thus the decoherence rate in momentum is continuously limited by the rate of the dynamical evolution of the system. This means that the only timescale relevant to decoherence in momentum is the dynamical timescale. This analysis provides the rigorous support for our initial intuitive explanation of the much slower decoherence rate in momentum space when compared to the decoherence rate in position space.

What are the resulting preferred states (i.e., the pointer states) singled out by the interaction with the environment? To answer this question, one can compute the change in purity $\varsigma = \operatorname{Tr} \hat{\rho}_S^2$ [see (2.30)] of the reduced density matrix $\hat{\rho}_S$ of the harmonic-oscillator system. We can calculate this change from the Born–Markov master equation (5.25). We shall here neglect the influence of the momentum-damping term γ, because one can show that this term always leads to an unphysical increase in purity over time that is completely insensitive to the form of the density matrix [16]. Omitting this term can be well justified in the limit of weak damping, i.e., in the case of an "underdamped" harmonic oscillator. We shall also neglect the anomalous-diffusion term, which is appropriate in the high-temperature limit (see Sect. 5.2.4). With these assumptions, the *instantaneous* change in purity is found to be [86]

$$\frac{\mathrm{d}}{\mathrm{d}t}\varsigma(t) = -4D\Delta X^2(t), \tag{5.81}$$

where $\Delta X^2(t) = \langle \hat{X}^2 \rangle_t - \langle \hat{X} \rangle_t^2$ is the dispersion in position space. Thus we see that the instantaneous decrease in purity, and thus the amount of decoherence, is minimized if this spatial dispersion is minimized. That is, the instantaneous pointer states would be those perfectly localized in position. We may have anticipated this result based on the fact that the environment directly monitors the position of the system.

However, the intrinsic evolution of the harmonic-oscillator system as given by the self-Hamiltonian (5.23) also plays an important role. In fact, as is well known for a harmonic oscillator, over the course of an oscillation period $T = 2\pi/\Omega$ position and momentum interchange their roles. This symmetry of the Hamiltonian of the system with respect to the position and momentum variables means that a state that is asymmetrically localized in phase space (say, well localized in position but completely delocalized in momentum) would be strongly altered under the time evolution generated by the Hamiltonian of the system.

To make this argument more precise, instead of studying the instantaneous change (5.81) in purity, let us now look at the change in purity averaged over one full period $t = 0 \ldots T$ of the harmonic-oscillator system. This *average* change can be shown to be given by [15, 84]

$$\varsigma(T) - \varsigma(0) = -2D \left(\Delta X^2 + \frac{\Delta P^2}{M^2 \Omega^2} \right), \tag{5.82}$$

where $\Delta X^2 \equiv \Delta X^2(0)$ and $\Delta P^2 \equiv \Delta P^2(0)$ are the dispersions in position and momentum at the initial time $t = 0$. Note that the anomalous-diffusion term f does not appear in (5.82) even if we do not neglect it *a priori* [as in (5.81)], because it turns out that the influence of this term averages out over one oscillation period.

Equation (5.82) shows that the average decrease in purity, and thus the amount of decoherence, is smallest if ΔX^2 and ΔP^2 are chosen such as to minimize the term in brackets on the right-hand side of the equation, constrained by the requirement $\Delta X \Delta P \geq 1/2$ as mandated by the uncertainty relation (recall that we have set $\hbar \equiv 1$ throughout this chapter). The choice for ΔX^2 and ΔP^2 that minimizes the decrease (5.82) in purity is then found to be

$$\Delta X^2 = \frac{1}{2M\Omega}, \qquad \Delta P^2 = \frac{M\Omega}{2}. \tag{5.83}$$

Thus the environment-superselected pointer states in quantum Brownian motion are minimum-uncertainty Gaussians (coherent states) that are well localized in both position and momentum. This was first shown by Zurek, Habib, and Paz [86] (see also [84, 223–226]).

Quantum mechanics, of course, forbids perfect localization in phase space, but the coherent states selected by the interaction with the environment represent phase-space localization that is "as good as it gets." Thus we may view these coherent states as the quantum version of the idealized concept of a classical point in phase space. By reidentifying over time the (improper) ensembles of minimum-uncertainty Gaussian wave packets described by the reduced density matrix, we can recover quasi-Newtonian trajectories of particles in phase space.

Strictly speaking, the above derivation of coherent states as the pointer states in quantum Brownian motion was based on the assumption of a weakly damped harmonic oscillator. However, our conclusions are not significantly altered in the strong-damping regime, the main difference being a preference for stronger localization in position than momentum [227]. Moreover, Eisert [106] showed that the environment-induced superselection of minimal-uncertainty

Gaussian states is completely generic for the case of free quantum particles coupled to a thermal bath, in the sense that exact decoherence to such coherent states occurs for arbitrary bath temperatures, system–environment coupling strengths, and spectral densities of the environment.

Finally, one may also investigate the environment-induced superselection of preferred states in quantum Brownian motion for the limiting case of very small frequencies of the environment relative to the intrinsic frequency of the system. This limit of a "slow" environment is opposite to the situation of the Caldeira–Leggett model of Sect. 5.2.4 and corresponds to the "quantum limit" of decoherence (see Sect. 2.8.2). Paz and Zurek [103] considered a variant of the standard quantum Brownian motion model (namely, a particle interacting with a scalar field) and found that in this limit the decrease of purity of the reduced density matrix is minimized for energy eigenstates of the system (i.e., eigenstates of the self-Hamiltonian), which thus emerge as the pointer states for the system. This result is in agreement with our general discussion in Sect. 2.8.

5.2.6 Limitations of the Quantum Brownian Motion and Caldeira–Leggett Models

In Sect. 5.2.4, we showed how the Caldeira–Leggett model leads to the very simple equation (5.61) for estimating the rate of spatial decoherence. In particular, we found that the ratio of the rate of spatial decoherence to the rate of dissipation was given by the ratio of the coherent separation to the thermal de Broglie wavelength. As demonstrated in Sect. 2.11, in typical situations of mesoscopic and macroscopic objects described by superpositions of two positions separated by similarly mesoscopic or macroscopic distances, this ratio can be extremely large. This suggests that in such situations typical decoherence rates will be many orders of magnitude larger than the dissipation rates.

One should be somewhat guarded, however, against too a literal interpretation of such estimates. The assumptions that have led to the Caldeira–Leggett equation (5.57), and thus to the expression (5.61) for the decoherence rate, are often not fulfilled in physically realistic cases of interest [162,228,229]. It would therefore be inappropriate to overgeneralize (5.61) by interpreting it as a universally valid expression for the decoherence rate of spatial superpositions. Moreover, the general decoherence (and dissipation) dynamics arising from the Caldeira–Leggett equation (5.57) should be considered in the context and within the scope of this model, and one should be cautious about drawing overly generic conclusions from results obtained from this model.

For example, the decoherence rate as given by (5.61) scales quadratically with the coherent separation $\Delta X = |X - X'|$ in a position-space superposition state. Of course, this is intuitively reasonable: The larger the separation, the more nonclassical the superposition state, and the faster we expect decoherence to be. However, on physical grounds there should exist an upper limit to this decoherence rate that cannot be exceeded even if we increase the separation further.

The reason for this limit is easy to see by recalling our results for the case of environmental scattering discussed in Chap. 3. There, we distinguished two cases: The situation in which the typical wavelength of the scattered environmental particles is short with respect to the coherent separation ΔX (the "short-wavelength limit," see Sect. 3.3.1), and the opposite case in which this wavelength is long in comparison with ΔX (the "long-wavelength limit," see Sect. 3.3.2). In the short-wavelength limit, a single scattering event is able to completely resolve the separation ΔX, whereas in the long-wavelength limit many such events are needed to transmit an appreciable amount of which-path information to the environment. As discussed in Sect. 3.3.2, the decoherence rate obtained in the short-wavelength limit therefore imposes an upper bound on the decoherence rate: The environment cannot possibly obtain any *more* which-path information, and thus the decoherence rate cannot exceed the value calculated in the short-wavelength limit.

A similar argument should be expected to hold true for the case of the model for quantum Brownian motion discussed here. That is, we would anticipate that the environment of harmonic oscillators possesses a certain maximum coherence length. Then, for separations ΔX exceeding this coherence length, we would enter the equivalent of the short-wavelength regime considered for the case of environmental scattering, leading to a saturation of the decoherence rate. Evidently, such saturation does not occur in the Caldeira–Leggett model. There is no place in which a quantity equivalent to a coherence length of the environment would enter into the model, and the decoherence rate (5.61) grows without bounds as we increase the separation ΔX.

This example clearly shows that the results of the Caldeira–Leggett model cannot be extrapolated to arbitrary physical situations and parameters. To investigate this issue further, one can consider the general model of a massive particle coupled to a massless scalar field. This model was investigated in detail by Unruh and Zurek [109]. It was shown that the model for quantum Brownian motion emerges as a limiting case of this particle–field model if one assumes that the dominant wavelengths of the environmental field are much longer than the typical coherent separation ΔX of the central particle (this corresponds to a dipole-type approximation) [15, 109].

This indicates that an *implicit long-wavelength assumption* underlies the model for quantum Brownian motion—not only the Caldeira–Leggett model but also the more general model described by the master equation (5.25). This in turn explains why the decoherence rate predicted by the Caldeira–Leggett model does not saturate. In fact, in the general particle–field model the decoherence rate is found to exhibit an approximately quadratic dependence on the separation ΔX for values of ΔX small in comparison with the dominant wavelengths of the field, and to approach a constant value as ΔX grows much larger [15, 109, 229]. The lack of saturation of the decoherence rate as evident in (5.61) is therefore merely an artifact of the underlying model [228, 229].

Another limitation of the Markovian model for quantum Brownian motion is due to the use of the time-independent coefficients (5.26). As mentioned at the end of Sect. 5.2.2, if the more realistic treatment of time-dependent coefficients is used [see (5.42)], the normal-diffusion coefficient D is found to exhibit a temperature-independent transient during which D grows from an initial value of zero before settling into its final value (5.48). The timescale of this transient is given by the high-frequency cutoff Λ, see (5.44), which therefore imposes a limit on the decoherence rate. In our treatment, we have completely neglected this transient, and instead assumed that the coefficient D is constant at all times.

Furthermore, the role of the assumption (4.23) of an initially completely uncorrelated system–environment state in inducing rather spurious decoherence dynamics has been discussed in detail by Anglin and Zurek [229]. Finally, the Caldeira–Leggett model is based on the assumption of a high-temperature environment, and hence it may be inadequate for representing decoherence in low-temperature experimental settings such as those involving, say, superconductors. In particular, at low temperatures the anomalous-diffusion term (5.40) in the full master equation (5.25) may become relevant, approaching a similar magnitude as the normal-diffusion term (5.34) and thus significantly influencing the dynamics of the system [15, 220].

Thus, in summary, conclusions drawn from a specific decoherence model should not be misinterpreted as universally valid statements about generic properties of decoherence in nature. As we have seen, some seemingly physically relevant features observed in models of decoherence may often simply be due to artifacts of the model and the assumptions made in deriving the model.

However, despite these cautionary remarks, it should not go unmentioned that the estimates for decoherence rates obtained from the Caldeira–Leggett model have been found to yield surprisingly robust estimates of decoherence rates. That is, when these estimates are compared to predictions derived from more general and realistic models, good agreement is found even in cases where one may have anticipated greater discrepancies. For example, for environments where the high-temperature assumption underlying the Caldeira–Leggett model is not valid, the estimates of decoherence rates obtained from this—now strictly speaking incorrect—model are often surprisingly close to those derived from more complicated non-Markovian models. (See the analysis by Paz, Habib, and Zurek [184] for a more detailed comparison and discussion.) This feature, together with the simplicity of arriving at such estimates [see (5.61)], makes the Caldeira–Leggett model an excellent starting point for the treatment of many decoherence problems.

5.2.7 Exact Master Equation

The model for linear quantum Brownian motion has the remarkable feature of being exactly solvable. Maybe even more astonishing, as first shown by Hu, Paz, and Zhang in 1992 [219] following preliminary work by other authors [109,221,230–232], the resulting exact (non-Markovian) master equation is local in time for arbitrary spectral densities of the environment (see also our discussion in Sect. 4.4). This master equation takes exactly the same functional form as the Born–Markov master equation (5.25) derived in Sects. 5.2.1 and 5.2.2 above, namely,

$$\begin{aligned}\frac{\mathrm{d}}{\mathrm{d}t}\hat{\rho}_{\mathcal{S}}(t) = -\mathrm{i}\left[\hat{H}_{\mathcal{S}} + \frac{1}{2}M\widetilde{\Omega}^2(t)\hat{X}^2, \hat{\rho}_{\mathcal{S}}(t)\right] - \mathrm{i}\gamma(t)\left[\hat{X}, \left\{\hat{P}, \hat{\rho}_{\mathcal{S}}(t)\right\}\right] \\ - D(t)\left[\hat{X}, \left[\hat{X}, \hat{\rho}_{\mathcal{S}}(t)\right]\right] - f(t)\left[\hat{X}, \left[\hat{P}, \hat{\rho}_{\mathcal{S}}(t)\right]\right], \quad (5.84)\end{aligned}$$

with $\hat{H}_{\mathcal{S}} = \frac{1}{2M}\hat{P}^2 + \frac{1}{2}M\Omega^2\hat{X}^2$ representing the harmonic-oscillator self-Hamiltonian (5.23) of the system.

The only, and key, difference between the exact equation (5.84) and the Born–Markov version (5.25) is the fact that now the coefficients $\widetilde{\Omega}^2$, γ, D, and f are explicitly time-dependent and fairly complicated functions of other time-dependent coefficients, which in turn involve various integrals over the noise and dissipation kernels (5.16) and (5.17) (see [219] for details). All subtle features of the exact non-Markovian dynamics are encapsulated in these coefficients $\widetilde{\Omega}^2(t)$, $\gamma(t)$, $D(t)$, and $f(t)$ appearing in (5.84), while the interpretation of the different terms on the right-hand side of (5.84) is exactly the same as in the case of the Born–Markov master equation (5.25). As briefly mentioned in Sect. 4.3, the particular time dependences of the coefficients $\widetilde{\Omega}^2(t)$, $\gamma(t)$, $D(t)$, and $f(t)$ enforce the positivity of the reduced density matrix [see (4.42)], despite the fact that (not surprisingly) the exact master equation cannot be written in the positivity-ensuring Lindblad form (4.44).

The fact that the reduced dynamics in the model for linear quantum Brownian can be determined exactly and are given by a time-local master equation is a fairly unique feature. Not many other physically interesting models seem to allow for the calculation of the exact master equation. One such example that permits an exact solution is that of a two-level system interacting with an environment of harmonic oscillators described by an interaction Hamiltonian of a particular form (which is an example of a spin–boson model; see the next Sect. 5.3). As shown by Garraway [233, 234], the resulting exact master equation is local in time and has time-dependent coefficients, just as the master equation (5.84) for quantum Brownian motion.

5.3 The Spin–Boson Model

The spin–boson model corresponds to a single two-level system interacting with a large reservoir of bosonic field modes, i.e., a spin-$\frac{1}{2}$ particle coupled to an environment of harmonic oscillators. This model has been widely studied in the context of decoherence and dissipation in quantum systems. The role of two-level (qubit) systems in quantum computing (Chap. 7) and in experiments on macroscopic quantum coherence (Chap. 6) has led to additional interest in the spin–boson model. Recently, the model has even been used to analyze the role of quantum decoherence in biological systems [235].

The seminal review paper by Leggett et al. [209] discusses the extremely rich dynamics of the spin–boson model in great detail. Since the environment consists of harmonic oscillators, it is possible to derive the exact Feynman–Vernon influence functional [211] for the model and thereby to effectively eliminate the individual bath degrees of freedom from the problem (see Sect. IV of [209]). Thus, despite its complex dynamics, it is in principle possible to solve the spin–boson model exactly. In particular, one can formally write down exact expressions for the quantities of interest for the central spin, such as the expectation value $\langle \hat{\sigma}_z(t) \rangle$ and correlation functions.

Unfortunately, these exact expressions are rather complicated and very cumbersome to evaluate in practical situations of interest. One therefore often employs certain assumptions and approximations. One important example is the so-called "non-interacting blip approximation" that has been studied in detail by Leggett et al. [209]. This approximation not only greatly simplifies the task of determining the dynamics of the central spin, but the solutions obtained from it also turn out to be very close to the exact results over most of the parameter space. In fact, in certain parameter regimes they even coincide with the exact results.

Here we shall not discuss these more advanced methods and instead approach the spin–boson model in the following two steps. First, we shall consider a simplified version of the model in which the Hamiltonian of the spin system does not contain a tunneling term. For this model, the exact system–environment dynamics can be determined in a fairly straightforward manner without the use of the master-equation formalism. This simplified spin–boson model neatly exhibits characteristic features of decoherence, making it an ideal candidate for the study of decoherence in two-level systems. In fact, when quantum information and quantum computation started to attract attention in the 1990s (see Chap. 7), the model was used to gain some first insights into the decoherence of a single qubit in the presence of a thermal environment [89, 236]. Since then, it has been frequently revisited (see, e.g., [96, 237]).

In Sect. 5.3.2, we shall then generalize our model to the full spin–boson model by including a tunneling term in the Hamiltonian. We will derive the explicit form of the Born–Markov master equation (4.8) for this model, which will give us another opportunity to see how the general master-equation formalism described in Sect. 4.2 is applied to a concrete example.

5.3.1 Simplified Spin–Boson Model Without Tunneling

In the following, let us consider the simplified spin–boson model described by the Hamiltonian

$$\hat{H} = \hat{H}_{\mathcal{S}} + \hat{H}_{\mathcal{E}} + \hat{H}_{\text{int}}. \tag{5.85}$$

Here,

$$\hat{H}_{\mathcal{S}} = \frac{1}{2}\omega_0 \hat{\sigma}_z \tag{5.86}$$

is the self-Hamiltonian of the system. Let us denote the eigenstates of $\hat{\sigma}_z$ by $|0\rangle$ and $|1\rangle$. The asymmetry energy ω_0 is the difference in energy between the basis states $|0\rangle$ and $|1\rangle$ of the system. By contrast with the general spin–boson model, we have not included a tunneling term $-\frac{1}{2}\Delta_0\hat{\sigma}_x$ [see (5.1)] in $\hat{H}_{\mathcal{S}}$ that would generate intrinsic dynamics of the central spin.

In (5.85), $\hat{H}_{\mathcal{E}}$ denotes the familiar self-Hamiltonian (5.3) of the environment of harmonic oscillators,

$$\hat{H}_{\mathcal{E}} = \sum_i \left(\frac{1}{2m_i}\hat{p}_i^2 + \frac{1}{2}m_i\omega_i^2\hat{q}_i^2 \right). \tag{5.87}$$

As usual, the ith harmonic oscillator in the bath is described by its natural frequency ω_i, mass m_i, and position and momentum operators $\hat{q}_i$ and $\hat{p}_i$, respectively, and it defines a bosonic mode i. The interaction Hamiltonian

$$\hat{H}_{\text{int}} = \hat{\sigma}_z \otimes \sum_i c_i \hat{q}_i \tag{5.88}$$

describes the linear coupling of the $\hat{\sigma}_z$ coordinate of the system to the positions coordinates $\hat{q}_i$ of each harmonic oscillator in the environment, with coupling strengths c_i. Finally, let us also define $\hat{H}_0$ as the sum of the self-Hamiltonians of the system and the environment, $\hat{H}_0 \equiv \hat{H}_{\mathcal{S}} + \hat{H}_{\mathcal{E}}$.

Equivalently, we may recast the Hamiltonian (5.85) in terms of bosonic creation and annihilation operators $\hat{a}_i^\dagger$ and $\hat{a}_i$ for each mode i. This formulation leads to a more compact notation which, among other things, does no longer require us to explicitly include the oscillator masses m_i in the Hamiltonian. Using the standard relations

$$\hat{q}_i = \sqrt{\frac{1}{2m_i\omega_i}} \left(\hat{a}_i + \hat{a}_i^\dagger \right), \tag{5.89a}$$

$$\hat{p}_i = -\mathrm{i}\sqrt{\frac{m_i\omega_i}{2}} \left(\hat{a}_i - \hat{a}_i^\dagger \right), \tag{5.89b}$$

and dropping, for simplicity, the vacuum-energy term $\sum_i \frac{\omega_i}{2}$, we can rewrite (5.85) as

$$\hat{H} = \frac{1}{2}\omega_0\hat{\sigma}_z + \sum_i \omega_i \hat{a}_i^\dagger \hat{a}_i + \hat{\sigma}_z \otimes \sum_i \left(g_i \hat{a}_i^\dagger + g_i^* \hat{a}_i \right), \tag{5.90}$$

with the obvious relations between the c_i and g_i.[2] The operators $\hat{a}_i^\dagger$ and $\hat{a}_i$ obey the usual bosonic commutation relations,

$$\left[\hat{a}_i, \hat{a}_j^\dagger \right] = \delta_{ij}. \tag{5.91}$$

Let us first note that, similarly to the spin–spin model described in Sect. 2.10,

$$\left[\hat{H}, \hat{\sigma}_z \right] = 0. \tag{5.92}$$

There exists neither an intrinsic tunneling term proportional to $\hat{\sigma}_x$ nor any coupling between an environmental coordinate and $\hat{\sigma}_x$ that could induce any transitions between the two basis states $|0\rangle$ and $|1\rangle$ of the system. Thus the populations of the two levels of the system are conserved quantities. There is no energy exchange between the system and the environment, and we therefore deal with a model of decoherence without dissipation. As discussed in Sect. 2.11, while in realistic systems usually both dissipation and decoherence are present, the timescale for decoherence is typically many orders of magnitude shorter than the timescale for thermal relaxation. Thus our model can be regarded as a good representation of such rapid decoherence processes during which the amount of dissipation is negligible.

Solving the Model

To solve the model defined by the Hamiltonian (5.90), let us now switch to the interaction picture (see Appendix). Here the evolution of the interaction-picture Hamiltonian $\hat{H}_{\text{int}}(t)$ is given by the free part $\hat{H}_0$,

$$\hat{H}_{\text{int}}(t) = \mathrm{e}^{\mathrm{i}\hat{H}_0 t} \hat{H}_{\text{int}} \mathrm{e}^{-\mathrm{i}\hat{H}_0 t} = \hat{\sigma}_z \otimes \sum_i \left(g_i \hat{a}_i^\dagger \mathrm{e}^{\mathrm{i}\omega_i t} + g_i^* \hat{a}_i \mathrm{e}^{-\mathrm{i}\omega_i t} \right). \tag{5.93}$$

This equation is easily proved by noting that the Heisenberg equations of motion for the operators $\hat{\sigma}_z$, $\hat{a}_i^\dagger$, and $\hat{a}_i$ read, using (5.91),

[2] Note that, although the c_i are real numbers and are related to the g_i by a real-valued factor, we have used the complex conjugate of g_i in (5.90). This complex notation, introduced *a posteriori*, is a generalization that should be of no concern to the reader here.

$$\frac{\mathrm{d}\hat{\sigma}_z}{\mathrm{d}t} = \mathrm{i}\left[\hat{H}_0, \hat{\sigma}_z\right] = 0, \tag{5.94a}$$

$$\frac{\mathrm{d}\hat{a}_i^\dagger}{\mathrm{d}t} = \mathrm{i}\left[\hat{H}_0, \hat{a}_i^\dagger\right] = \mathrm{i}\omega_i\left[\hat{a}_i^\dagger \hat{a}_i, \hat{a}_i^\dagger\right] = i\omega_i \hat{a}_i^\dagger, \tag{5.94b}$$

$$\frac{\mathrm{d}\hat{a}_i}{\mathrm{d}t} = \mathrm{i}\left[\hat{H}_0, \hat{a}_i\right] = i\omega_i\left[\hat{a}_i^\dagger \hat{a}_i, \hat{a}_i\right] = -\mathrm{i}\omega_i \hat{a}_i. \tag{5.94c}$$

Next, we write down the evolution operator in the interaction picture, which reads

$$\hat{U}(t) = \mathcal{T}_{\leftarrow} \exp\left[-\mathrm{i}\int_0^t \mathrm{d}t'\, \hat{H}_{\mathrm{int}}(t')\right], \tag{5.95}$$

where $\mathcal{T}_{\leftarrow}$ denotes the time-ordered product of operators (i.e., operators are arranged such that their time arguments increase from right to left).[3] We can formally expand (5.95) in terms of a Dyson series [58] as

$$\hat{U}(t) = \sum_{n=0}^{\infty} \frac{(-\mathrm{i})^n}{n!} \int_0^t \mathrm{d}t_1 \int_0^{t_1} \mathrm{d}t_2 \cdots \times \int_0^{t_{n-1}} \mathrm{d}t_n\, \mathcal{T}_{\leftarrow} \left\{\hat{H}_{\mathrm{int}}(t_1)\hat{H}_{\mathrm{int}}(t_2)\cdots \hat{H}_{\mathrm{int}}(t_n)\right\}. \tag{5.96}$$

Since the commutator $\left[\hat{H}_{\mathrm{int}}(t), \hat{H}_{\mathrm{int}}(t' \neq t)\right]$ will in general be equal to another (time-dependent) operator, the required time ordering means that we usually cannot evaluate this series in closed form. However, a great simplification occurs in the present case. Since the commutator of the operators $\hat{a}_i^\dagger$ and $\hat{a}_i$ is a c-number (instead of another operator), the commutator of $\hat{H}_{\mathrm{int}}(t)$ and $\hat{H}_{\mathrm{int}}(t' \neq t)$, with $\hat{H}_{\mathrm{int}}(t)$ given by (5.93), is simply a function of c-numbers,

$$\left[\hat{H}_{\mathrm{int}}(t), \hat{H}_{\mathrm{int}}(t')\right] = -2\mathrm{i}\sum_i |g_i|^2 \sin\omega_i(t-t'). \tag{5.97}$$

The explicit evaluation of (5.95) using (5.97) is somewhat cumbersome, so we refer readers interested in the derivation to [237] for details. The key point is that (5.97) enables us to write $\hat{U}(t)$ as a product of a global time-dependent phase factor and the ordinary (not time-ordered) equivalent of (5.95),

$$\hat{U}(t) = \mathrm{e}^{\mathrm{i}\phi(t)} \exp\left[-\mathrm{i}\int_0^t \mathrm{d}t'\, \hat{H}_{\mathrm{int}}(t')\right] \equiv \mathrm{e}^{\mathrm{i}\phi(t)}\hat{V}(t). \tag{5.98}$$

In our subsequent discussion, the global phase factor will play no role (although it may become relevant in other settings; see [96] for an example).

[3] The necessity of a time ordering of the operators has sometimes been overlooked in discussions of this model in the literature (see, e.g., [89] and the correction in [96] and [237]).

Therefore, for our purposes, instead of the time-ordered evolution operator (5.95) we may simply use $\hat{V}(t)$, which can be directly evaluated by carrying out the integration in the argument of the exponential. This yields

$$\hat{V}(t) = \exp\left[\frac{1}{2}\hat{\sigma}_z \otimes \sum_i \left(\lambda_i(t)\hat{a}_i^\dagger - \lambda_i^*(t)\hat{a}_i\right)\right], \tag{5.99}$$

where we have defined the time-dependent coefficients

$$\lambda_i(t) \equiv 2\frac{g_i}{\omega_i}\left(1 - \mathrm{e}^{\mathrm{i}\omega_i t}\right). \tag{5.100}$$

What time evolution does $\hat{V}(t)$, see (5.99), induce? For the remainder of this section, let us make the usual assumption that there exist no correlations between the system and the environment at $t = 0$,

$$|\Psi(0)\rangle = (a|0\rangle + b|1\rangle)|\Phi_\mathcal{E}\rangle. \tag{5.101}$$

We then readily obtain the time evolution

$$\begin{aligned} |\Psi(t)\rangle &= \hat{V}(t)|\Psi(0)\rangle \\ &= a|0\rangle \prod_i \hat{D}(\lambda_i(t)/2)|\Phi_\mathcal{E}\rangle + b|1\rangle \prod_i \hat{D}(-\lambda_i(t)/2)|\Phi_\mathcal{E}\rangle \\ &\equiv a|0\rangle|\mathcal{E}_+(t)\rangle + b|1\rangle|\mathcal{E}_-(t)\rangle. \end{aligned} \tag{5.102}$$

Here we have introduced the operator

$$\hat{D}(\lambda_i(t)) \equiv \exp\left[\lambda_i(t)\hat{a}_i^\dagger - \lambda_i^*(t)\hat{a}_i\right], \tag{5.103}$$

which generates the evolution of the ith environmental oscillator.

The reader may immediately recognize the similarity of the time evolution described by (5.102) to that encountered in the simple static model of Sect. 2.10, see (2.105). In both cases, the interaction establishes quantum correlations between the basis states $|0\rangle$ and $|1\rangle$ of $\mathcal{S}$ and the corresponding relative states $|\mathcal{E}_+(t)\rangle$ and $|\mathcal{E}_-(t)\rangle$ of the environment.

What concrete form do these environmental states $|\mathcal{E}_+(t)\rangle$ and $|\mathcal{E}_-(t)\rangle$ take? To answer this question, we must first choose a particular initial state $|\Phi_\mathcal{E}\rangle$ of the environment [see (5.101)], which we shall now do for two examples of interest. First, we shall consider the case in which the environment is in the ground state. Second, we will look at the situation of the environment's being in thermal equilibrium.

Case Study #1: Environment in the Ground State

Here we shall assume that each harmonic oscillator in the environment is initially in the energy ground state $|E_0\rangle$,

$$|\Phi_{\mathcal{E}}\rangle = \prod_i |E_0\rangle_i \,, \tag{5.104}$$

where the index i runs over all environmental oscillators. For a quantum harmonic oscillator, a coherent state $|\lambda\rangle$ of amplitude λ is defined via the relation [238–241]

$$|\lambda\rangle \equiv \exp\left[\lambda \hat{a}^\dagger - \lambda^* \hat{a}\right] |E_0\rangle \,. \tag{5.105}$$

Comparing this definition with the expression for the operator $\hat{D}\left(\lambda_i(t)\right)$, see (5.103), we immediately see that $\hat{D}\left(\lambda_i(t)\right)$ is simply the generator of a coherent state of amplitude $\lambda_i(t)$ for the ith environmental oscillator. Thus the two final states $|\mathcal{E}_\pm(t)\rangle$ are products of coherent states with amplitudes $\pm\lambda_i(t)/2$,

$$|\mathcal{E}_\pm(t)\rangle = \prod_i |\pm\lambda_i(t)/2\rangle \,. \tag{5.106}$$

Therefore we deal with a measurement-like process. If the system is in the state $|0\rangle$ ($|1\rangle$), the state of the environment gets shifted into the state $|\mathcal{E}_+(t)\rangle$ ($|\mathcal{E}_-(t)\rangle$). If the states $|\mathcal{E}_\pm(t)\rangle$ are sufficiently distinguishable, i.e., if their overlap $\langle\mathcal{E}_-(t)|\mathcal{E}_+(t)\rangle$ is small, they act as "pointers" capable of discriminating between the states $|0\rangle$ and $|1\rangle$. In this case quantum coherence between $|0\rangle$ and $|1\rangle$ will be lost (or, rather, delocalized) from the system, and the off-diagonal elements in the reduced density matrix of the system expressed in the $\{|0\rangle\,, |1\rangle\}$ basis will decay.

To quantify this decoherence process, let us compute the time evolution of the overlap $r(t) \equiv \langle\mathcal{E}_-(t)|\mathcal{E}_+(t)\rangle$ (which plays the role of a decoherence factor). For two coherent states $|\lambda\rangle$ and $|\mu\rangle$ we have

$$\langle\lambda|\mu\rangle = \exp\left[-\frac{1}{2}|\lambda|^2 - \frac{1}{2}|\mu|^2 + \lambda^*\mu\right] . \tag{5.107}$$

Thus the decoherence factor $r(t)$ is given by

$$\begin{aligned} r(t) = \langle\mathcal{E}_-(t)|\mathcal{E}_+(t)\rangle &= \prod_i \exp\left[-\frac{1}{2}|\lambda_i(t)|^2\right] \\ &= \exp\left[-\sum_i 4\frac{|g_i|^2}{\omega_i^2}\left(1-\cos\omega_i t\right)\right] . \end{aligned} \tag{5.108}$$

For times t much smaller than the dynamical timescales ω_i^{-1} of the environmental modes, i.e., $\omega_i t \ll 1$, we can approximate the cosine in (5.108) by

$$\cos\omega_i t \approx 1 - \frac{1}{2}\omega_i^2 t^2 . \tag{5.109}$$

In this case the decoherence factor is

$$r(t) = \langle\mathcal{E}_-(t)|\mathcal{E}_+(t)\rangle \approx \exp\left[-\sum_i 2|g_i|^2 t^2\right] , \tag{5.110}$$

i.e., we obtain Gaussian decay of coherence between the states $|0\rangle$ and $|1\rangle$.

Deviations from this Gaussian behavior occur when t becomes comparable to the typical timescales of the environment, $\omega_i t \approx 1$, and thus the approximation (5.109) starts to fail. Fig. 5.5 shows an example for the time evolution of the decoherence factor $r(t)$ also for such larger times, with the purely Gaussian decay (5.110) pictured for comparison.

As already pointed out in Sect. 2.10, it is important to keep in mind that for finite numbers of environmental oscillators, $r(t)$ will return to its initial value of one after a finite (albeit for large environments very long) time. We can see this point directly from recognizing that the argument in the exponential of (5.108) is a sum of functions that are periodic in time, and thus the sum itself is periodic in time. Therefore, for finite-sized environments, the coherence initially localized at the level of the system will always be restored after a finite time span.

Case Study #2: Environment in Thermal Equilibrium

Let us now turn to the more general case in which each environmental oscillator i is in a thermal state,

$$\hat{\rho}_{\mathcal{E}_i} = \frac{1}{Z_i} \mathrm{e}^{-\omega_i \hat{a}_i^\dagger \hat{a}_i / k_\mathrm{B} T} \equiv \frac{1}{Z_i} \mathrm{e}^{-\hat{H}_{\mathcal{E}_i} / k_\mathrm{B} T}. \tag{5.111}$$

Here

$$Z_i = \mathrm{Tr}_{\mathcal{E}_i}\, \mathrm{e}^{-\hat{H}_{\mathcal{E}_i} / k_\mathrm{B} T} \tag{5.112}$$

is the partition function for the ith mode. Thus the initial state of the composite system $\mathcal{SE}$ is given by

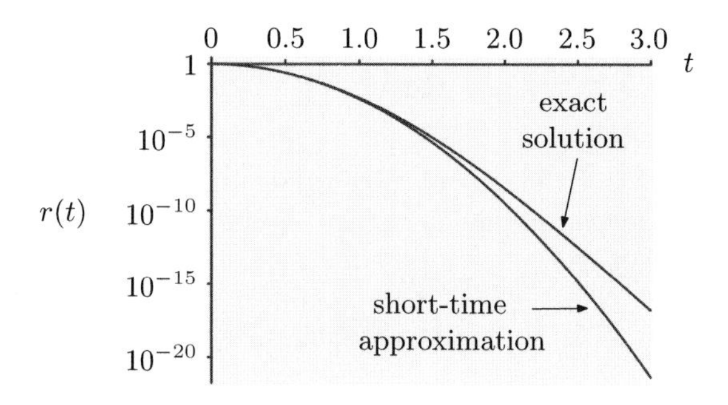

Fig. 5.5. Time evolution of the decoherence factor $r(t) = \langle \mathcal{E}_-(t) | \mathcal{E}_+(t) \rangle$ (i.e., of the overlap of the relative environmental states). Both the exact evolution (5.108) and the short-time approximation (5.110) are shown for an environment consisting of $N = 20$ harmonic oscillators. The couplings g_i and frequencies ω_i were chosen randomly from the interval $[0, 1]$.

$$\hat{\rho}(0) = \hat{\rho}_{\mathcal{S}}(0) \bigotimes_i \frac{1}{Z_i} \mathrm{e}^{-\hat{H}_{\mathcal{E}_i}/k_{\mathrm{B}}T}. \tag{5.113}$$

The time evolution of the reduced density matrix $\rho_{\mathcal{S}}(t)$ for the system is then obtained in the usual manner via

$$\hat{\rho}_{\mathcal{S}}(t) = \mathrm{Tr}_{\mathcal{E}} \left[\hat{V}(t) \rho_{\mathcal{S}}(0) \hat{V}^{-1}(t) \right], \tag{5.114}$$

with $\hat{V}(t)$ given by (5.99). Since we already know from (5.92) that the diagonal elements $\rho_{\mathcal{S}}^{(ii)}(t)$ of $\hat{\rho}_{\mathcal{S}}(t)$ expressed in the $\{|0\rangle, |1\rangle\}$ basis are constant in time,

$$\rho_{\mathcal{S}}^{(ii)}(t) = \rho_{\mathcal{S}}^{(ii)}(0), \qquad i = 0, 1, \tag{5.115}$$

we only need to compute the off-diagonal elements,

$$\begin{aligned} \rho_{\mathcal{S}}^{(01)}(t) &= \langle 0| \, \mathrm{Tr}_{\mathcal{E}} \left[\hat{V}(t) \hat{\rho}(0) \hat{V}^{-1}(t) \right] |1\rangle \,, \\ \rho_{\mathcal{S}}^{(10)}(t) &= \langle 1| \, \mathrm{Tr}_{\mathcal{E}} \left[\hat{V}(t) \hat{\rho}(0) \hat{V}^{-1}(t) \right] |0\rangle \,. \end{aligned} \tag{5.116}$$

Using the definitions above, these expressions are readily computed as follows,

$$\begin{aligned} \rho_{\mathcal{S}}^{(01)}(t) &= \rho_{\mathcal{S}}^{(01)}(0) \, \mathrm{Tr}_{\mathcal{E}} \left\{ \exp \left[\sum_i \left(\lambda_i(t) \hat{a}_i^\dagger - \lambda_i^*(t) \hat{a}_i \right) \right] \bigotimes_i \hat{\rho}_{\mathcal{E}_i} \right\} \\ &= \rho_{\mathcal{S}}^{(01)}(0) \prod_i \mathrm{Tr}_{\mathcal{E}_i} \left\{ \hat{D}\left(\lambda_i(t)\right) \hat{\rho}_{\mathcal{E}_i} \right\} \\ &= \rho_{\mathcal{S}}^{(01)}(0) \prod_i \left\langle \hat{D}\left(\lambda_i(t)\right) \right\rangle_{\hat{\rho}_{\mathcal{E}_i}} = \left[\rho_{\mathcal{S}}^{(10)}(t) \right]^* . \end{aligned} \tag{5.117}$$

In other words, the decoherence factor $r(t)$ is given by

$$r(t) = \prod_i \left\langle \hat{D}\left(\lambda_i(t)\right) \right\rangle_{\hat{\rho}_{\mathcal{E}_i}} = \prod_i \left\langle \exp \left[\lambda_i(t) \hat{a}_i^\dagger - \lambda_i^*(t) \hat{a}_i \right] \right\rangle_{\hat{\rho}_{\mathcal{E}_i}}, \tag{5.118}$$

i.e., by the product of the expectation values of the operators $\hat{D}\left(\lambda_i(t)\right)$, see (5.103), for the thermal modes $\hat{\rho}_{\mathcal{E}_i}$ of the environment.

It now turns out that

$$\chi(\lambda_i, t) \equiv \left\langle \exp \left[\lambda_i(t) \hat{a}_i^\dagger - \lambda_i^*(t) \hat{a}_i \right] \right\rangle_{\hat{\rho}_{\mathcal{E}_i}} \tag{5.119}$$

is just the symmetrically ordered characteristic function for a single harmonic oscillator (a single field mode) in thermal equilibrium, a function that is well-known from quantum optics (see, e.g., [241]). This function is given by [18, 150]

$$\chi(\lambda_i, t) = \exp \left[-\frac{1}{2} \left| \lambda_i(t) \right|^2 \coth\left(\omega_i / 2 k_{\mathrm{B}} T \right) \right]. \tag{5.120}$$

Hence

$$r(t) = \exp\left[-4\sum_i \frac{|g_i|^2}{\omega_i^2}\left(1-\cos\omega_i t\right)\coth\left(\omega_i/2k_\mathrm{B}T\right)\right] \equiv \mathrm{e}^{\Gamma(t)}, \qquad (5.121)$$

where we have denoted the argument of the exponential by $\Gamma(t)$.

Time evolution of the decoherence factor

After these rather formal manipulations, let us now try to get a feel for the time dependence of $r(t)$ by studying $\Gamma(t)$. First, let us assume that the environment is sufficiently large such that we can assume a continuous density of environmental modes. Just as in our discussion of quantum Brownian motion in Sect. 5.2, we therefore go from our expression (5.121) in terms of a sum over the discrete couplings g_i to a continuous description by means of a spectral density $J(\omega)$,

$$\sum_i |g_i|^2 \longrightarrow \int_0^\infty \mathrm{d}\omega\, J(\omega). \qquad (5.122)$$

With this replacement, $\Gamma(t)$ appearing in (5.121) becomes

$$\Gamma(t) = -\int_0^\infty \mathrm{d}\omega\, \frac{4J(\omega)}{\omega^2}\left(1-\cos\omega t\right)\coth\left(\omega/2k_\mathrm{B}T\right). \qquad (5.123)$$

Of course, if we choose

$$J(\omega) = \sum_i |g_i|^2\, \delta(\omega-\omega_i), \qquad (5.124)$$

the right-hand side of (5.123) will be trivially identical to our original expression (5.121). Instead, let us now represent $J(\omega)$ by a suitably chosen smooth function of ω.

As in the case of quantum Brownian motion (see Sect. 5.2.3), we shall consider a spectral density that is ohmic for sufficiently small frequencies, $J(\omega) \propto \omega$, and that has a smooth high-frequency cutoff quantified by Λ. Instead of the Lorentz–Drude form (5.44) used in our model for quantum Brownian motion, let us here choose an exponential cutoff of the form $\mathrm{e}^{-\omega/\Lambda}$. Thus our spectral density takes the form

$$J(\omega) = 4J_0\omega\mathrm{e}^{-\omega/\Lambda}, \qquad (5.125)$$

where J_0 is a dimensionless constant. The factor of four appearing in (5.123) has been absorbed into (5.125). The spectral density is plotted in Fig. 5.6. As desired, $J(\omega)$ increases approximately linearly for frequencies $\omega < \Lambda$ and decreases for $\omega > \Lambda$.

With the functional form (5.125) for $J(\omega)$, (5.123) reads

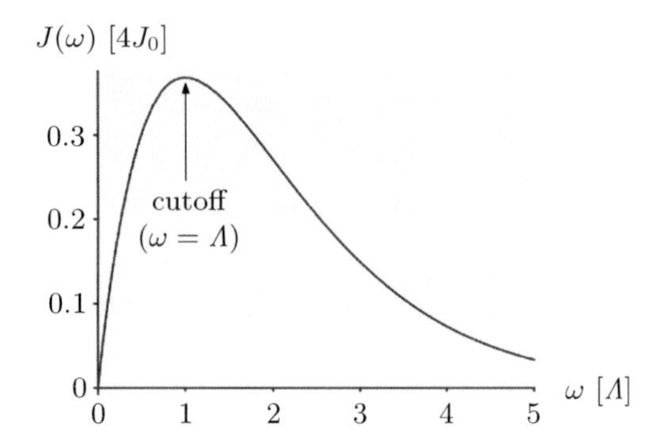

Fig. 5.6. Spectral density $J(\omega)$ given by (5.125), shown in units of $4J_0$. The spectral density is chosen to be approximately ohmic (i.e., $J(\omega) \propto \omega$) for frequencies well below the cutoff frequency Λ. At frequencies $\omega > \Lambda$, the ohmic character is suppressed by a term $\mathrm{e}^{-\omega/\Lambda}$. The frequency ω is measured in units of the cutoff Λ.

$$\Gamma(t) = -J_0 \int_0^{\infty} \mathrm{d}\omega \, \mathrm{e}^{-\omega/\Lambda} \frac{1 - \cos \omega t}{\omega} \coth\left(\omega / 2k_{\mathrm{B}} T\right). \tag{5.126}$$

If we now make the reasonable assumption that the thermal energy $k_{\mathrm{B}}T$ of the environment is small in comparison with the cutoff frequency Λ, $k_{\mathrm{B}}T \ll \Lambda$, this integral can be solved exactly. The result can be decomposed as

$$\Gamma(t) = \Gamma_{\text{fluc}}(t) + \Gamma_{\text{therm}}(t). \tag{5.127}$$

Here

$$\Gamma_{\text{fluc}}(t) = -\frac{1}{2} \ln\left(1 + \Lambda^2 t^2\right) \tag{5.128}$$

is the (exact) contribution to $\Gamma(t)$ due to purely quantum vacuum fluctuations (note that this term is independent of the bath temperature), and

$$\Gamma_{\text{therm}}(t) \approx -\ln\left[\frac{\sinh(\pi k_{\mathrm{B}} T t)}{\pi k_{\mathrm{B}} T t}\right] \tag{5.129}$$

is the contribution due to thermal fluctuations in the bath.

A plot of $\Gamma(t)$ is shown in Fig. 5.7. We see that $\Gamma(t)$ decreases from its initial value of one. We can therefore conclude that coherence between the components $|0\rangle$ and $|1\rangle$ indeed becomes locally damped, since the relevant decoherence factor (5.118) is given by $r(t) = \mathrm{e}^{\Gamma(t)}$. As we shall see in the following, we can distinguish several characteristic timescales for this decoherence process.

Decoherence timescales

Let us investigate the relevant timescales over which either one of the two contributions $\Gamma_{\text{fluc}}(t)$ and $\Gamma_{\text{therm}}(t)$ dominates the time evolution of $\Gamma(t)$.

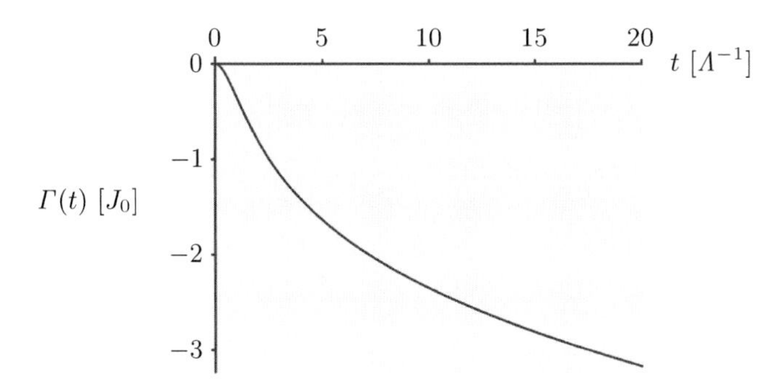

Fig. 5.7. Time dependence of the exponent $\Gamma(t)$ (in units of J_0), see (5.126), with time measured in units of the cutoff period Λ^{-1}. Here we have chosen the temperature of the bath such that $k_B T = \Lambda/60$, i.e., the thermal correlation time is assumed to be 60 times longer than the cutoff period Λ^{-1}.

From (5.128) and (5.129) we see that these timescales will be defined by the magnitude of t relative to the cutoff period Λ^{-1} and the thermal correlation time $(k_B T)^{-1}$.

First, let us assume that we are in the extreme short-time regime with $t \ll \Lambda^{-1}$ and $t \ll (k_B T)^{-1}$. Then we can terminate the Taylor expansion of the natural logarithm in the expression (5.128) for $\Gamma_{\text{fluc}}(t)$ after the second term (i.e., we expand $\Gamma_{\text{fluc}}(t)$ to first order in $\Lambda^2 t^2$),

$$\Gamma_{\text{fluc}}(t) \approx -\frac{1}{2}\Lambda^2 t^2, \tag{5.130}$$

and therefore $\Gamma_{\text{fluc}}(t) \ll 1$. The thermal contribution $\Gamma_{\text{therm}}(t)$ plays no role at all in this regime. Thus the decoherence factor $r(t)$, see (5.118), is close to one, with any (weak) decay being entirely due to quantum vacuum fluctuations.

Second, suppose t is larger than the characteristic period set by the cutoff frequency, $t \gg \Lambda^{-1}$, but still much smaller than the thermal correlation time, $t \ll (k_B T)^{-1}$. Then we can again neglect the contribution of $\Gamma_{\text{therm}}(t)$ to $\Gamma(t)$, and we can approximate $\Gamma_{\text{fluc}}(t)$ as

$$\Gamma_{\text{fluc}}(t) \approx -\frac{1}{2}\ln\left(\Lambda^2 t^2\right) = -\ln \Lambda t. \tag{5.131}$$

Thus $\Gamma_{\text{fluc}}(t)$ is of magnitude larger than one, and quantum vacuum fluctuations are a dominant source of decoherence in this regime.

Finally, let us assume the long-time regime for which t is much larger than the typical thermal fluctuation time, i.e., $t \gg (k_B T)^{-1}$. Then, using that

$$\sinh(x) = \frac{1}{2}\left(e^x - e^{-x}\right) \approx \frac{1}{2}e^x \qquad \text{if } x \gg 1, \tag{5.132}$$

we obtain

$$\Gamma_{\text{therm}}(t) \approx -\pi k_{\text{B}}Tt + \ln(2\pi k_{\text{B}}Tt) \approx -\pi k_{\text{B}}Tt \gg \Gamma_{\text{fluc}}(t), \tag{5.133}$$

and thus

$$\Gamma(t) \approx -\pi k_{\text{B}}Tt. \tag{5.134}$$

Inserting this expression for $\Gamma(t)$ into (5.121), we thus see that the decoherence factor $r(t)$ decays exponentially on a timescale set by the thermal correlation time $(k_{\text{B}}T)^{-1}$,

$$r(t) \approx \mathrm{e}^{-(\pi k_{\text{B}}T)t}. \tag{5.135}$$

5.3.2 Born–Markov Master Equation for the Spin–Boson Model

In the simplified spin–boson model discussed in the previous Sect. 5.3.1, the total Hamiltonian (5.90) was completely diagonal in the $\hat{\sigma}_z$ eigenbasis of the central spin. This means that the spin system has no intrinsic dynamics in the $\{|0\rangle, |1\rangle\}$ basis: If the system did not couple to any environment, the probability of finding the system in the state $|0\rangle$ (or $|1\rangle$) upon measurement would not change in time.

Let us now generalize this simplified model to include intrinsic dynamics by adding a tunneling term $-\frac{1}{2}\Delta_0\hat{\sigma}_x$ to the Hamiltonian (5.90), where Δ_0 denotes the tunneling matrix element [see (5.1)]. Switching back to the $(\hat{q}_i, \hat{p}_i)$ representation for the environmental coordinates, this yields the general spin–boson model defined by the Hamiltonian

$$\hat{H} = \frac{1}{2}\omega_0\hat{\sigma}_z - \frac{1}{2}\Delta_0\hat{\sigma}_x + \sum_i \left(\frac{1}{2m_i}\hat{p}_i^2 + \frac{1}{2}m_i\omega_i^2\hat{q}_i^2\right) + \hat{\sigma}_z \otimes \sum_i c_i\hat{q}_i. \tag{5.136}$$

Leggett et al. [209] showed in detail how this form of the Hamiltonian can be rigorously motivated from physical considerations. For our purposes, the introductory remarks of Sect. 5.1 on the role of spin-system and oscillator-environment models will be considered sufficient to motivate our interest in the Hamiltonian (5.136).

Despite the fact that this Hamiltonian seems only marginally more complicated than the no-tunneling Hamiltonian (5.90) considered in Sect. 5.3.1, the presence of the tunneling term renders the general spin–boson model defined by (5.136) much more complex, both in terms of its mathematical solution and of the dependences of the resulting decoherence dynamics on the relative magnitudes of the parameters. The aforementioned review by Leggett et al. [209] discusses in great detail exact and approximate solutions of the model.

Here, we shall focus on the derivation of the Born–Markov master equation for the spin–boson model. Recall that this master equation is appropriate if the system is weakly coupled to a large environment. To facilitate the subsequent calculations, we shall introduce one minor simplification in the

Hamiltonian (5.136). Namely, we will assume that the asymmetry energy ω_0 is equal to zero, corresponding to the central system moving in a symmetric double-well potential. With this assumption, the first $\hat{\sigma}_z$ term in (5.136) vanishes, such that our model is now defined by the Hamiltonian

$$\hat{H} = -\frac{1}{2}\Delta_0\hat{\sigma}_x + \sum_i \left(\frac{1}{2m_i}\hat{p}_i^2 + \frac{1}{2}m_i\omega_i^2\hat{q}_i^2 \right) + \hat{\sigma}_z \otimes \sum_i c_i\hat{q}_i. \tag{5.137}$$

This slightly simplified form of the Hamiltonian when compared with (5.136) makes the subsequent derivation of the Born–Markov master equation more transparent and will also allow us to more closely establish interesting analogies with quantum Brownian motion.

Fortunately, in our derivation we can proceed in a very similar manner as in the case of quantum Brownian motion (see Sect. 5.2.1). In fact, since we deal with the same type of environment and a bilinear system–environment coupling, many results can be taken over directly. For example, the bulk of the derivation of the master equation for quantum Brownian motion had been devoted to the calculation of the environment self-correlation functions (5.15). Since these functions are independent of the representation of the central system, we can directly use the results obtained in Sect. 5.2.1 in the following derivation.

In particular, we can take over the form of the master equation (5.22) for quantum Brownian motion and simply replace the position operator $\hat{X}$ by the Pauli z-spin operator $\hat{\sigma}_z$,

$$\begin{aligned}\frac{\mathrm{d}}{\mathrm{d}t}\hat{\rho}_S(t) = -\mathrm{i}\left[\hat{H}_S, \hat{\rho}_S(t)\right] - \int_0^\infty \mathrm{d}\tau \, \{\nu(\tau)\left[\hat{\sigma}_z, [\hat{\sigma}_z(-\tau), \hat{\rho}_S(t)]\right] \\ -\mathrm{i}\eta(\tau)\left[\hat{\sigma}_z, \{\hat{\sigma}_z(-\tau), \hat{\rho}_S(t)\}\right]\}.\end{aligned} \tag{5.138}$$

The noise and the dissipation kernels $\nu(\tau)$ and $\eta(\tau)$ appearing in (5.138) have exactly the same forms (5.16) and (5.17) as in the case of quantum Brownian motion, namely,

$$\nu(\tau) = \frac{1}{2}\sum_i c_i^2 \left\langle \{\hat{q}_i(\tau), \hat{q}_i\} \right\rangle_{\hat{\rho}_\mathcal{E}} \equiv \int_0^\infty \mathrm{d}\omega J(\omega) \coth\left(\frac{\omega}{2k_\mathrm{B}T}\right) \cos(\omega\tau), \tag{5.139}$$

$$\eta(\tau) = \frac{\mathrm{i}}{2}\sum_i c_i^2 \left\langle [\hat{q}_i(\tau), \hat{q}_i] \right\rangle_{\hat{\rho}_\mathcal{E}} \equiv \int_0^\infty \mathrm{d}\omega J(\omega) \sin(\omega\tau). \tag{5.140}$$

The time dependence of the operator $\hat{\sigma}_z$ in the interaction picture can be computed by solving the Heisenberg equations of motion with respect to the system Hamiltonian $\hat{H}_S = -\frac{1}{2}\Delta_0\hat{\sigma}_x$. The result is

$$\hat{\sigma}_z(\tau) = \mathrm{e}^{\mathrm{i}\hat{H}_S\tau}\hat{\sigma}_z\mathrm{e}^{-\mathrm{i}\hat{H}_S\tau} = \hat{\sigma}_z\cos(\Delta_0\tau) + \hat{\sigma}_y\sin(\Delta_0\tau). \tag{5.141}$$

Note the formal similarity of this expression to the result (5.24) for the interaction-picture position operator $\hat{X}(\tau)$ used in the context of the model for quantum Brownian motion.

We now insert (5.141) into our master equation (5.138). Using the standard commutation relations for the Pauli spin operators, we obtain after some straightforward rearrangements the final expression for the Born–Markov master equation for the spin–boson model as defined by the Hamiltonian (5.137),

$$\frac{\mathrm{d}}{\mathrm{d}t}\hat{\rho}_{\mathcal{S}}(t) = \underbrace{-\mathrm{i}\left[\hat{H}'_{\mathcal{S}}, \hat{\rho}_{\mathcal{S}}(t)\right]}_{\text{unitary evolution}} \underbrace{- \widetilde{D}\left[\hat{\sigma}_z, \left[\hat{\sigma}_z, \hat{\rho}_{\mathcal{S}}(t)\right]\right]}_{\text{decoherence}} + \underbrace{\zeta\hat{\sigma}_z\hat{\rho}_{\mathcal{S}}(t)\hat{\sigma}_y + \zeta^*\hat{\sigma}_y\hat{\rho}_{\mathcal{S}}(t)\hat{\sigma}_z}_{\text{decay}} . \tag{5.142}$$

Here, the renormalized ("Lamb-shifted") Hamiltonian of the system is

$$\hat{H}'_{\mathcal{S}} = \left[-\frac{1}{2}\Delta_0 - \zeta^*\right]\hat{\sigma}_x . \tag{5.143}$$

We thus see that the tunneling matrix element $\Delta_0/2$ is changed by an amount ζ^* due to the influence of the environment. The coefficient ζ is given by

$$\zeta = \int_0^\infty \mathrm{d}\tau\, \left[\nu(\tau) - \mathrm{i}\eta(\tau)\right] \sin\left(\Delta\tau\right) \equiv \widetilde{f} - \mathrm{i}\widetilde{\gamma}. \tag{5.144}$$

The coefficients $\widetilde{D}$, $\widetilde{f}$, and $\widetilde{\gamma}$ appearing in (5.142) and (5.144) are determined by expressions analogous to those for, respectively, the normal-diffusion and anomalous-diffusion coefficients D and f [see (5.26c) and (5.26d)] and the damping coefficient γ [see (5.26b)] in the case of quantum Brownian motion,

$$\widetilde{D} \equiv \int_0^\infty \mathrm{d}\tau\, \nu(\tau) \cos\left(\Delta\tau\right), \tag{5.145a}$$

$$\widetilde{f} \equiv \int_0^\infty \mathrm{d}\tau\, \nu(\tau) \sin\left(\Delta\tau\right), \tag{5.145b}$$

$$\widetilde{\gamma} \equiv \int_0^\infty \mathrm{d}\tau\, \eta(\tau) \sin\left(\Delta\tau\right). \tag{5.145c}$$

Accordingly, once a specific form of the spectral density of the environment is chosen, the coefficients (5.145) can be evaluated in much the same way as discussed for the case of quantum Brownian motion (see Sect. 5.2.3).

The interpretation of the different terms appearing on the right-hand side of the master equation (5.142) is also similar to the case of quantum Brownian motion. The term

$$\widetilde{D}\left[\hat{\sigma}_z, \left[\hat{\sigma}_z, \hat{\rho}_{\mathcal{S}}(t)\right]\right] = \widetilde{D}\left(\frac{1}{2}\hat{\rho}_{\mathcal{S}}(t) - 2\hat{\sigma}_z\hat{\rho}_{\mathcal{S}}(t)\hat{\sigma}_z\right) \tag{5.146}$$

is of the standard Lindblad double-commutator form (4.45), representing the direct monitoring of the system observable $\hat{\sigma}_z$ by the environment, as already evident from the structure of the interaction Hamiltonian, see (5.136). It thus describes decoherence in the basis $\{|0\rangle, |1\rangle\}$ at a rate given by the coefficient $\widetilde{D}$.

This can also be seen explicitly by expressing (5.146) as a 2×2 matrix in the basis $\{|0\rangle, |1\rangle\}$. If we denote the matrix elements $\langle i| \hat{\rho}_\mathcal{S}(t) |j\rangle$, $i \in \{0, 1\}$, of the reduced density matrix by $\rho_\mathcal{S}^{(ij)}(t)$, we get

$$\widetilde{D} [\hat{\sigma}_z, [\hat{\sigma}_z, \hat{\rho}_\mathcal{S}(t)]] \doteq \widetilde{D} \begin{pmatrix} 0 & \rho_\mathcal{S}^{(01)}(t) \\ \rho_\mathcal{S}^{(10)}(t) & 0 \end{pmatrix}. \tag{5.147}$$

Thus the time evolution of the off-diagonal matrix elements of the reduced density matrix (expressed in the eigenbasis of $\hat{\sigma}_z$) due to the influence of the $\widetilde{D}$ term (5.146) alone (neglecting all other terms appearing in the master equation) is

$$\begin{aligned} \frac{\mathrm{d}}{\mathrm{d}t} \rho_\mathcal{S}^{(01)}(t) &= -\widetilde{D} \rho_\mathcal{S}^{(01)}(t), \\ \frac{\mathrm{d}}{\mathrm{d}t} \rho_\mathcal{S}^{(10)}(t) &= -\widetilde{D} \rho_\mathcal{S}^{(10)}(t). \end{aligned} \tag{5.148}$$

This implies exponential decay of the off-diagonal elements at a rate given by $\widetilde{D}$. The diagonal elements $\rho_\mathcal{S}^{(00)}(t)$ and $\rho_\mathcal{S}^{(11)}(t)$ (i.e., the occupation probabilities) remain unaffected by the dynamics generated by the term (5.146), indicating a pure decoherence process without damping. This, again, is analogous to the case of quantum Brownian motion, where we found that the normal-diffusion term induced exponential damping of spatial coherences [see (5.35)].

Note that for the case of a vanishing tunneling matrix element, $\Delta_0 = 0$, we also have $\zeta = 0$. Then the master equation (5.142) reduces to the "pure-decoherence" form

$$\frac{\mathrm{d}}{\mathrm{d}t} \hat{\rho}_\mathcal{S}(t) = -\mathrm{i} \left[\hat{H}_\mathcal{S}, \hat{\rho}_\mathcal{S}(t) \right] - \widetilde{D} [\hat{\sigma}_z, [\hat{\sigma}_z, \hat{\rho}_\mathcal{S}(t)]], \tag{5.149}$$

which does not contain any terms that correspond to damping or frequency shifts. This is simply the Lindblad equation (4.45) with Lindblad operator $\hat{L} = \hat{\sigma}_z$, describing the environmental monitoring of the value of z-spin of the system.

Finally, the last two terms on the right-hand side of the master equation (5.142) describe the decay of the two-level system, quantified by an interplay between the coefficients $\widetilde{f}$ and $\widetilde{\gamma}$, see (5.145b) and (5.145c). This can be seen quite easily, e.g., by writing out these terms in matrix form. (Since our focus is on decoherence, we shall not show the proof here.)

In summary, we have seen how the Born–Markov master equation for the spin–boson model can be obtained in a manner quite analogous to the derivation of the master equation for quantum Brownian motion. The next step would consist of studying the spin–boson model beyond the Born–Markov approximations, which often break down in physical situations where the central two-level system and its environment are at very low temperatures (see the discussion in Sect. 4.4). Prominent examples for such low-temperature systems are the superconducting qubit systems operated at temperatures close to absolute zero (see Sect. 6.3).

As mentioned above, the rich non-Markovian dynamics of the spin–boson model have been analyzed in great detail by Leggett et al. [209]. As expected, the dynamics are found to exhibit strong dependences on the various parameters in the model, such as the temperature of the environment, the form of the spectral density (subohmic vs. ohmic vs. supraohmic), and the overall system–environment coupling strength. For each parameter regime, a characteristic dynamical behavior emerges: Localization, exponential or incoherent relaxation, exponential decay, and strongly or weakly damped coherent oscillations (see Table I of [209] for an overview of the different parameter regimes and the resulting dynamics).

We shall here refrain from discussing these rather complex dynamics and instead consider ourselves satisfied with the two models studied above, namely, the simplified spin–boson model and the spin–boson model in the framework of the Born–Markov approximations. These models already exhibit the main features relevant to a study of decoherence in models of the spin–boson type. Readers interested in an in-depth analysis of the non-Markovian dynamics of dissipation and decoherence in the spin–boson model will find a vast amount of information in the article by Leggett et al. [209] and the book by Weiss [218].

5.4 Spin-Environment Models

Let us now switch our focus from the study of environments composed of harmonic oscillators to the analysis of models in which the environment is represented by a collection of spin-$\frac{1}{2}$ particles (i.e., quantum two-level systems). In Sect. 5.1.2, we already discussed the physical motivation for the consideration of such spin environments and mentioned some key differences between spin and oscillator environments.

A central system represented by a spin particle linearly coupled to a collection of other spins comprises the physically most relevant (and most studied) spin-environment model. The typical physical situation for which such a "spin–spin model" is appropriate is that of a single two-level system, such as a superconducting qubit, strongly coupled to a low-temperature environment (see our discussion in Sect. 5.1). The other "canonical" possibility, a central harmonic oscillator interacting with a spin-type environment, has only recently begun to attract attention in the context of studying decoherence and dissipation in so-called quantum-electromechanical systems.

In this section, we shall investigate different aspects of such spin-environment models. We shall structure our discussion of spin–spin models in a similar manner as in the case of the spin–boson model. First, in Sect. 5.4.1, we will discuss a spin–spin model that is a simple generalization of the decoherence model studied in Sect. 2.10. Next, in Sect. 5.4.2, we will discuss spin-environment models in the limit of weak linear system–environment couplings. We will show how in this case the spin environment can be mapped onto an oscillator environment with a modified spectral density, as originally suggested by Feynman and Vernon [211] (see also Sect. 5.1.2). This will allow us to take over the formal results from our study of quantum Brownian motion, that is, we can derive a Born–Markov master equation of the form (5.22) for a central system weakly interacting with a spin bath.

In Sect. 5.4.3, we will discuss features of more general spin–spin models. We will focus on giving an overview of the decoherence dynamics arising in such models, without going into the mathematics required to solve the models. Readers interested in a more detailed account of the theory of spin-environment models will find plenty of material in the review article by Prokof'ev and Stamp [199]. A short and accessible summary is given in [200].

5.4.1 A Simple Dynamical Spin–Spin Model

We already encountered a member of the class of spin–spin models in Sect. 2.10 when we discussed a prototype model for illustrating basic features of environmental decoherence and superselection. We were able to solve the global dynamics of this model exactly in a straightforward fashion, because we had neglected the self-Hamiltonians of both the central spin and the spin environment, and because we had assumed that the interaction Hamiltonian was already diagonal in the $\hat{\sigma}_z$ eigenbasis of both the central spin and the collection of environmental spins [see (2.100)]. Clearly, this model was rather artificial, as it did not model the internal dynamics of the system and environment and thus corresponded to the limit of strong system–environment interactions which completely dominate the dynamics (the quantum-measurement regime).

We shall now generalize this static spin–spin model by adding a tunneling term of the form $-\frac{1}{2}\Delta_0\hat{\sigma}_x$ [see (5.1)] to the original Hamiltonian (2.100). Thus we consider the model described by the total Hamiltonian

$$\hat{H} = \hat{H}_{\mathcal{S}} + \hat{H}_{\text{int}} = -\frac{1}{2}\Delta_0\hat{\sigma}_x + \frac{1}{2}\hat{\sigma}_z \otimes \sum_{i=1}^{N} g_i \hat{\sigma}_z^{(i)} \equiv -\frac{1}{2}\Delta_0\hat{\sigma}_x + \frac{1}{2}\hat{\sigma}_z \otimes \hat{E}. \quad (5.150)$$

We will show that the inclusion of the tunneling term in (5.150) will in general lead to significantly different dynamics compared to the original static model defined by the Hamiltonian (2.100). The last term on the right-hand side of (5.150) corresponds to the familiar linear interaction between the z-spin coordinate of the central system and the z-spin coordinate of each environmental spin, with coupling strengths g_i.

Solving the Model

Despite the presence of the tunneling term, the global dynamics of the model defined by (5.150) can be solved exactly, as shown by Dobrovitski et al. [242] and Cucchietti, Paz, and Zurek [118]. We shall closely follow the latter approach [118] in our subsequent derivation.

A key simplifying feature of the Hamiltonian (5.150) arises from the fact that this Hamiltonian still contains only $\hat{\sigma}_z$ environment operators, just as in the static case (2.100). Thus, as in the model discussed in Sect. 2.10, the eigenstates $|n\rangle$ of the environment part $\hat{E}$ of the Hamiltonian are given by products of the eigenstates $|\uparrow\rangle_i$ and $|\downarrow\rangle_i$ of the ith environment operator $\hat{\sigma}_z^{(i)}$ [see (2.101)]. Explicitly,

$$\hat{E}\,|n\rangle = \epsilon_n\,|n\rangle\,, \quad n = 0, \ldots, 2^N - 1, \tag{5.151}$$

with eigenvalues [see (2.102)]

$$\epsilon_n = \sum_{i=1}^{N} (-1)^{n_i} g_i, \tag{5.152}$$

where $n_i = 0$ if the ith environmental spin is in the “up” state, and $n_i = 1$ if this spin is in the “down” state.

Consider now a pure product system–environment state given by

$$|\Psi_n\rangle = |\psi_S\rangle\,|n\rangle\,, \tag{5.153}$$

where $|\psi_S\rangle$ denotes an arbitrary state vector of the system. The action of the Hamiltonian (5.150) on this state is

$$\hat{H}\,|\Psi_n\rangle = \left(-\frac{1}{2}\Delta_0\hat{\sigma}_x + \frac{1}{2}\epsilon_n\hat{\sigma}_z\right)|\Psi_n\rangle\,. \tag{5.154}$$

In other words, for every environment state $|n\rangle$ we can consider a corresponding effective system Hamiltonian of the form

$$\hat{H}_S^{(n)} \equiv -\frac{1}{2}\Delta_0\hat{\sigma}_x + \frac{1}{2}\epsilon_n\hat{\sigma}_z. \tag{5.155}$$

The time evolution of an arbitrary initial product state of the form

$$|\Psi(0)\rangle = |\psi_S(0)\rangle \left(\sum_{n=0}^{2^N-1} c_n\,|n\rangle\right) \tag{5.156}$$

can be written as

$$|\Psi(t)\rangle = \sum_{n=0}^{2^N-1} c_n \left[\hat{U}_n(t) |\psi_S(0)\rangle\right] |n\rangle , \tag{5.157}$$

and thus

$$\hat{\rho}_S(t) = \mathrm{Tr}_{\mathcal{E}} |\Psi(t)\rangle\langle\Psi(t)| = \sum_{n=0}^{2^N-1} |c_n|^2 \hat{U}_n(t)\hat{\rho}_S(0)\hat{U}_n^\dagger(t), \tag{5.158}$$

where $\hat{\rho}_S(0) = |\psi_S(0)\rangle\langle\psi_S(0)|$. Here $\hat{U}_n(t)$ is the system evolution operator corresponding to the effective Hamiltonian (5.155). This operator is given by [118]

$$\hat{U}_n(t) = \hat{I}\cos(\Omega_n t) - \frac{\mathrm{i}}{\Omega_n}\left(\epsilon_n\hat{\sigma}_z - \frac{1}{2}\Delta_0\hat{\sigma}_x\right)\sin(\Omega_n t), \tag{5.159}$$

with $\Omega_n = \sqrt{\epsilon_n^2 + \Delta_0^2/4}$.

Following the approach of [118], let us now formally express the reduced density matrix $\hat{\rho}_S(t)$ in terms of the so-called polarization vector $\boldsymbol{p}(t) \equiv (p_x(t), p_y(t), p_z(t))$ as

$$\hat{\rho}_S(t) = \frac{1}{2}\left(\hat{I} + \boldsymbol{p}(t)\cdot\hat{\boldsymbol{\sigma}}\right) = \frac{1}{2}\left(\hat{I} + p_x(t)\hat{\sigma}_x + p_y(t)\hat{\sigma}_y + p_z(t)\hat{\sigma}_z\right). \tag{5.160}$$

We can interpret $\boldsymbol{p}$ as follows. Broadly speaking, the direction of $\boldsymbol{p}$ tells us into what set of eigenstates the density matrix (5.160) decomposes. For example, if $p_x = p_y = 0$,

$$\hat{\rho}_S = \frac{1}{2}\begin{pmatrix} 1+p_z & 0 \\ 0 & 1-p_z \end{pmatrix}, \tag{5.161}$$

and thus the eigenstates of $\hat{\rho}_S$ are given by the eigenstates $|0\rangle$ and $|1\rangle$ of $\hat{\sigma}_z$. On the other hand, if $p_y = p_z = 0$,

$$\hat{\rho}_S = \frac{1}{2}\begin{pmatrix} 1 & p_x \\ p_x & 1 \end{pmatrix}, \tag{5.162}$$

and the eigenstates of this density matrix are the eigenstates of $\hat{\sigma}_x$.

The dominating component of the polarization vector $\boldsymbol{p}(t \longrightarrow \infty)$ will tell us in which "direction" (i.e., in which basis) the decoherence process is most effective. Thus the polarization-vector description allows for an intuitive analysis of the preferred pointer basis arising in our model. For example, if $p_x(t)$ and $p_y(t)$ tend toward zero while $p_z(t)$ approaches a finite asymptotic value, we can conclude that the eigenbasis $\{|0\rangle, |1\rangle\}$ of $\hat{\sigma}_z$ is the pointer basis of the system.

Therefore our next goal will be to calculate the components $p_x(t)$, $p_y(t)$, and $p_z(t)$ of the polarization vector corresponding to the reduced density matrix (5.158). This will then allow us to study the resulting dynamics of the system. Explicit expressions $\left(p_x^{(n)}(t), p_y^{(n)}(t), p_z^{(n)}(t)\right)$ corresponding to each term $\hat{U}_n(t)\hat{\rho}_S(0)\hat{U}_n^\dagger(t)$ in the sum on the right-hand side of (5.158) are obtained rather easily by using the explicit expression (5.159) for $\hat{U}_n(t)$ (see [118] for the results).

The next step is then to calculate the sums over all individual terms of the form $|c_n|^2 p_s^{(n)}(t)$, $s \in \{x, y, z\}$, to obtain the full polarization vector [see (5.158) and (5.160)]. Instead of working with discrete indices n, let us now take the usual approach of rewriting (5.158) in an equivalent integral form as

$$\hat{\rho}_S(t) = \int \mathrm{d}\epsilon\, J(\epsilon)\hat{U}_\epsilon(t)\hat{\rho}_S(0)\hat{U}_\epsilon^\dagger(t), \tag{5.163}$$

where we have introduced the spectral density describing the distribution of the environment energies ϵ as

$$J(\epsilon) = \sum_{n=0}^{2^N-1} |c_n|^2 \,\delta(\epsilon - \epsilon_n). \tag{5.164}$$

As shown by Cucchietti, Paz, and Zurek [118], this distribution approaches a Gaussian for large numbers N of environmental spins,

$$J(\epsilon) \xrightarrow{N\longrightarrow\infty} \frac{1}{\sqrt{2\pi s_N^2}} \exp\left(-\frac{\epsilon^2}{2s_N^2}\right). \tag{5.165}$$

The width s_N of the Gaussian (5.165) quantifies the typical range of the environmental energies ϵ [see (5.152)]. We proved the result (5.165) in Sect. 2.10 for the simplified situation in which all environment couplings and expansion coefficients were assumed to be equal. However, (5.165) not only holds for much more general choices of coefficients, but is also a good (continuum) approximation to the exact expression (5.164) for already very modest values of N [118].

Using the Gaussian spectral density (5.165), the components $p_x(t)$, $p_y(t)$, and $p_z(t)$ of the polarization vector are then given by [118]

$$p_x(t) = \int \mathrm{d}\epsilon\, J(\epsilon) \frac{\Delta_0^2 + \epsilon^2 \cos(2\Omega_\epsilon t)}{\Omega_\epsilon^2} p_x(0), \tag{5.166a}$$

$$p_y(t) = \int \mathrm{d}\epsilon\, J(\epsilon) \left[\cos(2\Omega_\epsilon t)\, p_y(0) - \frac{\Delta_0 \sin(2\Omega_\epsilon t)}{\Omega_\epsilon} p_z(0)\right], \tag{5.166b}$$

$$p_z(t) = \int \mathrm{d}\epsilon\, J(\epsilon) \left[\frac{\epsilon^2 + \Delta_0^2 \cos(2\Omega_\epsilon t)}{\Omega_\epsilon^2} p_z(0) + \frac{\Delta_0 \sin(2\Omega_\epsilon t)}{\Omega_\epsilon} p_y(0)\right], \tag{5.166c}$$

where $\Omega_\epsilon = \sqrt{\epsilon^2 + \Delta_0^2/4}$.

Dynamics of the Model

Let us now evaluate the above integrals (5.166) for two important limiting situations. The first case, $\Delta_0 \ll s_N$, corresponds to weak intrinsic dynamics of the system. In this limit, the dynamics of the system will hardly be influenced by the presence of the self-Hamiltonian of the system. Thus we expect our model to approach the results of the static model discussed in Sect. 2.10. Following the study of Cucchietti, Paz, and Zurek [118], the evolution of the components (5.166) of the polarization vector for this case is shown as the top graph in Fig. 5.8. The polarization vector at time $t = 0$ was chosen to be $p_x(0) = p_y(0) = p_z(0) = 1/\sqrt{3}$ so as to avoid initial bias toward the eigenstates of any one of the operators $\hat{\sigma}_x$, $\hat{\sigma}_y$, or $\hat{\sigma}_z$.

We see that the components $p_x(t)$ and $p_y(t)$ rapidly decay from their initial values. The component $p_y(t)$ converges toward zero, while $p_x(t)$ approaches a constant positive value close to zero. By contrast, $p_z(t)$ hardly decays at all. Instead, it remains close to its initial value. The components $p_x(t)$ and

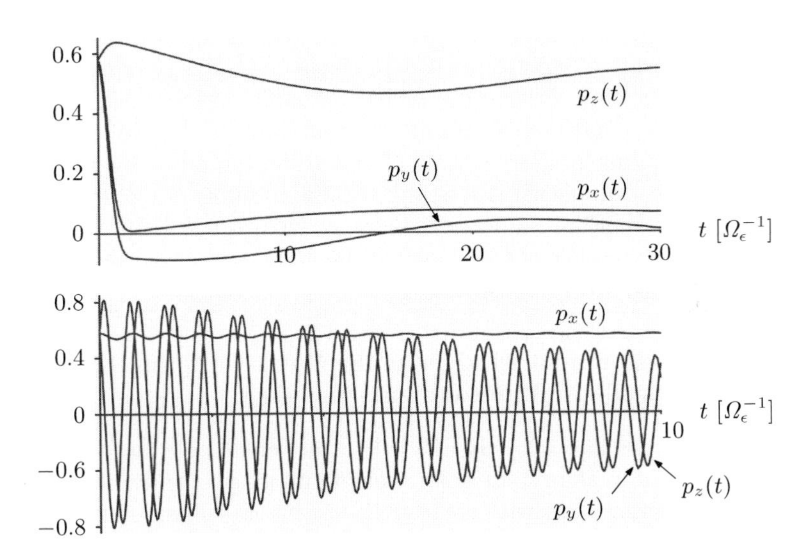

Fig. 5.8. Time dependence of the components $p_x(t)$, $p_y(t)$, and $p_z(t)$ of the polarization vector, see (5.166), as studied by Cucchietti, Paz, and Zurek [118]. This polarization vector describes the evolution of the reduced density matrix [see (5.160)] for the spin–spin model defined by the Hamiltonian (5.150). The time t is measured in units of the characteristic timescale Ω_ϵ^{-1} for the intrinsic evolution of the system. The top plot shows the case of weak intrinsic dynamics ($\Delta_0 = 0.1 s_N$), which leads to the selection of approximate eigenstates of $\hat{\sigma}_z$ as the preferred basis. The bottom plot depicts the opposite situation of strong self-dynamics of the system and low-energy ("slow") modes of the environment ($\Delta_0 = 5 s_N$). Now approximate eigenstates of $\hat{\sigma}_x$, and thus of the self-Hamiltonian, emerge as pointer states.

$p_z(t)$ settle into final values whose ratio is similar to the (small) ratio Δ_0/s_N. We thus see that the preferred pointer states selected by the environment are close to the eigenstates of $\hat{\sigma}_z$, with a small added contribution from $\hat{\sigma}_x$.

Let us now consider the second limiting case in which the highest energies available in the environment are smaller than the energy spacing Δ_0 of the system, i.e., $s_N \ll \Delta_0$. The bottom graph in Fig. 5.8 depicts the corresponding time evolution of the components of the polarization vector. We observe that $p_y(t)$ and $p_z(t)$ exhibit an oscillatory time dependence with a quickly decaying envelope, whereas $p_x(t)$ rapidly settles into a stationary value close to 1/2. This implies that the dynamically selected preferred basis of the system is now close to the eigenbasis of the x-spin operator $\hat{\sigma}_x$ and thus to the eigenbasis of the self-Hamiltonian. That is, the dynamics select energy eigenstates as the pointer basis for the system.

Our model nicely illustrates the dependence of the preferred basis on the relative strengths of the self-Hamiltonian of the system and the interaction Hamiltonian. The preferred basis emerges as the local basis that is most robust under the *total* Hamiltonian. As we have seen, in the quantum-measurement limit, i.e., when the interaction Hamiltonian dominates over the self-Hamiltonian ($\Delta_0 \ll s_N$), the preferred states are eigenstates of the interaction Hamiltonian, in agreement with the commutativity criterion (2.89) discussed in Sect. 2.8.1. Conversely, when the modes of the environment are "slow" ($s_N \ll \Delta_0$) and the self-Hamiltonian dominates the evolution of the system, the resulting preferred basis of the system is given by "local" energy eigenstates, i.e., eigenstates of the self-Hamiltonian. As emphasized in Sect. 2.8.2, this quantum limit of decoherence [103] is not simply equivalent to neglecting the presence of the environment. Rather, the environment leads to the decoherence of superpositions of energy eigenstates.

5.4.2 Spin-Environment Models in the Weak-Coupling Limit: Mapping to Oscillator Environments

In several places we have already alluded to the remarkable fact that, under the assumption of weak system–environment couplings, the dynamics of the reduced system interacting with *some* environment can be represented by this system interacting with an equivalent environment of harmonic oscillators [211, 212]. The mapping of a spin environment was first studied by Caldeira, Castro Neto, and de Carvalho [208].

We emphasize that this mapping does not mean that the dynamics of the system generated by an oscillator environment will be identical to those induced by the spin environment. In fact, the resulting dynamics will in general be very different. Instead, by "mapping" we mean to convey the insight that the reduced dynamics in the presence of the spin environment with a given spectral density are identical to those in the presence of an oscillator environment described by a spectral density that is suitably modified from the spectral density of the original spin environment. In other words, the Born–Markov master equation for the spin environment is given by the Born–Markov master equation (5.22) for the oscillator environment evaluated with a modified spectral density. We shall now show how this is done.

The Model

We decompose the Hamiltonian for our spin-environment model in the usual form as

$$\hat{H} = \hat{H}_{\mathcal{S}} + \hat{H}_{\mathcal{E}} + \hat{H}_{\text{int}}. \tag{5.167}$$

Here, $\hat{H}_{\mathcal{S}}$ is a general self-Hamiltonian of the system. We have also included a self-Hamiltonian

$$\hat{H}_{\mathcal{E}} = -\sum_i \frac{\Delta_i}{2} \hat{\sigma}_x^{(i)} \equiv \sum_i \hat{H}_{\mathcal{E}}^{(i)} \tag{5.168}$$

of the environment, where Δ_i denotes the tunneling matrix element for the ith environmental spin. This self-Hamiltonian lends intrinsic dynamics to the environment, in contrast with the more simplified spin-environment models considered in Sects. 2.10 and 5.4.1. For simplicity, we assume here that the asymmetry energy for each spin vanishes, i.e., that each environmental "particle" moves in a symmetric double-well potential.[4]

Finally, the interaction Hamiltonian $\hat{H}_{\text{int}}$ takes the usual form [compare (5.150)]

$$\hat{H}_{\text{int}} = \hat{s} \otimes \sum_i g_i \hat{\sigma}_z^{(i)} \equiv \hat{s} \otimes \hat{E}, \tag{5.169}$$

i.e., a general coordinate $\hat{s}$ of the system interacts bilinearly with each environmental spin, with coupling strengths quantified by the g_i. Later, we will specialize on the cases $\hat{s} = \hat{\sigma}_z$ for a spin system (the "spin–spin model") and $\hat{s} = \hat{X}$ for a harmonic-oscillator system.

Computing the Spin-Environment Self-Correlation Function

We assume the limit of weak system–environment couplings and take the environment to be in thermal equilibrium at temperature T. We would now like to derive the Born–Markov master equation (see Sect. 4.2) for this spin-bath model. To do so, the first goal will be the derivation of the environment self-correlation functions (4.10). Just as in the case of quantum Brownian motion [compare (5.5)], the interaction Hamiltonian (5.169) contains only a single

[4] The case of nonvanishing asymmetry energies, and thus the inclusion of terms proportional to $\hat{\sigma}_z^{(i)}$ in the self-Hamiltonian (5.168), can be treated in much the same way as outlined in this section. The only difference is that now the time dependence of the environmental z-spin operators in the interaction-picture becomes slightly more complicated, leading to the appearance of an additional (constant) term in the environment self-correlation function. However, the main conclusions of this section remain unaltered by this generalization.

term, and thus we need to calculate only one environment self-correlation function, namely,

$$\mathcal{C}(\tau) = \left\langle \hat{E}(\tau)\hat{E} \right\rangle_{\hat{\rho}_{\mathcal{E}}}. \tag{5.170}$$

Here $\hat{E}(\tau) = \mathrm{e}^{\mathrm{i}\hat{H}_{\mathcal{E}}\tau}\hat{E}\mathrm{e}^{-\mathrm{i}\hat{H}_{\mathcal{E}}\tau}$ denotes the collective environment operator $\hat{E}$ [defined in (5.169)] in the interaction picture.

Since the environmental spins do not interact with each other and thus $\left[\hat{\sigma}_x^{(i)}, \hat{\sigma}_z^{(j)}\right] = 0$ for $i \neq j$, the exponential $\mathrm{e}^{\mathrm{i}\hat{H}_{\mathcal{E}}\tau}$ factors into single-spin terms,

$$\mathrm{e}^{\mathrm{i}\hat{H}_{\mathcal{E}}\tau} \equiv \mathrm{e}^{\mathrm{i}\sum_i \hat{H}_{\mathcal{E}}^{(i)}\tau} = \prod_i \mathrm{e}^{\mathrm{i}\hat{H}_{\mathcal{E}}^{(i)}\tau}. \tag{5.171}$$

Therefore we can write the environment self-correlation function (5.170) as

$$\mathcal{C}(\tau) = \sum_{ij} g_i g_j \left\langle \mathrm{e}^{\mathrm{i}\hat{H}_{\mathcal{E}}^{(i)}\tau}\hat{\sigma}_z^{(i)}\mathrm{e}^{-\mathrm{i}\hat{H}_{\mathcal{E}}^{(i)}\tau}\hat{\sigma}_z^{(j)} \right\rangle_{\hat{\rho}_{\mathcal{E}}} \equiv \sum_{ij} g_i g_j \left\langle \hat{\sigma}_z^{(i)}(\tau)\hat{\sigma}_z^{(j)} \right\rangle_{\hat{\rho}_{\mathcal{E}}}. \tag{5.172}$$

We now use again the fact that the environmental spins do not directly interact with each other and are therefore uncorrelated, which implies [compare (5.7) for the analogous argument in the model for quantum Brownian motion]

$$\left\langle \hat{\sigma}_z^{(i)}(\tau)\hat{\sigma}_z^{(j)} \right\rangle_{\hat{\rho}_{\mathcal{E}}} = \left\langle \hat{\sigma}_z^{(i)}(\tau) \right\rangle_{\hat{\rho}_{\mathcal{E}}} \left\langle \hat{\sigma}_z^{(j)} \right\rangle_{\hat{\rho}_{\mathcal{E}}} \tag{5.173}$$

for $i \neq j$. We may therefore decompose the sum in (5.172) as

$$\mathcal{C}(\tau) = \sum_i g_i \left\langle \hat{\sigma}_z^{(i)}(\tau) \right\rangle_{\hat{\rho}_{\mathcal{E}}} \sum_{j\neq i} g_j \left\langle \hat{\sigma}_z^{(j)} \right\rangle_{\hat{\rho}_{\mathcal{E}}} + \sum_i g_i^2 \left\langle \hat{\sigma}_z^{(i)}(\tau)\hat{\sigma}_z^{(i)} \right\rangle_{\hat{\rho}_{\mathcal{E}}}. \tag{5.174}$$

We now make the nonrestrictive assumption that the average of the "quantum force" due to the spin bath vanishes at $t = 0$ (if this is not the case, we can always add a constant to achieve this goal), that is,

$$\left\langle \hat{E} \right\rangle_{\hat{\rho}_{\mathcal{E}}} = \sum_i g_i \left\langle \hat{\sigma}_z^{(i)} \right\rangle_0 = 0. \tag{5.175}$$

This implies that the term $\sum_{j\neq i} g_j \left\langle \hat{\sigma}_z^{(j)} \right\rangle_{\hat{\rho}_{\mathcal{E}}}$ appearing in the first sum on the right-hand side of (5.174) is also (approximately) zero. This allows us to greatly simplify (5.172) by neglecting the crossterms $i \neq j$,

$$\begin{aligned} \mathcal{C}(\tau) &= \sum_i g_i^2 \left\langle \hat{\sigma}_z^{(i)}(\tau)\hat{\sigma}_z^{(i)} \right\rangle_{\hat{\rho}_{\mathcal{E}}} \\ &= \sum_i g_i^2 \,\mathrm{Tr}_{\mathcal{E}} \left\{ \left[\frac{1}{Z} \prod_i \mathrm{e}^{-\hat{H}_{\mathcal{E}}/k_{\mathrm{B}}T} \right] \hat{\sigma}_z^{(i)}(\tau)\hat{\sigma}_z^{(i)} \right\}. \end{aligned} \tag{5.176}$$

In the last line we have explicitly written out the expression for the thermal average over the initial state $\hat{\rho}_{\mathcal{E}}$ of the bath, with $Z = \mathrm{Tr}_{\mathcal{E}}\, \mathrm{e}^{-\hat{H}_{\mathcal{E}}/k_{\mathrm{B}}T}$. First of all, since the bath spins are uncorrelated, the thermal average over the initial state $\hat{\rho}_{\mathcal{E}}$ involving all bath spins reduces to single-spin averages,

$$\mathcal{C}(\tau) = \sum_i g_i^2 \frac{1}{Z_i} \mathrm{Tr}_{\mathcal{E}_i} \left\{ \mathrm{e}^{-\hat{H}_{\mathcal{E}}^{(i)}/k_{\mathrm{B}}T} \hat{\sigma}_z^{(i)}(\tau) \hat{\sigma}_z^{(i)} \right\}, \tag{5.177}$$

where Z_i is defined as in (5.112). Next, we need to compute the time evolution of the interaction-picture operator $\hat{\sigma}_z^{(i)}(\tau)$ given by

$$\hat{\sigma}_z^{(i)}(\tau) = \mathrm{e}^{-\mathrm{i}\hat{H}_{\mathcal{E}}^{(i)}\tau} \hat{\sigma}_z^{(i)} \mathrm{e}^{-\mathrm{i}\hat{H}_{\mathcal{E}}^{(i)}\tau}. \tag{5.178}$$

Denoting the eigenbasis of $\hat{\sigma}_z^{(i)}$ by $\{|\uparrow\rangle_i, |\downarrow\rangle_i\}$, the explicit matrix representation of the environment Hamiltonian $\hat{H}_{\mathcal{E}}^{(i)}$ is

$$\hat{H}_{\mathcal{E}}^{(i)} = \begin{pmatrix} 0 & -\Delta_i/2 \\ -\Delta_i/2 & 0 \end{pmatrix}. \tag{5.179}$$

Diagonalizing this matrix yields the eigenvalues $E_\pm = \pm(-\Delta_i/2)$ and corresponding eigenvectors $|\pm\rangle_i = (|\uparrow\rangle_i \mp |\downarrow\rangle_i)/\sqrt{2}$. In this *new* basis $\{|+\rangle_i, |-\rangle_i\}$, the operator $\hat{\sigma}_z^{(i)}$ takes the form

$$\hat{\sigma}_z^{(i)} = (|+\rangle\langle-|)_i + (|-\rangle\langle+|)_i. \tag{5.180}$$

Thus the time evolution in the interaction picture is

$$\hat{\sigma}_z^{(i)}(\tau) = \mathrm{e}^{-\mathrm{i}\Delta_i\tau}(|+\rangle\langle-|)_i + \mathrm{e}^{\mathrm{i}\Delta_i\tau}(|-_i\rangle\langle+|)_i. \tag{5.181}$$

With these explicit expressions, the argument in curly brackets on the right-hand side of (5.177) reads

$$\mathrm{e}^{-\hat{H}_{\mathcal{E}}^{(i)}/k_{\mathrm{B}}T} \hat{\sigma}_z^{(i)}(\tau) \hat{\sigma}_z^{(i)} = \mathrm{e}^{\Delta_i/2k_{\mathrm{B}}T - \mathrm{i}\Delta_i\tau}(|+\rangle\langle+|)_i + \mathrm{e}^{-\Delta_i/2k_{\mathrm{B}}T + \mathrm{i}\Delta_i\tau}(|-\rangle\langle-|)_i. \tag{5.182}$$

Taking the trace over this expression (for simplicity in the $\{|+\rangle_i, |-\rangle_i\}$ basis) and using that

$$Z_i = \mathrm{Tr}_{\mathcal{E}_i}\, \mathrm{e}^{-\hat{H}_{\mathcal{E}}^{(i)}/k_{\mathrm{B}}T} = \mathrm{e}^{\Delta_i/2k_{\mathrm{B}}T} + \mathrm{e}^{-\Delta_i/2k_{\mathrm{B}}T}, \tag{5.183}$$

we obtain the explicit form of the environment self-correlation function (5.177),

$$\begin{aligned} \mathcal{C}(\tau) &= \sum_i g_i^2 \frac{1}{Z_i} \mathrm{Tr}_{\mathcal{E}} \left\{ \mathrm{e}^{-\hat{H}_{\mathcal{E}}^{(i)}/k_{\mathrm{B}}T} \hat{\sigma}_z^{(i)}(\tau) \hat{\sigma}_z^{(i)} \right\} \\ &= \sum_i g_i^2 \left\{ \cos(\Delta_i\tau) - \mathrm{i} \tanh\left(\frac{\Delta_i}{2k_{\mathrm{B}}T} \right) \sin(\Delta_i\tau) \right\}. \end{aligned} \tag{5.184}$$

Finally, let us follow the usual procedure and replace the discrete sum involving the environmental self-energies Δ_i by an integral using the spectral density

$$J(\Delta) \equiv \sum_i g_i^2 \delta(\Delta - \Delta_i). \tag{5.185}$$

This establishes our final expression for the environment self-correlation function,

$$\mathcal{C}(\tau) = \int \mathrm{d}\Delta \, J(\Delta) \left\{ \cos(\Delta\tau) - \mathrm{i} \tanh\left(\frac{\Delta}{2k_\mathrm{B}T}\right) \sin(\Delta\tau) \right\}. \tag{5.186}$$

This equation is our main result, which we shall now use to see how the spin bath is mapped onto an equivalent bath of oscillators.

Mapping Spins onto Oscillators

Let us contrast our expression (5.186) for the spin-environment self-correlation function with the environment self-correlation function (5.15) obtained for the case of an harmonic-oscillator bath, which reads

$$\begin{aligned} \mathcal{C}_\mathrm{osc}(\tau) &= \int \mathrm{d}\omega \, J_\mathrm{osc}(\omega) \left\{ \coth\left(\frac{\omega}{2k_\mathrm{B}T}\right) \cos(\omega\tau) - \mathrm{i} \sin(\omega\tau) \right\} \\ &\equiv \nu(\tau) - \mathrm{i}\eta(\tau), \end{aligned} \tag{5.187}$$

where

$$J_\mathrm{osc}(\omega) \equiv \sum_i \frac{g_i^2}{2m_i\omega_i} \delta(\omega - \omega_i) \tag{5.188}$$

is the spectral density of the harmonic-oscillator environment. Here, we have added the subscript "osc" in order distinguish the above expressions from their counterparts in the spin-bath case.

Comparing expressions (5.186) and (5.187), we see that the explicitly temperature-dependent noise kernel $\nu(\tau)$ [i.e., the real part of $\mathcal{C}_\mathrm{osc}(\tau)$] in the oscillator-bath case has lost this temperature dependence in the case of the spin bath: The real part of $\mathcal{C}(\tau)$ in the expression (5.186) does not contain any explicit temperature dependence. On the other hand, the imaginary part of $\mathcal{C}(\tau)$ for the spin bath contains a temperature-dependent factor $\tanh(\Delta/2k_\mathrm{B}T)$, whereas the corresponding dissipation kernel $\eta(\tau)$ appearing in $\mathcal{C}_\mathrm{osc}(\tau)$ does not exhibit such an explicit temperature dependence.

The only case for which the two expressions (5.186) and (5.187) agree is at zero temperature, since $\tanh(\Delta/2k_\mathrm{B}T) \longrightarrow 1$ and $\coth(\omega/2k_\mathrm{B}T) \longrightarrow 1$ as $T \longrightarrow 0$. This behavior can be understood as follows. At zero temperature, both the spins and harmonic oscillators in the respective environments will be effectively confined to their ground states and cannot occupy any excited states. Thus the characteristic difference between spins and oscillators—namely, that the former are restricted to a two-dimensional state space, while the latter can occupy a continuum of energy levels—fades away in the limit $T \longrightarrow 0$, and the spin and oscillator baths behave similarly.

However, even at finite temperatures, we can *map* the expression (5.186) for the spin bath onto the expression (5.187) for the oscillator bath by using a little trick. Suppose we introduce a new effective spectral density $J_{\text{eff}}(\Delta, T)$ defined as

$$J_{\text{eff}}(\Delta, T) \equiv J(\Delta) \tanh\left(\frac{\Delta}{2k_{\text{B}}T}\right), \tag{5.189}$$

where $J(\Delta)$ is the original spectral density (5.185) of the spin bath. Let us now replace the oscillator-bath spectral density $J_{\text{osc}}(\omega)$ appearing in the environment self-correlation function (5.187) by this explicitly temperature-dependent expression $J_{\text{eff}}(\Delta, T)$. Changing notation $\omega \longrightarrow \Delta$ in the integral on the right-hand side of (5.187), we get

$$\begin{aligned} \mathcal{C}_{\text{osc}}(\tau) &= \int \mathrm{d}\Delta\, J(\Delta) \tanh\left(\frac{\Delta}{2k_{\text{B}}T}\right) \left\{\coth\left(\frac{\Delta}{2k_{\text{B}}T}\right) \cos(\Delta\tau) - \mathrm{i}\sin(\Delta\tau)\right\} \\ &= \int \mathrm{d}\Delta\, J(\Delta) \left\{\cos(\Delta\tau) - \mathrm{i}\tanh\left(\frac{\Delta}{2k_{\text{B}}T}\right) \sin(\Delta\tau)\right\}. \end{aligned} \tag{5.190}$$

But this is precisely the expression (5.186) for the environment self-correlation function of the spin bath!

We have therefore discovered a simple route for taking over all of our previous results derived in Sect. 5.2 for the oscillator bath to the spin-bath case. All we need to do is to evaluate the oscillator-bath expressions that depend on the spectral density of the bath with the modified "surrogate" spectral density (5.189), which is given by the true spectral density $J(\Delta)$ of the spin bath multiplied by the additional temperature-dependent and frequency-dependent factor $\tanh(\Delta/2k_{\text{B}}T)$.

In particular, the kernels $\nu(\tau)$ and $\eta(\tau)$ for the spin bath are given by

$$\nu(\tau) = \int_0^\infty \mathrm{d}\Delta\, J(\Delta) \cos(\Delta\tau), \tag{5.191}$$

$$\eta(\tau) = \int_0^\infty \mathrm{d}\Delta\, J(\Delta) \tanh\left(\frac{\Delta}{2k_{\text{B}}T}\right) \sin(\Delta\tau), \tag{5.192}$$

obtained by inserting the new effective spectral density (5.189) into (5.16) and (5.17). The formal differences between these expressions for the kernels and their counterparts (5.16) and (5.17) for the oscillator bath may not seem dramatic at a first glance. However, the effect on the reduced dynamics can be enormous. Let us now study two explicit examples for the mapping procedure.

Example #1: Mapping of the Spin–Spin Model onto the Spin–Boson Model

We consider the case of the central system's being represented by a spin-$\frac{1}{2}$ particle bilinearly coupled to the spin bath through $\hat{\sigma}_z$. Following our above mapping technique, the Born–Markov master equation can then be immediately written down by substituting $\hat{X} \longrightarrow \hat{\sigma}_z$ in the general expression (5.22) for the master equation for quantum Brownian motion,

$$\frac{\mathrm{d}}{\mathrm{d}t}\hat{\rho}_{\mathcal{S}}(t) = -\mathrm{i}\left[\hat{H}_{\mathcal{S}}, \hat{\rho}_{\mathcal{S}}(t)\right] - \int_0^{\infty} \mathrm{d}\tau \left\{\nu(\tau)\left[\hat{\sigma}_z, \left[\hat{\sigma}_z(-\tau), \hat{\rho}_{\mathcal{S}}(t)\right]\right] - \mathrm{i}\eta(\tau)\left[\hat{\sigma}_z, \left\{\hat{\sigma}_z(-\tau), \hat{\rho}_{\mathcal{S}}(t)\right\}\right]\right\}. \quad (5.193)$$

The kernels $\nu(\tau)$ and $\eta(\tau)$ are now given by the spin-bath expressions (5.191) and (5.192).

If we take the self-Hamiltonian of the system to be of the form $\hat{H}_{\mathcal{S}} = -\frac{1}{2}\Delta_0\hat{\sigma}_x$, as in our discussion of the spin–boson model (see Sect. 5.3.2), the explicit time evolution of the interaction-picture operator $\hat{\sigma}_z(\tau)$ is

$$\hat{\sigma}_z(\tau) = \mathrm{e}^{\mathrm{i}\hat{H}_{\mathcal{S}}\tau}\hat{\sigma}_z\mathrm{e}^{-\mathrm{i}\hat{H}_{\mathcal{S}}\tau} = \hat{\sigma}_z\cos(\Delta_0\tau) + \hat{\sigma}_y\sin(\Delta_0\tau). \quad (5.194)$$

Using this expression in (5.193) yields the master equation

$$\frac{\mathrm{d}}{\mathrm{d}t}\hat{\rho}_{\mathcal{S}}(t) = -\mathrm{i}\left[\hat{H}'_{\mathcal{S}}, \hat{\rho}_{\mathcal{S}}(t)\right] - \widetilde{D}\left[\hat{\sigma}_z, \left[\hat{\sigma}_z, \hat{\rho}_{\mathcal{S}}(t)\right]\right] + \zeta\hat{\sigma}_z\hat{\rho}_{\mathcal{S}}(t)\hat{\sigma}_y + \zeta^*\hat{\sigma}_y\hat{\rho}_{\mathcal{S}}(t)\hat{\sigma}_z. \quad (5.195)$$

This equation is formally identical to the spin–boson master equation (5.142). This result is of course not surprising: After all, it was our goal to map the spin bath onto an equivalent oscillator (i.e., bosonic) bath. The quantities appearing in (5.195) are defined just as in the spin–boson case, see (5.143), (5.144), and (5.145a), but are now evaluated with the kernels (5.191) and (5.192).

Example #2: Mapping of the Oscillator–Spin model onto the Model for Quantum Brownian Motion

As another example for the mapping approach, let us discuss the case of a central harmonic oscillator (weakly) coupled to a spin bath. A potential application of this (weak-coupling) "oscillator–spin model" is the modeling of decoherence and dissipation in quantum-electromechanical systems, which will be discussed in more detail in Sect. 6.4.2. Let us consider the rather general oscillator–spin model described by the total Hamiltonian

$$\hat{H} = \hat{H}_{\mathcal{S}} + \hat{H}_{\mathcal{E}} + \hat{H}_{\text{int}} = \hat{H}_{\mathcal{S}} - \sum_i \frac{\Delta_i}{2}\hat{\sigma}_x^{(i)} + \hat{X} \otimes \sum_i g_i\hat{\sigma}_z^{(i)}, \quad (5.196)$$

where

$$\hat{H}_{\mathcal{S}} = \frac{\hat{P}^2}{2M} + \frac{M\Omega^2}{2}\hat{X}^2 \quad (5.197)$$

is the usual harmonic-oscillator Hamiltonian describing the intrinsic dynamics of the central system. The environment self-Hamiltonian $\hat{H}_{\mathcal{E}}$ has the familiar form (5.168). The interaction Hamiltonian $\hat{H}_{\text{int}}$ describes the bilinear coupling of the position coordinate of the central oscillator to the z-spin coordinate of each environmental spin.

Again following our strategy for mapping spin baths onto oscillator baths (assuming the weak-coupling limit applies), we can now readily describe the reduced dynamics of the oscillator system by the Born–Markov master equation (5.25) for quantum Brownian motion of a particle confined to a harmonic-oscillator potential,

$$\frac{\mathrm{d}}{\mathrm{d}t}\hat{\rho}_{\mathcal{S}}(t) = -\mathrm{i}\left[\hat{H}_{\mathcal{S}} + \frac{1}{2}M\widetilde{\Omega}^2\hat{X}^2, \hat{\rho}_{\mathcal{S}}(t)\right] - \mathrm{i}\gamma\left[\hat{X}, \left\{\hat{P}, \hat{\rho}_{\mathcal{S}}(t)\right\}\right] - D\left[\hat{X}, \left[\hat{X}, \hat{\rho}_{\mathcal{S}}(t)\right]\right] - f\left[\hat{X}, \left[\hat{P}, \hat{\rho}_{\mathcal{S}}(t)\right]\right]. \quad (5.198)$$

The frequency shift $\widetilde{\Omega}^2$, momentum-damping rate γ, and the diffusion coefficients D and f are given by the same expressions (5.26) as in the case of quantum Brownian motion, but with the kernels $\nu(\tau)$ and $\eta(\tau)$ appearing these expressions replaced by the "surrogate" spin-bath kernels (5.191) and (5.192).

Let us consider the example of an ohmic spectral density for the spin bath,

$$J(\Delta) = \frac{2M\gamma_0}{\pi}\Delta\frac{\Lambda^2}{\Lambda^2 + \Delta^2}. \quad (5.199)$$

This spectral density is formally identical to the ohmic spectral density (5.44) considered in the context of quantum Brownian motion. Following the same approach as detailed in Sect. 5.2.3, the choice of an explicit form for the spectral density now allows us to evaluate the coefficients appearing in the master equation (5.198).

As before, the coefficients γ and D are given by double Fourier sine and cosine transforms, respectively (see the discussion in Sect. 5.2.3). Thus we can readily write down the resulting expressions [compare (5.46) and (5.48)],

$$\gamma = \frac{\pi}{2}\frac{1}{M\Delta}J_{\text{eff}}(\Delta, T) = \gamma_0\frac{\Lambda^2}{\Lambda^2 + \Omega^2}\tanh\left(\frac{\Delta}{2k_{\text{B}}T}\right), \quad (5.200)$$

$$D = \frac{\pi}{2}J_{\text{eff}}(\Delta, T)\coth\left(\frac{\Delta}{2k_{\text{B}}T}\right) = M\gamma_0\Delta\frac{\Lambda^2}{\Lambda^2 + \Omega^2}. \quad (5.201)$$

We see that, in contrast with the momentum-damping coefficient in the oscillator-bath case, the momentum-damping coefficient γ for the ohmic spin bath is explicitly temperature-dependent. The factor $\tanh(\Delta/2k_{\text{B}}T)$, and thus γ, *decrease* with increasing temperature. This may seem counterintuitive at a first glance, since we would expect a "warmer" environment to either have a stronger effect on the system, or, at least, not induce *less* damping of the system than a "colder" environment.

However, the temperature dependence of the damping coefficient is rather easily explained. For an oscillator bath, each oscillator possesses an infinity of energy levels, so there is no upper limit to the amount of energy that can be absorbed by the bath. By contrast, a spin-$\frac{1}{2}$ particle has only two possible energy levels, so the spin bath as a whole saturates very quickly as the temperature (and thus the amount of energy that can be absorbed from the system) is increased. It follows that, since the dissipation rate of the oscillator bath (in the model for quantum Brownian motion) is temperature-independent [see (5.46)], the ability of the spin bath to exert a dissipative effect on the system must *decrease* with increasing temperature. Indeed, the temperature dependence of the damping rate as predicted by (5.200) has been observed experimentally [243] in physical systems such as glasses, where dissipation is due to the interaction of phonon (i.e., oscillator) modes with two-level systems [244].

The normal-diffusion coefficient (5.201), and thus the rate of spatial decoherence, is seen to have no explicit temperature dependence in the case of an ohmic spin bath. This is in contrast with the normal-diffusion coefficient (5.48) in quantum Brownian motion, which scales as $\coth(\hbar\omega_0/2k_{\mathrm{B}}T)$, making spatial decoherence stronger as we go to higher temperatures of the environment. In fact, in the high-temperature limit of the Caldeira–Leggett model, we had found a linear increase of D with temperature [see (5.51)].

Once again, this difference between the spin-bath and oscillator-bath settings can be understood from the point of view of the size of the state spaces available to each particle in the environment. The ability of environmental oscillators to occupy a significant number of increasingly excited energy levels with growing temperature translates into shorter and shorter characteristic wavelengths present in the environment. The shorter the typical environmental wavelengths, the better the position of the central system can be resolved and thus the more which-path information can be obtained, leading to stronger spatial decoherence (see also the discussion in Sect. 5.2.6).

By contrast, spin-$\frac{1}{2}$ particles can only occupy two energy levels. The range of typical environmental wavelengths, and therefore the rate of spatial decoherence, will saturate very quickly as we increase the temperature. Within the context of our particular model, this even leads to a completely temperature-independent decoherence rate. In fact, in this model we had assumed that the environmental two-level systems move in a symmetric double-well potential. Thus the two states of the corresponding spin particle are here associated with the same energy, restricting the state space even further. It is easy to show that if this symmetry assumption is dropped and an asymmetry term of the form $\sum_i \frac{\omega_i}{2}\hat{\sigma}_z^{(i)}$ is included in the bath Hamiltonian, the rate of spatial decoherence is increased by a constant (but still temperature-independent) value. This increase reflects the larger state space now available to the bath particles, which corresponds to the presence of shorter environmental wavelengths.

5.4.3 Beyond Markov: Solving General Spin-Environment Models

Finally, let us conclude our discussion of spin-environment models by giving the reader a brief idea of how more general models of this type may be solved without resorting to the limiting Born–Markov approximations. As we discussed in Sect. 5.1.2, the interesting regime for spin-environment models is typically that of low temperatures and strong system–environment couplings. This leads to significant non-Markovian memory effects in the environment and usually renders simplifying weak-coupling approximations inapplicable. It follows that spin-environment models frequently cannot be treated by means of an approximate Born–Markov master equation (as we did in the previous Sect. 5.4.2), whose derivation is based on weak couplings and negligible environment self-correlation times. This renders the mathematical solution of such non-Markovian spin-environment models often somewhat cumbersome.

In most cases, the relevant decoherence dynamics arising from general spin-environment models—for example, in terms of the central-spin expectation value $\langle \hat{\sigma}_z(t) \rangle$—can be calculated analytically using advanced techniques such as the instanton formalism [199]. In this context, the problem arises of how to properly "average" over the degrees of freedom of a strongly coupled spin environment. Prokof'ev and Stamp [199] showed that this task can be facilitated by breaking down the general spin-environment model into four limiting cases, each of which emphasizes a different parameter regime and thus disregards certain terms in the general expression for the total Hamiltonian. Each of these cases can then be solved separately, and each yields an expression for an averaging integral over certain quantities in the model. As shown in [199], the correct procedure for averaging over the spin environment in the general case—and thus for obtaining the reduced dynamics of the central system—is then simply given by the combined application of these four separate averages. Here, we shall not go into the details of this approach and instead refer interested readers to [199] for details. Further discussions of the physical and mathematical foundations of spin-environment models can be found in [200, 210, 214, 216].

5.5 Summary

This chapter on decoherence models has been somewhat lengthy and technical. Let us therefore summarize some of the main results.

Mapping onto Canonical Models

A large class of physical systems of interest can be represented either by spin-$\frac{1}{2}$ particles (for systems whose state space is effectively reduced to two different basis states), or by particles that are described by the continuous canonical coordinates position and momentum and that move in some potential.

Environments can be modeled either as a collection of harmonic oscillators, corresponding to a quasicontinuum of delocalized bosonic field modes, or as a collection of localized modes represented by spins. Spin environments are often the appropriate model for low-temperature environments strongly coupled to the central system.

Quantum Brownian Motion

This is the model of a central particle moving in some potential in real space and bilinearly interacting with an environment of noninteracting harmonic oscillators. We focused on the case of the central system represented by a harmonic oscillator.

We derived the Born–Markov master equation for this model [see (5.25)],

$$\frac{\mathrm{d}}{\mathrm{d}t}\hat{\rho}_S(t) = -\mathrm{i}\left[\hat{H}_S + \frac{1}{2}M\widetilde{\Omega}^2\hat{X}^2, \hat{\rho}_S(t)\right] - \mathrm{i}\gamma\left[\hat{X}, \left\{\hat{P}, \hat{\rho}_S(t)\right\}\right] - D\left[\hat{X}, \left[\hat{X}, \hat{\rho}_S(t)\right]\right] - f\left[\hat{X}, \left[\hat{P}, \hat{\rho}_S(t)\right]\right]. \quad (5.202)$$

The first term on the right-hand side of this equation describes the unitary dynamics (with a frequency shift quantified by $\widetilde{\Omega}^2$). The second term corresponds to momentum damping with a rate proportional to γ. The third term describes decoherence of spatial coherences over a distance ΔX at a rate $D(\Delta X)^2$. The fourth term also represents decoherence but can often be neglected, especially at higher temperatures. All coefficients appearing in the master equation (5.202) are strongly dependent on the spectral density of the environment, which encapsulates the physical characteristics of the environment.

In the high-temperature limit, the reduced dynamics of the central system are in many cases well approximated by a simplified form of the master equation (5.202) known as the Caldeira–Leggett master equation, given by [see (5.57)]

$$\frac{\mathrm{d}}{\mathrm{d}t}\hat{\rho}_S(t) = -\mathrm{i}\left[\hat{H}'_S, \hat{\rho}_S(t)\right] - \mathrm{i}\gamma_0\left[\hat{X}, \left\{\hat{P}, \hat{\rho}_S(t)\right\}\right] - 2M\gamma_0 k_\mathrm{B}T\left[\hat{X}, \left[\hat{X}, \hat{\rho}_S(t)\right]\right]. \quad (5.203)$$

Again, the first term on the right-hand side represents the unitary evolution (governed by the environment-renormalized self-Hamiltonian $\hat{H}'_S$), the second term describes dissipation, and the third term corresponds to decoherence of spatial superpositions at a rate that increases linearly with the temperature of the bath. Let us summarize the assumptions that have entered into the derivation of the Caldeira–Leggett master equation (5.57):

- The Born–Markov approximations hold (see Sect. 4.2.1).
- The environment of harmonic oscillators is described by an ohmic spectral density $J(\omega) \propto \omega$ with a high-frequency cutoff Λ.
- The environment is at a sufficiently high temperature, such that its thermal energy is much larger than the energy scale set by the natural frequency of the system ($k_\mathrm{B}T \gg \Omega$).
- The natural frequency of the system is much smaller than the high-frequency cutoff Λ of the environment ($\Omega \ll \Lambda$).

We found that the interplay of environmental monitoring and intrinsic dynamics leads to the emergence of pointer states that are minimum-uncertainty Gaussians and that are therefore well-localized in both position and momentum, approximating classical points in phase space (see Sect. 5.2.6). These states arise as a compromise: They are the most robust under the combined influence of the direct monitoring of the position coordinate by the environment on the one hand, and under the intrinsic dynamics of the system (whose self-Hamiltonian is symmetric in both position and momentum coordinates) on the other hand. The intrinsic dynamics, through their creation of spatial superpositions from superpositions of momentum, lead to decoherence in momentum, even though the momentum coordinate is not directly monitored by the environment.

This difference between "direct" decoherence in position and "indirect" decoherence in momentum is also reflected in the fact that the timescales for decoherence in position and decoherence in momentum are distinctly different. Decoherence in momentum occurs on the dynamical timescale set by the intrinsic evolution of the system. By contrast, the timescale for decoherence in position is essentially independent of (and typically much shorter than) this dynamical timescale.

Spin–Boson Model

This model consists of a central two-level system (represented by a spin-$\frac{1}{2}$ particle) bilinearly coupled to an environment of noninteracting harmonic oscillators.

For a simplified version of the model (without a tunneling term in the Hamiltonian of the system) and for an ohmic spectral density of the environment, we found that interferences between the "spin up" and "spin down" states of the system are exponentially damped at a rate which increases linearly with the temperature T of the environment (see Sect. 5.3.1).

We also derived the Born–Markov master equation for the general spin–boson model, which reads [see (5.142)]

$$\frac{\mathrm{d}}{\mathrm{d}t}\hat{\rho}_{\mathcal{S}}(t) = -\mathrm{i}\left[\hat{H}'_{\mathcal{S}}, \hat{\rho}_{\mathcal{S}}(t)\right] - \widetilde{D}\left[\hat{\sigma}_z, \left[\hat{\sigma}_z, \hat{\rho}_{\mathcal{S}}(t)\right]\right] + \zeta\hat{\sigma}_z\hat{\rho}_{\mathcal{S}}(t)\hat{\sigma}_y + \zeta^*\hat{\sigma}_y\hat{\rho}_{\mathcal{S}}(t)\hat{\sigma}_z. \tag{5.204}$$

The first term on the right-hand side represents the unitary evolution under the environment-shifted self-Hamiltonian $\hat{H}'_{\mathcal{S}}$, the second term corresponds to decoherence in the $\hat{\sigma}_z$ eigenbasis of the system, and the last two terms describe the decay of the two-level system.

Spin-Environment Models

We used the model of a tunneling central spin whose $\hat{\sigma}_z$ coordinate is linearly coupled to an environment of other spins to illustrate the selection of different preferred pointer bases by the environment. We also showed how, in the weak-coupling limit, spin environments can be mapped onto oscillator environments. That is, the reduced dynamics of the system weakly coupled to a spin environment can be described by this system coupled to an equivalent oscillator environment described by an explicitly temperature-dependent spectral density of the form

$$J_{\mathrm{eff}}(\omega, T) \equiv J(\omega)\tanh\left(\frac{\omega}{2k_{\mathrm{B}}T}\right), \tag{5.205}$$

where $J(\omega)$ is the spectral density of the spin environment. We can therefore take over all our previous expressions from the oscillator-environment case (i.e., from the quantum Brownian motion and spin–boson models), but evaluate them with the effective spectral density (5.205). We discussed the mapping of the spin–spin model onto the spin–boson model, and the mapping of the oscillator–spin model onto the model for quantum Brownian motion.

Master Equations are General;
Coefficients and Spectral Densities are Specific

The canonical models are described by fairly general master equations that, under certain assumptions, may even take similar formal structures for different canonical models. For example, in Sect. 5.4.2 we have seen that, in the weak-coupling limit, spin-environment models can be mapped onto oscillator-environment models in the sense that the former can be described by master equations that are formally identical to the master equations derived for the latter models. In fact, the oscillator-environment equations turn out to be *universal* in the case of weak coupling [211].

However, what encapsulates the differences between environments and their physical properties is the specific form of the coefficients multiplying the terms in the master equation. These coefficients are in turn dependent on the spectral density of the environment. For example, the master equation for quantum Brownian motion evaluated with a standard ohmic spectral density will in general describe reduced dynamics that are significantly different from the reduced dynamics obtained from the master equation evaluated with the temperature-dependent "effective" spectral density (5.205).

The specific functional form of the coefficients also plays an important role. For example, in our treatment of quantum Brownian motion we have mainly focused on Markov-type coefficients without explicit time dependence [see (5.26)]. However, we have also shown how an explicitly time-dependent version of the coefficients [see (5.42)] can be used instead to achieve a more realistic treatment of the short-time dynamics of decoherence. Finally, one can use the exact expressions for the coefficients, yielding the complete non-Markovian dynamics (see Sect. 5.2.7). In each case, the master equation takes the same form (5.202), but the resulting dynamics may differ greatly between the different choices for the coefficients, even when using the same spectral density.

Thus we conclude that the actual physics of a master equation for a decoherence model is mainly contained in the particular form of the coefficients and spectral density used to evaluate the master equation.

Generality

Needless to say, models are only simplified representations of physical systems. We discussed several important assumptions and approximations that have entered in some of our models:

- The Born–Markov approximations.
- The assumption of bilinear system–environment couplings (made in all models discussed here).
- Implicit long-wavelength assumptions for the environment, leading to an unrealistic lack of the saturation of the decoherence rate in quantum Brownian motion.
- The approximation of time-dependent coefficients in the master equation by their long-time asymptotic values [compare (5.26) and (5.42)].
- The assumption of a high-temperature environment in deriving the Caldeira–Leggett master equation.

Any of these assumptions may be completely innocuous in the context of the modeling of one physical system, but may lead to unrealistic results for another system. Therefore, whenever a concrete experimental situation is to be modeled by a particular canonical model for decoherence, special care must be taken to properly understand the assumptions underlying each model and the extent to which these assumptions may hold in the experimental setting of interest.

6 Of Buckey Balls and SQUIDs: Observing Decoherence in Action

Until the mid-1990s, decoherence was mainly studied through theoretical models. As described in Chap. 3, in 1985 Joos and Zeh [7] showed that spatial superposition states of even minuscule objects, such as dust grains or large molecules, are rapidly decohered by the scattering of only minimal environments. Furthermore, the rule-of-thumb expression (2.113) for spatial decoherence rates derived by Zurek in 1984 [12] suggested that on macroscopic scales decoherence would be overwhelmingly more rapid than dissipation. These results led to the common notion that decoherence is extremely efficient and fast, and to the general prediction that any nonclassical superposition in the everyday world would be immediately decohered.

That much was known (or at least suspected) in the 1980s. What was sorely missing, though, was some kind of experiment that would demonstrate the *dynamics* of decoherence, by showing how superposition states becomes gradually unobservable due to the action of decoherence. Instead of simply stating that decoherence is so strong and rapid as to preclude the observation of Schrödinger-cat states at all length scales relevant to the world of our experience, such experiments would enable us to see directly how this quantum-to-classical transition happens. We could observe how the smooth action of decoherence carries away quantum features and shuttles our object of interest safely into the classical domain—and maybe even how quantum coherence can subsequently be restored, thus showing that the quantum-to-classical transition really is a two-way process. We could change the dynamics and properties of this process by manipulating experimental parameters, thereby challenging the Copenhagen view of a fundamental quantum–classical boundary (see Sect. 8.1). Finally, we could test our theoretical models for decoherence to find out how much they really capture the effects of environmental interactions in an actual experimental setting.

Now that readers have hopefully been convinced of how useful and exciting such experiments would be, let us look at the difficulties that we immediately face in realizing experiments on decoherence. What do we need to do in order to observe the gradual action of decoherence? First, we ought to be able to prepare a system in a nonclassical superposition state. Of course, the system should have some minimal "size" to be sufficiently susceptible to decoherence—it should not just be a single electron, but preferably some mesoscopic system such as a few entangled photons or a molecule. But here lies the catch. We know from theoretical studies that superposition states in such systems would typically be decohered much faster than we could resolve in an actual experiment. Therefore the experiment must be designed very cleverly to ensure that decoherence does not happen too quickly and that it can also be sufficiently well controlled. Then, once the superposition state is created, we must have means available to continuously monitor the state of the system, but without introducing too much additional decoherence. Given all these desired properties, the realization of experiments on decoherence constitutes a formidable task which requires very special systems and observation techniques.

In this chapter, we shall describe some of the recent experiments that have made it possible to generate "Schrödinger kittens" in the laboratory and to observe how they are turned into effectively classical systems through decoherence, in impressive agreement with theoretical predictions. In Sect. 6.1, we will discuss the first experiment that allowed for the observation of gradual decoherence processes. This experiment was based on superpositions of distinct states of an electromagnetic field. In Sect. 6.2, we will describe experiments in which interference patterns (and their decoherence-induced disappearance) were observed for rather massive C_{70} molecules. In Sect. 6.3, we will focus on experiments involving superconducting two-level ("qubit") systems. Finally, in Sect. 6.4, we will outline two other experimental domains—Bose–Einstein condensation and quantum-electromechanical systems—that are promising candidates for future studies of decoherence.

6.1 The First Milestone: Atoms in a Cavity

The year 1996 saw several experimental breakthroughs in the creation and verification of mesoscopic Schrödinger-cat states. The basic mechanism of these experiments is rooted in a process similar to that originally envisioned by Schrödinger in his cat paradox (see Chap. 1) and can be formally described by the von Neumann measurement scheme (see Sect. 2.5.1). We let a microscopic system $\mathcal{S}$, which is rather easily prepared in a superposition state $(|s_1\rangle + |s_2\rangle)/\sqrt{2}$, interact with a mesoscopic or macroscopic apparatus $\mathcal{A}$, such that the latter "measures" the state of $\mathcal{S}$ in the $\{|s_1\rangle, |s_2\rangle\}$ basis (in the general von Neumann sense of the formation of quantum correlations; see Sect. 2.5.1). This results in an entangled joint state [compare (2.54)],

$$|\Psi\rangle = \frac{1}{\sqrt{2}} (|s_1\rangle |a_1\rangle + |s_2\rangle |a_2\rangle) . \tag{6.1}$$

Thus the combined mesoscopic or macroscopic system $\mathcal{SA}$ is now described by the superposition ("cat") state $|\Psi\rangle$.

Monroe and coworkers [245] used this principle to generate a superposition of two mesoscopically separated but spatially well-localized coherent-state wave packets of a single trapped $^9\text{Be}^+$ ion, where each wave packet was correlated with distinct (microscopic) internal electronic states of the ion. Thus the ion is described by an entangled cat state of the form (6.1), namely,

$$|\Psi\rangle = \frac{1}{\sqrt{2}} \left(|\boldsymbol{x}_1\rangle |\uparrow\rangle + |\boldsymbol{x}_2\rangle |\downarrow\rangle \right). \tag{6.2}$$

Here the $|\boldsymbol{x}_i\rangle$ denote the two wave packets localized around $\boldsymbol{x}_i$, and $|\uparrow\rangle$ and $|\downarrow\rangle$ are the two relevant electronic states of the ion (Monroe et al. used a pair of hyperfine ground states). In the notation introduced above, these electronic states would correspond to the system $\mathcal{S}$, whereas the two distinct spatial regions of the wave packets act as the mesoscopic "meter" $\mathcal{A}$.

In the experiment, the centers of the wave packets were about 80 nm apart, much more than the width of each individual wave packet (which was about 7 nm). The existence of the superposition was verified through the explicit observation of an interference pattern. While Monroe et al. pointed out some possible techniques for studying decoherence in the experiment, they did not actually experimentally investigate this problem.

This ambitious goal was accomplished during the same year in a remarkable experiment by Michel Brune and coworkers at the Ecole Normale Supérieure in Paris [246] (see also [247] for an accessible review of the experiment). Instead of creating superpositions of a matter particle, the researchers generated a mesoscopic cat state of radiation fields with classically distinguishable phases and then "watched" how this superposition was gradually destroyed by decoherence. This experiment comes even closer to the spirit of the original Schrödinger-cat setting than the experiment by Monroe et al., as the microscopic system $\mathcal{S}$ and the "meter" $\mathcal{A}$ are here realized as physically distinct systems.

The basic idea of the experiment is as follows. An atom is prepared in a superposition of two energy eigenstates $|g\rangle$ and $|e\rangle$ and then allowed to transverse a cavity containing a coherent state $|\alpha\rangle$ of an electromagnetic field. The field effectively measures the state of the atom in such a way that, if the atom is in the state $|e\rangle$, the coherent state of the field undergoes a phase shift ϕ, $|\alpha\rangle \longrightarrow |\mathrm{e}^{i\phi}\alpha\rangle$, whereas nothing happens if the atom is in the state $|g\rangle$. Due to the linearity of the time evolution, the superposition of the atom is imprinted on the combined atom–field state,

$$\frac{1}{\sqrt{2}} \left(|g\rangle + |e\rangle \right) |\alpha\rangle \longrightarrow \frac{1}{\sqrt{2}} \left(|g\rangle |\alpha\rangle + |e\rangle \left|\alpha \mathrm{e}^{\mathrm{i}\phi}\right\rangle \right). \tag{6.3}$$

Since in the experiment the coherent state consists of about 10 photons and the achieved phase shift ϕ is on the order of π, the relative states $|\alpha\rangle$ and $|\mathrm{e}^{i\phi}\alpha\rangle$ of the field are mesoscopically distinct. Therefore the right-hand side of (6.3) represents a truly mesoscopic cat state. A cleverly designed method, to be described below, is then used to verify the existence of the superposition state and to observe the gradual decoherence of this superposition. We will explain this experiment in more detail in the following.

6.1.1 Atom–Field Interactions and Rabi Oscillations

In the experiment, the microscopic system $\mathcal{S}$ (which serves as the "seed" of the superposition) is realized in form of a rubidium atom prepared in a superposition of two distinct energy eigenstates $|g\rangle$ and $|e\rangle$ corresponding to two circular Rydberg states. These are states with a large quantum number n (Brune et al. [246] used $n = 50$ and 51) and maximum orbital angular momentum along the quantization axis, i.e., $|m| = n - 1$. The corresponding valence-electron probability density is torus-shaped, such that the valence electron moves in a quasiclassical circular planar orbit perpendicular to the quantization axis. Such states have many interesting properties. Most important to this experiment, they interact strongly with even weak microwave radiation due to their large magnetic dipole moment. Also, the electrons are far away from the nucleus and thus their acceleration is comparably small, leading to long radiative decay times. Both features allow for the preparation of long-lived atom–field correlations necessary to an observation of the gradual action of decoherence.

Since the interaction between an atom and an external oscillating field is the key component of the experiment, let us briefly review so-called *Rabi oscillations*. Suppose that we can make the assumption that only the two energy levels $|g\rangle$ (with energy E_g) and $|e\rangle$ (with energy E_e) of the atom are relevant. Since the circular Rydberg levels used in the experiment interact very strongly with radiation, this is a good approximation. The state of the atom can then be written as

$$|\psi_{\text{atom}}\rangle = c_g(t)\,|g\rangle + c_e(t)\,|e\rangle\,, \tag{6.4}$$

where $c_g(t)$ and $c_e(t)$ are arbitrary complex coefficients such that $|c_g(t)|^2 + |c_e(t)|^2 = 1$. Without the field, the Hamiltonian for the atom is

$$\hat{H}_0 = E_g|g\rangle\langle g| + E_e|e\rangle\langle e|. \tag{6.5}$$

Let us now turn on an external sinusoidal oscillating field with frequency ω. We will treat this field classically, which turns out to be a perfectly legitimate approximation in our experiment for the following reason. The particular cavity used in the experiment that contains the field is very dissipative, i.e., it interacts strongly with the environment. Thus energy and coherence of the field continuously leak out of the cavity, making the field effectively classical (another decoherence process!) [248]. This is despite the fact that the average number of photons in the cavity in the experiment is of order one, such that a quantum treatment may have appeared necessary.

When the field is turned on, the Hamiltonian of the atom becomes (setting $\hbar \equiv 1$)

$$\hat{H} = \hat{H}_0 + \frac{\epsilon}{2}\mathrm{e}^{\mathrm{i}\omega t}|g\rangle\langle e| + \frac{\epsilon}{2}\mathrm{e}^{-\mathrm{i}\omega t}|e\rangle\langle g|, \tag{6.6}$$

where the *Rabi frequency* ϵ measures the coupling between the atom and the field. Thus the external field can induce transitions between the atomic levels $|g\rangle$ and $|e\rangle$. Suppose the atom is initially in the ground state $|g\rangle$ (i.e., $c_g(0) = 1$, $c_e(0) = 0$), then a simple calculation yields the probability of the atom to be found in state $|e\rangle$ after a time t,

$$|c_e(t)|^2 = \frac{\epsilon^2}{\Omega^2} \sin^2 \frac{\Omega t}{2}. \tag{6.7}$$

Here $\Omega \equiv \sqrt{\Delta^2 + \epsilon^2}$ is called the *generalized Rabi frequency*, where $\Delta = \omega - \omega_{eg}$ quantifies the detuning between the frequency ω of the field and the atomic transition frequency $\omega_{eg} \equiv (E_e - E_g)$. We see that the larger the detuning Δ, the smaller the maximum amplitude of the component $|e\rangle$ in the state $|\psi_{\text{atom}}\rangle$.

At resonance, i.e., when the frequency ω of the field is tuned to the atomic transition frequency ω_{eg} and thus $\Delta = 0$, we have $\Omega = \epsilon$. Then (6.7) becomes

$$|c_e(t)|^2 = \sin^2 \frac{\epsilon t}{2} = \frac{1}{2} (1 - \cos \epsilon t), \tag{6.8}$$

i.e., the atom cycles between the two energy levels $|g\rangle$ and $|e\rangle$ at the Rabi frequency ϵ. In particular, if the atoms starts out in the state $|g\rangle$ (or $|e\rangle$) and the field is switched on for a time $\tau = \pi/2\epsilon$ (a so-called "resonant $\pi/2$ pulse"), $|c_e(\tau)|^2 = 1/2$, i.e., the resulting state of the atom is the equal-weight superposition

$$|\psi_{\text{atom}}\rangle \propto |g\rangle + e^{i\Phi} |e\rangle, \tag{6.9}$$

where Φ is some (real) phase. This is exactly how the initial superposition state of the rubidium atom is achieved in the experiment.

We shall also note that the technique of Ramsey interferometry outlined in Sect. 2.2.2 is directly based on the application of two $\pi/2$ Rabi pulses. The first $\pi/2$ pulse creates the superposition of the two basis states, as in (2.4) and (6.9), whereas the second $\pi/2$ pulse, applied after a certain delay time, leads to the final state (2.6) in which the magnitudes of the expansion coefficients fluctuate as a function of the delay time.

6.1.2 Creating the Cat State

How precisely is the superposition state of radiation fields achieved in the experiment? For the following discussion, the reader may refer to Fig. 6.1 for a schematic overview of each step of the process. Let us start by having a look at the experimental arrangement sketched in Fig. 6.2. Rubidium atoms, emitted from the source O and prepared in the initial state $|e\rangle$, pass through a series of three cavities, labeled R_1, C, and R_2. The purpose of the first cavity R_1 is to create the superposition state of the rubidium atoms in the manner described above. Namely, a $\pi/2$ pulse with frequency equal to the frequency of the atomic transition $|g\rangle \longleftrightarrow |e\rangle$ is applied to the rubidium atoms, creating the state (6.9) with $\Phi = 0$, i.e.,

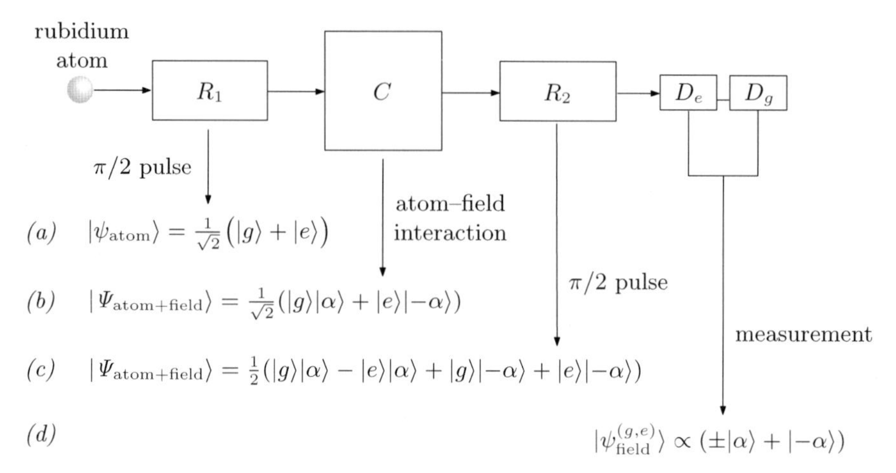

Fig. 6.1. Sequence of events leading to the generation the cat state. *(a)* A rubidium atom is prepared in the superposition (6.10) of the states $|g\rangle$ and $|e\rangle$ by application of a $\pi/2$ pulse in cavity R_1. *(b)* When passing through cavity C containing a coherent field $|\alpha\rangle$, the atom imparts a dispersive phase shift $|\alpha\rangle \longrightarrow |-\alpha\rangle$ on the field if the atom is in state $|e\rangle$. Since the atom is in a superposition of $|g\rangle$ and $|e\rangle$, an entangled atom–field state [see (6.15)] is created. *(c)* In the cavity R_2, another $\pi/2$ pulse is applied to the atom, leading to the atom–field state (6.17). *(d)* The energy state of the atom ($|g\rangle$ or $|e\rangle$) is measured by the detector pair D_e and D_g, projecting the state of cavity C on the nonentangled cat state $\left|\Psi_{\text{field}}^{(g,e)}\right\rangle \propto (\pm|\alpha\rangle + |-\alpha\rangle)$.

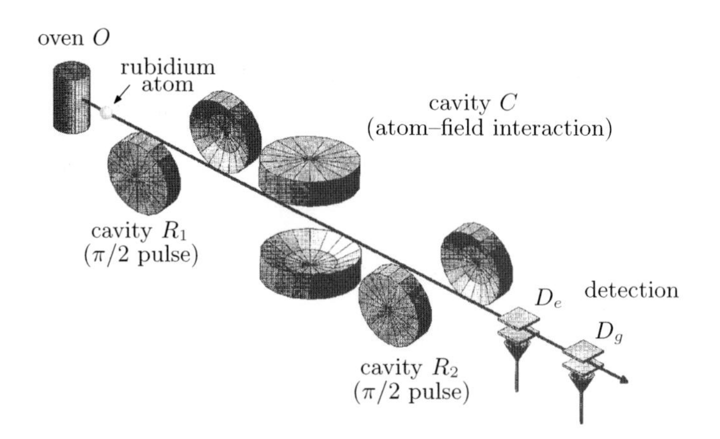

Fig. 6.2. Setup used in the experiment by Brune et al. [246] for creating mesoscopic Schrödinger-cat states of radiation fields and for monitoring the gradual decoherence of these states. Rubidium atoms emitted from the oven O pass through a sequence of cavities R_1, C, and R_2 containing photon fields and are detected by the ionization chambers D_e and D_g. Figure adapted with permission from [249]. Copyright 1996 by the American Physical Society.

$$|\psi_{\text{atom}}\rangle = \frac{1}{\sqrt{2}}\left(|g\rangle + |e\rangle\right). \tag{6.10}$$

(If the initial state of the atom had been $|g\rangle$, the final state would be $|\psi_{\text{atom}}\rangle = \frac{1}{\sqrt{2}}\left(|g\rangle - |e\rangle\right)$.) Next, the atoms enter the cavity C, which is cooled down to the superconducting regime close to absolute zero in order to minimize the thermal radiation present in the cavity. In contrast to cavity R_1, C has a very large so-called "Q factor" (Q stands for "quality"). This means that dissipative losses are minimized, such that photons in the cavity have rather long lifetimes. The cavity is made of two parallel mirrors about 3 cm apart. However, the distance between these mirrors can also be adjusted, and thereby the frequency of the radiation inside the cavity can be varied. This radiation consists of a coherent field containing a few photons.

It is very important to note that the frequency in the cavity is far off the resonance frequency for any transitions from level $|g\rangle$, so nothing happens to the state $|g\rangle$ interacting with the field, but the frequency is fairly close to that of a transition from $|e\rangle$ involving yet another level $|i\rangle$ (namely, the circular Rydberg state with $n = 52$). However, the field increases and decreases in strength so slowly along the trajectory of the atom that the atom–field interaction is nearly adiabatic, i.e., no energy quanta are exchanged between the atom and the field and no actual changes in the amplitude $|c_e(t)|$ for the component $|e\rangle$ in the state (6.4) occur. What happens instead is most interesting. The atom entering the cavity C acts as a refractive medium for the field, i.e., it leads to a shift $\delta\omega$ in the frequency of the field. It turns out that this shift is only significant if the atom is in the state $|e\rangle$, and the shift is negligible if the atom is in the state $|g\rangle$. By fine-tuning the interaction time τ between the atom and the field, the experiment achieves a phase shift $\phi = \tau\delta\omega$ of the field with a value close to π if the atom is in state $|e\rangle$.

What happens to the coherent state $|\alpha\rangle$ if such a phase shift ϕ is applied to each photon? A coherent state of a photon field, with (complex) amplitude α, is defined as a superposition of photon-number eigenstates $|n\rangle$ [238–241],

$$|\alpha\rangle = \sum_{n=0}^{\infty} c_n |n\rangle, \tag{6.11}$$

with

$$c_n = \mathrm{e}^{-|\alpha|^2/2} \frac{\alpha^n}{\sqrt{n!}}. \tag{6.12}$$

Thus the distribution of photon-number states follows a Poisson distribution with average photon number $\bar{n} = |\alpha|^2$,

$$\varrho(n) \equiv |c_n|^2 = \mathrm{e}^{-\bar{n}} \left(\frac{\bar{n}^n}{n!}\right). \tag{6.13}$$

Now, if every photon suffers a phase shift ϕ, the state $|\alpha\rangle$ becomes

$$|\alpha\rangle \longrightarrow \mathrm{e}^{-|\alpha|^2/2} \sum_{n=0}^{\infty} \mathrm{e}^{\mathrm{i}n\phi} \frac{\alpha^n}{\sqrt{n!}} |n\rangle = \mathrm{e}^{-|\alpha|^2/2} \sum_{n=0}^{\infty} \frac{(\mathrm{e}^{\mathrm{i}\phi}\alpha)^n}{\sqrt{n!}} |n\rangle = \left|\mathrm{e}^{\mathrm{i}\phi}\alpha\right\rangle . \quad (6.14)$$

Hence for a phase shift $\phi = \pi$, as in the experiment, the coherent field state $|\alpha\rangle$ is transformed into the state $|-\alpha\rangle$. This implies that, after the atom has passed through the cavity C, the composite atom–field system is described by the state [see (6.3)]

$$|\Psi_{\text{atom+field}}\rangle = \frac{1}{\sqrt{2}} \left(|g\rangle |\alpha\rangle + |e\rangle |-\alpha\rangle \right) . \quad (6.15)$$

Let us now have a look back at the experimental setup sketched in Fig. 6.2. We see that there is in fact a third cavity, R_2, which is identical to R_1 and positioned after cavity C along the atomic trajectory. Inside R_2, a second $\pi/2$ pulse is applied to the atom emerging from the cavity C. Recall that this pulse implements the transformations [see (6.10)]

$$\begin{aligned} |g\rangle &\longrightarrow \frac{1}{\sqrt{2}} \left(|g\rangle - |e\rangle \right) , \\ |e\rangle &\longrightarrow \frac{1}{\sqrt{2}} \left(|g\rangle + |e\rangle \right) . \end{aligned} \quad (6.16)$$

Thus the entangled state (6.15) becomes

$$\begin{aligned} |\Psi_{\text{atom+field}}\rangle &= \frac{1}{2} \left(|g\rangle |\alpha\rangle - |e\rangle |\alpha\rangle + |g\rangle |-\alpha\rangle + |e\rangle |-\alpha\rangle \right) \\ &= \frac{1}{2} \left(|\alpha\rangle + |-\alpha\rangle \right) |g\rangle + \frac{1}{2} \left(- |\alpha\rangle + |-\alpha\rangle \right) |e\rangle . \end{aligned} \quad (6.17)$$

After passage through R_2, the atom enters two ionization chambers D_e and D_g, which measure whether the atom is in the state $|g\rangle$ or $|e\rangle$. Given the state (6.17) and the usual projection postulate of quantum mechanics, this means that the state of the field after the measurement will be either

$$\left|\psi_{\text{field}}^{(g)}\right\rangle = \frac{1}{N} \left(|\alpha\rangle + |-\alpha\rangle \right) \quad (6.18)$$

if the atom is found in the state $|g\rangle$, or

$$\left|\psi_{\text{field}}^{(e)}\right\rangle = \frac{1}{N} \left(- |\alpha\rangle + |-\alpha\rangle \right) , \quad (6.19)$$

if the atom is found in the state $|e\rangle$. Here, N is a normalization factor that is very close to $\sqrt{2}$ for larger values of $|\alpha|^2$. For simplicity, let us assume $N = \sqrt{2}$ in the following.

From (6.18) and (6.19) we see that we have now indeed prepared a superposition of two coherent field states with distinct phases. The "catness" of these states increases with the average number $\bar{n} = |\alpha|^2$ of photons in the cavity C. We note that these superposition states differ from the entangled state $|\Psi_{\text{atom+field}}\rangle$, see (6.15). The latter state does not allow one to speak of the field itself being in a superposition—only the *combined* atom–field system is described by a superposition. By contrast, the measurement of the state of the atom after passage through the cavity R_2 disentangles the atom and the field, and thus (6.18) and (6.19) represent true superposition states of the radiation field.

One important question remains: How can we distinguish the superpositions (6.18) and (6.19) from the corresponding mixture? This is not only important for claiming success of the above scheme for the creation of mesoscopic Schrödinger cats, but also for achieving our main goal, namely, for observing the action of decoherence. The experiment solves this problem through a clever trick, which will be explained in the next section.

6.1.3 Observing the Gradual Action of Decoherence

Note that the passage and measurement of the first atom has left behind information in the cavity C about the state of the atom, since the final state of the field in C is either $\left|\psi^{(g)}_{\text{field}}\right\rangle$ or $\left|\psi^{(e)}_{\text{field}}\right\rangle$, depending on the outcome of the measurement. As we can see from (6.18) and (6.19), this information is encoded in the relative sign between the components $|\alpha\rangle$ and $|-\alpha\rangle$ of the superposition. We know that decoherence tends to delocalize such phase information into the environment and to thus make it inaccessible at the level of the system. How can we monitor the resulting decay of coherence?

The trick is to send a *second* atom, acting as a probe, through the sequence $R_1 \to C \to R_2$ of cavities. If this is done quickly enough after passage of the first atom, the information about the first atom will still be well-encoded in the state of the field in the cavity C, and the second atom will become correlated with this information. The remarkable result is that, in absence of decoherence (and for photon numbers $\bar{n} = |\alpha|^2 \gg 1$), after detection at D_e or D_g the second atom will always be found in the same state as the first atom.

That this is indeed true can be easily seen from the following simple calculation (see Fig. 6.3). Let us assume atom 1 has been detected in the state $|e\rangle$, such that the field state is $\left|\psi^{(e)}_{\text{field}}\right\rangle$. After traversing cavity R_1, atom 2 is described by the superposition $|\psi_{\text{atom2}}\rangle = \frac{1}{\sqrt{2}}(|g\rangle + |e\rangle)$ [see (6.10)] (the basis states $|g\rangle$ and $|e\rangle$ now refer to atom 2). Thus the initial uncorrelated state of this atom and the field is

$$|\Psi_{\text{atom2+field}}\rangle = \frac{1}{2}(|g\rangle + |e\rangle)(-|\alpha\rangle + |-\alpha\rangle). \tag{6.20}$$

What happens when atom 2 passes through cavity C? Recall that the field state suffers a phase shift $|\pm\alpha\rangle \longrightarrow |\mp\alpha\rangle$ if the atom is in the state $|e\rangle$. Therefore the combined state after passage through C is now

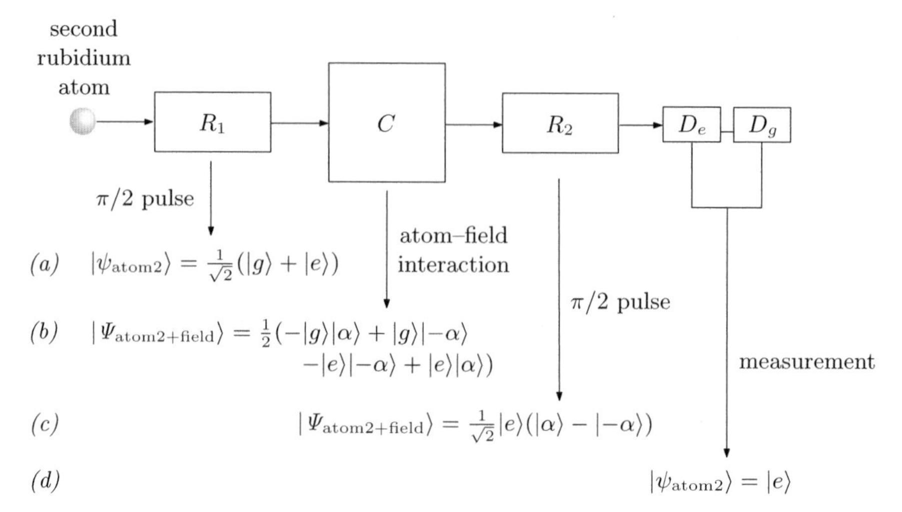

Fig. 6.3. Sequence of events for detecting the superposition state of fields in the cavity C and for monitoring its decoherence, assuming that the initial atom has been detected in the state $|e\rangle$. *(a)* A $\pi/2$ pulse in cavity R_1 prepares a second rubidium atom in the superposition (6.20) of the states $|g\rangle$ and $|e\rangle$. *(b)* The atom enters cavity C, which has been left in a superposition state after passage of the first atom. Atom 2 imparts another phase shift on the field in C, leading to an entangled atom–field state [see (6.21)]. *(c)* The $\pi/2$ pulse in cavity R_2 disentangles the field and the atom [see (6.22)]. *(d)* If the field states $|\pm\alpha\rangle$ are (approximately) orthogonal, this means that atom 2 will always be detected in state $|e\rangle$, identical to the state of atom 1. Thus the two measurement outcomes are perfectly correlated. Decay of this correlation implies decoherence of the cat state in the cavity C.

$$|\Psi_{\text{atom2+field}}\rangle = \frac{1}{2}\left(-|g\rangle|\alpha\rangle + |g\rangle|-\alpha\rangle - |e\rangle|-\alpha\rangle + |e\rangle|\alpha\rangle\right). \qquad (6.21)$$

Next, the atom passes through cavity R_2, which transforms $|g\rangle$ and $|e\rangle$ into superpositions of these states [see (6.16)]. Thus the state (6.21) becomes

$$\begin{aligned}|\Psi_{\text{atom2+field}}\rangle &= \frac{1}{2\sqrt{2}}\big(-|g\rangle|\alpha\rangle + |e\rangle|\alpha\rangle + |g\rangle|-\alpha\rangle - |e\rangle|-\alpha\rangle \\ &\qquad - |e\rangle|-\alpha\rangle - |g\rangle|-\alpha\rangle + |e\rangle|\alpha\rangle + |g\rangle|\alpha\rangle\big) \\ &= \frac{1}{\sqrt{2}}|e\rangle\left(|\alpha\rangle - |-\alpha\rangle\right). \end{aligned} \qquad (6.22)$$

Assuming that $|\alpha\rangle$ and $|-\alpha\rangle$ are orthogonal—an approximation that gets better as the number of photons $\bar{n} = |\alpha|^2$ in the cavity C increases—the reduced density operator for atom 2 corresponding to the state (6.22) is

$$\hat{\rho}_{\text{atom2}} = \text{Tr}_{\text{field}}\,|\Psi_{\text{atom2+field}}\rangle\langle\Psi_{\text{atom2+field}}| = |e_2\rangle\langle e_2|. \qquad (6.23)$$

It immediately follows that the probability P_{ee} of finding atom 2 in the state $|e\rangle$, given the detection of atom 1 in the state $|e\rangle$, is $P_{ee} = 1$. A similar calculation can be done under the assumption that atom 1 was detected in the state $|g\rangle$. This establishes our above assertion of perfect two-atom correlations.

Thus all we need to do to observe the decoherence of our cat state is to measure the decay of correlations $P_{ee}(T)$ [or $P_{gg}(T)$] between measurement outcomes for pairs of atoms as we increase the time interval T between sending off the first and the second atom. If T is much smaller than the characteristic decoherence time of the field superposition, $P_{ee}(T)$ will be close to one, reflecting the perfect correlation between the measurement outcomes when the superposition has not yet been appreciably decohered. As we make the delay time T longer, the superposition in the cavity will become increasingly decohered, until all relative-phase information (and thus information about the state of the first atom) has been delocalized into the environment and the field is described by the incoherent statistical mixture

$$\hat{\rho}_{\text{field}} = \frac{1}{2}\left(|\alpha\rangle\langle\alpha| + |-\alpha\rangle\langle-\alpha|\right). \tag{6.24}$$

This will then yield completely uncorrelated measurement results between the first and the second atom, i.e., the outcome of the measurement on the second atom will be random and independent of the result of the measurement on the first atom. Thus this decoherence process will be observable as the decay of $P_{ee}(T)$ down to a value of 0.5.

Fig. 6.4 shows a theoretical prediction for $P_{ee}(T)$ [249], which was found to match very well the experimental observations [250]. We clearly see how $P_{ee}(T)$ quite rapidly decreases from its initial value of one down to a plateau of $P_{ee}(T) = 0.5$, corresponding to full decoherence. The reader may wonder why $P_{ee}(T)$, after remaining around 0.5 over a range of delay times T, suddenly continues to decay toward a value of zero for larger values of T. The explanation is quite simple. The photons in the cavity C have an only finite lifetime, so eventually there are no photons left in C. Thus this cavity ceases to have an effect on the atoms passing through it. In this case, the atoms experience only the two $\pi/2$ pulses in the cavities R_1 and R_2, whose combined effect it is to transform the initial state $|e\rangle$ into $|g\rangle$ [see (6.16)],

$$|e\rangle \xrightarrow{\pi/2} \frac{1}{\sqrt{2}}\left(|g\rangle + |e\rangle\right) \tag{6.25}$$

$$\xrightarrow{\pi/2} \frac{1}{\sqrt{2}}\left[\frac{1}{\sqrt{2}}\left(|g\rangle - |e\rangle\right) + \frac{1}{\sqrt{2}}\left(|g\rangle + |e\rangle\right)\right] = |g\rangle\,. \tag{6.26}$$

Thus the atoms are never detected in the state $|e\rangle$, and thus $P_{ee}(T) \longrightarrow 0$ as $T \longrightarrow \infty$.

From the graph in Fig. 6.4, we can read off the characteristic decoherence time of the field superposition. The experiment showed that the decoherence time depends on two key parameters:

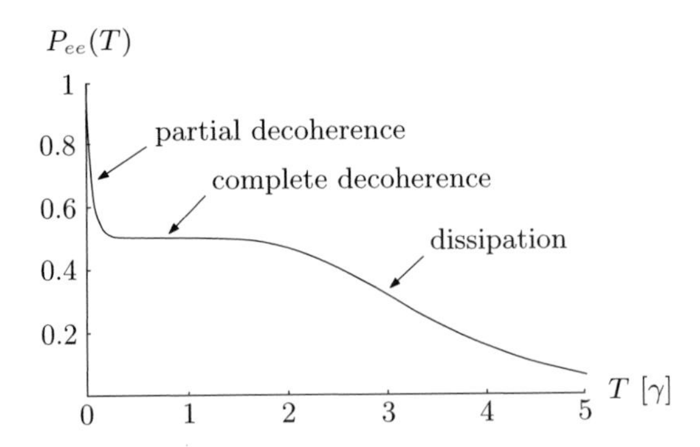

Fig. 6.4. Dependence of the joint probability $P_{ee}(T)$ on the delay time T between the passage of a pair of atoms, as given by Davidovich et al. [249]. The initial fast decay of $P_{ee}(T)$ corresponds to decoherence of the cat state in the cavity C; $P_{ee}(T) = 0.5$ represents complete decoherence. For even larger values of T, $P_{ee}(T)$ decreases further due to photon leakage from the cavity. The delay time T is measured in units of the characteristic dissipation period γ of the cavity. Figure adapted with permission from [249]. Copyright 1996 by the American Physical Society.

1. *The phase shift ϕ of the field induced by atoms in the state $|e\rangle$ passing through the cavity C.* It is found that, the larger ϕ, the shorter the decoherence time of the field superposition. This behavior is reasonable: The "catness" of the superposition state depends on how distinguishable the components of the superposition are. If ϕ is small, then the field states $|\alpha\rangle$ and $|e^{i\phi}\alpha\rangle$ [see (6.14)] have a large overlap. On the other hand, as ϕ approaches its maximal value of π (the value that we had assumed in our above discussion and that comes close to the phase shift achieved in the experiment), $|\alpha\rangle$ and $|e^{i\phi}\alpha\rangle$ become orthogonal and thus maximally distinct.
2. *The mean number $\bar{n} = |\alpha|^2$ of photons in the cavity C.* Again, this makes sense: The more photons, the more mesoscopic (and ultimately macroscopic) the states $\left|\Psi^{(g,e)}_{\text{field}}\right\rangle$, see (6.18), and thus the more strongly these states are subject to decoherence. Also, the larger $\bar{n}$, the smaller the overlap between the components $|\alpha\rangle$ and $|e^{i\phi}\alpha\rangle$ for a given phase shift ϕ, and thus the more distinct these components become.

In their experiment, Brune and coworkers varied these parameters and observed how the decoherence rate changed accordingly. Remarkably, the results were found to be in astonishing agreement with theoretical predictions of Davidovich et al. [249]. It was the first time that a mesoscopic "Schrödinger kitten" was generated, its existence verified, and its decoherence observed in a controlled way.

The type of experiment carried out by Brune and coworkers has since inspired a number of further experiments that have confirmed and extended the results described above. For example, by increasing the number of photons in the cavity C (which requires improving the Q factor of the cavity), larger and larger cat states can be created (recent experiments have used several tens of photons [251]). This class of experiments is commonly referred to as "cavity quantum electrodynamics," or "cavity QED" for short. Apart from allowing for the generation of Schrödinger-cat states and the observation of their decoherence, cavity QED could potentially be used to implement quantum computers. Readers interested in learning more about cavity QED will find a good and entertaining introduction in the book by Dutra [252].

6.1.4 Bringing Schrödinger Cats Back to Life

Despite the fact that decoherence is a completely unitary (and thus in principle reversible) process, we are accustomed to thinking of the quantum-to-classical transition as a "one-way street." Once quantum coherence is lost, the object behaves classically, and nothing seems to bring it back to the quantum world. Of course, from our general discussion of decoherence in Sect. 2.7, we know the reason why decoherence is considered irreversible for all practical purposes. To actually "relocalize" the superposition at the level of the system (i.e., to effectively time-reverse the process of decoherence), we would need to have appropriate control over the environment, which is usually impossible to achieve in practice.

In Sect. 2.13, we pointed out that a truly reversible decoherence process (an example of virtual decoherence) could be studied by letting the system interact with an appropriately designed "artificial" environment over which we can maintain full control. In 1997, the Brune group [253] proposed a modified version of the cavity-QED experiment described in the previous sections that contains such an engineered environment. As of 2007, the proposal has not yet been experimentally realized, mainly because it requires a reversible coupling between two superconducting cavities that is difficult to implement in practice. Nonetheless, since the idea behind this experiment is very intriguing and chances are good that the technical obstacles can be overcome in the near future, we shall outline the proposal in this section.

The experimental setup is essentially the same as described above (see Fig. 6.2). The key difference is that now the superconducting cavity C is linearly coupled (e.g., by means of a superconducting wave guide) to a second, identical cavity C_2 that is initially empty. (We shall denote the first cavity C by C_1 in the following to distinguish it clearly from the additional cavity C_2.) Through the coupling, C_1 and C_2 can exchange energy (i.e., photons). Recall that after a rubidium atom has passed through the apparatus and its state has been measured by the detectors D_e and D_g, the cavity C_1 is left in a superposition of the two field states $|\alpha\rangle$ and $|-\alpha\rangle$ [see (6.18) and (6.19)], and the relative sign between these two components encodes information about the measured state of the atom.

Now suppose that this phase information is transferred to the "reservoir" C_2 by the coupling between the two cavities. Then the state of C_1 must accordingly be described by the mixture (6.24). The idea behind the experimental proposal is that, if the coupling between C_1 and C_2 is reversible to a sufficiently large degree, the direction of this information transfer can subsequently be reversed, such that the initial superposition state in C_1 is restored while C_2 is returned to its initial empty state. Thus we can simulate a reversible decoherence process, in which the cavity C_2 plays the role of a controllable single-mode environment for C_1 that leads to decoherence and subsequent reappearance of the Schrödinger-cat state in C_1.

Let us make this argument a little bit more precise. Suppose energy is exchanged between the two cavities C_1 and C_2 in an oscillatory manner with frequency Ω_C. This frequency is chosen such that the time required for the preparation of the cat state in C_1 is much shorter than the time it takes for one cycle of energy exchange. Thus it is assumed that the coupling between C_1 and C_2 does not influence the preparation of the superposition state. Since the coupling is linear, a coherent field $|\alpha(t)\rangle$ in C_1 gives rise to another coherent field $|\beta(t)\rangle$ in C_2. Taking into account the fact that C_2 contains no field at $t = 0$, we can write $\alpha(t) = \alpha_0 \cos(\Omega_C t/2)$ and $\beta(t) = \alpha_0 \sin(\Omega_C t/2)$. Therefore the initial cat state in C_1 [see (6.18) and (6.19)]

$$\left|\Psi_{C_1}^{(g,e)}\right\rangle = \frac{1}{N}\left(\pm|\alpha\rangle + |-\alpha\rangle\right), \tag{6.27}$$

evolves into an entangled state for the combined C_1+C_2 system,

$$\left|\Psi_{C_1+C_2}^{(g,e)}\right\rangle = \frac{1}{N}\left(\pm|\alpha(t)\rangle|\beta(t)\rangle + |-\alpha(t)\rangle|-\beta(t)\rangle\right). \tag{6.28}$$

What happens now can easily be understood from the fact that the overlap of the two coherent-field states $|\pm\alpha\rangle$ decreases as the mean number of photons $\bar{n} = |\alpha|^2$ increases. After a time $t = \pi/\Omega_C$ has passed, $\beta(t) = \alpha_0$, so all photons are concentrated in cavity C_2. Thus the states $|\pm\beta\rangle$ have attained their maximum degree of distinctness, whereas the distinguishability of the states $|\pm\alpha\rangle$ is at its minimum. This implies that cavity C_2 now holds a maximum amount of information about the relative sign between the components in the superposition, at the expense of cavity C_1. In other words, there is no measurement that we could perform on C_1 alone that would tell us anything about the relative sign between the components. C_2 has carried away the phase information, and C_1 is formally described by an incoherent mixture of states $|\pm\alpha\rangle$, see (6.24).

The reader may recognize this state of affairs as exactly analogous to our usual description of decoherence processes where the states associated with the environment rapidly attain orthogonality, leading to a decay of relative-phase information in the density matrix of the system (see Sect. 2.7). This analogy is expected, since cavity C_2 precisely plays the role of the environment in this context. However, the key difference of the proposal discussed here is that, after a characteristic time $\tau = 2\pi/\Omega_C$, which is the period of energy exchange between the two cavities C_1 and C_2, the process has been completely reversed: Now the states $|\pm\alpha\rangle$ are again maximally distinct, as at $t = 0$. All information about the relative sign can be retrieved from cavity C_1, and the coherence of the superposition of field states in C_2 has been restored.

Now we can observe the gradual decoherence and recoherence in exactly the same way as before, namely, by sending a second atom through the apparatus and measuring the dependence of the joint probability $P_{ee}(T)$ on the delay time T between the passages of the two atoms. If sources of decoherence other than the coupling between the cavities C_1 and C_2 are neglected, we expect that $P_{ee}(T)$ will periodically return to its initial value of one at times $T_n = n\tau = 2n\pi/\Omega_C$, with $n = 1, 2, \ldots$. At these times T_n the decoherence of the cat state in C_1 has been completely undone. Conversely, at intervals $n\tau/2 = n\pi/\Omega_C$, the state of C_1 is described by an incoherent mixture, and thus the outcomes of the measurements on atom 1 and 2 are fully uncorrelated, yielding $P_{ee}(n\tau/2) = 0.5$.

This behavior is sketched in Fig. 6.5. Instead of directly plotting the joint probability $P_{ee}(T)$, we have here used the two-atom correlation function $\eta(T)$ as originally introduced by Raimond, Brune, and Haroche [253]. This function takes a value of 0.5 in the case of perfect correlations, and a value of zero in the case of complete decoherence. We see that for delay times equal to integer multiples of the characteristic time $\tau = 2\pi/\Omega_C$, the initial state is restored and we thus obtain a revival of coherence.

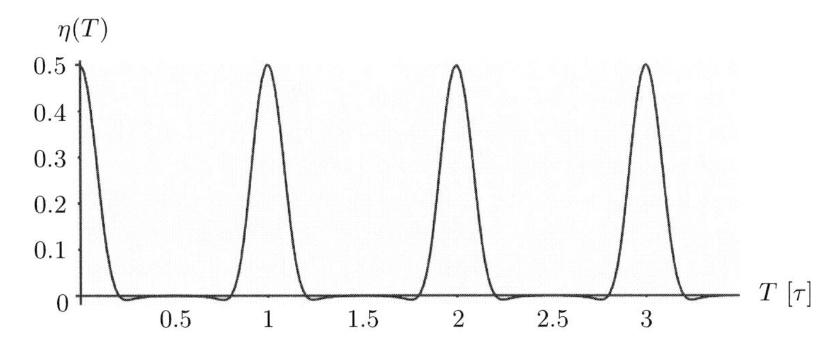

Fig. 6.5. Dependence of the two-atom correlation function $\eta(T)$ on the delay time T in the recoherence experiment proposed by Raimond, Brune, and Haroche [253]. Coherence is continuously removed from, and returned to, the cavity C_1 by a second cavity C_2 (the reservoir). This leads to a periodical revival of the cat state in C_1 and thus to reappearance of the maximum-correlation signal $\eta(T) = 0.5$. The delay time T is expressed in units of the characteristic period $\tau = 2\pi/\Omega_C$ with which energy (and thus phase information) is exchanged between the two cavities C_1 and C_2. The initial field in C_1 was taken to contain an average of five photons.

Of course, in a realistic experimental setting, this revival process will not go on forever, since the cavities C_1 and C_2 are subject to other sources of decoherence and dissipative effects such as photon loss on the mirror surfaces. It has been shown [254] that these processes would lead to an exponential decay of the amplitude of the revival peaks shown in Fig. 6.5, and thus it is unlikely that more than a few peaks could be observed in practice. Nonetheless, the observation of a single revival peak would already be an impressive experimental illustration of the fact that decoherence is, at least in principle, completely reversible.

6.2 Interferometry with C_{70} Molecules

In this section, we will describe a series of experiments carried out by the group of Anton Zeilinger at the University of Vienna. In essence, these experiments are a sophisticated version of the well-known double-slit experiment. However, instead of light or microscopic entities such as electrons, interference patterns are here observed for massive molecules. These experiments are remarkable for two reasons. First, they demonstrate the quantum "wave nature" of objects that would normally clearly fall into the "matter" category. Second, they show directly how the continuous action of decoherence gradually takes away the "quantumness" of these objects and transforms them into the familiar classical objects of our experience. Before we describe these experiments in more detail, let us first sketch in a few paragraphs a predecessor, namely, the double-slit experiment carried out with single electrons.

6.2.1 The Double-Slit Experiment with Electrons

When, in the year of 2002, the readers of *Physics World* were asked to nominate the "most beautiful experiment in physics" of all times, the winner turned out to be the double-slit experiment with single electrons [255]. Feynman had famously remarked [256] that the double-slit experiment is a phenomenon "which has in it the heart of quantum mechanics; in reality it contains the *only* mystery" of the theory. Indeed, it is probably difficult to fathom another experiment that embodies so simply and completely the strange features of quantum mechanics.

Interestingly, it was not until the year 1961 that the experiment in its original form—a double slit traversed by electrons—had actually been carried out by Claus Jönsson, a student of Gottfried Möllenstedt's at the University of Tübingen [257,258]. Möllenstedt had previously experimentally demonstrated electron interference using a different device, called the electron biprism—in essence, a very thin wire that splits the electron beam in half. This method bears similarities to the slip of paper the English scientist Thomas Young had used at the beginning of the 19th century to split a beam of sunlight in experiments demonstrating the ability of light to exhibit wave-like constructive and destructive interference effects [259].

Perhaps even more surprising, the experiment that ensured that only a single electron was present in the apparatus at any time was carried out only in 1989 by Akira Tonomura and coworkers at Hitachi using an electron biprism, just as in Möllenstedt's experiments [260]. The resulting interference pattern is shown in Fig. 6.6. This experiment clearly reflects the "mysteries" of quantum mechanics, as it shows that the interference fringes are not induced by interactions between electrons, but are rather due to the wave nature of *individual* electrons. Decoherence in the electron-biprism interferometer was recently investigated by Sonnentag and Hasselbach [261], who studied the decay of the visibility of the interference pattern induced by Coulomb interactions between the electrons and a macroscopic and dissipative electron-gas and phonon-gas environment.

6.2.2 Experimental Setup

Let us now outline the experimental setup used by the Zeilinger group for demonstrating the wave nature of massive molecules. The original experiment, reported in 1999 in a *Nature* article by Markus Arndt and coworkers [262], used C_{60} molecules, often nicknamed "buckey balls" because of their shape (see Fig. 6.7). Subsequent experiments, results of which were first reported in a 2002 paper by the Zeilinger group [263], were carried out using even bigger C_{70} molecules (in the shape of an elongated buckey ball) and a different experimental setup [263–266]. Our following discussion will be based on this later series of experiments.

Let us first get a sense of how truly massive these C_{70} molecules are (relative to microscopic entities such as electrons, that is). Since each carbon

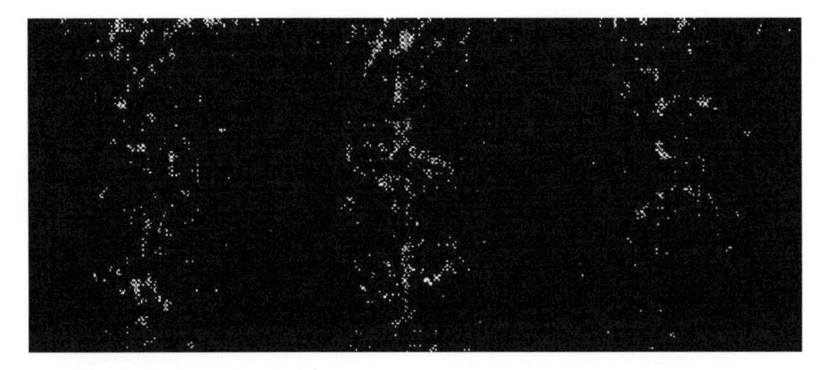

Fig. 6.6. Interference pattern from single electrons obtained in the double-slit experiment by Tonomura and coworkers [260]. Figure reprinted with permission from [260]. Copyright 1989 American Institute of Physics.

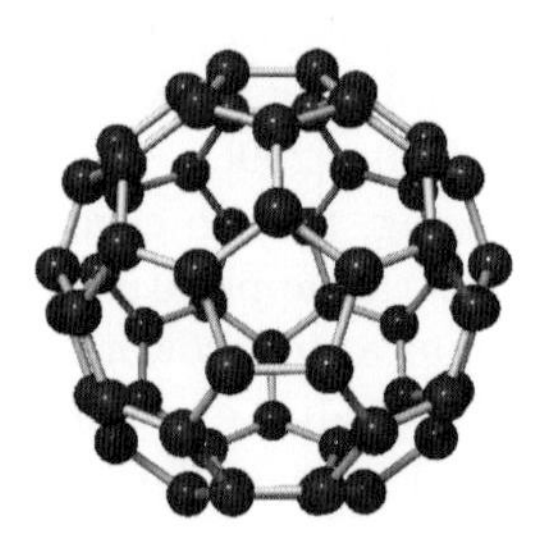

Fig. 6.7. Schematic illustration of a C_{60} molecule used in the original matter-wave interference experiment of Arndt et al. [262]. Sixty carbon atoms are arranged in the shape of a buckey ball with a diameter of about 1 nm.

atom contains six protons, six neutrons, and six electrons, a C_{70} molecule is composed of a total of over 1,000 microscopic constituents. Each such molecule possesses a very large number of highly excited internal rotational and vibrational degrees of freedom. This makes it possible to assign a finite temperature to each individual molecule. Also, emission of thermal radiation (i.e., of photons) from the molecules can be observed [266]. Thus, C_{70} molecules seem to clearly fall into the "particle" category.

Why is it so difficult to observe interference effects for such massive molecules? The relevant quantity is the matter de Broglie wavelength $\lambda = h/mv$, where m and v are the mass and velocity, respectively, of the particle. Thus the more massive and faster the particle, the shorter this wavelength. To observe interference effects with particles in a double slit–type experiment, the separation between the slits (and thus even more so the width of each slit) must be on the order of the wavelength. For typical velocities of C_{70} molecules in the experiment (around 100 m/s), the de Broglie wavelength of these molecules is a mere few picometers. The reader may easily imagine the impossibility of manufacturing such narrow, and narrowly spaced, slits.

How can this difficulty be overcome? The experiments of the Zeilinger group solve the problem through a neat trick that is based on the so-called *Talbot–Lau effect.* The basic principle of this effect is the following (see Fig. 6.8). Suppose a plane wave is traveling in the $+z$ direction, $f(z) = \mathrm{e}^{2\pi \mathrm{i} z/\lambda}$. Now we place a grating, composed of an array of parallel slits, at a right angle to the incoming wave. Let us denote the spacing of adjacent slits by d. One can then show (which we will not do here—see [267] for a short derivation) that at integer multiples of the distance

$$L_\lambda = \frac{d^2}{\lambda} \tag{6.29}$$

behind the grating the plane wave will in fact be *equal to the pattern of the grating.* In other words, if we place a screen at a distance nL_λ, $n = 1, 2, \ldots$, behind the grating and carry out the experiment with light, the screen will

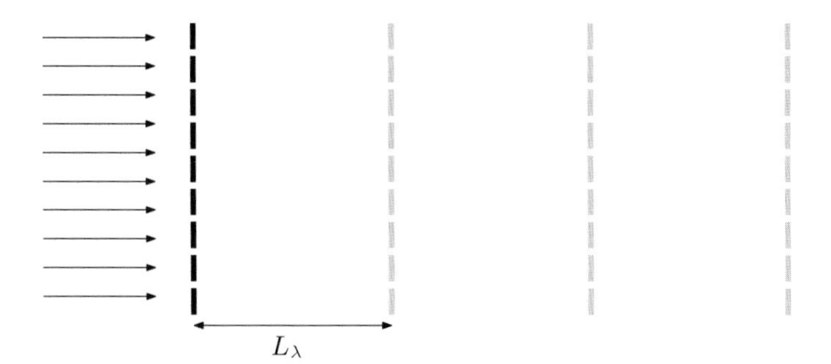

Fig. 6.8. Illustration of the Talbot–Lau effect. A plane wave traversing the grating on the left will generate an image of the grating at distances nL_λ, $n = 1, 2, \ldots$, from the grating.

show dark and light bands spaced a distance d apart, creating an image of the grating itself. This Talbot–Lau effect is a true interference phenomenon. Hence, if we carry out this experiment with matter particles and observe a density pattern equal to the grating pattern at multiples of the "Talbot length" L_λ, we have demonstrated the wave nature of these particles. To be sure, the fringes may in principle also be due to a simple blocking of rays by the grating (the so-called Moiré effect). However, the Moiré pattern is independent of the wavelength λ. Thus, to make sure that we actually deal with the Talbot–Lau effect, we can simply observe a variation in the pattern while varying λ. In the matter-wave case, this is achieved by changing the mean velocity v of the particles, since $\lambda = h/mv$.

A simple estimate will immediately convince the reader of the advantage of the Talbot–Lau method for detecting interference effects with C_{70} molecules. In the experiments of the Zeilinger group, the spacing d of the grating is about one micrometer, and the wavelength of the molecules equals a few picometer. Thus $L_\lambda \sim (10^{-6}\,\mathrm{m})^2/10^{-12}\,\mathrm{m} = 1\,\mathrm{m}$, i.e., all we need to do to observe the interference pattern is to place the "screen," i.e., a particle detector, at the macroscopic distance L_λ behind the grating. (The actual distance in the experiment is $L_\lambda = 38$ cm.) If we then measure a periodic variation of the particle density (with period d) that exhibits the characteristic wavelength dependence as we change the velocity of the molecules, we have confirmed the wave nature of these molecules.

The setup in the experiments of the Zeilinger group is schematically shown in Fig. 6.9. The C_{70} molecules are emitted from a source and then pass through a total of three identical free-standing gold gratings, each containing on the order of a thousand slits with period $d = 990\,\mathrm{nm}$. The purpose of the first grating is to induce a certain degree of coherence in the beam such that a diffraction pattern can be observed. The center grating acts as the actual diffraction grating. Finally, a third grating is placed at a distance equal to the

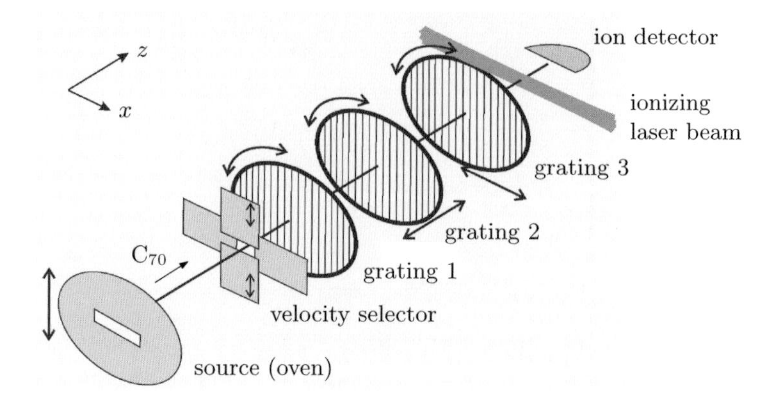

Fig. 6.9. Setup for observing matter-wave interference with C_{70} molecules in the experiments of the Zeilinger group [263–266]. Figure adapted with permission from [263]. Copyright 2002 by the American Physical Society.

Talbot length L_λ behind the diffraction grating. Recall that this means that, if the C_{70} molecules indeed possess a wave nature, an interference pattern will be obtained that, at the position of the third grating, will be an image of the diffraction grating itself. In other words, the molecular density would have periodic maximums and minimums along the x direction, with the peaks spaced apart by the slit spacing d. The purpose of the third grating is to act as a scanning mask for this molecular-density pattern. By moving the grating along x, the number of molecules registered behind the grating will fluctuate between a maximum and minimum value if an interference pattern is present. This fluctuation can then be easily measured by counting the number of molecules behind the third grating as a function of the x position of the grating.

The mean velocities of the molecules passing through the apparatus can be adjusted between about 90 m/s and 220 m/s by changing the position of the molecular source such that the fixed-height delimiter selects different portions of the beam, which follows a free-flight parabola. These velocities correspond to de Broglie wavelengths between 2 and 6 pm. The entire apparatus is contained in a vacuum chamber to minimize collisions between the C_{70} molecules and other particles.

6.2.3 Confirming the Wave Nature of Massive Molecules

The result of a typical run of the experiment is shown in Fig. 6.10. The horizontal axis corresponds to the position of the scanning mask along the x axis (see Fig. 6.9), and the vertical axis shows the molecular density behind the grating for each position. We clearly see the oscillatory fluctuations in the density of C_{70} molecules, which confirms the existence of spatial coherence

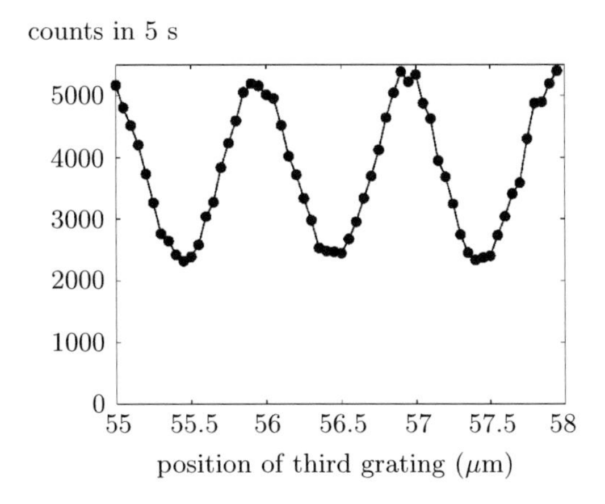

Fig. 6.10. Interference fringes for C_{70} molecules observed in the experiments of the Zeilinger group. The solid line is a fitted sine curve based on theoretical predictions. Figure adapted with permission from [263]. Copyright 2002 by the American Physical Society.

and interference effects. By varying the velocity of the molecules and observing the change in the density pattern (in agreement with the Talbot–Lau theory), it was also explicitly confirmed that this pattern is indeed due to the Talbot–Lau effect.

Also, the interference effect is a strict one-particle effect and not due to any interference between *different* molecules, for two reasons. First, inter-particle interference would require that the interacting molecules are in the same state. However, since each C_{70} molecule has a very large number of internal degrees of freedom, the chances for this to be the case are effectively zero. Second, the distance between two molecules is much larger than the range of any intermolecular forces, so essentially only a single molecule is passing through the apparatus at any given time.

6.2.4 Which-Path Information and Decoherence

Let us now turn to the main theme of this chapter, namely, the experimental observation of decoherence. In the C_{70}-interference experiments, decoherence is dominantly due to collisions between C_{70} molecules and molecules in the background gas (which is always present in the apparatus to some degree, since the vacuum is never perfect). We described such collisional decoherence in detail in Chap. 3 and showed that it constitutes a ubiquitous source of decoherence in nature. However, as mentioned in the opening paragraphs of this chapter, for mesoscopic and macroscopic objects this decoherence is typically so fast as to make it effectively impossible to observe its gradual action.

This brings us to a key advantage of the C_{70}-interference experiments. Here, the amount of collisional decoherence can be precisely tuned by changing the density of the background gas. Fig. 6.11 displays the experimentally measured interference pattern for the C_{70} molecules for two different pressures (i.e., densities) of the background gas, as reported by Lucia Hackermüller et al. [267] (see also [264]). The left plot shows clearly visible interference fringes. As the pressure of the background gas is increased, the interference fringes become less pronounced (right plot). Let us quantify the visibility $V(p)$ of the interference pattern at a given background-gas pressure p as the relative difference between the maximum amplitude $c_{\max}(p)$ and minimum amplitude $c_{\min}(p)$ of the interference pattern,

$$V(p) = \frac{c_{\max}(p) - c_{\min}(p)}{c_{\max}(p) + c_{\min}(p)}. \tag{6.30}$$

In the idealized case of no decoherence and no other experimental imperfections, we would have $c_{\min} = 0$ and therefore full visibility. In the completely decohered case, no interference pattern would be observable and thus $c_{\max} = c_{\min}$, which implies zero visibility.

The experimental data [264, 267] for the change of the visibility (6.30) with gas pressure p is shown in Fig. 6.12. The visibility is found to decrease exponentially with p (note that the vertical axis is in logarithmic units). This exponential decay is in remarkable agreement with theoretical calculations [160, 268].

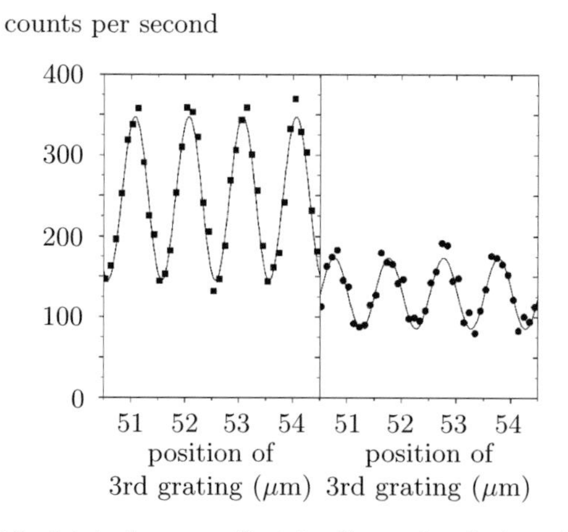

Fig. 6.11. Diminished interference effect in C_{70}-molecule interferometry due to decoherence induced by collisions with background-gas molecules, as observed in the experiments of the Zeilinger group [267]. The visibility of the interference fringes decreases when the pressure of the gas is increased from the left to the right panel. Figure adapted, with kind permission from Springer Science and Business Media, from [267].

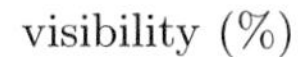

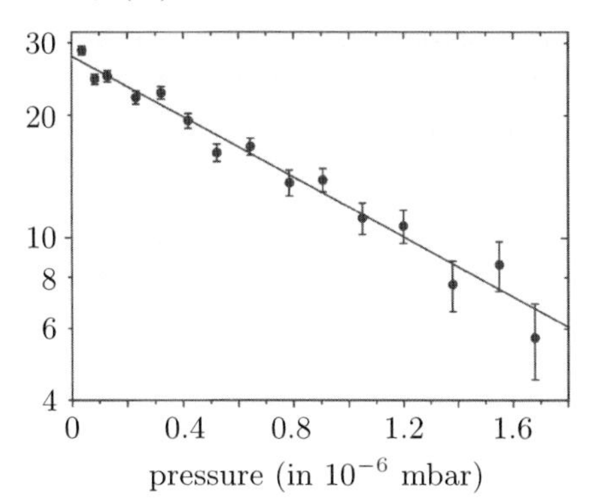

Fig. 6.12. Dependence of the visibility (6.30) of the interference pattern on the pressure of the background gas, as reported by the Zeilinger group [267]. The measured values (circles) are seen to agree well with the theoretical prediction (solid line), which describes an exponential decay of the visibility with pressure. Figure adapted, with kind permission from Springer Science and Business Media, from [267].

Thus these experiments provide impressive direct evidence for how the interaction with the environment gradually delocalizes the quantum coherence required for interference effects to be observed. This decoherence process occurs in a completely controlled way: By changing the amount of which-path information obtained by the environment (e.g., by altering the density of the background gas), the amount of decoherence and thus the loss of quantum features can be precisely tuned. So we can smoothly navigate and explore the quantum–classical boundary, and we find our observations to be in excellent agreement with theoretical predictions.

6.2.5 Decoherence Due to Emission of Thermal Radiation

What are the most important decoherence mechanisms for macroscopic bodies? Certainly, as we have seen, collisions with other particles (photons, air molecules, etc.) are ubiquitous and play a crucial role. However, another fundamental source of decoherence on these length scales is the emission of thermal radiation. Every "large" object is able to store energy in its many internal degrees of freedom. Macroscopic bodies essentially continuously absorb and emit photons. Each emitted photon carries away information about the path of the body, leading to decoherence in the position basis and therefore to spatial localization of the object.

For macroscopic objects, thermal emission of radiation is typically so strong as to completely preclude the possibility of observing spatial interference effects. Interestingly, as we shall see below, C_{70} molecules perfectly navigate the boundary between the microscopic and macroscopic regimes with respect to thermal decoherence. They are small enough to allow for visible interference patterns to emerge, but they are also sufficiently complex that, when heated to a temperature of several thousand Kelvin, thermal decoherence leads to a complete disappearance of interference patterns. Thus these molecules are ideally suited for an explicit observation of thermal decoherence effects. All we need to do is to gradually heat up the molecules before they pass through the apparatus and then record the decay of the visibility of the interference fringes as a function of the molecular temperature.

This idea has been realized in a remarkable experiment that was carried out by members of the Zeilinger group and reported in 2004 by Hackermüller et al. [266]. Before passage through the gratings, the C_{70} molecules were heated by a laser beam from their source temperature of about 900 K to temperatures up to around 3,000 K. The observed interference pattern for four different values of laser power (and thus mean temperature of the molecules) is shown in Fig. 6.13.

We see that the visibility decreases with laser power due to thermal emission of radiation from the molecules, as expected. The results from a series of runs at various laser heating powers are shown in Fig. 6.14, and we observe that the experimental data is in good agreement with predictions obtained

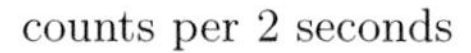

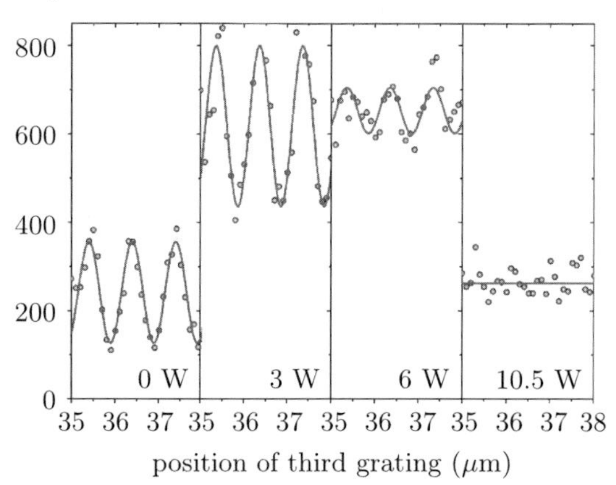

Fig. 6.13. Interference patterns for C_{70} molecules for different laser heating powers (inducing different mean temperatures of the molecules), as measured in the experiment by Hackermüller et al. [266]. The visibility of the interference fringes decreases as the average temperature of the molecules is increased. The variation in the overall average count rate at different values of laser heating power is due to variations of the efficiency of the ion-counting detector with the energy of the molecules and due to ionization and fragmentation effects in the heating stage. Figure reprinted from [266] by permission from Macmillan Publishers Ltd: Nature, copyright 2004.

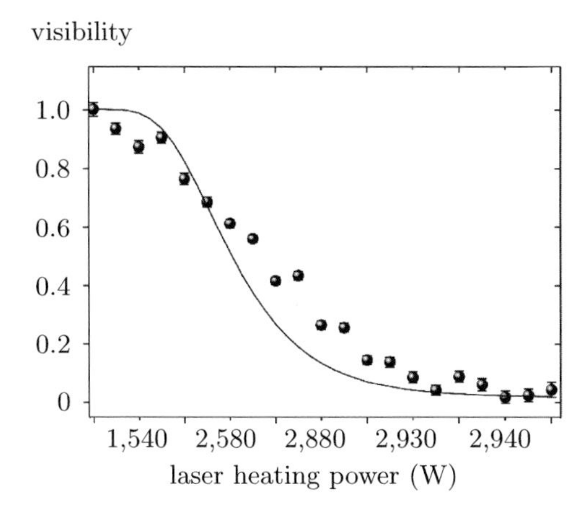

Fig. 6.14. Decay of the visibility of the C_{70} interference pattern as a function of laser heating power and mean molecular temperature in the experiment of Hackermüller et al. [266]. The circles represent experimental data, which agrees well with the theoretical prediction represented by the solid line. Figure reprinted from [266] by permission from Macmillan Publishers Ltd: Nature, copyright 2004.

from theoretical calculations [266]. A further detailed theoretical analysis of the experiment can be found in [269, 270].

The results demonstrate that for C_{70} molecules below temperatures of about 1,000 K, thermal decoherence is sufficiently weak for an interference pattern to be observed. At temperatures above about 2,000 K, thermal decoherence significantly reduces the visibility of the fringes, while around 3,000 K the interference pattern disappears completely.

This behavior can be explained as follows. For an emitted photon to transmit a sufficient amount of which-path information to resolve the path of the C_{70} molecule, its wavelength must be comparable to the separation of the different paths of the molecule, i.e., to the spacing between the maxima of the wave packets corresponding to passage of the molecule through the different slits in the diffraction grating. This separation is on the order of the spacing of the slits themselves, $d \approx 1\,\mu$m, so thermal decoherence requires the emission of photons of wavelength $\lambda \lesssim 1\ \mu$m. It turns out that only for temperatures above 2,000 K there is a significant probability of a C_{70} molecule to emit a photon of this wavelength. For temperatures around 3,000 K, the molecule typically emits several such photons. This transfers a sufficient amount of which-path information to the environment to entirely destroy the interference pattern.

6.2.6 Beyond Buckey Balls

Now that single-particle interference fringes have been experimentally observed for C_{70} molecules, we can naturally go further and attempt to demonstrate such quantum effects for larger and larger molecules. In fact, the Zeilinger group has already taken the next step into this direction in an experiment reported by Hackermüller et al. in 2002 [265]. Using an experimental setup similar to that for C_{70} molecules, interference patterns were observed for the even more massive fluorinated fullerene $C_{60}F_{48}$ (mass $m = 1632$ amu, 108 atoms) and for the biomolecule tetraphenylporphyrin $C_{44}H_{30}N_4$ (mass $m = 614$ amu, width over 2 nm). The structure of these molecules is sketched in Fig. 6.15, and the resulting interference patterns are displayed in Fig. 6.16. Again, well-defined interference fringes are observed even for these comparably massive and complex molecules, and the measured visibilities were found to be in good agreement with theoretically predicted values.

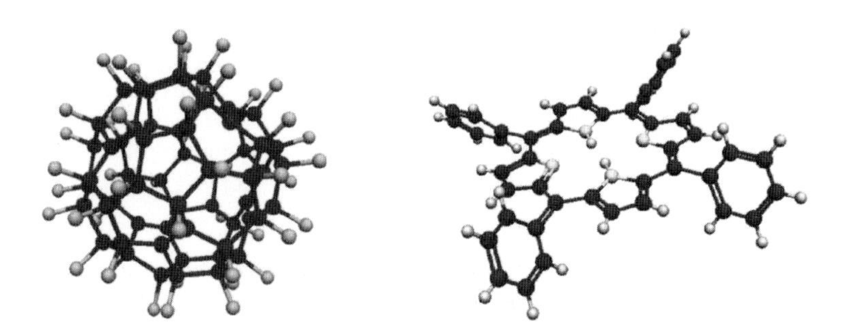

Fig. 6.15. Three-dimensional structure of the $C_{60}F_{48}$ molecule (left) and the biomolecule tetraphenylporphyrin $C_{44}H_{30}N_4$ (right) used in the interference experiments by Hackermüller et al. [265]. Figure reprinted with permission from [265]. Copyright 2003 by the American Physical Society.

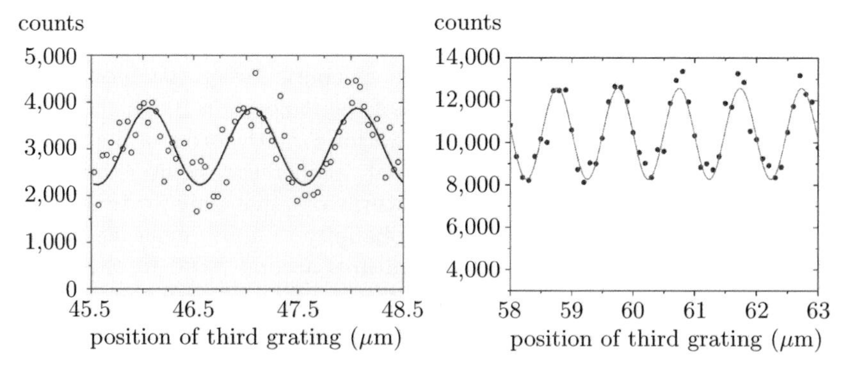

Fig. 6.16. Interference fringes for the $C_{60}F_{48}$ molecule (left) and the biomolecule tetraphenylporphyrin ($C_{44}H_{30}N_4$) (right) observed in the experiment by Hackermüller et al. [265]. The solid line is a fitted sine curve. Figure adapted with permission from [265]. Copyright 2003 by the American Physical Society.

One may now speculate about the feasibility of interference experiments that use even larger particles, for instance, biomolecules such as proteins and viruses [267, 271] or carbonaceous aerosols [270]. Such experiments will be limited by three main factors:

1. *Collisional decoherence.* In order to observe interference fringes for these molecules, the vacuum in the apparatus must be very good. Using a decoherence model for the visibility of interference fringes as a function of the density of the background gas, the quality of the vacuum in the apparatus required for the observation of interference effects has been estimated for several biomolecules up to the size of a rhinovirus [267,271] (see Fig. 6.17). The estimates assume the availability of an elongated Talbot–Lau interferometer with a spacing of one meter between the gratings (which is about three times more than the currently used distance), which has not yet been experimentally realized. Given this design, the required low background-gas densities are within the reach of current technology [264, 267].
2. *Thermal decoherence.* This effect will become increasingly relevant on larger scales. For example, for a simple double-slit setup (as opposed to the Talbot–Lau interferometer) it has been estimated that a small virus (mass 5×10^7 amu) would need to be cooled down to about 40 K to exhibit a sufficiently visible interference pattern [270]. Obviously, it would be quite difficult to produce a beam of such "cold viruses" in the laboratory. For objects larger than the C_{70} molecules, it may thus well turn out that

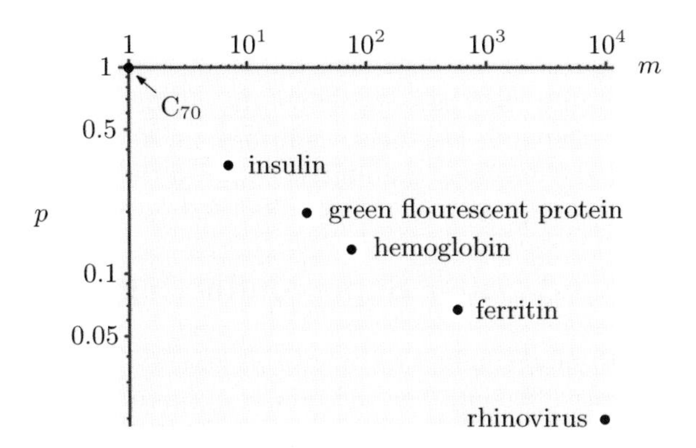

Fig. 6.17. Estimates for the maximum background-gas pressure p that would still permit the observation of interference fringes for several biomolecules in an elongated Talbot–Lau interferometer. The horizontal axis shows the molecular weight m relative to the mass of a C_{70} molecule. The pressure, shown on the vertical axis, is normalized by the pressure required to observe interference with C_{70} molecules. Plot generated using data reported by Hackermüller et al. [267].

it will be experimentally much more challenging to mitigate the effect of thermal decoherence than that of collisional decoherence [270].

3. *Dephasing due to inertial forces and vibrations.* While not a decoherence effect *per se*, the visibility of the interference pattern is also affected by noise due to gravitational, rotational, and acoustic perturbations, which are induced by the mass and rotation of the earth and by vibrations of the diffraction gratings [272]. These effects can severely limit the observability of the interference fringes, so the experimenters have taken much care to minimize these sources of noise in the Talbot–Lau setup described here. Still, for the particles listed in Fig. 6.17, vibrational noise is predicted to be a major obstacle to observing interference [272]. In currently available setups, this noise is likely to diminish the visibility of the interference pattern to a larger extent than collisional and thermal decoherence.

The important point to bear in mind is, however, that the conditions for an observation of interference effects can be precisely specified and quantified using theoretical decoherence models. We no longer have to limit ourselves to the assumption of a vaguely defined, fundamental divide between the quantum and classical realms as postulated by the Copenhagen interpretation. Instead, by treating the molecules of interest as open quantum systems interacting with their environments, we can understand what we need to do in order to observe interference effects, and why interference fringes are unobservable in a particular experimental setting.

The C_{70} molecules have the perfect amount of susceptibility to decoherence to allow for an observation of the gradual action of decoherence due to collisions with surrounding particles and due to the emission of thermal radiation. Thus it is difficult to imagine a more accessible and intuitive experiment for demonstrating the direct and controllable influence of decoherence that drives the system into the classical regime. It will be exciting to follow the development of research in this field, as future interference experiments are expected to push the envelope even further toward the macroscopic realm.

6.3 SQUIDs and Other Superconducting Qubits

Let us now turn to superconducting quantum two-state (qubit) systems. Over the past decade, these systems have become key players in experimental studies of macroscopic coherence and decoherence. They have gained additional importance as potential building blocks of superconducting quantum computers [273]. In the following, we shall focus on so-called superconducting quantum interference devices, or SQUIDs for short. We will also mention some related types of superconducting qubit systems such as Cooper-pair boxes.

A SQUID is a macroscopic quantum system that can exhibit a variety of fascinating quantum effects. In SQUIDs the state of billions of Bose-condensed electron pairs (see below) is described by a single collective macroscopic variable whose evolution is governed by the Schrödinger equation. The potential of SQUIDs for explorations of the quantum domain and of possible limits to quantum mechanics has been recognized early. In 1980, Leggett [274] proposed a *Gedankenexperiment* involving the superposition of macroscopic flux states in a SQUID to demonstrate the possibility of quantum-coherent behavior in a macroscopic system. However, it took two decades for this proposal to mature into successful experiments [275, 276].

6.3.1 Superconductivity and Supercurrents

The phenomenon of superconductivity was first discovered in the early 1900s by Heike Kamerlingh Onnes. Onnes used liquid helium to cool down solid mercury and found that at a temperature of several Kelvin above absolute zero, the electrical resistance of the mercury disappeared. He published his results in 1911 in a paper aptly entitled "On the sudden rate at which the resistance of mercury disappears" [277]. Two years later, he was awarded the Nobel Prize for his work. However, a theoretical understanding of superconductivity at the microscopic level was not reached until 1957, when Bardeen, Cooper, and Schrieffer [278] explained superconductivity as the formation of so-called *Cooper pairs* of electrons of opposite spin, where the mutual attraction is mediated by phonons in the crystal lattice of the material. Cooper pairs effectively act as bosons and can therefore exhibit properties that are strikingly different from those of single electrons, which are fermions.

Why can Cooper pairs traverse the material without encountering any resistance? To answer this question, we note that each Cooper pair corresponds to a low-energy ground state that is separated by an energy gap ΔE from the first excited state. If the thermal vibrational energy of the crystal lattice is less than ΔE, the lattice cannot induce any transitions in the state of the Cooper pair, and thus the pair remains unaffected by collisions with the lattice. This is in contrast with a single (free) electron that can absorb and emit arbitrary amounts of energy, leading to dissipation and thus to electrical resistance.

A key feature of superconductivity is the fact that, since the Cooper pairs all assume the same low-energy quantum state, a number of Cooper pairs can form a resistanceless current ("supercurrent") in which the collective center-of-mass motion of the pairs is represented by a single macroscopic wave function. Thanks to the peculiar features of quantum mechanics, the flow of such a current does not even require the application of an external voltage. All we need to do is to insert a very thin insulating barrier between two pieces of superconducting material. Cooper pairs then tunnel through the barrier, leading to the flow of a supercurrent at zero voltage. This experimentally well-confirmed, purely quantum-mechanical phenomenon is one of the different incarnations of the so-called *Josephson effect*, and the barrier is correspondingly referred to as the *Josephson junction* (for reviews, see [279, 280]).

6.3.2 Basic Physics of SQUIDs

A SQUID is directly based on the Josephson effect. It consists of a ring of superconducting material interrupted by one (rf-SQUID) or several (dc-SQUID) Josephson junctions (Fig. 6.18). In this section, we shall focus on the rf-SQUID with a single junction. The quantum-mechanical tunneling of Cooper pairs through the junction leads to the flow of a supercurrent around the loop, which creates a magnetic flux threading the SQUID ring. Additionally, the SQUID is immersed into an external magnetic field whose strength can be adjusted, giving rise to an additional flux Φ_{ext} through the loop. The purpose of this external flux will be explained below.

Quantum mechanics requires that the wave function describing the supercurrent be continuous around the loop, and thus an integer multiple of the wavelength must equal the circumference of the loop. The total change in phase around the loop is given by $2\pi\Phi/\Phi_0$. Here $\Phi_0 = h/2e$ is the so-called flux quantum (with e being the electron charge). Φ denotes the total trapped flux through the loop, composed of the flux created by the flow of supercurrent around the loop and the flux Φ_{ext} due to the external magnetic field. If the loop did not contain a Josephson junction, then the requirement of continuity of the wave function would translate into the condition

$$\Phi/\Phi_0 = k, \qquad k = 1, 2, \ldots \tag{6.31}$$

However, the presence of the Josephson junction introduces an additional phase shift $\Delta\phi_{\text{J}}$, which is completely determined by the physical properties of the junction. The complete condition for phase continuity then reads

$$\Delta\phi_{\text{J}} + 2\pi\Phi/\Phi_0 = 2\pi k, \qquad k = 1, 2, \ldots. \tag{6.32}$$

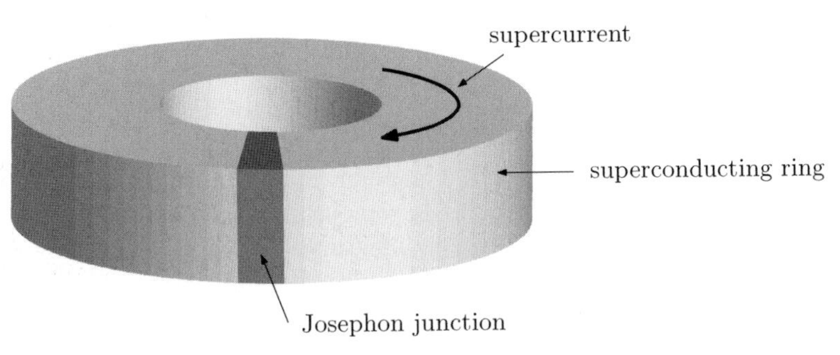

Fig. 6.18. Schematic illustration of a SQUID. A ring of superconducting material is interrupted by one or several Josephson junctions, which induce the flow of a dissipationless "supercurrent" around the loop. This supercurrent creates a magnetic flux through the loop.

Since the quantities $\Delta\phi_J$ and Φ_0 are fixed, the only free parameter in (6.32) is the total flux Φ. Thus in SQUIDs the dynamics of a macroscopic quantity of matter (namely, on the order of 10^9 Cooper pairs) are collectively determined by the quantum-mechanical evolution of a single macroscopic coordinate Φ. Because flux is the key variable in SQUIDs, these devices are often referred to as superconducting flux qubits. (As we shall describe in Sect. 6.3.4, there exist other types of superconducting qubit systems in which physical quantities such as charge or phase are the key variables [280].)

Let us now discuss the dynamics of Φ. It turns out that the evolution of Φ is governed by an effective Hamiltonian of the form

$$H = \frac{P_\Phi^2}{2C} + U(\Phi) = \frac{\hbar^2}{2C}\frac{\mathrm{d}^2}{\mathrm{d}\Phi^2} + \left[\frac{(\Phi - \Phi_{\mathrm{ext}})^2}{2L} - \frac{I_c\Phi_0}{2\pi}\cos\left(2\pi\frac{\Phi}{\Phi_0}\right)\right]. \quad (6.33)$$

Here C denotes the total capacitance, L is the self-inductance of the loop, and I_c is the critical current of the Josephson junction. We shall not derive the structure of this Hamiltonian here (interested readers may find the derivation in [218]). Instead, let us get a feel for what kind of dynamics this Hamiltonian implies.

We can interpret the Hamiltonian as determining the motion of a fictitious "particle" of "mass" C in a potential given by $U(\Phi)$. The role of the usual canonical variables position and momentum is here played by the total trapped flux Φ and the conjugate quantity $P_\Phi = -i\hbar\mathrm{d}/\mathrm{d}\Phi$, which is often referred to as the total displacement current and has units of charge. The analog of the kinetic energy of a particle moving in real space is in our case the electrostatic energy.

If the external magnetic field is adjusted such that the induced flux Φ_{ext} is in the vicinity of the bias point $\Phi_{\mathrm{ext}} = \Phi_0/2$, $U(\Phi)$ takes the form of a tilted one-dimensional double-well potential (Fig. 6.19). This is the experimentally relevant situation, which we shall assume from now on. The amount of tilt is determined by the flux Φ_{ext}, and at $\Phi_{\mathrm{ext}} = \Phi_0/2$ the double well becomes exactly symmetric (this case will be discussed in more detail below). Thus we can change the shape of the potential by adjusting the strength of the externally applied magnetic field.

Broadly speaking, the two wells of the potential $U(\Phi)$ correspond to the two possible directions (clockwise and counterclockwise) of the supercurrent around the loop. Each well contains a number of energy eigenstates $|k\rangle$ of the Hamiltonian (6.33) that are (provided the damping induced by the Josephson junction is weak) localized well below the barrier separating the two wells. Thus we have a set of left-well (right-well) energy eigenstates that correspond, in an approximate manner, to a classical persistent clockwise (counterclockwise) current around the loop.

More precisely, the corresponding wave functions $\psi_k(\Phi) \equiv \langle\Phi|k\rangle$ are locally of the s-wave type. This means that the amplitudes of these wave functions are narrowly peaked around the bottom of either one of the wells of

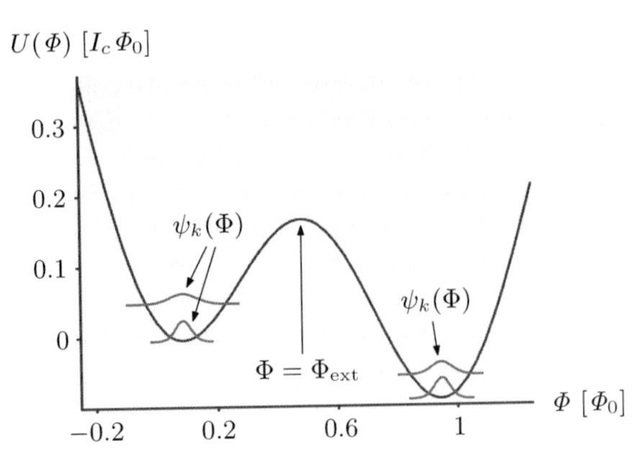

Fig. 6.19. Effective double-well potential $U(\Phi)$ of the SQUID (in units of $I_c\Phi_0$) in the vicinity of the bias point $\Phi_{\rm ext} = \Phi_0/2$. Low-lying energy eigenstates $\psi_k(\Phi) \equiv \langle\Phi|k\rangle$ are well-localized near the bottom of each well and are therefore also approximate eigenstates of the flux operator. In the plot, the Φ-axis is measured in units of the flux quantum Φ_0, and we chose $\Phi_{\rm ext} = \frac{3}{5}\Phi_0$.

$U(\Phi)$, with very little spread in flux space. Thus each state $|k\rangle$ corresponds to a relatively narrow range of values of the total flux Φ and can therefore (at least for sufficiently small k) also be regarded as a "fuzzy" eigenstate of the flux operator, with the possible discrete flux values determined by the continuity condition (6.32). Accordingly, the states $|k\rangle$ are often referred to as "k-fluxoid states."

Classically, the barrier separating the two wells of the potential $U(\Phi)$ is impenetrable for the low-lying k-fluxoid states $|k\rangle$, since their energy is less than the barrier height. However, quantum-mechanically, resonant tunneling between the wells can occur [281,282]. For example, one can apply a radiation field to excite the system from a low-lying state to a higher-energy state, localized in the same well, and this state can subsequently decay to a lower (or equal) energy level in the other well via quantum tunneling. Such tunneling processes lead to a macroscopic change of the flux Φ threading the ring, which can in turn be observed in form of a macroscopic change in the magnetic moment of the system.

One typically engineers the SQUID in such a way as to ensure that the ground state $|0\rangle$ and first excited state $|1\rangle$ (localized at the bottom of opposite wells) are well-separated in energy from higher-lying states. Then the SQUID can be effectively described as a macroscopic two-state (qubit) system. As basis states, we choose the two classical persistent-current states $|\circlearrowright\rangle$ and $|\circlearrowleft\rangle$ corresponding to the supercurrent flowing in, respectively, definite clockwise and anticlockwise directions around the loop. Formally, we may then write the effective Hamiltonian of this two-state SQUID as

$$\hat{H} = -\epsilon\hat{\sigma}_z - \Delta\hat{\sigma}_x. \tag{6.34}$$

Here the eigenstates of the Pauli spin operator $\hat{\sigma}_z$ are $|\circlearrowright\rangle$ and $|\circlearrowleft\rangle$, with $2\epsilon = 2I_p\,(\Phi_{\text{ext}} - \Phi_0/2)$ denoting the asymmetry energy between these states, and $I_p \approx I_c$. Δ is the matrix element for tunneling between $|\circlearrowright\rangle$ and $|\circlearrowleft\rangle$. The two lowest-lying energy eigenstates $|0\rangle$ and $|1\rangle$ are then linear combinations of $|\circlearrowright\rangle$ and $|\circlearrowleft\rangle$ given by

$$|0\rangle = \cos(\theta)\,|\circlearrowright\rangle + \sin(\theta)\,|\circlearrowleft\rangle\,, \tag{6.35a}$$

$$|1\rangle = -\sin(\theta)\,|\circlearrowright\rangle + \cos(\theta)\,|\circlearrowleft\rangle\,, \tag{6.35b}$$

with $\tan 2\theta = \Delta/\epsilon$ and energies $E_{0,1} = \mp\sqrt{\Delta^2 + \epsilon^2}$. The energy-level splitting between $|0\rangle$ and $|1\rangle$ is therefore $\Delta E = 2E_1 = 2\sqrt{\Delta^2 + \epsilon^2}$. Away from the bias point $\Phi_{\text{ext}} = \Phi_0/2$, $\epsilon \gg \Delta$, and thus the states $|\circlearrowright\rangle$ and $|\circlearrowleft\rangle$ are essentially identical to the zero-fluxoid and one-fluxoid energy eigenstates $|0\rangle$ and $|1\rangle$, in agreement with our discussion above. The expectation value of the supercurrent corresponding to the two states $|0\rangle$ and $|1\rangle$ can be shown to be equal to $\pm I_p \cos 2\theta$ (where the positive sign indicates clockwise direction of the current).

6.3.3 Superposition States and Coherent Oscillations in SQUIDs

Let us now consider what happens if the SQUID is at the bias point $\Phi_{\text{ext}} = \Phi_0/2$. Now the double well becomes symmetric and we have $\epsilon = 0$ in (6.34), i.e., the energy levels in the left and right wells line up. In this case, classically, the energy levels associated with the states $|0\rangle$ and $|1\rangle$ would cross and therefore these states would become degenerate. Quantum-mechanically, however, this degeneracy is lifted by the continued existence of a tunneling barrier Δ between the wells [see (6.34)], which leads to a "repulsion" of the energy levels, producing an *anticrossing* of these levels (Fig. 6.20). Since now $\theta = \pi/4$ in (6.35a) and (6.35b), the energy ground state $|0\rangle$ and the first excited state $|1\rangle$ are given by symmetric and antisymmetric superpositions of the persistent-current states $|\circlearrowright\rangle$ and $|\circlearrowleft\rangle$,

$$|0\rangle = \frac{1}{\sqrt{2}}\left(|\circlearrowright\rangle + |\circlearrowleft\rangle\right), \tag{6.36a}$$

$$|1\rangle = \frac{1}{\sqrt{2}}\left(-\,|\circlearrowright\rangle + |\circlearrowleft\rangle\right). \tag{6.36b}$$

The energy splitting ΔE between these two new eigenstates is typically very small and is solely determined by the capacitance C of the junction, scaling as $\Delta E \propto \mathrm{e}^{-\sqrt{C}}$.

Since $|\circlearrowright\rangle$ and $|\circlearrowleft\rangle$ are localized in, respectively, the left and right wells, superpositions (6.36a) and (6.36b) are delocalized across the two wells. The existence of these superpositions in a SQUID was experimentally confirmed in 2000 by Friedman et al. [275] and independently by van der Wal et al. [276]. In the experiment by Friedman et al., the states $|\circlearrowright\rangle$ and $|\circlearrowleft\rangle$ corresponded

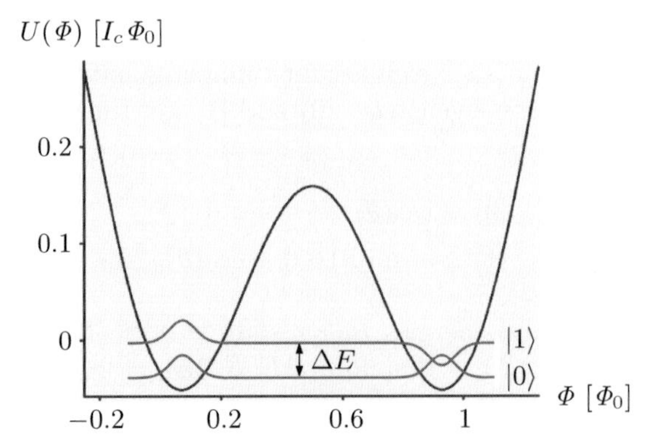

Fig. 6.20. Effective potential $U(\Phi)$ of the SQUID at bias $\Phi_{\text{ext}} = \Phi_0/2$. At this bias point the double well becomes symmetric. The presence of the tunneling barrier prevents the energy levels in the two wells from crossing. Instead, level anticrossing occurs via the formation of delocalized energy eigenstates $|0\rangle = \frac{1}{\sqrt{2}}(|\circlearrowright\rangle + |\circlearrowleft\rangle)$ (the new symmetric ground state) and $|1\rangle = \frac{1}{\sqrt{2}}(-|\circlearrowright\rangle + |\circlearrowleft\rangle)$ (the new antisymmetric first excited state). These eigenstates are superpositions of the "classical" persistent-current states $|\circlearrowright\rangle$ and $|\circlearrowleft\rangle$. Their presence has been confirmed experimentally via spectroscopic measurements [275, 276] and via the observation of coherent Rabi oscillations [283, 284].

to opposite-direction supercurrents of several μA, each current composed of billions of Cooper pairs. This amounts to a macroscopic difference of more than $\Phi_0/4$ in the flux generated by these two currents, equal to about $10^{10}\mu_{\text{B}}$ (where μ_{B} is the Bohr magneton) in local magnetic moment. The states (6.36a) and (6.36b) therefore represent superpositions of macroscopically distinct states of the Schrödinger-cat type, just as envisioned in Leggett's original thought experiment [274] twenty years earlier. Friedman et al. verified the existence of the superposition states by indirect means via a static spectroscopic measurement of the energy difference ΔE between these two states. Their result for ΔE turned out to be in excellent agreement with theoretical predictions.

There is another way of demonstrating the existence of such superposition states in SQUIDs, namely, through the observation of coherent oscillations between the persistent-current states $|\circlearrowright\rangle$ and $|\circlearrowleft\rangle$. These oscillations are of the Rabi type described in Sect. 6.1.1. First, the external flux threading the SQUID is tuned slightly away from the degeneracy point $\Phi_{\text{ext}} = \Phi_0/2$, such that the energy eigenstates $|0\rangle$ and $|1\rangle$ are close to degeneracy. The SQUID is initialized in the ground state $|0\rangle$ by allowing it to relax. Then a pulse of microwave radiation is applied whose frequency is tuned to the (small) energy difference ΔE between the states $|0\rangle$ and $|1\rangle$. Following our discussion of Rabi oscillations in Sect. 6.1.1, this will cause the qubit to oscillate between the states $|0\rangle$ and $|1\rangle$ and thus also between the persistent-current states $|\circlearrowright\rangle$ and $|\circlearrowleft\rangle$ [see (6.35)]. This effect manifests itself in an oscillation of the macroscopic supercurrent in the SQUID between clockwise and counterclockwise directions around the loop (see Fig. 6.21). The oscillation occurs at the Rabi frequency, which is determined by the strength of the coupling between the qubit and the microwave field and is linearly dependent on the microwave amplitude. Thus the relative probability of the SQUID's being in each of the two states $|\circlearrowright\rangle$ and $|\circlearrowleft\rangle$ depends on the amplitude and length of the microwave pulse [see (6.8)].

In 2003 Chiorescu and coworkers [283] experimentally demonstrated such Rabi oscillations in a superconducting flux qubit. The full experimental setup consisted here of a μm-sized superconducting loop interrupted by three Josephson junctions (Fig. 6.22). The use of three junctions does not significantly alter the basic physics of this type of qubit compared to the single-junction rf-SQUID described above, but it facilitates the correct tuning of the qubit. The qubit was in turn coupled to a dc-SQUID (with two Josephson junctions), which acted as an extremely sensitive read-out device for the flux in the qubit.

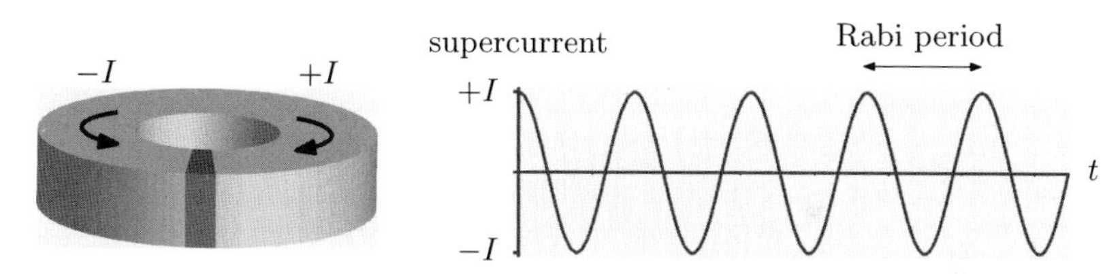

Fig. 6.21. Quantum-coherent tunneling in a SQUID (tuned close to the degeneracy point $\Phi_{\text{ext}} = \Phi_0/2$) can be observed via the coherent oscillation of the macroscopic supercurrent between clockwise and counterclockwise directions.

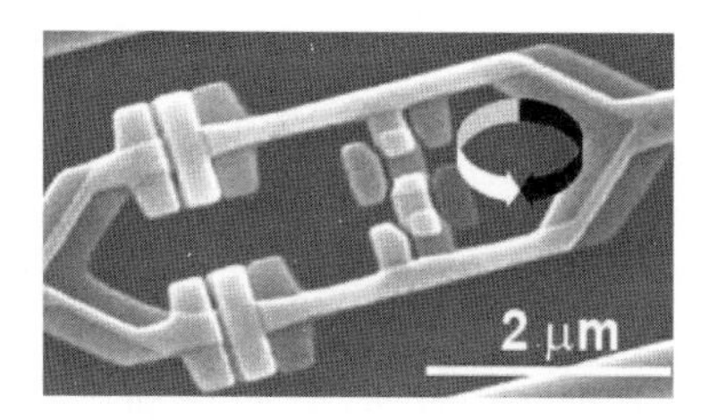

Fig. 6.22. Superconducting flux qubit used in the experiment by Chiorescu et al. [283]. The actual qubit, shown on the right, consists of a tiny superconducting loop with three Josephson junctions. The two possible directions of the persistent current around the loop are indicated by white and black arrows. The change in magnetic flux in the qubit is detected by a SQUID composed of a larger superconducting loop (partially shared with the qubit) interrupted by two Josephson junctions. Figure reprinted with permission from [283]. Copyright 2003 by AAAS.

Chiorescu et al. applied a microwave pulse of given length and amplitude and then performed a projective measurement on the flux qubit in the persistent-current basis $\{|\circlearrowright\rangle, |\circlearrowleft\rangle\}$ (corresponding to the experimentally accessible observable). By carrying out this procedure several thousand times, they obtained the experimental probability of finding the qubit in (say) the state $|\circlearrowright\rangle$, corresponding to the supercurrent flowing in clockwise direction. By repeating this experiment over a range of pulse lengths of the microwave radiation (while keeping the microwave amplitude fixed), they were able to trace out hundreds of damped Rabi oscillations with a surprisingly long decay time of about 150 ns (Fig. 6.23). To confirm that these oscillations were indeed quantum-coherent oscillations of the Rabi type, the researchers successfully verified that the observed oscillation frequencies scaled linearly with the amplitude of the microwave pulse, which is a key signature of the Rabi process. An alternative continuous read-out scheme for the detection of coherent oscillations was implemented in a subsequent experiment by Ilichev et al. [284].

The Rabi oscillations observed by Chiorescu et al. [283] and Ilichev et al. [284], together with the earlier spectroscopic results of Friedman et al. [275]

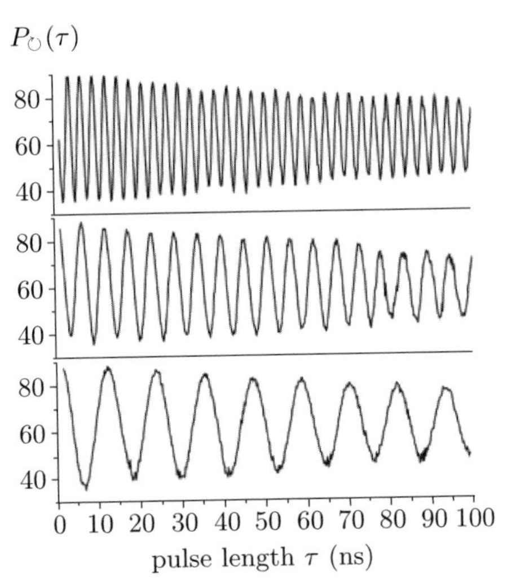

Fig. 6.23. Rabi oscillations between the macroscopic opposite-direction supercurrent states $|\circlearrowright\rangle$ and $|\circlearrowleft\rangle$ in a superconducting flux qubit as observed in the experiment by Chiorescu et al. [283]. These oscillations confirm the existence of coherent superpositions of the states $|\circlearrowright\rangle$ and $|\circlearrowleft\rangle$. Results for three different amplitudes of the resonant microwave radiation are shown (with the amplitude decreasing from top to bottom). The vertical axis denotes the percentage $P_{\circlearrowright}(\tau)$ of runs in which the qubit was found in the particular definite-current state $|\circlearrowright\rangle$ (say), as a function of the length τ of the microwave pulse. Figure adapted with permission from [283]. Copyright 2003 by AAAS.

and van der Wal et al. [276], impressively demonstrate the presence of coherent superpositions involving macroscopically distinct supercurrent states. SQUIDs are particularly good candidates for the creation of such superposition states, because the relevant macroscopic variable (the trapped macroscopic flux through the SQUID ring) can be controlled via microscopic energy differences in the Josephson junction [285]. The height Δ of the tunneling barrier is dominantly determined by properties of the junction, independently of the size of the loop. This means that the scaling of the achievable "size" of the superpositions of macroscopically distinct flux states is only weakly dependent on the amount of distinctness itself (i.e., on the difference in flux between the persistent-current states).

This feature of SQUIDs allows for the creation of superpositions of "classical" states that have a significantly larger degree of macroscopic distinctness than those achieved in other experiments. For example, in the matter-wave diffraction experiments with C_{70} molecules (Sect. 6.2), the spacing of the diffraction grating must decrease as $1/\sqrt{N}$ with the number N of atoms in the molecule in order for a diffraction pattern to be observed. Thus, in this case, our ability to experimentally produce interference patterns for molecules of a particular size scales with the size of the molecules itself.

6.3.4 Observing and Quantifying Decoherence

It is important to note that the characteristic damping time of the Rabi oscillations derived from the decay of the envelope of the oscillatory occupation probability (see Fig. 6.23) does not immediately give access to the intrinsic decoherence timescale of the persistent-current superpositions (6.36a) and (6.36b). The reason lies in the fact that the Rabi oscillations are induced by an external driving field, namely, by the applied microwave pulse. To measure the decoherence timescale, however, we must consider the delocalization of relative phase information between the components in the superposition under the *free* (nondriven) evolution of the qubit interacting with its uncontrolled environment.

This goal can be accomplished by using the technique of Ramsey interferometry [51]. The following procedure is a variation of the "resonant" Ramsey technique described in Sect. 2.2.2. First, with the qubit biased close to the degeneracy point $\Phi_{\text{ext}} = \Phi_0/2$, we again start by initializing the qubit in the energy ground state $|0\rangle$. We then apply a $\pi/2$ microwave pulse to the qubit that is off-resonance by an amount δf from the transition frequency $\Delta E/2\pi$ between the two levels $|0\rangle$ and $|1\rangle$. This transforms the state of the qubit into a coherent superposition of the states $|0\rangle$ and $|1\rangle$, which is then allowed to evolve freely for a certain period of time τ. During this time the relative phase between the components $|0\rangle$ and $|1\rangle$ in the superposition increases by $2\pi(\delta f)\tau$ (in the frame rotating with the applied microwave frequency). Then a second off-resonant $\pi/2$ pulse is applied. In the resulting superposition, the magnitudes of the coefficients multiplying the components $|0\rangle$ and $|1\rangle$

will exhibit an oscillatory dependence on the delay time τ, with a frequency equal to $\delta f/2$. It follows that the occupation probabilities of the classical persistent-current states $|\circlearrowright\rangle$ and $|\circlearrowleft\rangle$ will oscillate as a function of τ with frequency δf.

In practice, we expect the amplitude of this oscillation to decay, as decoherence diminishes the amount of coherence between the states $|0\rangle$ and $|1\rangle$ (and thus also between the persistent-current states $|\circlearrowright\rangle$ and $|\circlearrowleft\rangle$) during the duration τ of the free qubit evolution. As we have explained in Sect. 2.2.2, if the superposition becomes completely decohered, then we arrive at an incoherent mixture of the states $|\circlearrowright\rangle$ and $|\circlearrowleft\rangle$ for which the occupation probabilities are independent of the delay time τ.

This eventual disappearance of the oscillations of the occupation probabilities with increasing delay time τ allows us to observe and quantify the gradual effect of decoherence. Chiorescu et al. [283] measured the relevant occupation probabilities of the persistent-current states over a range of delay times τ. They traced out a damped oscillation of this probability, from which the characteristic decoherence timescale could be read off (Fig. 6.24).

We see the expected oscillatory dependence of the occupation probability on the delay time τ. The frequency of the oscillation was found to match the detuning δf of the microwave radiation, in excellent agreement with the predictions of the Ramsey theory. From the decay of the envelope of the oscillations, the authors obtained a decoherence time of about 20 ns, which represents the characteristic timescale for an environment-induced loss of phase coherence between the classical persistent-current states $|\circlearrowright\rangle$ and $|\circlearrowleft\rangle$ during the free evolution of the qubit in this experiment.

In more recent experiments involving superconducting flux qubits, significantly longer decoherence times up to 4 μs have been observed [286]. These promising results have nourished hopes that superconducting qubits might be good candidates for the implementation of solid-state quantum-computing

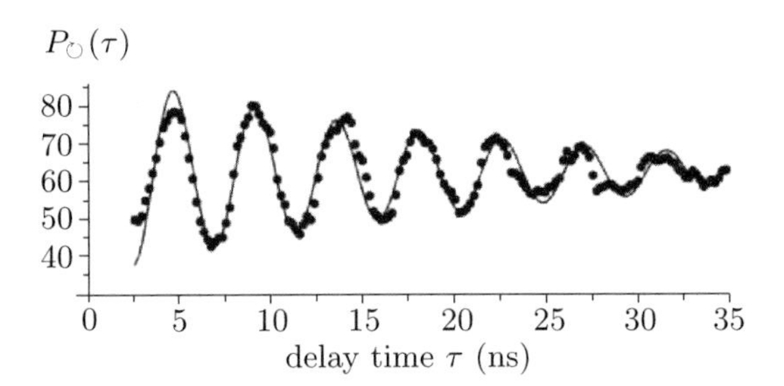

Fig. 6.24. Dependence of the occupation probability $P_{\circlearrowright}(\tau)$ on the delay time τ between the application of two $\pi/2$ pulses in a superconducting flux qubit as observed in the experiment of Chiorescu et al. [283]. The damping of the oscillation amplitude corresponds to the gradual loss of coherence from the system. Figure adapted with permission from [283]. Copyright 2003 by AAAS.

devices (see Chap. 7). Here, gate operations via the application of control pulses can be performed very fast and single-qubit gate operations can be as short as about 2 ns. One may therefore be able to perform on the order of a thousand gate operations before coherence is appreciably lost. This number, however, is still rather far away from the $O(100,000)$ gate operations required to implement even simple quantum algorithms.

Quantum-coherent oscillations and decoherence have also been observed in superconducting devices whose key variable is either charge or phase, instead of the flux variable used in the case of SQUIDs. Generally, the formal description of the physics of superconducting charge and phase devices is very similar to that of the SQUID given above.

Superconducting charge qubits, commonly known as Cooper-pair boxes, consist of a tiny superconducting "island" onto which Cooper pairs can tunnel from a reservoir through a Josephson junction. The two qubit basis states then correspond to two different charge states of the island, differing by at least one Cooper pair. Cooper-pair boxes were the first superconducting qubit systems that allowed for the experimental observation of Rabi oscillations between charge states, which was accomplished in 1999 by Nakamura, Pashkin, and Tsai [287]. In 2002, Vion et al. [288] reported the observation of many thousands of coherent oscillations with a decoherence time of 0.5 μs.

Coherent oscillations of the Rabi type have also been observed in phase qubits. Phase qubits contain a single Josephson junction, similar to the rf-SQUID. However, instead of the flux threading the loop, the key variable is here the phase difference between the electrodes of the junction. Yu et al. [289] measured Rabi oscillations between macroscopically distinct phase states and found decoherence times up to several μs. Similar results were also reported by Martinis et al. [290].

Current experimental evidence (see, e.g., [290, 291]) indicates that the main source of decoherence in superconducting qubits is the presence of intrinsic defects in the Josephson junctions and the superconductor, rather than (as one may have expected) interactions with the external circuit used to control the loop–junction setup. Often, these defects can be modeled as effective two-level systems. As briefly mentioned in Sect. 5.1.2, it has been well known for a while that decoherence and dissipation in low-temperature (e.g., superconducting) systems is mainly due to such intrinsic two-level systems [199, 200]. Since the superconducting qubit itself is a two-level system, the appropriate model for a theoretical description of decoherence in such qubits is likely to be of the spin–spin type discussed in Sect. 5.4 [199].

For example, Martinis et al. [291] applied this model to a study of decoherence in superconducting phase qubits due to dielectric loss from two-level states present both in the bulk insulating material as well as in the Josephson junction. Insights gathered from these theoretical studies were successfully translated into significantly improved experimental structures. Other theoretical studies of decoherence in superconducting qubits have focused on the influence of noise processes, such as fluctuations in the bias current controlling a current-biased Josephson junction [292]. However, in view of our general discussion in Sect. 2.12 (see also Sect. 7.3) it is important to clearly distinguish such noise-induced ensemble dephasing from "true" decoherence created by environmental entanglement, even though both processes may lead to similar results from the phenomenological perspective of local measurement statistics.

6.4 Other Experimental Domains

In the previous sections, we have described experiments that have successfully generated and detected superposition states of mesoscopically and macroscopically distinct states and that have allowed us to observe the gradual decoherence of such superpositions. In these experiments, decoherence has provided a qualitative and quantitative explanation of the quantum-to-classical phenomena observed in the laboratory.

In a similar vein—but now taking a different point of view—decoherence also helps us understand what *prevents* us from being able to generate and observe superposition states in certain experiments. In this section, we will describe two such experimental domains, which are promising candidates for the future creation and observation of even larger nonclassical superposition states than those available to date. As we shall see, decoherence allows us to describe qualitatively and quantitatively why it is so difficult to create and observe such superpositions in these experiments. Furthermore, decoherence suggests methods for improving the experimental procedures, such that we can come closer to achieving our goal of generating the desired superposition states in the laboratory.

6.4.1 Decoherence in Bose–Einstein Condensates

Bose–Einstein condensation is a quantum-mechanical phenomenon in which a macroscopic number of atoms (up to $O(10^7)$ in some experiments) undergoes a quantum phase transition into a condensate in which all atoms lose their individuality and occupy the same quantum state. Thus the condensate can be fully described by a single quantum-mechanical N-particle wave function with a phase. This phase of the wave function is experimentally accessible. For example, if two condensates are allowed to overlap, an interference pattern is observed due to the phase difference between the two condensates [293, 295–298] (Fig. 6.25). Recently it has also become possible to construct an interferometer in which a single condensate is coherently split and then recombined, resulting in the observation of interference fringes [294] (see again Fig. 6.25).

In spite of these successes in demonstrating the quantum properties of Bose–Einstein condensates, experimental attempts to produce Bose–Einstein

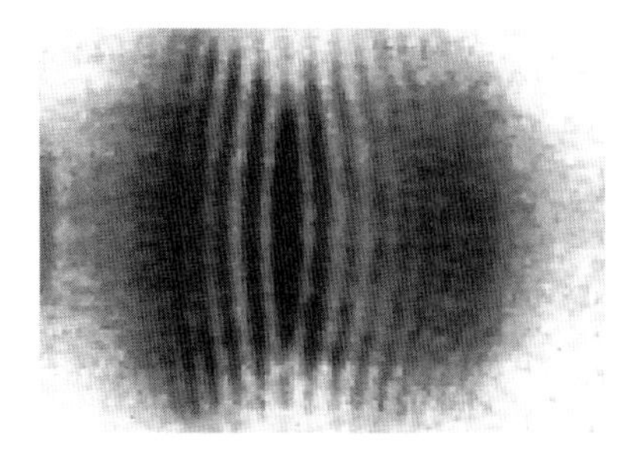

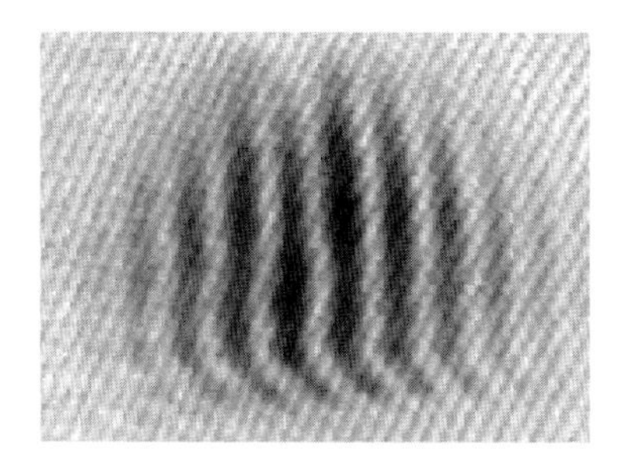

Fig. 6.25. Interference patterns of the atomic gas density in Bose–Einstein condensates. *Left:* Pattern resulting from the phase difference between two independent condensates that were let to overlap. Figure reprinted with permission from [293]. Copyright 1997 by AAAS. *Right:* Interference fringes obtained in a "double-slit" interferometer for Bose–Einstein condensates. A single condensate was coherently split (mimicking the role of the slits in the standard double-slit experiment) and subsequently recombined. Figure reprinted with permission from [294]. Copyright 2004 by the American Physical Society.

condensates in a superposition of states involving macroscopically distinguishable numbers of particles have thus far failed, despite several promising theoretical proposals [299–305]. However, this failure can be well understood, both qualitatively and quantitatively, by taking into account the relevant decoherence mechanisms. In the case of superposition states of Bose–Einstein condensates, decoherence is found to be dominantly due to elastic and inelastic scattering between condensate and noncondensate atoms.

Detailed models of such scattering processes have been developed (see, e.g., [102]), providing us with realistic predictions for decoherence timescales as a function of the key parameters of the experiment (such as the size of the condensate). Insights gained from these decoherence models have also led to concrete proposals for how to minimize decoherence through clever experimental design. Examples of such suggested improvements include the construction of modified condensate traps that allow for a faster evaporation of the decoherence-inducing thermal cloud of noncondensate atoms [102]; the creation of macroscopic superpositions of relative-phase (instead of number-difference) states, thereby taking into account the environment-selected pointer basis in order to minimize decoherence [302]; the implementation of active environment engineering (Sect. 7.5.3) to decrease the effective size of the thermal cloud [102]; and the faster generation of the superposition state [305]. Each of these proposals is derived from an analysis of the various decoherence mechanisms involved in a particular experimental setup. Given such a setup, models for decoherence then tell us the location of the quantum–classical boundary and what we can and need to do to push this boundary in order to realize the desired cat states.

6.4.2 Decoherence in Quantum-Electromechanical Systems

Quantum-electromechanical systems, or QEMS for short, are a relative recent addition to the zoo of experiments that have the potential to generate even larger nonclassical superposition states [306] (see also [307, 308] for an introduction to QEMS). We already mentioned QEMS in Sect. 5.4.2 in the context of the "oscillator–spin model" for decoherence and dissipation. QEMS show great promise for the demonstration of truly mechanical "Schrödinger kittens" involving billions of atoms in a superposition of two well-distinguishable positions in space. Given the rapid experimental progress in the field of QEMS, it is quite likely that over the next few years these systems will become a major player in explorations of the quantum-to-classical transition.

QEMS are a result of the recent revolution in nanotechnology. The main component of a QEMS is a miniature mechanical double-clamped beam or single-clamped cantilever, manufactured from crystalline materials such as silicone and with typical dimensions on the order of nanometers to micrometers (Fig. 6.26). The beam (or cantilever) acts as tiny mechanical resonator. If a suitable driving force is applied for a brief period of time, the resonator is set into an oscillatory motion.

Because of the extremely small dimensions of the resonator, the oscillation frequencies can be remarkably high, given the mechanical nature of such systems. Recent experiments have achieved frequencies up to one gigahertz, i.e., frequencies in the microwave range [310]. Furthermore, dissipative effects in the resonators can be made relatively weak, resulting in tens of thousands of free oscillations before a significant decrease in the oscillation amplitude occurs. The oscillatory motion of the resonator is detected by an electronic

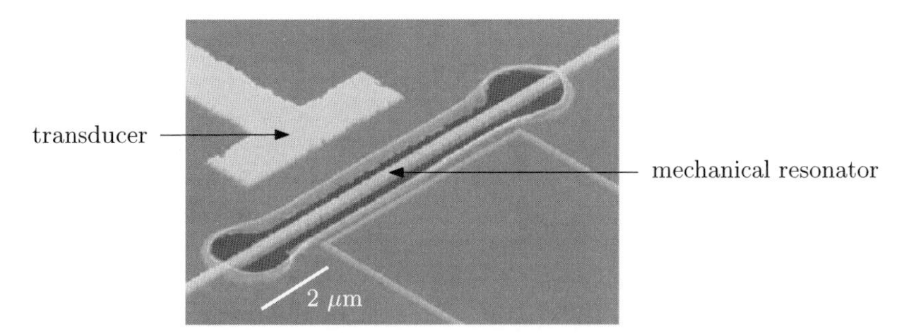

Fig. 6.26. Experimental realization of a quantum-electromechanical system, constructed in the laboratory of Keith Schwab at Cornell University [309]. A beam of width 200 nm and length 8 μm acts as a nanomechanical resonator with a natural frequency of 19.7 MHz. The vibrational motion of the beam is translated into an electrical signal by means of an electronic transducer. Figure reprinted with permission from [309]. Copyright 2004 by AAAS.

transducer and converted into a corresponding electrical signal that is subsequently read out.

Remarkably, most degrees of freedom of the resonator are either "frozen out" at the typical low (millikelvin) temperatures at which QEMS operate, or they couple only negligibly to the transducer. Thus, in most cases, the macroscopic resonator can be effectively treated as a single quantum-mechanical harmonic oscillator (corresponding to the lowest fundamental flexural mode of the beam or cantilever). The nanomechanical resonators in QEMS are macroscopic mechanical objects composed of billions of atoms and perturbed by a host of intrinsic defects, imperfections, etc. Nonetheless, as we shall discuss now, they can exhibit distinctly quantum-mechanical effects.

One key feature of QEMS is their extreme sensitivity to spatial displacements. In this context, an important quantity is the quantum zero-point displacement uncertainty Δx_{zp} of the harmonic oscillator, which represents the quantum limit to position detection. For the nanomechanical resonators currently manufactured in the laboratory, Δx_{zp} is on the order of only a tenth of a picometer [306]. Nonetheless, in recent experiments (involving, for example, the QEMS shown in Fig. 6.26) it has been possible to resolve spatial displacements of the resonator down to a few times of Δx_{zp} [309]. Thus experiments on QEMS are very close to achieving the goal of quantum-limited position measurements.

A Scheme for Creating Mechanical "Schrödinger Kittens" Using QEMS

Researchers have also considered the possibility of putting the resonator into a coherent superposition of two different spatial displacements and to observe the gradual decoherence of such states. What makes this prospect particularly exciting is the fact that QEMS are purely mechanical structures that are much more closer to the objects of the everyday world around us than, say, superconducting qubits (Sect. 6.3) or Bose–Einstein condensates (Sect. 6.4.1). Mechanical "Schrödinger kittens" in QEMS, if realized, would represent superpositions of two distinct positions of a macroscopic "rough" ordinary-matter mechanical object. At this point, no such superposition states have been realized in the laboratory, but concrete proposals exist [311] and theoretical studies of the relevant decoherence and dissipation mechanisms are underway [312,313]. The idea of entangling several nanomechanical resonators that are separated by macroscopic distances has also been investigated theoretically [314]. One may even conceive of the possibility of using QEMS as mechanical qubits in quantum computers [315].

Let us now briefly outline the scheme for the generation of spatial superposition states in QEMS as proposed by Armour, Blencowe, and Schwab [311] (see also [306]). The basic idea consists of using a two-state quantum system whose basis states $|0\rangle$ and $|1\rangle$ couple to two different displacements of the resonator and which is rather easily prepared in a superposition of $|0\rangle$ and

$|1\rangle$. In [311], the authors chose a Cooper-pair box (see Sect. 6.3.4). Here the two basis states $|0\rangle$ and $|1\rangle$ correspond to different values of electrical charge present on the superconducting island. The electrostatic interaction between the Cooper-pair box and the cantilever will result in two different displacements of the resonator, depending on whether the Cooper-pair box is in the state $|0\rangle$ or in the state $|1\rangle$ (see Fig. 6.27). Let us denote the center-of-mass position states of the resonator corresponding to these two displacements by $|P_0\rangle$ and $|P_1\rangle$.

Suppose now we prepare the Cooper-pair box in a superposition of the form $\frac{1}{\sqrt{2}}(|0\rangle + |1\rangle)$. As shown in [311], if a particular sequence of control pulses is applied to the box, we can obtain an entangled box–resonator superposition state of the form

$$|\Psi\rangle = \frac{1}{\sqrt{2}}\left(|0\rangle |P_0\rangle + |1\rangle |P_1\rangle\right). \tag{6.37}$$

Provided the spatial separation between the center-of-mass states $|P_0\rangle$ and $|P_1\rangle$ is sufficiently macroscopic, we have therefore generated a state of the Schrödinger-cat type. Here, the resonator plays the role of the cat, while the Cooper-pair box corresponds to the unstable atom in Schrödinger's original setting.

The existence of the superposition state (6.37) can be confirmed using the technique of Ramsey interferometry (see Sect. 2.2.2). By applying an additional control pulse to the Cooper-pair box after a certain delay time, the entangled state (6.37) is transformed in such a way that the probabilities

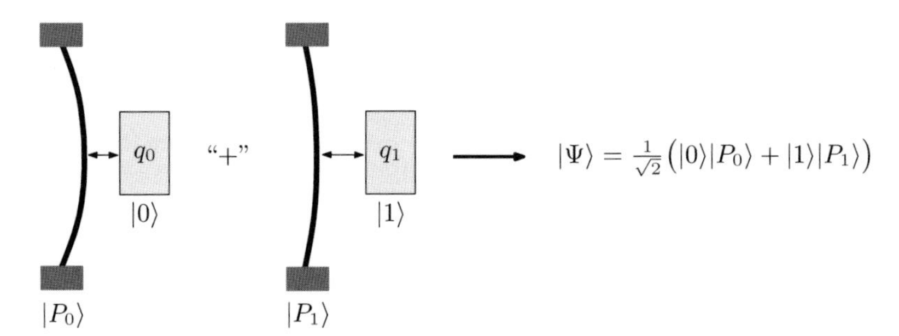

Fig. 6.27. Scheme for the generation of a superposition state involving two distinct positions of the nanomechanical resonator, as proposed by Armour, Blencowe, and Schwab [311]. A Cooper-pair box, described by basis states $|0\rangle$ and $|1\rangle$ (corresponding to different electric charges q_0 and q_1 on the superconducting island of the box), is electrostatically coupled to the nanomechanical resonator. If the Cooper-pair box is prepared in a coherent superposition $\frac{1}{\sqrt{2}}(|0\rangle + |1\rangle)$ and a suitable sequence of control pulses is applied, an entangled box–resonator superposition state $\frac{1}{\sqrt{2}}(|0\rangle |P_0\rangle + |1\rangle |P_1\rangle)$ results, where $|P_0\rangle$ and $|P_1\rangle$ correspond to the two distinguishable center-of-mass positions of the resonator.

of finding the box in the state $|0\rangle$ or $|1\rangle$ exhibit an oscillatory dependence on the delay time. This dependence would be absent if the box–resonator system were in a classical mixture of the component states of the superposition. As discussed in Sect. 6.3.4, the Ramsey method also yields information about the characteristic decoherence time of the superposition.

Decoherence in QEMS

To evaluate the practical viability of the superposition scheme described above, it will be important to understand the main sources of decoherence affecting such superpositions states in QEMS. First of all, the Cooper-pair box itself will of course be subject to decoherence. However, as mentioned in Sect. 6.3.4, relatively long decoherence times in the microsecond regime are within experimental reach [288]. We can therefore focus on the decoherence due to the environmental interactions of the resonator itself.

Interestingly, current experimental evidence [316, 317] (see also earlier results in, e.g., [318, 319]) indicates that the dominant source of dissipation (and therefore also decoherence) in QEMS may be the interaction of the lowest fundamental flexural mode of the resonator with localized defects present in the resonator itself. These defects are effective two-level systems that may assume various physical realizations, such as charge traps, impurity atoms, elastic centers, and dangling bonds created by the disruption of the crystal structure at the surface of the resonator.

At this point, the precise physical details underlying these defects and their interactions with the flexural mode of the resonator are rather poorly understood at the level of both experiment and theory. Since the spatial superposition states in QEMS described above have not yet been experimentally realized, there exists no direct experimental data on decoherence in QEMS. However, recent experiments have yielded valuable information on the characteristic properties of dissipation in QEMS [316, 317].

Since dissipation is generally accompanied by decoherence (but not vice versa), such measurements can also provide us with insights into the decoherence properties of QEMS. For example, we may be able to extract an effective spectral density of the environment composed of the defects from the data on dissipation. Provided we have available a suitable model for the interaction between the resonator and the defects, we can compute the relevant decoherence coefficients and dynamics using this spectral density. This will then allow us to estimate characteristic decoherence rates for the above QEMS superposition states and thus to assess the viability of proposed schemes for the generation of such states. One overarching goal is therefore the development of realistic models for dissipation and decoherence in QEMS. Achieving this goal would not only enhance our theoretical understanding of QEMS, but it would also enable us to make progress toward the experimental realization of superpositions states in QEMS and would suggest general improvements in the implementation of QEMS.

In their original proposal [311], Armour, Blencowe, and Schwab estimated the decoherence time of the spatial box–resonator superposition state (6.37) using the well-explored model for quantum Brownian motion described in Sect. 5.2. Here, the central harmonic oscillator represents the fundamental flexural mode of the resonator, which is in turn weakly coupled to an environment of other harmonic oscillators. However, as we have outlined above, a more realistic representation of the environment of a nanomechanical resonator at millikelvin temperatures may be a bath of two-level systems representing the defects. If we model these two-level systems as quantum spin-$\frac{1}{2}$ particles, we arrive at the oscillator–spin model. We derived the weak-coupling master equation for this model in Sect. 5.4.2 [see (5.198)].

However, the temperature dependence of the damping rate derived from this model [see (5.200)] has been found to disagree with data obtained from experiments on QEMS, in which the dissipation rate has been observed to *increase* with temperature [316,317]. This discrepancy indicates that at least some of the two-level defects in the resonator should be modeled as spin-$\frac{1}{2}$ particles that are strongly coupled to an additional decohering and dissipative environment (see Fig. 6.28). This environment renders the tunneling of the spins incoherent and also leads to relaxation of the spins, thereby inducing defect dynamics akin to those of classical two-level fluctuators. By absorbing the excess energy from the spin bath, it prevents this bath from saturating. The more realistic model for decoherence and dissipation in QEMS may therefore be that of the central harmonic oscillator coupled to a number of independent spin–boson models (i.e., spins interacting with an oscillator bath). QEMS thus offer the opportunity to experimentally distinguish and test different models for decoherence and dissipation (e.g., oscillator vs. spin baths, spin baths coupled to a dissipative bath, etc.).

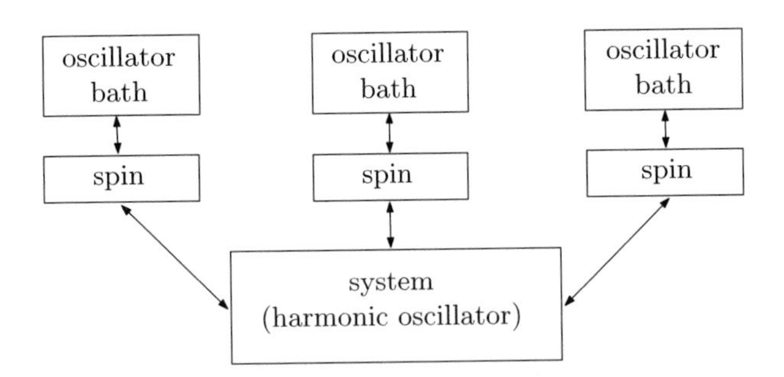

Fig. 6.28. Schematic illustration of a possible model for decoherence and dissipation in QEMS. The central system consists of a harmonic oscillator, which represents the fundamental flexural mode of the nanomechanical resonator. The oscillator interacts with a collection of two-level defects modeled as spin-$\frac{1}{2}$ particles. In turn, each spin is further coupled to a dissipative bath of oscillators that absorbs energy from the (otherwise easily saturated) spin environment.

6.5 Outlook

When Schrödinger first sketched his cat paradox in 1935, he regarded it as a pure thought experiment, as an example for the evidently "absurd" consequences that arise if one applies quantum mechanics to the macroscopic world of our experience. Yet, over the past years immense progress has been made in breeding (and observing) comparably small, but nonetheless macroscopic Schrödinger cats in the laboratory. In this chapter, we have described experiments that have achieved the generation of superpositions of two macroscopically distinguishable states involving dozens of photons (Sect. 6.1), the observation of interference patterns for massive C_{70} molecules (Sect. 6.2), and the verification of the existence of superpositions of two μA currents flowing in opposite directions (Sect. 6.3).

In each of these experiments, researchers have also managed to explicitly observe the gradual action of decoherence, i.e., to measure how the superposition becomes eventually unobservable due to the interaction with the environment. We have thus clearly seen the smoothness of the quantum-to-classical transition: There is no fixed barrier between the quantum and classical domains that would manifest itself as a discontinuous transformation of a quantum state into a classical state. By manipulating the interactions of the system with its environment, we can directly change the dynamics and timescales of the transition.

The experiments described in this chapter therefore show that the scope of decoherence is much broader than to simply refer to the effectively instantaneous "localization" of macroscopic objects around us. Instead, decoherence explains the emergence of effective classicality as a gradual and controllable process that can be directly observed and tested in the laboratory and is accessible to a rigorous theoretical description in terms of the standard quantum-mechanical formalism.

The interplay between theory and experiment is here a very close one. Measurements of decoherence timescales and other properties of the decoherence dynamics enable us to directly test our decoherence models in the laboratory. We thereby learn to what extent these models are realistic and physically adequate. In turn, the decoherence models provide us with valuable information about existing and future experiments. They tell us at which stage of "macroscopicity" the quantum-to-classical transition can be expected to occur in a particular experimental setting, what type of "Schrödinger kittens" we might be able to create and observe given a certain set of experimental parameters, and how we may be able to improve our experiments to generate increasingly "cat-like" superpositions. In this chapter, we have described two experimental domains—Bose–Einstein condensation (Sect. 6.4.1) and quantum-electromechanical systems (Sect. 6.4.2)—that have not yet achieved the creation of certain superposition states proposed in theoretical schemes. Nonetheless, decoherence models, combined with measured data, allow us to identify the physically relevant decoherence mechanisms in these systems that prevent us from "seeing" the desired superposition states, and they suggest strategies for mitigating these detrimental decoherence effects.

The natural question to ask is where the experiments on superpositions of macroscopically distinct states will be headed in the future, and what the resulting implications for our understanding of quantum mechanics and for the construction of large-scale quantum systems, such as quantum computers, will be. There are many answers to this question, so let us just mention a few key points.

First of all, on the scales investigated thus far, the existing experimental evidence does not point to any fundamental limitations of the superposition principle. It appears that, as long as we are able to sufficiently control decoherence effects (and this limit is open to precise quantitative analysis) via a proper experimental design which minimizes unwanted interactions with the environment, any number of superpositions of "classical" states could be generated and observed.

Thus it is safe to say that, over the next few years, we will witness many experiments that will push the envelope for the achievable "size" of superpositions of macroscopically distinct states further and further. We may soon be able to observe interference patterns for even larger biomolecules such as viruses, realize superposition states of Bose–Einstein condensates and nanomechanical resonators, and achieve impressively increased decoherence times in superconducting qubits. Such experiments will also be motivated by the continuing search for possible breakdowns of the superposition principle. However, since decoherence leads to an apparent breakdown from the view of the local observer, it will remain very difficult to design an experiment that would be sufficiently shielded from decoherence but reasonably susceptible to the mechanism inducing the deviation from unitary evolution (see Sect. 8.4 and also [285, 320]).

It is important to emphasize that the ability to generate and observe superpositions of increasingly distinct macroscopic states does not solve any of the fundamental interpretive questions associated with such superpositions in particular, and with decoherence and quantum mechanics in general. The problem of the proper interpretation and physical meaning of quantum states (Sect. 2.1), especially of superposition states (Sect. 2.2.1), is already evident in microscopic settings, such as the Stern–Gerlach experiment described in Sect. 2.2.2. While the recent experiments on superpositions in the macroscopic domain may add a flavor of counterintuitivity of the Schrödinger-cat type to the problem, they neither solve nor exacerbate the fundamental problems of interpretation.

In particular, the question of how to understand the persistence of coherence in the global quantum state involving the environment, and thus the issue of how to properly relate the improper mixtures of quantum states arising in the density-matrix description of decoherence to some form of underlying quantum reality, remains a matter of foundational debate. Experiments of the type discussed here will certainly continue to enlighten the discussion, if only by further challenging our preconceived notions about the "nature of reality." Once experiments may be able to conclusively demonstrate (or rule out) deviations from purely unitary dynamics, we may gain—or be forced into—new insights into quantum mechanics and the structure of quantum reality. For now, though, the interpretive issues are here to stay, even in view of the stunning experimental achievements described in this chapter.

7 Decoherence and Quantum Computing

Quantum information theory and quantum computing have received rapidly growing attention over the past decade and have become one of the focal points of theoretical and experimental research. Quantum information theory views many fundamental concepts of quantum mechanics through a new lens. It has also inspired proposals for quantum technologies and applications, including the possibility of constructing a quantum computer in the future. Quantum computers that would implement useful algorithms are still far from any experimental realization, but many ideas have been worked out and some components that could serve as the basic building blocks of quantum computers have already been developed in the laboratory.

At the same time, research on quantum information theory and quantum computing has enormously increased the interest in studying and controlling decoherence effects. From a fundamental point of view, the basic formal description of decoherence as a von Neumann measurement interaction with the environment (see Sects. 2.5.1 and 2.6.3) can be recast in quantum information–theoretical terms using the formalism of conditional quantum dynamics [16, 321] (see also Sect. 7.4.1 below). Furthermore, in previous chapters of this book we have already alluded to the connection between decoherence and a (rather vague) notion of "information." Specifically, we have made this link in discussing how environmental entanglement carries away which-state information, leading to decoherence (see Sect. 2.6). From a practical point of view, the functioning of any quantum computer is crucially dependent on the ability to maintain quantum coherence. Decoherence induces errors into the computation and is therefore the key obstacle to implementations of quantum computers.

In this chapter, we shall outline the role played by decoherence in quantum computing. In Sect. 7.1, we will review basic aspects of quantum computation, including the key differences between quantum and classical computation and the historical development of the core ideas on which quantum computation is based. In Sect. 7.2, we will discuss the trade-off between decoherence and controllability that is at the heart of any implementation of quantum computers. Next, in Sect. 7.3, we will make some further remarks on the differences between decoherence and noise (see also Sect. 2.12), tailored to the topic of quantum computation. Sect. 7.4 will describe methods of mitigating detrimental decoherence effects in quantum computers by means of so-called quantum error correction. Finally, in Sect. 7.5, we will discuss quantum computation using the concept of decoherence-free (i.e., pointer) subspaces introduced in Sect. 2.8.1. Readers who would like to learn more about quantum information theory and quantum computing in general will find plenty of material, for example, in the book by Nielsen and Chuang [56].

7.1 A Brief Overview of Quantum Computing

We all know that a computer bought today will be hopelessly outdated only a few years later. The rate at which computers have become faster over the past decades is nothing short of astounding. Accordingly, the complexity of the circuits that make up the computer's processor (measured, for example, by the number of transistors per chip) continues to increase rapidly. Readers may have heard of Moore's famous "law," which states that circuit complexity roughly doubles every two years, a prediction that has proved surprisingly accurate up to date.

As a consequence, a few researchers in the 1970s and 1980s realized that at some point, the individual processing structures (e.g, transistors) on a chip would have to reach atomic scales at which quantum effects would become important.[1] They were thus led to speculations about whether such quantum effects could in fact be harnessed in order to enhance computing power. Or, put even more boldly, could one build a computer whose power *depends specifically* on a clever use of quantum-mechanical principles?

7.1.1 The Power of Quantum Computing

Let us think for a second why a computer that is inherently quantum in nature would be so much more powerful than a standard ("classical") computer. A classical computer encodes information in the elementary unit of a logical bit, which can take values of either "0" or "1." A key mantra of quantum information theory is that "information is inevitably physical" (an insight probably first emphasized by Landauer [323]). Information must always be encoded by some physical system, be it by a pencil stroke on a sheet of paper or by the presence of electrons in a transistor, and as such, the encoding of information itself is subject to the laws of physics.

Accordingly, let us consider a *physical system* that can be in two states corresponding to the bit values of "0" and "1," and that therefore is capable of encoding one bit of information. In the spirit of our above argument of shrinking component sizes in computers, this system may well be a quantum system with two basis states $|0\rangle$ and $|1\rangle$ that are classically distinguishable

[1] In a 1997 speech, Moore himself predicted this to happen around the year 2017 [322].

and correspond to the two possible values of the classical bit. Such "qubit" systems have already been discussed in Sect. 2.10 and Chap. 5. In principle, any quantum system with an effectively two-dimensional state space can be used as a qubit. Examples include photons with their two polarization degrees of freedom, ionic or nuclear spins, and atoms with two dominant energy levels. We shall describe some of these physical realizations in Sect. 7.1.5 below.

Quantum mechanics then tells us that this system could, at least in principle (i.e., provided decoherence in the $\{|0\rangle, |1\rangle\}$ basis can be neglected), also be found in a coherent superposition

$$|\psi\rangle = \alpha |0\rangle + \beta |1\rangle . \tag{7.1}$$

Sloppily speaking, this means that a qubit can simultaneously encode *both* logical bits 0 and 1, in contrast with a classical bit that has a value of either 0 or 1. Since there are infinitely many ways to choose the coefficients α and β, there exist infinitely many possible states of a single qubit, as opposed to the two states of a classical bit. A single qubit in a given state $|\psi\rangle$ can encode two real numbers, whereas a classical qubit represents only a single binary digit. This can be seen from the fact that $|\psi\rangle$ can always be written in the form

$$|\psi\rangle = |\alpha| \, e^{i\phi_0} |0\rangle + |\beta| \, e^{i\phi_1} |1\rangle . \tag{7.2}$$

If we disregard an irrelevant overall phase factor, this may be expressed as

$$|\psi\rangle = |\alpha| \, |0\rangle + |\beta| \, e^{i(\phi_1 - \phi_0)} |1\rangle . \tag{7.3}$$

The normalization requirement $|\langle\psi|\psi\rangle|^2 = 1$ then implies that we may rewrite this equation as

$$|\psi\rangle = \cos\left(\frac{\theta}{2}\right) |0\rangle + \sin\left(\frac{\theta}{2}\right) e^{i\phi} |1\rangle , \tag{7.4}$$

with $\phi \equiv \phi_1 - \phi_0$ and an appropriate choice of the angle θ $(0 \leq \theta \leq \pi)$. We note that the form (7.4) allows for a nice visualization of such a state in terms of the Bloch-sphere picture (see Fig. 7.1). We have thus shown that any of the infinitely many possible single-qubit states encodes two real numbers (here given by the angles θ and ϕ).

Suppose now that our system is composed of N qubits. The Hilbert space of this system has dimension 2^N, since it is given by the tensor product $\mathcal{H} = \mathcal{H}_1 \otimes \mathcal{H}_2 \otimes \cdots \otimes \mathcal{H}_N$ of the single-qubit Hilbert spaces $\mathcal{H}_i$, $1 \leq i \leq N$. Thus there are 2^N mutually orthogonal basis states, which are typically chosen to be product states of the original single-qubit basis states $|0\rangle_i$ and $|1\rangle_i$ of the ith qubit $(1 \leq i \leq N)$,

$$\begin{aligned} |00\cdots 0\rangle &\equiv |0\rangle_1 |0\rangle_2 \cdots |0\rangle_N , \\ |00\cdots 1\rangle &\equiv |0\rangle_1 |0\rangle_2 \cdots |1\rangle_N , \\ &\vdots \\ |11\cdots 1\rangle &\equiv |1\rangle_1 |1\rangle_2 \cdots |1\rangle_N . \end{aligned} \tag{7.5}$$

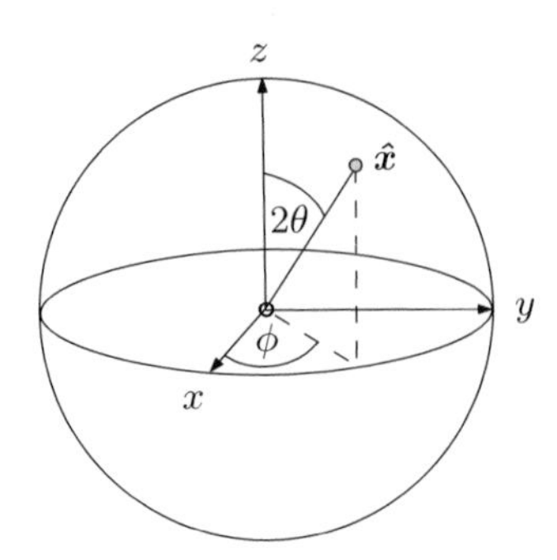

Fig. 7.1. Geometrical representation of the state of a qubit as a point on the Bloch sphere, a unit sphere in $\mathbb{R}^3$. An arbitrary qubit state can be described by two real parameters θ and ϕ [see (7.4)], where 2θ and ϕ play the role of the familiar spherical coordinates. That is, the state (7.4) is represented by the unit vector $\hat{\boldsymbol{x}} = (\sin 2\theta \cos\phi, \sin 2\theta \sin\phi, \cos 2\theta)$.

This basis is referred to as the *computational basis* for the N-qubit system. Any N-qubit state can also be described by a $2^N \times 2^N$ density matrix. If we diagonalize this matrix, we see that this state can simultaneously encode 2^N independent real numbers, since there are 2^N real-valued entries (representing the populations in the eigenbasis of the density operator) on the matrix diagonal. This is to be compared to N classical bits which encode only a binary string of length N. This shows the first advantage of quantum computers: The storage capacity of arrays of qubits is *exponentially* larger than that of classical bits, and a qubit can be in any of the infinitely many coherent superposition states (7.1).

The second advantage is related to the different ways in which classical and quantum computers process information. In a classical computer, we apply logical gates (such as AND, NOT, and XOR) that act on one or two bits at a time. In the quantum case, implementation of gates corresponds to the application of unitary transformations $\hat{U}$ that act on the Hilbert space of the N-qubit system.[2]

[2]We note that there exists an alternative approach to quantum computing in which the entire quantum computation proceeds in form of a time-ordered series of projective measurements on a qubit system which has initially been prepared in a highly entangled state (the so-called *cluster state*) [324–326]. It can be shown that such measurement-based (or cluster-state) quantum computation is able to perform exactly the same computational tasks as conventional quantum computation based on unitary gate operations. In other words, a measurement-based quantum computer is just as universal as a "standard" quantum computer. This is a rather remarkable result, since the nonunitary nature of quantum measurement would intuitively lead one to believe that it is incompatible with the simulation of arbitrary (and therefore also unitary) quantum dynamics. Since cluster-state quantum computation requires the preparation of a highly entangled N-qubit state that must

The key point is now that any such transformation $\hat{U}$, even if it acts on only a single qubit, is capable of changing the *global* state of the N-qubit system. For a maximally entangled state, for example, a unitary operation applied to a single qubit will in general alter the global state of all N qubits. It thus operates simultaneously on all 2^N basis states of the total qubit system. As a simple example, suppose a two-qubit entangled state of the Bell form (2.7a),

$$|\Phi^+\rangle = \frac{1}{\sqrt{2}} (|00\rangle + |11\rangle) , \tag{7.6}$$

is subject to a unitary transformation $\hat{U}$ that acts only on the first qubit, e.g.,

$$\hat{U} = (-|0\rangle\langle 1| + |1\rangle\langle 0|)_1 \otimes \hat{I}_2. \tag{7.7}$$

Then the transformed state reads

$$\hat{U} |\Phi^+\rangle = \frac{1}{\sqrt{2}} (|10\rangle - |01\rangle) . \tag{7.8}$$

We may view this feature as a consequence of the nonlocal, holistic nature of entangled multi-particle states. Again, this is to be contrasted with the classical situation, in which we would either need to perform the same operation 2^N times, or have 2^N "processors" operating in parallel, each of which applies one classical logic gate.[3] We remark here that, in practice, one needs to consider only quantum gates that act on one or two qubits at a time, since one can show that a certain set of such one-qubit and two-qubit gates is sufficient to perform any quantum computation [56,329–332]. This is a very useful result, since it is in practice difficult to engineer interactions between more than two qubits that would be required to physically implement multi-qubit gates.

7.1.2 Reading Out a Quantum Computer

In the case of a classical computer, we obtain the result of the computation simply by reading out the final values of all bits. We can do this at any time without altering the state of the computer. The corresponding procedure in

remain sufficiently robust to decoherence over time, its general decoherence properties are likely to be similar to those of gate-based quantum computation. For an accessible introduction to cluster-state quantum computation, we refer the reader to [327].

[3] In analogy with this observation, quantum computers are often said to exhibit massive "quantum parallelism" (this feature was pointed out early by Zurek under the name of "quantum Monte Carlo" [328]). However, this term should be interpreted with care. Parallelism creates the impression of many independent processes performed in parallel, whereas in the quantum case we really should think of a single quantum-mechanical "whole" (namely, the global quantum state) subject to a single operation (the quantum gate).

quantum computing consists of a projective measurement, carried out in some chosen basis, of some (or all) qubits at the end of the computation. According to the standard theory of quantum measurement, this will collapse the global state of the N-qubit system onto one of the states in the measured basis and destroy any existing entanglement between each individually measured qubit and the other qubits. The key difference to the classical case is related to the fact that measurement in quantum mechanics is probabilistic. That is, if we apply the same series of quantum gates to a collection of identically prepared N-qubit systems, the measurement outcomes will in general vary from system to system—we can only specify the probability of the system to be found in a certain state upon measurement (see Sect. 2.1).

This fact may seem to render quantum computation essentially useless, since we would in general not be able to obtain a single consistent result at the conclusion of a set of identical quantum computations. Thus the quantum computer will typically at least in some cases not yield the correct answer to the computational problem in question. Fortunately, in many situations it is rather easy to verify whether a given output is actually the correct result or not. For example, the famous algorithm for the factorization of integers on a quantum computer (see Sect. 7.1.4 below) will not always yield the correct answer. However, it is trivial to check whether a given decomposition of an integer into prime numbers is correct, and therefore we can simply manually discard any incorrect results from the quantum computer. Provided the probability of the final measured state to correspond to the correct solution is reasonably high (e.g., such that obtaining the correct answer does not require a number of runs of the quantum computer that grows exponentially with the size of the problem), we have still employed the power of the quantum computer to our advantage.

7.1.3 Simulating Physical Systems

Let us now ask the following question: Could a classical computer, at least in principle, always *simulate* a quantum system?[4] After all, one might think of emulating the evolution of, say, an N-qubit system by representing its quantum state by $2^N \times 2^N$ matrices and by applying other $2^N \times 2^N$ matrices to represent gate operations. There are, however, two caveats. First, as we

[4]At the risk of stating the obvious, we note that a quantum computer can also simulate any classical computer if states are prepared and quantum gates are applied such that each qubit is at any time described by either one of the states $|0\rangle$ and $|1\rangle$ but not by superpositions thereof. This point is directly relevant to our subsequent discussion of the role of decoherence in quantum computing. Decoherence tends to destroy any coherent superpositions of the computational-basis states (7.5) of an N-qubit system, thereby effectively transforming the qubits into a classical logical-bit register. The quantum computer then loses its "quantum power"—it still can perform the same functions as a classical computer, but nothing more.

have seen, this approach becomes prohibitively computationally expensive for larger N, since the size of the matrices and the number of matrix operations (and thus the demand for storage space and processing power) blows up exponentially with increasing N. This makes it impossible for all practical purposes to simulate a larger quantum system (in other words, a quantum algorithm) on a classical computer. Second, from a more fundamental point of view, Bell's theorem [30] shows that there exists no classical mechanism that could prepare two correlated distant systems whose correlations violate the Bell inequality. Thus such (EPR/Bell-type) correlations are a distinct feature of the quantum world and of quantum computing [333, 334].

The problem of how to efficiently simulate quantum systems lies at the roots of the quantum-computing program. In a seminal paper of 1982, Feynman [333] explored the issue of constructing a quantum system that could efficiently emulate the physical behavior of any other (quantum) system in nature. Consequently, such a system would not only represent a universal emulator, but also a universal quantum computer, since quantum computers must necessarily also be physical systems. Feynman modeled this universal emulator in terms of a lattice of spins with nearest-neighbor interactions. Famously, Feynman was able to demonstrate that this model would, in principle, be able to simulate *any* other quantum system that can be described by a finite-dimensional Hilbert space. This was a truly groundbreaking discovery of fundamental importance. However, Feynman's model still had a drawback. Each new computational problem would in general require one to choose a different form and strength of the spin–spin interactions. Obviously, this requirement is hard to implement in practice, and Feynman did not specify the physical mechanism that would enable the user to freely choose the governing Hamiltonian.

Therefore the next step consisted of devising a universal quantum computer that, based on a fixed Hamiltonian (i.e., on a specific physical architecture), would be able to solve any given problem simply by choosing different operations applied to the quantum computer. This was accomplished by David Deutsch in an important paper of 1985 [335]. Deutsch used a model based on a collection of two-state systems. He showed that a small set of unitary transformations—the aforementioned "quantum gates"—applied to the combined state of these systems could in principle implement any arbitrary unitary evolution and thus simulate any physical system. For a large class of possible evolutions, this quantum computer would also run *efficiently* [336], in the sense that the number of computational steps required to implement a particular evolution would grow less than exponentially with the size of the input (i.e., the number of two-state systems in Deutsch's model). This ability to efficiently simulate (virtually) any other physical system made Deutsch's proposal the first blueprint for a (virtually) universal quantum computer.

7.1.4 Examples of Famous Quantum Algorithms

In the years following Deutsch's paper, researchers busied themselves with finding computational problems that would demonstrate the superior power of a quantum computer over a classical computer [337–341]. Initial attempts were rather unfruitful, until in 1994 Peter Shor [342,343] presented a quantum algorithm that could solve the factoring problem (i.e., the task of decomposing a given integer into a product of prime numbers) in only polynomial time—namely, in time $O(N^2)$ for an N-digit number. By contrast, solving the factoring problem on a classical computer is widely believed (although not rigorously proved) to require exponential time and is therefore essentially impossible to accomplish on such a computer for any reasonably large number. The importance of Shor's discovery may not lie as much in its potential practical applications as in the fact that the algorithm plays, as Steane [334] put it, a "conceptual role similar to that of Bell's inequality, in defining something of the essential nature of quantum mechanics."

It is fair to say that Shor's algorithm was the root of much of the initial and continuing interest that surrounds the field of quantum computing. Since then, progress in finding new algorithms that would demonstrate the power of quantum computers has been rather slow (cf. [337, 340]). The probably most significant discovery since Shor's algorithm has been Grover's search algorithm [344,345]. This algorithm allows for highly efficient database search: It is capable of finding a particular entry in a completely unstructured list of N items in $O(\sqrt{N})$ steps. By contrast, on a classical computer this task would take on the order of N steps, since in average we will need to look through half of the list (i.e., perform $N/2$ steps) to find a specific entry. Thus a quantum computer that implements Grover's algorithm would be faster by a factor $O(\sqrt{N})$ than any classical computer.

7.1.5 Physical Realizations of Quantum Computers

Our above introduction to quantum computing has been deliberately kept rather abstract. We have simply assumed that we have a collection of qubits available that we can coherently manipulate through the application of unitary transformations (gates). We have specified neither the physical nature of the qubits nor how such gate operations are implemented in practice. In fact, the area of research on experimental realizations of prototype quantum computers (systems containing a few qubits that can be coherently controlled) is vast, and it would be completely outside of the scope of this book to attempt to give an overview of this rapidly evolving field.

Just to give the reader a first idea, in Table 7.1 we list some of the possible physical realizations of the qubits and control mechanisms (which implement gates and measurements) that may have the potential to mature into full-blown quantum computers at some point in the future. Each technology has its advantages and disadvantages, which can in general be traced back to

Table 7.1. Examples for the physical realization of the qubit system and the qubit control.

Scheme	Qubits realized as	Qubits controlled by
Optical	Photons	Optical media
Ion trap	Trapped ions	Laser beam
Cavity QED	Two-level atoms	Laser beam
NMR	Nuclear spins	Magnetic field
Solid state	Coupled quantum dots	Point contact
	Superconducting tunnel junctions	Bias current of junction

the fundamental trade-off problem between decoherence and controllability, which we shall discuss in the following Sect. 7.2. Current experiments using the components listed in Table 7.1 have achieved coherent control of only a few qubits. Implementation of any useful quantum algorithms will require many more qubits, and thus these experiments are still a far cry from practical realizations of quantum computers.

7.2 Decoherence Versus Controllability in Quantum Computers

As we have already indicated in several places, quantum computers derive their power from the presence of nonlocal coherent superpositions of the computational-basis states (7.5), which are product states of the individual qubit basis states $|0\rangle$ and $|1\rangle$ (which must be sufficiently distinguishable so we can manipulate and measure them by means of some apparatus controlling the quantum computer). Thus quantum computing is based on superpositions of mesoscopically or macroscopically distinct states, which are typically extremely sensitive to decoherence. It is fair to say that decoherence is the number-one enemy of quantum computers, and that much, if not most, of the research on the implementation of a quantum computer has revolved around the problem of how to control, minimize, and "undo" decoherence.

To address the problem of decoherence in quantum computers, one might be tempted to suggest that it will be sufficient to shield the qubits as much as possible from any environmental interactions. Indeed, isolating the qubits from their surroundings would minimize decoherence. However, such a strategy would also compromise the functionality of the quantum computer. As we have seen, a quantum computation is carried out through the application of a series of unitary operations to the qubits. Physically, such gate operations correspond to some interaction between the qubit system and an external apparatus. The specific physical form of the interaction depends on the particular implementation of the quantum computer (see Table 7.1). To reliably implement quantum gates, the interaction between the qubits and the control apparatus must be sufficiently strong.

Thus we are faced with two opposing demands. On the one hand, we would like to isolate the qubits in order to minimize unwanted environmental interactions and to thus protect the qubit superposition states. On the other hand, however, we need to keep the qubit system sufficiently open to allow for the qubits to be manipulated (in form of state preparations and gate operations) and read out by controlled interactions with a macroscopic apparatus. In turn, such qubit–apparatus interactions will in general induce decoherence of the qubits due to entanglement with the apparatus, with further perturbations introduced by classical noise (fluctuations) in the apparatus.

Therefore, the formidable challenge of designing a quantum computer consists of meeting both demands in a balanced way. Obviously, a quantum computer that has a very low environmental decoherence rate but cannot be manipulated in any reasonable way is as useless as a quantum computer that we are able to control well but whose environmental interactions are so strong as to immediately deteriorate the device into a classical computer. In short, we require the qubits to interact strongly with certain components of the environment (namely, the apparatus) but not with the remainder (the uncontrolled surroundings into which the qubits are immersed).

For example, optical quantum computing uses photons as physical qubits, with the two different orientations of polarization representing the computational-basis states. Photons are relatively insensitive to decoherence, since they are uncharged particles and exhibit usually only weak interactions with each other and with matter. In ordinary linear optical media photons do not interact with each other at all (and only extremely weakly in nonlinear media), posing the problem of how to implement two-qubit gate operations that would entangle a pair of photons. A possible way of out of this dilemma has been suggested by the famous Knill–Laflamme–Milburn scheme [346], which uses only linear optics, single-photon states, and projective measurements to entangle two photonic qubits. However, this scheme in turn introduces new challenges that need to be overcome for the construction of an optical quantum computer to become feasible. Similarly, qubits implemented as single spins inside of a nucleus have long decoherence times but are very difficult to control and measure.

7.3 Decoherence Versus Classical Fluctuations

As discussed in Sect. 2.12, decoherence should be understood as a distinctly quantum-mechanical effect with no classical analog. By contrast, in the literature on quantum computing readers will often find that the term "decoherence" is used in a more sloppy way as referring to *any* process that affects the qubits, including perturbations due to *classical* fluctuations and imperfections. Examples for sources of such classical noise in the context of quantum computing are the fluctuations in the intensity [347] and duration [348] of the laser beam incident on qubits in an ion trap, inhomogeneities in the magnetic fields in NMR quantum computing [349], and bias fluctuations in superconducting qubits [292].

Phenomenologically and formally the influence of such classical processes on the qubits may be described in a manner similar to the effect of environmental entanglement, namely, in terms of "qubit errors" (see Sect. 7.4.2 below). From a purely practical point of view, the distinction between decoherence due to entangling interactions with an environment on the one hand, and the loss of phase coherence due to classical noise (in an ensemble average) on the other hand, may therefore not be relevant and is often not made explicit. The distinction between classical noise and quantum decoherence has been further blurred by the field of quantum error correction (see Sect. 7.4), since the error-correcting schemes are insensitive to the physical origin of the qubit errors.

Furthermore, as mentioned in Sect. 2.12, the loss of phase coherence due to environmental entanglement is sometimes *simulated* by classical fluctuations perturbing the system, i.e., by the addition of certain time-dependent terms to the self-Hamiltonian of the system. This strategy was implemented, for example, in theoretical [347,350] and experimental [125,351] studies of the influence of fluctuating parameters on the phase coherence of qubits in ion-trap quantum computers. In the theoretical domain, Schneider and Milburn investigated "decoherence" due to random fluctuations both in the phase and intensity of the qubit-controlling laser beam [347] and in the parameters of the trap which confines the ions. In experiments carried out by a group at NIST [125, 351], the influence of different types of reservoirs on the decay of Schrödinger-cat states of a trapped ion (superpositions of coherent states and number eigenstates) was experimentally simulated by letting certain parameters of the harmonic trap, such as the location of the minimum or the trap frequency, fluctuate randomly.

As we have already discussed in Sect. 2.12, the key point is now that, if the influence of the environment is modeled in this way as classical noise, the decay of coherence can manifest itself only in an *ensemble*, i.e., in the form of an average over a large number of particular realizations of such noise processes. Since the system is not coupled to any external environment, in any individual realization of the noise process the dynamics of the system are completely unitary, and thus no coherence can have been lost from the system. By contrast, if the system becomes entangled with environmental degrees of freedom, at the very least we would need to perform a pair of measurements on the environment before and after the interaction with the system in order to gather enough information to reverse the effect of decoherence by application of an appropriate countertransformation. Moreover, as shown in the NIST experiment [125], these measurements would not always constitute a sufficient procedure for "undoing" decoherence (see also

Sect. IV.C of [16]). Thus this experiment illustrates nicely some of the key similarities and differences between classical noise and environmental decoherence.

This is of course not to say that classical noise is not important in quantum computing. To the contrary, noise due to fluctuations and imperfections in experimental implementations may be equally significant as the effect of decoherence induced by environmental entanglement. While we will continue to reserve the term "decoherence" to describe the consequences of environmental entanglement, the more general term "qubit error" will, if not specified further, refer to any change of the state of the qubit, regardless of whether the change is induced by classical noise or environmental entanglement.

7.4 Quantum Error Correction

We have available two main avenues toward combating decoherence effects in quantum computers. The first route consists of *minimizing* detrimental environmental influences before they even have a chance to severely corrupt the coherent superposition states of the qubits. In Sect. 7.5, we will discuss one possible method, namely, the encoding of logical qubits states in pointer subspaces that are immune to decoherence.

The second, complementary approach consists of trying to actively "undo" the effect that decoherence has imparted on the qubits. Since it is impossible in practice to exert full control over the environment, phase relations are typically irreversibly delocalized from the qubits into the environment (see Sect. 2.7). A true (time) reversal of the decoherence process, i.e., a complete relocalization of the phase coherence at the level of the qubits, is therefore out of reach. However, as discussed in Sect. 2.13, through the coupling of the system to an auxiliary system, we can *reconstruct* the original superposition state in a nonlocal fashion. In the context of quantum computing, such schemes have become known under the heading of *quantum error correction*, first developed independently by Steane [352] and Shor [353] in the mid-1990s (for accessible reviews, see, e.g., [56,354,355]). As we shall see, the basic idea of quantum error correction is similar to the process of quantum "erasure" discussed in Sect. 2.13. The main difference between the two schemes lies in the fact that quantum error correction requires the reconstruction of the exact initial superposition state. By contrast, quantum "erasure" simply recovers the possibility of interference between the path components, and the reconstructed superposition may be either the original or the phase-reversed superposition state [see (2.125)].

Let us emphasize that quantum error correction will be an integral, indispensable element of any foreseeable implementation of a quantum computer. Decoherence induced by interactions with the environment and the control apparatus as well as noise due to faulty gate operations will simply be too strong, according to current estimates, to allow for useful quantum computations to be carried out if no error-correcting methods are employed. A concrete example that demonstrates the importance of active quantum error correction for an even comparably basic computational task has been given by Miquel, Paz, and Zurek [348] and by Miquel, Paz, and Perazzo [356]. The authors modeled an ion-trap quantum computer composed of 18 ions (the qubit system) that implements Shor's factorization algorithm [342, 343] (see Sect. 7.1.4) through a series of about 15,000 gate operations, physically represented by laser pulses incident on the ions. Errors are introduced into the computation both by imperfect implementations of the gate operations [348] and by environmental decoherence of the qubit system [356]. It was demonstrated that, without error correction, the computational task of factorizing even small numbers (the authors studied the example of the number 15) would very rapidly go astray, making it effectively impossible to implement the algorithm in practice.

7.4.1 Classical Versus Quantum Error Correction

First of all, to get a feel for how error correction in the quantum setting demands new approaches, let us briefly review how error correction is typically implemented in a classical computer. Here, the key idea is that of *redundancy.* That is, instead of encoding one bit of information (a logical bit) in a single physical bit, we encode the logical bit in several physical bits. If now a small fraction of these physical bits gets corrupted, we can still extract the original logical bit by a simple majority vote. For example, we may encode the logical bits 0 and 1 in three physical bits,

$$0_{\mathrm{L}} \longleftrightarrow 000, \tag{7.9a}$$

$$1_{\mathrm{L}} \longleftrightarrow 111, \tag{7.9b}$$

where we have used the subscript "L" to denote the logical bit. Suppose now that some noise process flips the third physical bit of the encoded bit 0_{L},

$$000 \longrightarrow 001. \tag{7.10}$$

By a simple majority vote, we may now conclude that the uncorrupted state was likely to be "000." Of course, this protocol fails if more than one bit was flipped. Clearly, for our scheme to bring about any improvement over the completely uncorrected case, we must therefore require the probability of a single bit flip to be less than 50%. If this probability was larger than 50%, then the three-bit state would in average suffer more than one bit flip, and the error correction would, in average, yield the wrong state.

The approach of redundant encoding cannot be directly taken over to the quantum setting. The idea analog to the classical case would be to copy a single-qubit state

$$|\psi\rangle = \alpha |0\rangle + \beta |1\rangle \tag{7.11}$$

onto a set of $N-1$ other qubits, such that the total state of all N qubits would read

$$|\Psi\rangle = \prod_{i=1}^{N} \left(\alpha |0\rangle_i + \beta |1\rangle_i\right). \tag{7.12}$$

Thereby we would have redundantly encoded the same single-qubit state $|\psi\rangle$ in N physical qubits. However, it is a fundamental consequence of the linearity of quantum-mechanical time evolution that cloning of arbitrary quantum states is not possible. The proof of this famous *no-cloning theorem* [20, 21] is rather straightforward. Suppose we have some sort of device that is able to copy our two qubit basis states $|0\rangle_1$ and $|1\rangle_1$. That is, given a second qubit described by the initial state $|\psi_0\rangle_2$, the machine would implement the following process:

$$|0\rangle_1 |\psi_0\rangle_2 \longrightarrow |0\rangle_1 |0\rangle_2 , \tag{7.13a}$$

$$|1\rangle_1 |\psi_0\rangle_2 \longrightarrow |1\rangle_1 |1\rangle_2 . \tag{7.13b}$$

For the special case $|\psi_0\rangle_2 = |0\rangle_2$, this evolution is referred to as a so-called controlled-NOT (or CNOT) gate in quantum information theory. It simply corresponds to an evolution of the von Neumann type (2.52), where the change of the state of the second system is *conditional* on the state of the first system. That is, if the first system is in the state $|0\rangle_1$, the state of the second system remains unchanged, whereas the latter changes to $|1\rangle_2$ if the first system is in the state $|1\rangle_1$. This type of evolution is an example of so-called *conditional quantum dynamics*.

Since the evolution of the composite system consisting of the device and the two qubits is unitary and thus linear, it follows from (7.13) that the output for the input state $(|0\rangle_1 + |1\rangle_1)$ of the first qubit would be

$$(|0\rangle_1 + |1\rangle_1) |\psi_0\rangle_2 \longrightarrow |0\rangle_1 |0\rangle_2 + |1\rangle_1 |1\rangle_2 . \tag{7.14}$$

But this final state is evidently different from the desired output state, which is

$$(|0\rangle_1 + |1\rangle_1)(|0\rangle_2 + |1\rangle_2) = |0\rangle_1 |0\rangle_2 + |0\rangle_1 |1\rangle_2 + |1\rangle_1 |0\rangle_2 + |1\rangle_1 |1\rangle_2 . \tag{7.15}$$

This simple observation concludes our proof. Thus we can clone the single-qubit state (7.11) only if we had advance knowledge of the coefficients α and β. But this requirement would be extremely limiting, since we usually do not know these coefficients, except maybe for a very brief period of time directly after the preparation of the initial qubit state.

In fact, even if we were able to perfectly clone arbitrary quantum states and thus mimic the classical redundant encoding, it would not help us in detecting and correcting quantum errors. The reason for this problem lies in the second fundamental difference between the classical and quantum settings.

In classical computing, to detect the error that has occurred in a collection of bits, we can simply read out the state of each bit and then analyze the result in a suitable manner. By contrast, any measurement performed on the qubits will in general reveal some information about the state of the qubits. If we think of this measurement in the usual way as an interaction that entangles the qubits with some measurement device, then we know that this process will in turn lead to decoherence of the qubits.

The upshot of this fact is that, in order to successfully do error detection in the quantum setting, we must extract information about the error that has occurred in the qubit system *without* extracting any information about the state of the qubits themselves. Thus, even if we had managed to redundantly encode the state of a single qubit in a collection of physical qubits as in (7.12), we would have no direct way of measuring and comparing the states of each of the qubits without destroying the superposition and therefore the very quantum information that we were seeking to protect. We will see in the following sections how to overcome these obstacles and how to implement error correction in quantum computation.

7.4.2 Representing the Influence of Decoherence by Discrete Errors

While in the classical case we only need to correct discrete bit flips, in the quantum setting the qubit state is represented by a superposition described by two real numbers that can change by any continuous amount. It would therefore appear that we need to have available an infinite (in fact, uncountably infinite!) number of different error-correcting operations. Fortunately, this turns out not to be the case. As we shall see now, we can in fact *discretize* the continuum of possible errors into linear combinations of a very small number of errors, which we can then quite easily correct. This remarkable result is one of the cornerstones for making quantum error correction possible. Let us see how this discretization comes about.

For this purpose, consider a single qubit $\mathcal{S}$ interacting with some environment $\mathcal{E}$. Suppose $\mathcal{S}$ is initially in a pure state

$$|\psi\rangle = \alpha\,|0\rangle + \beta\,|1\rangle\,. \tag{7.16}$$

Then, following our discussion in Sect. 2.15.3, the initial state of the composite qubit–environment system can always be written in the product form $|\psi\rangle\,|e_\mathrm{r}\rangle$. A completely arbitrary interaction between $\mathcal{S}$ and $\mathcal{E}$ can be represented by the evolution

$$|0\rangle\,|e_\mathrm{r}\rangle \longrightarrow c_{00}\,|0\rangle\,|e_{00}\rangle + c_{01}\,|1\rangle\,|e_{01}\rangle\,, \tag{7.17a}$$

$$|1\rangle\,|e_\mathrm{r}\rangle \longrightarrow c_{10}\,|0\rangle\,|e_{10}\rangle + c_{11}\,|1\rangle\,|e_{11}\rangle\,, \tag{7.17b}$$

where the coefficients c_{ij} are here equal to either zero or one, and where the environmental states $|e_{ij}\rangle$ are not necessarily orthogonal or normalized.

Depending on the value of the c_{ij}, different physical interpretations may be attached to the general evolution (7.17). For example, the choice $c_{00} = c_{11} = 1$ and $c_{01} = c_{10} = 0$ would correspond to decoherence in the true sense, i.e., to a purely entangling (von Neumann–type) interaction with the environment [see (2.52)],

$$|0\rangle\,|e_{\mathrm{r}}\rangle \longrightarrow |0\rangle\,|e_{00}\rangle\,, \tag{7.18a}$$

$$|1\rangle\,|e_{\mathrm{r}}\rangle \longrightarrow |1\rangle\,|e_{11}\rangle\,, \tag{7.18b}$$

without any change of the component states $|0\rangle$ and $|1\rangle$.

Let us for now assume the most general case in which all coefficients c_{ij} in (7.17) are equal to one. Then the arbitrary pure initial state (7.16) of $\mathcal{S}$ evolves according to

$$|\psi\rangle\,|e_{\mathrm{r}}\rangle \longrightarrow \alpha\left[|0\rangle\,|e_{00}\rangle + |1\rangle\,|e_{01}\rangle\right] + \beta\left[|0\rangle\,|e_{10}\rangle + |1\rangle\,|e_{11}\rangle\right]. \tag{7.19}$$

It is elementary to rewrite this equation in the form

$$\begin{aligned} |\psi\rangle\,|e_{\mathrm{r}}\rangle \longrightarrow\; & (\alpha\,|0\rangle + \beta\,|1\rangle)\,\frac{1}{2}\left[|e_{00}\rangle + |e_{11}\rangle\right] \\ +\; & (\alpha\,|1\rangle + \beta\,|0\rangle)\,\frac{1}{2}\left[|e_{01}\rangle + |e_{10}\rangle\right] \\ +\; & (\alpha\,|0\rangle - \beta\,|1\rangle)\,\frac{1}{2}\left[|e_{00}\rangle - |e_{11}\rangle\right] \\ +\; & (\alpha\,|1\rangle - \beta\,|0\rangle)\,\frac{1}{2}\left[|e_{01}\rangle - |e_{10}\rangle\right]. \end{aligned} \tag{7.20}$$

In fact, we can simplify this expression by noting that the part of each term in the sum corresponding to the qubit system can be written as either the identity operator $\hat{I}$ or one of the Pauli operators $\hat{\sigma}_x$, $\hat{\sigma}_y$, and $\hat{\sigma}_z$ acting on the initial state $|\psi\rangle$ of the qubit. Namely,

$$\begin{aligned} |\psi\rangle\,|e_{\mathrm{r}}\rangle \longrightarrow\; & \left(\hat{I}\,|\psi\rangle\right)\frac{1}{2}\left[|e_{00}\rangle + |e_{11}\rangle\right] \\ +\; & (\hat{\sigma}_x\,|\psi\rangle)\,\frac{1}{2}\left[|e_{01}\rangle + |e_{10}\rangle\right] \\ +\; & (\hat{\sigma}_z\,|\psi\rangle)\,\frac{1}{2}\left[|e_{00}\rangle - |e_{11}\rangle\right] \\ +\; & (\hat{\sigma}_y\,|\psi\rangle)\,\frac{1}{2}\left[|e_{01}\rangle - |e_{10}\rangle\right]. \end{aligned} \tag{7.21}$$

We can summarize this result as follows. Consider a single qubit $\mathcal{S}$, initially described by a pure state $|\psi\rangle$ and interacting with an environment $\mathcal{E}$. Then an arbitrary evolution of the combined qubit–environment state can always be written in the form

$$|\psi\rangle\,|e_{\mathrm{r}}\rangle \longrightarrow \hat{I}\,|\psi\rangle\,|e_I\rangle + \sum_{s=x,y,z} (\hat{\sigma}_s\,|\psi\rangle)\,|e_s\rangle\,. \tag{7.22}$$

Here the $\hat{\sigma}_s$ act on the Hilbert space $\mathcal{H}_\mathcal{S}$ of $\mathcal{S}$, and $|e_I\rangle$ and $\{|e_s\rangle\}$ are environmental states that are not necessarily orthogonal or normalized.

Let us appreciate the meaning of this result. Equation (7.22) shows that a completely arbitrary influence of the environment on the qubit can be expressed simply in terms of a weighted sum of the discrete Pauli operators and the identity operator acting on the original state of the qubit. This discretization of the continuum of possible changes of the state into a superposition of only four operators can be done because $\{\hat{I}, \hat{\sigma}_x, \hat{\sigma}_y, \hat{\sigma}_z\}$ forms a complete set of operators for the Hilbert space $\mathcal{H}_\mathcal{S}$ of $\mathcal{S}$. We will see below that this discretization is crucial in detecting and correcting errors in quantum computers.

Because of their effect on the state of a qubit, the effect of $\hat{\sigma}_x$ and $\hat{\sigma}_z$ is often referred as a "bit-flip error" and "phase-flip error," respectively. This can be seen by comparing (7.20) and (7.21), or directly from the fact that

$$\hat{\sigma}_x (\alpha |0\rangle + \beta |1\rangle) = \alpha |1\rangle + \beta |0\rangle , \tag{7.23a}$$

$$\hat{\sigma}_z (\alpha |0\rangle + \beta |1\rangle) = \alpha |0\rangle - \beta |1\rangle . \tag{7.23b}$$

In other words, the operator $\hat{\sigma}_x$ swaps the two bits $|0\rangle$ and $|1\rangle$, while $\hat{\sigma}_z$ rotates the phase of the bit $|1\rangle$ by 180° (which amounts to a sign flip). Note that the phase flip operation has no classical analogue, since a classical bit is in either one of the two states 0 and 1. Therefore phase relations between these two states cannot play a role. Furthermore, the reader may easily verify that the operator $\hat{\sigma}_y$ is simply a combination of a bit-flip and a phase-flip error, since $\hat{\sigma}_x \hat{\sigma}_z = -\mathrm{i}\hat{\sigma}_y$.

It should be emphasized, though, that despite this suggestive labeling of the Pauli operators the physical interpretation of (7.20) and (7.21) in terms of a sequence of bit-flip and phase-flip errors is only justified if the corresponding relative states of the environment are orthogonal. Only in this case the Pauli operators represents mutually exclusive "error possibilities" (with associated probabilities) that could be distinguished in a measurement.

For our purpose of describing environmental entanglement and the resulting decoherence effects, a simplified version of the general result (7.22) will suffice. From (7.18) and (7.21) we see that the influence of environmental decoherence on a single qubit can be represented as a combination of the identity operator and the Pauli operator $\hat{\sigma}_z$. That is,

$$|\psi\rangle |e_\mathrm{r}\rangle \longrightarrow \frac{1}{\sqrt{2}} \Big(\hat{I} |\psi\rangle |e_+\rangle + \hat{\sigma}_z |\psi\rangle |e_-\rangle \Big) , \tag{7.24}$$

where we have defined new (conjugate) relative environmental states $|e_\pm\rangle$ as

$$|e_\pm\rangle \equiv \frac{1}{\sqrt{2}} (|e_{00}\rangle \pm |e_{11}\rangle) . \tag{7.25}$$

This result may not come as a surprise, since decoherence is associated with the delocalization of local phase relations, and we may thus intuitively expect the effect of decoherence to be related to phase-flip errors.

The evolution (7.22) can be generalized to a system of N qubits interacting with an environment. One can show that the dynamics of the composite system composed of these qubits and the environment can be written as

$$|\psi\rangle |e_{\mathrm{r}}\rangle \longrightarrow \sum_i \left(\hat{E}_i |\psi\rangle\right) |e_i\rangle . \tag{7.26}$$

Here $|\psi\rangle$ is the initial N-qubit state, and the $\hat{E}_i$ are tensor products of N operators involving identity and Pauli operators, with each operator acting on the Hilbert space of a single qubit. For example,

$$\hat{E}_i = \hat{\sigma}_y^{(1)} \otimes \hat{I}^{(2)} \otimes \hat{\sigma}_x^{(3)} \otimes \cdots \otimes \hat{\sigma}_z^{(N)} \qquad \text{etc.,} \tag{7.27}$$

where the superscripts indicate which qubit each operator acts on.

For the case of a single qubit, we found that an entangling interaction with the environment can be described as a combination of no error (represented by the identity operator) and a phase-flip error (represented by the Pauli operator $\hat{\sigma}_z$) [see (7.24)]. The analogous result holds true also for the case of N qubits. Now the relevant error operators $\hat{E}_i$ are products of N single-qubit operators, each of which is either the identity operator or the operator $\hat{\sigma}_z$ acting on the Hilbert space of a single qubit.[5] For example,

$$\hat{E}_i = \hat{\sigma}_z^{(1)} \otimes \hat{I}^{(2)} \otimes \hat{\sigma}_z^{(3)} \otimes \cdots \otimes \hat{I}^{(N)} \qquad \text{etc.} \tag{7.28}$$

To establish a more transparent notation, let us denote these N-qubit error operators (7.28) in the form $\hat{Z}_{j_1 j_2 \cdots j_N}$, where each j_k is either equal to 0 if the identity operator $\hat{I}^{(k)}$ acts on the kth qubit, or equal to 1 if the phase-flip operator $\hat{\sigma}_z^{(k)}$ acts on this qubit. For example, in this notation the operator $\hat{E}_i$ written out in (7.28) would read $\hat{Z}_{101\cdots 0}$. Then we can express a purely entangling system–environment interaction as

$$|\psi\rangle |e_{\mathrm{r}}\rangle \longrightarrow \left\{ \sum_{j_1=0}^{1} \sum_{j_2=0}^{1} \cdots \sum_{j_N=0}^{1} \hat{Z}_{j_1 j_2 \cdots j_N} |\psi\rangle \right\} |\widetilde{e}_{j_1 j_2 \cdots j_N}\rangle , \tag{7.29}$$

where the $|\widetilde{e}_{j_1 j_2 \cdots j_N}\rangle$ are some relative states of the environment.

We remark that the most general form of (7.29) corresponds, errorwise, to the "worst-case scenario" of environmental entanglement: In principle, 2^N different error operators $\hat{Z}_{j_1 j_2 \cdots j_N}$ will need to be taken into account. We say that phase-flip errors *up to weight* N will have to be considered, where

[5] There are at least two equivalent different ways of proving this result. First, we may try an explicit proof along the lines of the one-qubit case considered above. This proof, although quite straightforward, is notationally somewhat cumbersome. The second (and more elegant) route consists of making use of the so-called Jamiolkowski isomorphism [357]. Since this approach requires the introduction of new formal elements, we shall not further consider it here.

the weight of a general error operator $\hat{E}_i$ is defined as the number of the constituent single-qubit operators that are different from the identity.

Fortunately, in many cases of interest relevant to quantum error correction, simplified versions of this general setting can be used. One important case is that of *partial decoherence*. Here, only a small number $K < N$ of qubits becomes entangled with the environment during the time between two applications of an error-correcting mechanism. Then it will be sufficient to restrict our attention to the 2^K possible error operators of weight $\leq K$. Another case of interest is that of *independent qubit decoherence*. Here, we focus on error operators of weight equal to one only, i.e., we consider the case of a collection of independent phase-flip errors acting on single qubits. Thus each qubit undergoes phase flips independently of all other qubits: Phase-flip errors are not correlated among multiple qubits, but only occur locally at the site of an individual qubit. The assumption of independent qubit decoherence can be physically justified in situations where each qubit couples dominantly to its own environment and where these individual environments do not interact with each other. For example, this may be the case if the qubits are spatially separated and only couple to their immediate surroundings.

7.4.3 "Undoing" Decoherence in a Quantum Computer

Let us now employ the above results and insights to see how we can effectively "undo" the effects of decoherence on the system and return the qubit system to its original pre-decoherence state, thereby implementing quantum error correction. Instead of trying to find out which error to correct by measuring the system and then to directly manipulate the system–environment combination, we use an additional set of qubits, referred to as the *ancilla*, to help us in detecting and correcting errors. Such ancilla qubits play the role of an artificial "environment." They are in general of the same physical form as the computational qubits, but they do not take part in the actual execution of the quantum algorithm. Their only role is to help the quantum computer combat decoherence. Instead of manipulating the qubit–environment system (which is difficult, if not impossible), we operate on the qubit–ancilla system (over which we assume to have perfect control).

Recall that the influence of an entangling interaction between the qubit system $\mathcal{S}$ and its environment $\mathcal{E}$ can be written in the form[6] [see (7.26)]

$$|\psi\rangle\,|e_{\mathrm{r}}\rangle \longrightarrow \sum_i \left(\hat{E}_i\,|\psi\rangle\right)|e_i\rangle\,. \tag{7.30}$$

Our objective is now to recover, given the entangled state on the right-hand side of (7.30), the initial (and of course unknown!) state $|\psi\rangle$. To do so, let

[6] Although here we focus on decoherence-induced errors, the following scheme holds for arbitrary error operators $\hat{E}_i$ and is not restricted to error operators of the form $\hat{Z}_{j_1 j_2 \cdots j_N}$.

us now bring the aforementioned ancilla qubits (to be denoted by $\mathcal{A}$ in the following) into the game. One purpose of these ancilla qubits is to act as a diagnostic tool in the following sense. Suppose the initial state of the ancilla is described by the state $|a_r\rangle$. We now let the ancilla interact with the qubit system such that

$$|a_r\rangle \left[\sum_i \left(\hat{E}_i |\psi\rangle \right) |e_i\rangle \right] \longrightarrow \sum_i |a_i\rangle \left(\hat{E}_i |\psi\rangle \right) |e_i\rangle . \tag{7.31}$$

This equation describes a von Neumann-type interaction (see Sect. 2.5.1), with the ancilla playing the role of the "apparatus." To make the following argument more clear, let us assume that the ancilla states $|a_i\rangle$ appearing on the right-hand side of (7.31) are at least approximately mutually orthogonal, such that they can be distinguished by measurement. (We will elaborate on this important issue below.)

Now we can use the following procedure (see Fig. 7.2). First, we measure the observable

$$\hat{O}_{\mathcal{A}} = \sum_i a_i |a_i\rangle\langle a_i| \tag{7.32}$$

on the ancilla, with $a_i \neq a_j$ for $i \neq j$. The projective measurement will yield a particular outcome, say, a_k, and lead to the reduction of the entangled state on the right-hand side of (7.31) onto the kth component of the superposition, i.e.,

$$\sum_i |a_i\rangle \left(\hat{E}_i |\psi\rangle \right) |e_i\rangle \longrightarrow |a_k\rangle \left(\hat{E}_k |\psi\rangle \right) |e_k\rangle . \tag{7.33}$$

Since we know that the outcome of our measurement was a_k, we also immediately know which transformation to use to "undo" the effect of the environment, namely, $\hat{E}_k^{-1} = \hat{E}_k^\dagger$. (This feature motivates our earlier statement that the ancilla qubits act as a "diagnostic tool" for the "symptom" $\hat{E}_k$.) Thus we now apply the transformation $\hat{E}_k^{-1}$ to our qubit system, which yields

$$|a_k\rangle \left(\hat{E}_k |\psi\rangle \right) |e_k\rangle \xrightarrow{\hat{E}_k^{-1}} |a_k\rangle |\psi\rangle |e_k\rangle . \tag{7.34}$$

We have reached our desired result: We have successfully returned the qubit to its original state $|\psi\rangle$. In this way, we have indeed recovered from the detrimental effect of decoherence.

Note that, as required in order to avoid introducing additional decoherence in the computational basis of the qubit system, we have obtained no information whatsoever about the state of the qubits themselves. Instead, by quantum-correlating the composite qubit–environment system with the ancilla, we have encoded information about the various errors $\hat{E}_i$ in the different relative states $|a_i\rangle$ of the ancilla. We can safely measure the ancilla without destroying the quantum information contained in the qubits, since each of the states $|a_i\rangle$ of the ancilla is correlated with the full initial state $|\psi\rangle$ of

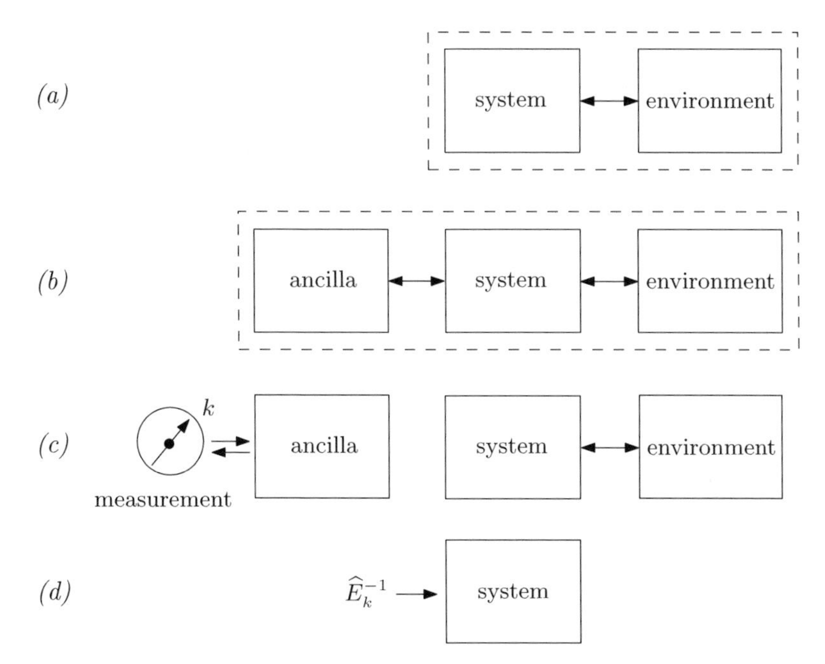

Fig. 7.2. Quantum error correction by means of ancilla qubits. *(a)* The interaction between the system $\mathcal{S}$ and the environment $\mathcal{E}$ entangles the two partners [see (7.30)]. *(b)* The ancilla qubits $\mathcal{A}$ are let to interact with $\mathcal{S}$, creating the tripartite entangled state (7.31) involving $\mathcal{S}$, $\mathcal{A}$, and $\mathcal{E}$. *(c)* A projective measurement on $\mathcal{A}$ reduces this state to a product state of the form (7.33), effectively disentangling the subsystems. The outcome of the measurement contains information about the error that needs to be corrected. *(d)* Application of an appropriate countertransformation $\hat{E}_k^{-1}$ then restores the original state of $\mathcal{S}$.

the qubits, albeit transformed by the error operators $\hat{E}_i$. The measurement of the ancilla then simultaneously reduces the total composite ancilla–qubit–environment state to a single one of these errors (it has "forced" the system to "decide" among the set $\{\hat{E}_i\}$ of possible errors), and tells us which error has been singled out. Thus the task of reconstructing the original superposition state has boiled down to reversing a single, and known, discrete error $\hat{E}_k$.

In this sense, the basic idea of quantum error correction is, as Preskill put it [358], to "fight entanglement with entanglement." By this statement we mean to convey the insight that we can combat the detrimental effect of environmental entanglement on our qubit system by entangling the qubit–environment combination with a third "artificial environment," namely, the ancilla, over which we have full control. Our procedure for error recovery has transferred entropy from the qubits to the ancilla, and the ancilla therefore acts a form of "sink" for the entropy introduced into the qubit system by the environmental interaction. Accordingly, we must "erase" any error information stored in the ancilla qubits (i.e., we must in turn couple the ancilla to some other sink for entropy) before we can reuse the ancilla in a subsequent cycle of error correction.

Our above scheme for quantum error correction has been highly idealized. In particular, three issues will make it challenging to implement this scheme in practice. First, as we shall discuss in Sect. 7.4.4 below, it is impossible to design an interaction between the computational qubits and the ancilla that would allow us to distinguish between all possible errors through a measurement of the ancilla. Second, in realistic settings the error operators $\hat{E}_i$ may be very complex, and it remains to be seen whether and how the corresponding countertransformations (whose physical implementation will require interactions of the computational qubits with some macroscopic device) can be applied to the quantum computer without introducing significant further decoherence. Third, the ancilla qubits are physically similar to the computational qubits and can therefore be expected to be equally prone to environmental interactions (and thus decoherence) as the computational qubits themselves. Such effective measurements of the ancilla qubits by the environment would then compromise the proper functioning of the quantum error-correction scheme. Since the inclusion of ancilla qubits increases the total number of qubits in the quantum computer, and since decoherence rates typically scale exponentially with the size of the system, it will require sophisticated experimental designs to ensure not only that quantum error correction works in practice, but also that it does not unintentionally make the problem of qubit decoherence even worse.

7.4.4 When Does an Error-Correcting Code Exist?

To keep the above outline of the basic principles of quantum error correction as simple as possible, we have assumed [see (7.31)] that there exists a unitary transformation acting on the Hilbert space of the composite ancilla–qubit system that leads to a correlated ancilla–qubit–environment state in such a way that the relative ancilla states $|a_i\rangle$ are all distinguishable and correspond to the different error types $\hat{E}_i$, i.e., $\langle a_i|a_j\rangle \approx \delta_{ij}$ for two different errors $\hat{E}_i \neq \hat{E}_j$. However, we did not elaborate at all on the question of how to show that this transformation exists and of how to determine it.

In fact, this question is at the heart of quantum error correction. It turns out that it is usually impossible to find a transformation that would account for all possible errors $\hat{E}_i$ that may in principle occur. Instead, the typical scope of a scheme for quantum error correction is restricted to transformations of the form (7.31) that can distinguish all errors $\hat{E}_i$ up to a certain weight $t < N$. Furthermore, error-correcting codes are usually not able to recover arbitrary initial qubit states, but only those states contained in a certain subspace $\mathcal{H}_{\text{code}}$ of the total Hilbert space $\mathcal{H}_{\mathcal{S}}$ of the qubit system. (The subspace $\mathcal{H}_{\text{code}}$ is often referred to as the *code subspace* for the quantum error-correcting scheme.)

As we have seen in our above outline, the basic strategy of quantum error correction is to project the qubit state into orthogonal subspaces labeled by the error operators $\hat{E}_i$. Then we can perform a measurement that reveals into which of these error subspaces the state has been projected and thus obtain information about which countertransformation to use in order to correct the error. Formally, a sufficient (albeit not necessary) condition for the existence of an error-correcting code that can correct errors up to a certain weight t is therefore given by requiring that the "error-projected" subspaces

$$\{\hat{E}_i \mathcal{H}_{\text{code}} \,|\, \text{weight}(\hat{E}_i) \leq t\} \tag{7.35}$$

are (at least approximately) mutually orthogonal for all i. Then, for any initial state $|\psi\rangle \in \mathcal{H}_{\text{code}}$, the different error terms $\hat{E}_i |\psi\rangle |e_i\rangle$ in the expansion of the perturbed state will be mutually orthogonal and thus distinguishable by a measurement, without revealing the quantum information encoded in the state $|\psi\rangle$ itself.

7.4.5 Importance of Redundant Encoding and the Three-Bit Code for Phase Errors

It turns out that, in order to fulfill condition (7.35), it is usually necessary to use redundant encoding. In Sect. 7.4.1, we explained how in classical computing redundant encoding helps recover the original, unperturbed state. We also mentioned that we cannot clone an unknown quantum state. However, note that (7.14) points to a method for how we can easily accomplish a redundant encoding in the "relative-state" form using the computational-basis states as the preferred basis. Namely, we simply apply a series of CNOT gates between the first and all other qubits, i.e.,

$$(\alpha |0\rangle + \beta |1\rangle) |00 \cdots 0\rangle \longrightarrow \alpha |000 \cdots 0\rangle + \beta |111 \cdots 1\rangle \,. \tag{7.36}$$

How may such redundant encoding help us in correcting quantum errors?

First of all, to see why we usually cannot get very far without such encoding, consider the example of a single-qubit state $|\psi\rangle = \alpha |0\rangle + \beta |1\rangle$ that becomes entangled with an environment. The final composite state can then be written as [see (7.24)]

$$|\Psi\rangle = \frac{1}{\sqrt{2}} \left[(\alpha |0\rangle + \beta |1\rangle) |e_+\rangle + (\alpha |0\rangle - \beta |1\rangle) |e_-\rangle \right] . \tag{7.37}$$

The assumption encapsulated in (7.31) is now that there exist two orthogonal ancilla states $|a_0\rangle$ and $|a_1\rangle$ and a unitary transformation acting on the Hilbert space of the composite ancilla–qubit system such that we can induce the evolution

$$|a_{\rm r}\rangle\,|\Psi\rangle \longrightarrow \frac{1}{\sqrt{2}}\left[|a_0\rangle\left(\alpha\,|0\rangle+\beta\,|1\rangle\right)|e_+\rangle+|a_1\rangle\left(\alpha\,|0\rangle-\beta\,|1\rangle\right)|e_-\rangle\right]. \quad (7.38)$$

But if we make the replacement $\beta \longrightarrow -\beta$, then the role of the ancilla states as "error indicators" will be reversed. Namely, the ancilla state $|a_0\rangle$ will now be correlated with a term of the phase-flip type, whereas the state $|a_1\rangle$ will correspond to the error-free term. Thus a measurement of the ancilla will not reveal any valuable information about the error that has occurred—unless, of course, if we knew something about the coefficients α and β (and, in turn, if the ancilla states were some function of these coefficients). But such knowledge can clearly not be assumed to exist, since the initial qubit state will in general be completely unknown.[7]

This difficulty marks the departure of quantum error correction from the more simple approach of quantum "erasure" (see Sect. 2.13). Note that the evolution described by (7.37) and (7.38) is essentially that of the "erasure" protocol (2.125), with the ancilla and environment corresponding to, respectively, the read-out device and the which-path detector. In quantum "erasure," the interference pattern is recovered in the sense of a conditioning of the outcome of the read-out measurement (2.124) on the observed position of the particle on the screen [130]. This conditioning is necessary because the states $|\psi_+\rangle$ and $|\psi_-\rangle$ in (2.125) correspond to, respectively, fringe and "antifringe" patterns on the screen, with the combined pattern being equal to the no-interference classical distribution. Thus the effective recovery of an interference pattern requires us to be able to *distinguish* the final relative states $|\Phi_\pm\rangle$ of the read-out device, but we do not need to infer from the read-out measurement the particular value of the relative phase between the components $|\psi_1\rangle$ and $|\psi_1\rangle$ in the states $|\psi_\pm\rangle$. By contrast, in quantum computing it is important to restore the original superposition state, and thus we will need to be able to detect and correct the relative phase between the computational-basis states, a requirement that (7.38) does not fulfill.

This problem can be solved through a redundant encoding of the single-qubit state. To illustrate this approach, we shall consider the example of a single logical qubit redundantly encoded in three physical qubits using the scheme (7.36),

$$(\alpha\,|0\rangle+\beta\,|1\rangle)\,|00\rangle \longrightarrow \alpha\,|000\rangle+\beta\,|111\rangle\,. \quad (7.39)$$

As we will see in the following, this encoding allows us to diagnose (and recover from) a phase-flip error affecting a single qubit without any knowledge about the initial qubit state.

First, we transform the encoded state (7.39) to the conjugate basis $\{|+\rangle\,,|-\rangle\}$ for each qubit, where $|\pm\rangle=(|0\rangle\pm|1\rangle)/\sqrt{2}$. This transformation

[7]This is of course the whole point of error correction in both the classical and quantum settings: We are given a perturbed state and would like to reconstruct the unknown initial state. If we knew from the outset what the unperturbed state should be, we could trivially just reprepare the system in this state.

can be performed by applying the so-called *Hadamard rotation*

$$H = \frac{1}{\sqrt{2}} \begin{pmatrix} 1 & 1 \\ 1 & -1 \end{pmatrix} \tag{7.40}$$

to the original basis states $|0\rangle$ and $|1\rangle$. Note that a phase-flip $\hat{\sigma}_z$ in the basis $\{|0\rangle, |1\rangle\}$ looks like a bit-flip error in the basis $\{|+\rangle, |-\rangle\}$, because

$$\hat{\sigma}_z |\pm\rangle = |\mp\rangle . \tag{7.41}$$

We have thus accomplished the encoding

$$\alpha |0\rangle + \beta |1\rangle \longrightarrow \alpha |+++\rangle + \beta |---\rangle . \tag{7.42}$$

The three qubits will interact with the environment and thus be subject to entanglement and decoherence, leading to phase errors $\hat{Z}_{ijk}$, $(i, j, k) \in \{0, 1\}$. Suppose we only consider error operators of weight ≤ 1 (namely, the operators $\hat{Z}_{000}$, $\hat{Z}_{100}$, $\hat{Z}_{010}$, and $\hat{Z}_{001}$). Then the composite qubit–environment state may be written as

$$\begin{aligned} |\Psi\rangle &= (\alpha |+++\rangle + \beta |---\rangle) |e_{000}\rangle \\ &+ (\alpha |-++\rangle + \beta |+--\rangle) |e_{100}\rangle \\ &+ (\alpha |+-+\rangle + \beta |-+-\rangle) |e_{010}\rangle \\ &+ (\alpha |++-\rangle + \beta |--+\rangle) |e_{001}\rangle . \end{aligned} \tag{7.43}$$

In the following, we shall refrain from writing down the environmental states. As we do not care about their precise form, we may simply imagine each of the qubit terms in the state to be correlated with some relative state of the environment.

Now we would like to diagnose the error. To do so, we first apply the Hadamard rotation (7.40) one more time to each qubit, which returns us to the original basis $\{|0\rangle, |1\rangle\}$. Then the state (7.43) becomes (omitting the environmental states)

$$\begin{aligned} |\Psi\rangle &= (\alpha |000\rangle + \beta |111\rangle) \\ &+ (\alpha |100\rangle + \beta |011\rangle) \\ &+ (\alpha |010\rangle + \beta |101\rangle) \\ &+ (\alpha |001\rangle + \beta |110\rangle) \end{aligned} \tag{7.44}$$

Next we introduce two ancilla qubits prepared in the state $|a_r\rangle = |00\rangle$.[8] To encode information about the qubit errors in the ancilla, we perform a series of CNOT gates in the following way:

[8] In fact, it can be shown that these two ancilla qubits are not strictly required to perform the error diagnosis and correction. The error syndrome could in principle also be extracted directly from the three encoded qubits themselves. However, inclusion of the ancilla allows our description of the three-bit code to follow more clearly the basic ideas of quantum error correction outlined above. Furthermore, it turns out that the ancilla qubits are required to implement so-called *fault-tolerant* schemes for error correction (see below).

1. First, we apply CNOT gates from the first and the second of the $\mathcal{S}$-qubits to the first ancilla qubit.
2. Then, we carry out CNOT gates from the first and the third of the $\mathcal{S}$-qubits to the second ancilla qubit.

The effect of this sequence of CNOT gates on the initial ancilla state $|00\rangle$ for each of the three-qubit computational basis states as the input is shown in Table 7.2. The combined state (7.44) of the ancilla and the three-qubit system after application of the above sequence of CNOT gates then reads

$$\begin{aligned} |\Psi\rangle\,|00\rangle \longrightarrow & (\alpha\,|000\rangle + \beta\,|111\rangle)\,|00\rangle \\ + & (\alpha\,|100\rangle + \beta\,|011\rangle)\,|11\rangle \\ + & (\alpha\,|010\rangle + \beta\,|101\rangle)\,|10\rangle \\ + & (\alpha\,|001\rangle + \beta\,|110\rangle)\,|01\rangle\,. \end{aligned} \tag{7.45}$$

We now see that we have correlated each of the four possibilities $\hat{Z}_{000}$, $\hat{Z}_{100}$, $\hat{Z}_{010}$, and $\hat{Z}_{001}$ with a distinct ancilla state $|ij\rangle$, $(i,j) \in \{0,1\}$. Since these states are mutually orthogonal, we can perfectly distinguish the four error types by measuring the two ancilla qubits.

Now our error-correction procedure is almost complete. Suppose that the ancilla measurement has yielded the outcome (i.e., the error syndrome) "01." This measurement has reduced the composite state (7.45) of the ancilla and the qubit system to

$$(\alpha\,|001\rangle + \beta\,|110\rangle)\,|01\rangle\,. \tag{7.46}$$

Since the ancilla is now completely disentangled from the qubit system, we can ignore the ancilla state in the following.

Based on the measurement outcome, we immediately know that we must apply the countertransformation $\hat{I}\otimes\hat{I}\otimes\hat{\sigma}_x$ to correct the error (see Table 7.3). This transformation changes the state (7.46) according to

Table 7.2. Evolution of the initial two-qubit ancilla state $|00\rangle$ after application of the first (center column) and the second (right column) CNOT gate, for the different qubit–environment input states listed in the left column.

Input	First CNOT	Second CNOT
$\|000\rangle$	$\|00\rangle$	$\|00\rangle$
$\|111\rangle$	$\|00\rangle$	$\|00\rangle$
$\|100\rangle$	$\|10\rangle$	$\|11\rangle$
$\|011\rangle$	$\|10\rangle$	$\|11\rangle$
$\|010\rangle$	$\|10\rangle$	$\|10\rangle$
$\|101\rangle$	$\|10\rangle$	$\|10\rangle$
$\|001\rangle$	$\|00\rangle$	$\|01\rangle$
$\|110\rangle$	$\|00\rangle$	$\|01\rangle$

Table 7.3. Three-bit code for correcting phase-flip errors. The table shows the four possible error syndromes obtained by measuring the two ancilla qubits, the corresponding errors in the encoded three-qubit state, and the countertransformations needed to restore the original three-qubit state.

Syndrome	Error	Recovery transformation
00	$\hat{Z}_{000}$ (no error)	None needed
11	$\hat{Z}_{100}$ (phase-flip of first qubit)	$\hat{\sigma}_x \otimes \hat{I} \otimes \hat{I}$
10	$\hat{Z}_{010}$ (phase-flip of second qubit)	$\hat{I} \otimes \hat{\sigma}_x \otimes \hat{I}$
01	$\hat{Z}_{001}$ (phase-flip of third qubit)	$\hat{I} \otimes \hat{I} \otimes \hat{\sigma}_x$

$$\alpha\,|001\rangle + \beta\,|110\rangle \xrightarrow{\hat{I}\otimes\hat{I}\otimes\hat{\sigma}_x} \alpha\,|000\rangle + \beta\,|111\rangle\,, \tag{7.47}$$

and we have therefore restored the initial, redundantly encoded qubit state (7.39). We can then, finally, apply the inverse of the encoding transformation (7.39) to arrive at the original, unperturbed single-qubit state $\alpha\,|0\rangle + \beta\,|1\rangle$.

This scheme for the correction of phase errors cannot always succeed if we include phase errors that affect more than one qubit (i.e., if we consider phase-error operators of weight greater than one). This can be easily seen from the fact that our two ancilla qubits can give us only two classical bits of information about the error. Therefore we can distinguish with certainty only up to four different error possibilities. If we include all possible phase-error operators for three qubits (i.e., all operators of weight between zero and three), we can recover the original state only with a certain probability. Since we now have to be able to correct for eight different errors, a specific error syndrome will correspond to two different errors. We can compute the probabilities of these two possibilities and then apply the countertransformation corresponding to the error that is more likely.

Assuming that the probability p of a single-bit error is less than 50% and that the errors are uncorrelated (such that the probabilities of two-qubit and three-qubit errors are $3p^2(1-p)$ and p^3, respectively), the most likely error types will be precisely the ones considered in our example above, namely, $\hat{Z}_{000}$, $\hat{Z}_{100}$, $\hat{Z}_{010}$, and $\hat{Z}_{001}$. More specifically, since the probability of our scheme to *fail* is $3p^2(1-p) + p^3$ (which is simply the probability of the occurrence of two-qubit and three-qubit errors), the probability of failure scales as $O(p^2)$ and therefore decreases quite rapidly as p becomes smaller. This is to be compared to the completely uncorrected case, in which the probability of failure scales as $O(p)$ (since our state will be perturbed any time a single error occurs, which, by definition, happens with probability p). Thus our error-correcting scheme described above will improve matters over the case of no error correction whenever $p < 50\%$. This situation is similar to the majority-voting scheme for classical error correction discussed in Sect. 7.4.1. There we showed that error recovery will yield an advantage only if in average no more than one bit flips (i.e., if the probability of single-bit flips is less than 50%).

Interestingly, the three-bit code for phase errors described here was the basis of the first experimental implementation of quantum error correction realized in 1998 [359].[9] The experiment was carried out using liquid-state NMR to represent an ensemble quantum-computing device [349]. It was found that the experimentally implemented quantum error-correction protocol indeed led to a slowdown of the theoretically predicted loss of coherence from the qubit system.

Since redundant encoding requires a significant increase in the number of physical qubits in the quantum computer over the unencoded case, redundant encoding has to be implemented very carefully so as to avoid introducing large amounts of further decoherence. (Above we had pointed out a similar problem in the context of the ancilla qubits.) The overarching goal is therefore to achieve a maximum of error-correcting abilities with the smallest possible ratio of physical to logical qubits.

7.4.6 Apparatus-Induced Decoherence and Fault Tolerance

Thus far we have considered interactions with the uncontrolled environment as the source of our phase errors. However, we really meant to convey a broader notion of "environment" in this context—namely, that of any physical system interacting with the qubit system. An important such interaction is the coupling of the qubit system to the apparatus that prepares, manipulates, and measures the qubits and thus processes quantum information. Any such interaction with the apparatus will in general lead to entanglement between the apparatus and the qubit system and thus to decoherence of the qubits. Each gate operation, for example, will then typically induce some qubit decoherence. In addition, the practical implementation of gate operations will always be imperfect, introducing classical noise. Quantum algorithms usually require the application of many thousands of gates to the qubit system, leading to the rapid accumulation of errors.

Therefore, we must not only periodically correct qubit errors due to the continuous interaction with the uncontrolled environment, but also due to the processing of quantum information. Of course, as mentioned above, since error correction itself is based on measurements and gate operations performed by some apparatus, it will in turn introduce decoherence effects into the qubits and cannot be executed with infinite accuracy in any realistic situation. This problem can be mitigated by so-called *fault-tolerant* methods for error detection and recovery [360–364], which aim at ensuring reliable operation of the quantum computer even in the realistic setting of each single

[9] We note, though, that this experiment did not make use of the two ancilla qubits introduced above. Instead the error syndrome was obtained directly from certain joint measurements on the qubit system.

gate and measurement operation being "noisy" (i.e., having a certain error probability).

While we shall not go into the details of fault-tolerant quantum computing (see, e.g., [56,365] for accessible reviews of this subject), we shall state an important result that is often referred to as the *threshold theorem* [358,366–369]. If one adopts a fault-tolerant implementation of quantum-error correction, then one can show that, provided the error rate per gate operation (in some physical implementation) is below a certain threshold value,[10] it should be possible, at least in principle, to efficiently perform arbitrarily large quantum computations. This important theorem can be proved given a small set of physically reasonable assumptions about the nature of the errors, and it suggests that there should be no fundamental obstacle to an implementation of quantum computers.

7.5 Quantum Computation on Decoherence-Free Subspaces

We introduced the general concept of decoherence-free subspaces (DFS) [89–98], or pointer subspaces [9], in Sect. 2.8.1. Recall that DFS are subspaces $\mathcal{H}_{\mathrm{DF}}$ of the Hilbert space $\mathcal{H}$ of the system such that *any* state $|\psi\rangle$ in this subspace remains (exactly or approximately) pure in spite of interactions with the environment. Applied to quantum computing, with the system represented by a collection of computational qubits, DFS allow us to encode quantum information in "quiet corners" of the Hilbert space so as to automatically protect this information from the detrimental influence of the environment. By contrast with quantum error correction, DFS prevent errors from happening in the first place and thus represent a strategy for *intrinsic error avoidance*.

We showed that [see (2.96)], given the general diagonal decomposition of the interaction Hamiltonian,

$$\hat{H}_{\mathrm{int}} = \sum_{\alpha} \hat{S}_{\alpha} \otimes \hat{E}_{\alpha}, \tag{7.48}$$

a necessary requirement for a DFS to emerge is the existence of a set of orthonormal degenerate eigenstates $|s_i\rangle$ of the operators $\hat{S}_{\alpha}$,

$$\hat{S}_{\alpha} |s_i\rangle = \lambda^{(\alpha)} |s_i\rangle \qquad \text{for all } \alpha \text{ and } i. \tag{7.49}$$

In Sect. 2.8.1, we also discussed that the condition (7.49) amounts to the presence of dynamical symmetries. Below, we will consider a model in which all qubits couple in exactly the same way to a single environment. In this

[10] Current estimates for this *accuracy threshold* vary between about 10^{-2} and 10^{-6} per computational step; see [370] for an analysis of the different estimates.

case the system–environment interaction is completely symmetric with respect to any permutations of the qubits, leading to a DFS of maximum size. What happens if this perfect symmetry is broken by additional small independent coupling terms? Clearly, we would like our DFS to be reasonably robust to such perturbations. Following initial studies by Lidar, Chuang, and Whaley [90], Bacon, Lidar, and Whaley [99] showed that, to first order in the perturbation strength, the *storage* of quantum information in DFS is stable to such perturbations to all orders in time. However, the *processing* of such quantum information encoded in DFS was found to be robust to symmetry-breaking perturbations only to *first* order in time. The authors also demonstrated how this insufficient robustness of quantum computation on DFS can be mitigated by combining DFS with active quantum error-correction schemes [94].

7.5.1 What Does a Decoherence-Free Subspace Look Like?

So far, we have only stated the general conditions for the existence of a DFS, without giving any concrete examples of its structure. Also, a DFS would obviously be of little use for quantum computation if it was so small as to effectively make it impossible to form the necessary higher-dimensional superposition states. The question of the size of the DFS relative to the dimension of the full Hilbert space of the system is therefore very important. Probably not surprisingly, we must adopt particular models for the system–environment interaction to give specific answers to this question. Here we will consider the model most commonly considered in the literature on DFS (see, e.g., [89,91,93,96,237]), namely, the spin–boson model discussed in Sect. 5.3.

Independent Versus Collective Decoherence

Let us distinguish two limiting cases for modeling decoherence in qubits. The first limit is that of independent decoherence, i.e., the assumption that each qubit couples independently to its own environment, without any interactions between these environments. We already mentioned this case in Sect. 7.4.2. The assumption of independent decoherence is often made in the context of quantum error correction. It implies that the error processes affecting the qubits are completely uncorrelated. Thus, if the probability of a particular error to affect one qubit is p, the probability of this error to occur in K qubits will be p^K. In this case, we may represent qubit errors simply as a linear combination of error operators of weight equal to one, i.e., as a combination of single-qubit errors with all other qubits unaffected. Many error-correcting schemes (such as the three-qubit code discussed in Sect. 7.4.5) are only efficient in correcting such single-qubit errors, and thus the assumption of independent decoherence frequently underlies these schemes.

However, this assumption is rather unrealistic when the qubits are located spatially close to each other (relative to the typical coherence length of the environment). In this case, all qubits "feel" the (approximately) same environment, and it is likely that errors will become correlated among multiple qubits. The limiting case corresponding to this situation is that of collective decoherence, in which all qubits couple to exactly the same environment.

It turns out that these two extremes, completely independent decoherence on the one hand and purely collective decoherence on the other hand, delineate the limits on the size of a DFS. For the spin–boson model, in the limit of collective decoherence (and many qubits), the dimension of the DFS will asymptotically approach the dimension of the original Hilbert space, thus allowing for optimal decoherence-free encoding. In the opposite limit of independent decoherence, however, we shall see that there exists no DFS that would allow for an encoding of logical qubits. We shall not reproduce the detailed models and proofs necessary to establish these results, but we nonetheless would like to gain some intuition into the relationship between the size of the DFS and the particular decoherence model. This will be done in the following.

A Simple Example

Let us first consider the case of collective decoherence in the spin–boson model [89, 91, 93, 96, 237]. DFS emerge here almost by definition. The interaction Hamiltonian for a collection of N qubits is given by a generalization of the interaction Hamiltonian in (5.90), i.e.,

$$\hat{H}_{\text{int}} = \sum_{i=1}^{N} \hat{\sigma}_z^{(i)} \otimes \sum_j \left(g_{ij} \hat{a}_j^\dagger + g_{ij}^* \hat{a}_j \right) \equiv \sum_{i=1}^{N} \hat{\sigma}_z^{(i)} \otimes \hat{E}_i. \tag{7.50}$$

Here the g_{ij} denote the coupling strength of the ith qubit to the jth environmental oscillator, and the $\hat{a}_j$ ($\hat{a}_j^\dagger$) are the annihilation (creation) operators for the jth mode of the bosonic field.

The assumption of collective decoherence means that all qubits couple to the *same* environment, which implies that the couplings g_{ij} (and thus the environment operators $\hat{E}_i$) must be independent of the index i. Then (7.50) becomes

$$\hat{H}_{\text{int}} = \left(\sum_i \hat{\sigma}_z^{(i)} \right) \otimes \hat{E} \equiv \hat{S}_z \otimes \hat{E}. \tag{7.51}$$

The precise form of the environment operator $\hat{E}$ is not really important here. The key point is that the interaction Hamiltonian has now been written in the form (7.51) with the sum replaced by a single term $\hat{S}_z \otimes \hat{E}$.

Now let us recall that a DFS is spanned by a degenerate set of eigenstates of the system operators $\hat{S}_\alpha$ of the interaction Hamiltonian [see (7.49)]. Thus in this case the DFS will be spanned by degenerate eigenstates of the collective spin operator $\hat{S}_z$ defined in (7.51). Clearly, any N-qubit product state expressed in the computational basis (7.5) will be an eigenstate of $\hat{S}_z$. There are $2N+1$ different possible integer eigenvalues m, ranging from $m = -N$ (corresponding to the basis state $|1\cdots 1\rangle$) to $m = +N$ (for the basis state $|0\cdots 0\rangle$). Generally, a computational-basis state with m_0 qubits in the "0" state has eigenvalue $m = 2m_0 - N$.

The largest number of mutually orthogonal computational-basis states with the same eigenvalue m of $\hat{S}_z$ is given by the set $\mathfrak{S}_0$ of basis states with $m = 0$, i.e., those with an equal number of qubits being in the "0" and "1" states. Formally, we may write

$$\mathfrak{S}_0 = \left\{ |i_1 \cdots i_N\rangle \mid i_j \in \{0,1\}, \sum_{j=1}^{N} i_j = 0 \right\}, \tag{7.52}$$

and there are

$$n_0 = \binom{N}{N/2} \tag{7.53}$$

states in this set. Therefore the states contained in $\mathfrak{S}_0$ span a DFS of dimension n_0. For large values of N, we can approximate the binomial coefficient in (7.53) using Stirling's formula,

$$\log_2 \binom{N}{N/2} \approx N - \frac{1}{2}\log_2(\pi N/2) \xrightarrow{N \gg 1} N. \tag{7.54}$$

Thus, as claimed above, in the limiting case of collective decoherence the dimension of our DFS approaches the dimension of the original Hilbert space, and the encoding efficiency approaches unity (i.e., we can, in the limit $N \longrightarrow \infty$, encode all 2^N qubit basis states of our original Hilbert space in a decoherence-free fashion).

For example, for $N = 4$ qubits, the set

$$\mathfrak{S}_0 = \{\, |0011\rangle\,, |0101\rangle\,, |0110\rangle\,, |1001\rangle\,, |1010\rangle\,, |1100\rangle\,\} \tag{7.55}$$

of computational-basis states spans a maximum-size DFS of dimension equal to six, to be compared with the dimension of the original Hilbert space, which is $2^4 = 16$. Thus, given the model for collective decoherence considered here, we can encode up to two logical qubits in our DFS. Three logical qubits would already require 2^3 computational basis states, which (7.55) evidently does not afford. Using four physical qubits, we can therefore carry out decoherence-free quantum computations that require no more than two logical qubits.

Conversely, in the case of purely independent decoherence, it is not possible to find any DFS of dimension greater than one. This can be seen from the following simple observation. The environment operators $\hat{E}_i$ appearing in the interaction Hamiltonian (7.50) will now in general all be different from one another. To find a DFS, we follow the usual strategy (7.49) of determining a set of orthonormal basis states $\{|s_i\rangle\}$ such that

$$\left[\hat{I}^{(1)} \otimes \cdots \otimes \hat{I}^{(j-1)} \otimes \hat{\sigma}_z^{(j)} \otimes \hat{I}^{(j+1)} \otimes \cdots \otimes \hat{I}^{(N)}\right] |s_i\rangle = \lambda^{(j)} |s_i\rangle \tag{7.56}$$

for all i and $1 \leq j \leq N$. As the reader may easily verify, there is only a single state that fulfills this eigenvalue problem, namely, the computational-basis state $|0 \cdots 0\rangle$. Since we need at least a two-dimensional subspace to encode a single logical qubit, the case of independent decoherence in the spin–boson model does not allow for the existence of a DFS for quantum computation. In the language of pointer subspaces, there is only a single exact pointer state, and this environment-superselected preferred state of the system will be the ground state $|0 \cdots 0\rangle$.

Of course, in realistic settings neither the assumption of purely independent decoherence nor the limit of entirely collective decoherence will be completely appropriate. Fortunately, however, we have now two powerful, complementary methods at our disposal for combating decoherence in each of these two limiting cases. We can use encoding in DFS to protect our qubit system from collective decoherence effects, and we can recover from single-qubit errors due to independent decoherence using the active error-correction methods described in Sect. 7.4. In fact, these two approaches can be tied together (or "concatenated" [94]) to allow for universal fault-tolerant quantum computation even when the restriction to single-qubit errors is dropped [95,371].

7.5.2 Experimental Realizations of Decoherence-Free Subspaces

The results of the first experiment that explicitly demonstrated the existence of a DFS were reported by Kwiat and coworkers in 2000 [372]. One should emphasize that the experiment was more a proof-of-principle demonstration than an actual realization of a DFS relevant to quantum computation. It was simply shown that a particular Bell state [see (2.7)] of a pair of photons was essentially immune to collective decoherence, while photon pairs in the other three Bell states suffered significant decoherence, in agreement with theoretical predictions. The experiment used an artificial environment, namely, two little pieces of quartz oriented at adjustable angles with respect to the paths of the two photons, to induce a controllable decohering interaction with the photons passing through the quartz. Aligning the two quartz pieces at identical angles (modulo 90°) induced collective decoherence. Changing the angle then corresponded to inducing decoherence in different bases. In essence, this experiment simply demonstrated the existence of a single pointer state.

The results of another experiment on DFS were reported in 2001 by Kielpinski et al. at NIST [373]. The authors used two-qubit Bell states similar to those employed in the experiment by Kwiat et al. [372] described in the previous paragraph. However, this time they encoded the single logical qubit into a DFS of a pair of two trapped and strongly interacting $^9\text{Be}^+$ ions. The experiment demonstrated that superpositions of the encoded basis states were indeed immune to the influence of collective decoherence. The observed decay of coherence was much slower than for an unencoded qubit state and could be attributed to sources other than collective decoherence (such as the degradation of the read-out pulses and the heating of the motional state of the ions). Thus the experiment successfully demonstrated the experimental construction of a two-dimensional subspace (using a four-dimensional Hilbert space of physical qubits) that was immune to collective decoherence.

Since then, experiments have aimed to create increasingly large DFS using different physical realizations of the qubits. For example, in 2001 Viola et al. succeeded in generating the first three-qubit DFS using NMR qubits [374].

7.5.3 Environment Engineering and Dynamical Decoupling

One may wonder to what extent the conditions for the existence of a DFS can be fulfilled, at least approximately, in situations of practical (i.e., experimental) interest. In the previous section we have described some recent experiments in which DFS have indeed been realized. But these DFS are still very small. It is clear that for reasonably large DFS to exist, the system–environment interaction must exhibit a sufficiently high degree of symmetry, and it seems quite unlikely that such conditions will arise naturally in a given experimental setting.

One possible way of overcoming this limitation is based on *environment engineering*. Here, one tries to actively create certain symmetries in the structure of the system–environment interactions. For obvious reasons, this strategy is sometimes also referred to as a "symmetrization of the environment." For example, Dalvit, Dziarmaga, and Zurek [102] showed how such an appropriately engineered symmetrization of the system–environment coupling could make superposition states in Bose–Einstein condensates to correspond to (approximate) degenerate eigenstates of the interaction Hamiltonian. Thus such superposition states would lie within a DFS, thereby significantly enhancing their longevity (see also Sect. 6.4.1).

As shown by Poyatos, Cirac, and Zoller [375], by changing the parameters in the effective system–environment interaction Hamiltonian for a trapped ion (with the controlling laser beam playing the role of the environment), one can also select different pointer subspaces and thereby actively control into which DFS the system (i.e., the trapped ion) is driven. This feature was subsequently experimentally demonstrated in ion-trap experiments, with application to the control and simulation of decoherence [125, 351] (see also [376]). This neatly lends direct experimental evidence to the pointer-basis concept. The form of the "classical" states is not predefined by some *a priori* criterion introduced into quantum mechanics from the outside, but is instead dynamically determined by the structure of the system–environment interaction Hamiltonian, as described in Sect. 2.8. By modifying this interaction Hamiltonian, various pointer bases emerge, corresponding to different quasiclassical states.

Another approach to the active creation of DFS has become known as *dynamical decoupling* [377–382]. Here the basic trick consists of introducing time-dependent modifications of the Hamiltonian of the system that effectively counteract the influence of the environment. These modifications take the form of sequences of rapid projective measurements or strong control-field pulses acting on the system only ("quantum bang-bang control" [377]). Even if the structure of the system–environment interaction Hamiltonian is not known at all, decoherence can be suppressed arbitrarily well in the limit of an infinitely fast rate of the decoupling control field, thus dynamically creating a DFS (which then represents a *dynamically decoupled subspace*). In the realistic case of a finite control rate, sufficient (albeit imperfect) protection from decoherence can be achieved via this decoupling technique provided the control rate is larger than the fastest timescale set by the rate of formation of environmental entanglement.

7.6 Summary and Outlook

The key obstacle to the realization of a quantum computer is the required trade-off between shielding and controllability. On the one hand, we need to construct qubit systems that are reasonably protected from unwanted interactions with the environment in order to minimize decoherence. On the other hand, these qubit systems must remain sufficiently open to allow for their coherent control in order to perform the quantum computation.

Given that we thus cannot simply isolate the qubits, the minimization of decoherence in quantum computation will always require a combination of techniques, such as active quantum error correction, decoherence-free subspaces, environment engineering, dynamical decoupling, etc. One can show that, once the (typically rather unrealistic) restriction to single-qubit errors is dropped, universal fault-tolerant quantum computation can only be achieved if one uses both DFS *and* quantum error correction [371]. The idea is then to first construct the quantum computer in such a way as to exploit (or create) dynamical symmetries which allow for an encoding in subspaces that are approximately decoherence-free to a sufficient degree. Effective DFS can also be created and maintained *during* the quantum computation via externally induced time-dependent modifications of the evolution of the qubit system. Since DFS exhibit a certain degree of robustness toward small perturbations from the ideal conditions, this approach will induce a certain stability of the quantum computation toward decoherence effects. Then, as a second step, schemes for quantum error correction can be used to try to actively recover from the influence of decoherence.

To date, it has been possible to coherently control only a handful of qubits, much too few to allow for any useful quantum computation to be carried out. It remains to be seen if, and when, it will possible to sufficiently prolong decoherence times to enable the construction of a working quantum computer. Regardless, the research into strategies for combating decoherence sparked by the interest in quantum computing has also proved fruitful for improving our understanding of decoherence itself, in both the theoretical and experimental domains.

In the early days of research on decoherence, the focus had been put on the fundamental implications of decoherence for quantum measurement and the quantum-to-classical transition in general, with the main feature of interest being the practically irreversible nature and extreme efficiency of decoherence. In the past decade, the field of quantum computation has significantly broadened this scope by focusing on the active control and engineering of decohering interactions. For instance, we have learned, and experimentally demonstrated, how to design and manipulate the structure of system–environment interactions in order to exploit and create niches in the Hilbert space that are less affected by decoherence. Quantum error correction has cast the influence and mitigation of decoherence effects into the language of discrete errors that can be effectively "undone" by exploiting the very effect that causes decoherence, namely, quantum entanglement. Many more examples, some of which have been discussed in this chapter, could be listed here. In this sense, our understanding of decoherence has received a significant boost from the recent work on quantum computing, and we can expect many more exciting insights into decoherence to result from this area of research.

8 The Role of Decoherence in Interpretations of Quantum Mechanics

Many decades have now elapsed since the famous Solvay conference of 1927 during which the elite of physicists engaged in a heated debate about the interpretation of quantum mechanics [383]. Yet, discussions about the meaning of quantum theory show no sign of abating. If one would like to go beyond a purely pragmatic "shut-up-and-calculate" approach to quantum mechanics[1] and relate the quantum formalism to a presumed physical reality "out there," it is virtually impossible not to get tangled up in interpretive questions. The existence of a variety of interpretations of quantum mechanics is therefore as old as quantum theory itself.

In this chapter, we shall discuss the following key question. What implications does decoherence have for these different interpretations of quantum mechanics, given that decoherence itself is related to interpretive problems of quantum mechanics? This question is of particular interest given that decoherence was "discovered" a fairly long time after many of the main interpretations of quantum mechanics—such as relative-state interpretations, modal interpretations, and the pilot-wave theory of de Broglie and Bohm—had already been formulated. (As recounted at the end of Chap. 1, it was only in the early 1990s that decoherence started to attract broader attention from the scientific community.) When the importance of decoherence for a realistic description of interpretation-related topics such as quantum measurement was finally realized, the question of the implications of decoherence for existing (and future) interpretations of quantum mechanics became acutely relevant. Since one of the main goals of interpretations was to make sense of the puzzling aspects of quantum measurement, and since decoherence was concerned (among other things) with a realistic description of such measurements, it became inevitable to revisit these interpretations in the context of the new insights gained from decoherence.

Let us outline some of the key questions that one encounters when investigating the implications of decoherence for interpretations of quantum mechanics:

---

[1] Although often attributed to Feynman, it appears that the nickname "shut-up-and-calculate interpretation" was actually coined by Mermin [384]. For an example of such an interpretive stance, see [44].

- Is decoherence by itself capable of solving certain foundational problems of quantum mechanics, thereby rendering some of the interpretive or formal additives to quantum mechanics introduced by particular interpretations superfluous? If this is indeed the case, what core foundational problems remain, and what form do they take?
- Can decoherence enforce the empirical adequacy of interpretations, that is, can it protect an interpretation from experimental disproof?
- In turn, can decoherence rule out certain interpretations by demonstrating that these interpretations lead to clashes with standard quantum mechanics or with our observations?
- Can decoherence provide physical motivations for some of the assumptions introduced by a particular interpretation and thereby also lend a more precise (physical) meaning to these assumptions?
- To what extent can decoherence act as an "amalgam" that may unify and simplify a spectrum of different interpretations, thereby showing that seemingly different interpretations can be reduced to a common core?

These and other questions will be addressed and analyzed in this chapter (parts of which are based on a review paper [1] of this author[2]), separately for each of the main interpretive strands of quantum mechanics. We will consider the standard and Copenhagen interpretations (Sect. 8.1), relative-state interpretations (Sect. 8.2), modal interpretations (Sect. 8.3), physical collapse theories (Sect. 8.4), and Bohmian mechanics (Sect. 8.5). By doing so, we will not only be able to assess the current status of the different interpretations in the light of decoherence, but also acquire new insights into the meaning and scope of decoherence itself.

8.1 The Standard and Copenhagen Interpretations

In this section, we shall discuss the standard interpretation of quantum mechanics as commonly presented (either explicitly or implicitly) in most textbooks on quantum mechanics. This interpretation is often also referred to as "orthodox" quantum mechanics. We deliberately distinguish the standard interpretation from the Copenhagen interpretation, which introduces the additional assumption of the necessity of fundamental, irreducible classical concepts in order to describe quantum phenomena, including measurements. We shall come back to this particular aspect of the Copenhagen interpretation in Sect. 8.1.3 below.

A defining feature of the standard interpretation is the collapse postulate, which we have already mentioned in many places of this book. This postulate can be decomposed into several parts. First, it states that every measurement performed on a quantum system induces a discontinuous break in the unitary

[2] Parts of this chapter adapted with permission from [1]. Copyright 2004 by the American Physical Society.

time evolution. Second, this break results in the "collapse" of the wave function, expanded in terms of the eigenstates of the measured observable, onto one of these terms. Thereby a single component in the superposition of terms is selected as the outcome of the measurement. Third, the probability of a particular outcome is given by the amplitude-squared overlap between the initial wave function and the quantum state corresponding to the outcome (Born's rule; see also footnote 9 on p. 35).

The standard interpretation does not give any explanations of the nature of the collapse.[3] Despite the prominent role of measurement and the collapse postulate in this interpretation, the concrete definition of "measurement" remains rather unclear. In principle, the standard interpretation does not *a priori* exclude the possibility of macroscopic superpositions of "classical" states. At the same time, it tells us that such superpositions could never be observed, since any such observation would amount to a collapse-inducing measurement interaction. However, the question of what precisely counts as a measurement is not answered: Is a human observer required to induce the collapse? When does a given interaction induces a collapse, and when does it allow the system to continue to evolve unitarily?

8.1.1 The Problem of Outcomes

We have already discussed the problem of outcomes in the context of the von Neumann measurement scheme (see Sect. 2.5.4). As mentioned there, in the standard interpretation the interpretive rule for the existence of outcomes (or definite "values") is given by the eigenvalue–eigenstate link, which prescribes that a system has a definite value of a physical quantity if and only if it is in an eigenstate of the observable corresponding to this quantity. This is an "objective" criterion since it allows us to infer the existence of a definite (quantum) state of the system to which a value of a physical quantity can be ascribed.

Within this interpretive framework (and without presuming the collapse postulate) decoherence cannot solve the problem of outcomes described in Sect. 2.5.4. This is so because phase coherence between macroscopically different pointer states is preserved in the state that includes the environment, and we can always enlarge the system so as to include (at least parts of) the environment. In other words, the superposition is only enlarged, with coherence being delocalized into the environment ("the interference terms still exist, but they are not *there*" [7, p. 224]), and the process of decoherence could in principle always be reversed. Therefore, if we assume the orthodox

[3]The Copenhagen interpretation, on the other hand, appears to emphasize an epistemic nature of the collapse. Since this interpretation regards the wave function as representing a probability amplitude only, the collapse has the character of a mere "increase of information" rather than that of an actual physical process. See also the discussion in Sect. 2.1.

eigenvalue–eigenstate link to establish the existence of determinate values of physical quantities, decoherence cannot ensure that the system actually ever is in a definite pointer state (unless, of course, the system is initially in an eigenstate of the pointer observable), or that measurements have outcomes at all.

Let us discuss this important issue in some more detail. First note that, with respect to the global system–environment quantum state vector, the interaction with the environment has only led to additional entanglement,

$$|\Psi(t=0)\rangle = \left(\sum_n c_n |s_n\rangle\right) |E_0\rangle \longrightarrow |\Psi(t)\rangle = \sum_n c_n |s_n\rangle |E_n(t)\rangle \,. \qquad (8.1)$$

In some sense, the entanglement brought about by the interaction with the environment could initially even be considered as making the measurement problem, as considered in the context of the von Neumann scheme (see Sect. 2.5), even worse. Bacciagaluppi [385, Sect. 3.2] puts it like this:

> Intuitively, if the environment is carrying out, without our intervention, lots of approximate position measurements, then the measurement problem ought to apply more widely, also to these spontaneously occurring measurements. (...) The state of the object and the environment could be a superposition of zillions of very well localised terms, each with slightly different positions, and which are collectively spread over a macroscopic distance, even in the case of everyday objects. (...) If everything is in interaction with everything else, everything is entangled with everything else, and that is a worse problem than the entanglement of measuring apparatuses with the measured probes.

The rapid decrease of the overlap $\langle E_n(t)|E_{m\neq n}(t)\rangle$ of the relative states of the environment, corresponding to an increase in the distinguishability of these states, becomes only relevant once we consider local measurements in the context of the usual measurement axioms of quantum mechanics, for example, by forming the reduced density matrix of the system corresponding to the state (8.1),

$$\hat{\rho}_S(t) = \sum_{mn} c_m c_n^* |s_m\rangle\langle s_n| \langle E_n(t)|E_m(t)\rangle. \qquad (8.2)$$

As discussed in Sect. 2.7, the rapidly decaying overlap $\langle E_n(t)|E_{m\neq n}(t)\rangle$ means that interference terms $|s_m\rangle\langle s_{n\neq m}|$ become damped. The reduced density matrix is transformed into an approximate, albeit improper, ensemble of the component states $|s_n\rangle$,

$$\hat{\rho}_S(t) \approx \sum_n |c_n|^2 |s_n\rangle\langle s_n|. \qquad (8.3)$$

Recall that the introduction of the reduced density matrix was motivated by the fact that this mathematical object completely encapsulates all measurement statistics of the local system. These statistics are given by the expectation values of all possible local observables $\hat{O}_S$, computed from the trace rule $\langle \hat{O}_S \rangle = \mathrm{Tr}(\hat{\rho}_S \hat{O}_S)$. For all practical purposes and for any local measurement performed on the system only, the statistics generated by the reduced density matrix (8.3) of the system will then be (approximately) the same as those generated by the corresponding proper mixture (ensemble) of pure states. In this purely operational sense, decoherence thus explains why certain interference effects are so difficult to observe, especially in the macroscopic domain (see also Sects. 2.7 and 2.5.3).

However, note that the trace operation is nonunitary and generally motivated by, and interpreted as, an averaging over different outcomes of measurements. As shown in Sect. 2.4.1, the identification of the formal expression $\mathrm{Tr}(\hat{\rho}\hat{O})$ as the expectation value of a quantity represented by the operator $\hat{O}$ relies on the mathematical fact that, when writing out this trace, it is found to be equal to a sum over the possible outcomes of the measurement represented by $\hat{O}$, weighted by the Born probabilities for the system to be "thrown" into a particular state corresponding to each of these outcomes in the course of a measurement. This interpretation, however, already *presumes* that measurements have outcomes and that the Born rule holds. Therefore the trace operation, and thus the concept and interpretation of reduced density matrices, is based on the statistical interpretation and the usual measurement axioms of quantum mechanics. In the words of Pessoa Jr. [64, p. 432], "taking a partial trace amounts to the statistical version of the projection postulate."

Since, in the standard interpretation of quantum mechanics, it is precisely this projection (or collapse) postulate that ensures the existence of outcomes and, consequently, defines when we can assign definite values of physical quantities to systems, this existence of outcomes cannot be *derived* from any formal structure that is obtained by means of the trace rule, such as the reduced density matrix. Once the measurement axioms (and thus the trace rule) are dropped, we are left with a global entangled system–environment state (8.1) that, according to the standard interpretation, does not allow us to say anything about the physical state of the system or to assign a particular outcome (i.e., a definite value of a physical quantity) to the system.

8.1.2 Observables, Measurements, and Environment-Induced Superselection

In the standard and Copenhagen interpretations, the assignment of physical properties is determined by an observable that represents the measurement of a physical quantity and that in turn defines the preferred basis. However, any Hermitian operator can play the role of an observable, and thus any given state has the potential for an infinite number of different properties, whose attribution is usually mutually exclusive unless the corresponding observables commute (in which case they share a common eigenbasis, which preserves the uniqueness of the preferred basis). What, then, determines the observable that is being measured? In the standard and Copenhagen interpretations, it is essentially the "user" who simply "chooses" the particular observable to be measured and thus determines which properties the system possesses.

This positivist point of view has led to a lot of controversy. It runs counter to the notion of an observer-independent reality, which has been at the heart of natural science since its beginning. Moreover, in practice, one certainly does not have the freedom to choose arbitrary observables and measure them. Instead, one has "instruments" (including one's senses) that are designed to measure a particular observable. For most (and maybe all) practical purposes, this will ultimately boil down to a single relevant observable, namely, position. But what makes the instruments designed for a particular observable?

Answering this crucial question essentially means abandoning the orthodox view of treating measurements as a "black-box" process that has little, if any, relation to the workings of actual physical measurements. The formalization of measurements as a formation of quantum correlations between a system and an apparatus goes back to the early years of quantum mechanics and is reflected in the von Neumann measurement scheme (see Sect. 2.5.1). However, as we have discussed in Sect. 2.5.2, it does not resolve the issue of how the choice of observables is made. The second element, the explicit inclusion of the environment in a description of the measurement process, was brought into quantum theory by the studies of decoherence. The stability criterion for pointer states introduced by Zurek [8] means that measurements must be of such a nature as to establish robust records, that is, the system–apparatus correlations ought to be preserved in spite of the inevitable interaction with the surrounding environment (see the discussion in Sect. 2.8). The "user" cannot choose the observables arbitrarily, but must design a measuring device whose interaction with the environment is such as to ensure stable records (which, in turn, defines a measuring device for this observable). In the reading of orthodox quantum mechanics, this can be interpreted as the environment determining the properties of the system.

In this sense, the decoherence program has embedded the rather formal concept of measurement as proposed by the standard and Copenhagen interpretations—with its vague notion of observables that are seemingly freely chosen by the observer—into a more realistic and physically motivated framework. This is accomplished via the specification of observer-free criteria for the selection of the measured observable by taking into account the physical structure of the measuring device and its interaction with the environment. In most cases, the environment is also needed to amplify the measurement record [76] and thereby to make it accessible to the external observer (see also Sect. 2.9).

8.1.3 The Concept of Classicality in the Copenhagen Interpretation

The Copenhagen interpretation additionally postulates that classicality is not to be derived from quantum mechanics, for example, as the macroscopic limit of an underlying quantum structure (as is in some sense assumed, but not explicitly derived, in the standard interpretation). Instead it prescribes that classicality ought to be viewed as an indispensable and irreducible element of a complete quantum theory—and, in fact, be considered as a concept prior to quantum theory. In particular, the Copenhagen interpretation assumes the existence of macroscopic measurement apparatuses that obey classical physics and that are not supposed to be described in quantum-mechanical terms (in stark contrast to the von Neumann measurement scheme). Such classical apparatuses are considered necessary in order to make quantum-mechanical phenomena accessible to us in terms of the "classical" world of our experience. This strict dualism between the system $\mathcal{S}$, to be described by quantum mechanics, and the apparatus $\mathcal{A}$, obeying classical physics, also entails the existence of a fundamental (albeit in principle movable) boundary between $\mathcal{S}$ and $\mathcal{A}$, which separates the microworld from the macroworld (the "Heisenberg cut").

In view of the insights gained from decoherence, it seems impossible to uphold the notion of the existence of a fundamentally classical realm. Environment-induced superselection and suppression of interference have demonstrated how quasiclassical robust states can emerge, or remain absent, using the quantum formalism alone and over a broad range of microscopic to macroscopic scales. Recent experiments (some of which we described in Chap. 6) have established the notion that the boundary between $\mathcal{S}$ and $\mathcal{A}$ is to a large extent movable toward $\mathcal{A}$. Similar results have been obtained from the general study of quantum nondemolition measurements (see, for example, Chap. 19 of [138]), which include the monitoring of a system by its environment. Also note that the Copenhagen view implies that the apparatus is macroscopic by definition, since it is described in classical terms. However, not every apparatus must be macroscopic; the actual "instrument" could well be microscopic. Only the "amplifier," which effectively assures the redundancy of the measurement records (as first emphasized by Zurek [76]; see also his discussion in [16, 321]), must be macroscopic. As an example, consider the decoherence model described in Sect. 2.10. Here the system can be associated with an "instrument" represented by a bistable atom, while the environment plays the role of the amplifier ("bit-by-byte measurement"). Similarly, certain macroscopic measurement devices may be described in ("microscopic") quantum-mechanical terms. Examples include the macroscopic detectors of gravitational waves [69] and the macroscopic resonators in quantum-electromechanical systems used to implement quantum-limited position measurements (see Sect. 6.4.2). Both devices may be treated as quantum-mechanical harmonic oscillators.

Based on the progress already achieved by the decoherence program, it is reasonable to anticipate that decoherence embedded in some additional interpretive structure could lead to a complete and consistent derivation of the (appearance of the) classical world from quantum-mechanical principles. This would make the assumption of intrinsically classical apparatuses (which have to be treated outside of the realm of quantum mechanics) appear as neither a necessary nor a viable postulate. In this way, we can simultaneously acknowledge the correctness of Bohr's notion of the necessity of a classical world and view the classical world as part of and as emerging from a purely quantum-mechanical substrate [67, 104, 385].

8.2 Relative-State Interpretations

The system–observer duality of orthodox quantum mechanics introduces into the theory external "observers" who are not described by the deterministic laws of quantum systems but instead follow a stochastic indeterminism. This approach obviously runs into problems when the universe as a whole is considered, since by definition there cannot be any external observers.

The central idea of relative-state interpretations, first described (but not worked out in detail) by Hugh Everett in the late 1950s [140] and subsequently further developed by many authors, is to abandon this system–observer duality. Instead, one (i) assumes the existence of a total quantum state $|\Psi\rangle$ representing the physical state of the entire universe and (ii) upholds the universal validity of the Schrödinger evolution. In addition, one (iii) postulates that at the completion of a measurement all terms in the expansion of the total state in the eigenbasis of the measured observable correspond to (physical) states in some sense, that is, no particular "outcome" is singled out, neither formally nor physically. Each of these (physical) states can be understood as relative (a) to the state of the other part in the composite system (as in Everett's original proposal; see also [141, 143]), (b) to a particular "branch" of a constantly "splitting" universe (the *many-worlds interpretation*, popularized by DeWitt [386] and Deutsch [387]), or (c) to a particular "mind" in the set of minds of the conscious observer (the *many-minds interpretation*; see, for example, [388]).

Relative-state interpretations face two main difficulties. First, the preferred-basis problem: If states are only relative, the question arises, relative to what? What determines the particular basis terms that are used to define the branches, which in turn specify the relative properties, worlds, or minds in the next instant of time? When precisely does the "splitting" occur? Which properties are made determinate in each branch, and how are they connected to the determinate properties of our experience? Second, what is the meaning of probabilities, given that every possible outcome "occurs" in some sense, and how can Born's rule be motivated in such an interpretive framework? As we will describe below, proponents of relative-state interpretations have frequently appealed to decoherence in solving these difficulties (see, for example, [387, 389–396]).

In turn, the pioneers of the decoherence program, such as Zeh and Zurek, have often made use of ideas of Everett's relative-state framework (see, for instance, [4, 5, 104, 397]), presumably because the Everett approach takes unitary quantum mechanics essentially "as is," with a minimum of added interpretive elements. This matches well the spirit of the decoherence program, which attempts to explain the emergence of classicality purely from the formalism of basic quantum mechanics. Zeh adheres to an Everett-style branching to which distinct observers are attached [397] (see also Sect. 9.4). Zurek has employed relative states as a useful concept to clearly bring out the preferred-basis problem [8] and to emphasize the importance and role of stable measurement records [84, 104] (see Zurek's "existential interpretation" described in Sect. 8.2.3 below).

8.2.1 Everett Branches and the Preferred-Basis Problem

Stapp [398, p. 1043] stated the requirement that "a many-worlds interpretation of quantum theory exists only to the extent that the associated basis problem is solved." In the context of relative-state interpretations, the preferred-basis problem is not only much more severe than in the orthodox interpretation, but also more fundamental for at least two reasons:

1. The branching occurs continuously and essentially everywhere. If measurements are understood in the general sense of the formation of quantum correlations, every newly formed correlation, whether it pertains to microscopic or macroscopic systems, corresponds to a branching.
2. The ontological implications are much more drastic, at least in those relative-state interpretations which assume an actual "splitting" of worlds or minds, since the choice of the basis determines the resulting "world" or "mind" as a whole.

The environment-induced superselection criteria of the decoherence program have frequently been employed to solve the preferred-basis problem of relative-state interpretations (see, for example, [104, 395, 396, 399]). A decoherence-based approach to selecting the preferred Everett bases has several advantages. First, no *a priori* existence of a preferred basis needs to be postulated, but instead the preferred basis arises dynamically from the physical criterion of robustness. Second, the selection will be empirically adequate, since the decoherence program is derived solely from the well-confirmed Schrödinger dynamics (modulo the possibility that robustness may not be the universally valid selection criterion). Lastly, the evolving decohered components of the wave function can be reidentified over time (forming "trajectories" in the preferred state spaces) and thus can be used to define stable, temporally extended Everett branches. Similarly, such trajectories can be associated with robust record states of observers and with environmental states that make information about the state of the system accessible to many observers (see Sect. 2.9 and also Sect. 8.2.3 below).

The approach of using environment-induced superselection and decoherence to define the Everett branches has been criticized on grounds of being "conceptually approximate," since environment-induced superselection generally leads to an only approximate specification of a preferred basis (see Sect. 2.8.3) and therefore cannot give an "exact" definition of the Everett branches (see, for example, the comment of Kent [400], and also Bell's essay [401]). Wallace [396, pp. 90–91] has argued against such an objection as

> (...) arising from a view implicit in much discussion of Everett-style interpretations: that certain concepts and objects in quantum mechanics must either enter the theory formally in its axiomatic structure, or be regarded as illusion. (...) [Instead] the emergence of a classical world from quantum mechanics is to be understood in terms of the emergence from the theory of certain sorts of structures and patterns, and ... this means that we have no need (as well as no hope!) of the precision which Kent [400] and others (...) demand.

Indeed, it is reasonable to assert that there is no *a priori* reason to doubt that an "approximate" criterion for the selection of the preferred basis can give a meaningful definition of the Everett branches—one that is empirically adequate and that accounts for our experiences. The environment-superselected basis emerges from the physically very reasonable criterion of robustness, together with the purely quantum-mechanical effect of decoherence. It would be rather difficult to imagine how an axiomatically introduced "exact" rule could be able to select preferred bases in a manner that is similarly physically motivated and capable of ensuring empirical adequacy.

Besides using the environment-superselected pointer states to describe the Everett branches, various authors have also directly used the instantaneous Schmidt decomposition of the composite state (or, equivalently, the set of orthogonal eigenstates of the reduced density matrix) to define the preferred basis (we discussed the Schmidt decomposition in Sect. 2.15.1). This approach is easier to implement than the explicit search for dynamically stable pointer states, since the preferred basis follows directly from a simple mathematical diagonalization procedure at each instant of time. Furthermore, it gives an "exact" rule for basis selection in relative-state interpretations. The quantum origin of the Schmidt decomposition, which matches well the "pure quantum mechanics" spirit of Everett's proposal (where the formalism of quantum mechanics supplies its own interpretation), has also been counted as an advantage [147]. In an earlier work, Deutsch [387] attributed a fundamental role to the Schmidt decomposition in relative-state interpretations as defining an "interpretation basis" which imposes the precise structure that is needed to give meaning to Everett's basic concept. However, as pointed out in Sect. 2.15.1, basis states obtained from the instantaneous Schmidt decomposition will frequently have properties that are very different from those selected by the stability criterion and that are undesirably nonclassical. For example, they may lack the spatial localization of the robustness-selected Gaussians [398].

The question to what extent the Schmidt basis states correspond to classical properties in Everett-style interpretations was investigated in detail by Barvinsky and Kamenshchik [147]. The authors compared the states selected by the Schmidt decomposition to coherent states (i.e., minimum-uncertainty Gaussians), where the latter were chosen as the "yardstick states" representing classicality. As we have seen in Sect. 5.2.5, such coherent states emerge as the preferred states, for example, in the model for quantum Brownian motion. For the investigated models, Barvinsky and Kamenshchik found that only subsets of the Everett branches defined by the Schmidt decomposition exhibit classicality in the sense of coherent states. Furthermore, the degree of classicality of these branches is very sensitive to the choice of the initial state and the interaction Hamiltonian, such that classicality emerges typically only temporarily, and the Schmidt basis generally lacks robustness under time evolution. Similar difficulties with the Schmidt-basis approach have been reported by Kent and McElwaine [148].

8.2.2 Probabilities in Relative-State Interpretations

The question of the origin and meaning of probabilities in a relative-state interpretation based solely on a deterministically evolving global quantum state, and the problem of how to consistently derive Born's rule in such a framework, has been the subject of much discussion and criticism directed at this type of interpretation. Early approaches that aimed at an understanding of probabilities in the relative-state framework in terms of relative frequencies (see, e.g., [140, 386, 402–404]) have been shown to be circular [400, 405, 406].

Initially, decoherence was thought to provide a natural account of the probability concept in a relative-state framework. The idea was to relate the diagonal elements of the decohered reduced density matrix to the collection of possible "events" and to interpret the corresponding coefficients as relative frequencies of branches [104, 407]. Since decoherence enables one to reidentify the individual localized components of the wave function over time (describing, for example, observers and their measurement outcomes attached to well-defined branches; see also Sect. 9.4), this leads to an interpretation of the Born probabilities as empirical frequencies. However, this argument cannot yield a noncircular derivation of the Born rule, since the formalism and interpretation of reduced density matrices presume this rule (see our discussion in Sects. 2.4.1, 2.4.6 and 8.1.1). Attempts to derive probabilities from reduced density matrices are therefore circular [16, 408].

A derivation that is based on the nonprobabilistic axioms of quantum mechanics and on elements of classical decision theory has been presented by Deutsch [407] (see also the critique by Barnum et al. [409] and the subsequent defense by Gill [410] and Wallace [411]; Saunders [412] embedded Deutsch's derivation into an operational framework). However, it is important to realize that such decision-theoretic approaches are subject to the same charge of circularity pointed out in the previous paragraph. This is so because these approaches first need to define the "classical" events (outcomes) to which probabilities are to be assigned. If one starts from pure states of the form $|\psi\rangle = \sum_n \mathrm{e}^{\mathrm{i}\varphi_n} |\phi_n\rangle$, as Deutsch's derivation does, one would need to (i) justify the identification of the states $|\phi_n\rangle$ with the possible events, and (ii) show that the phase relations φ_n between these states are irrelevant, i.e., do not influence the "decision" of the observer. The approach of Deutsch does not address these issues. Thus it tacitly uses environment-induced superselection (which selects the set of events that can be observed) and decoherence (which explains the irrelevance of phase relations from the view of the local observer), while it fails to supply a derivation of these processes in a manner that does not presume Born's rule.

The solution to the problem of understanding the meaning of probabilities and of deriving Born's rule in a relative-state framework must therefore be sought on a much more fundamental level of quantum mechanics. Quantum information theory has established the notion that quantum mechanics can be viewed as a description of what, and how much, "information" nature is willing to proliferate [56]. For example, a peculiar feature of quantum mechanics is that complete knowledge of a global pure bipartite quantum state $|\Psi\rangle = (|a_1\rangle |b_1\rangle + |a_2\rangle |b_2\rangle) / \sqrt{2}$ does not appear to contain any information about the "absolute" state of one of the subsystems. This hints at ways how a concept of "objective ignorance," and therefore of objective probabilities, may emerge directly from the quantum feature of entanglement without any classical counterpart. This idea has recently been developed under the heading of *environment-assisted invariance*, or *envariance* for short, in a series of papers by Zurek [16, 65–67]. The motivation and spirit of this approach is strongly based on decoherence. Envariance and Zurek's derivation of the Born rule have been discussed further by Schlosshauer and Fine [413], Barnum [414], and Mohrhoff [415]. Given a set of assumptions, envariance leads to a derivation of quantum probabilities and Born's rule. We shall outline this promising approach in the following.

Zurek's derivation is based on a study of the properties of a composite entangled state and therefore intrinsically requires the decomposition of the Hilbert space into subsystems and the usual tensor-product structure. Zurek considers a bipartite product Hilbert space $\mathcal{H}_\mathcal{A} \otimes \mathcal{H}_\mathcal{B}$ and a completely known composite pure state $|\Psi\rangle$ written in the diagonal Schmidt decomposition [see (2.127)]

$$|\Psi\rangle = \frac{1}{\sqrt{2}} \left(e^{i\varphi_1} |a_1\rangle |b_1\rangle + e^{i\varphi_2} |a_2\rangle |b_2\rangle \right). \tag{8.4}$$

Here $\{|a_1\rangle, |a_2\rangle\}$ and $\{|b_1\rangle, |b_2\rangle\}$ are sets of orthonormal basis vectors that span the Hilbert spaces $\mathcal{H}_\mathcal{A}$ and $\mathcal{H}_\mathcal{B}$, respectively. Then the core result to be established by Zurek's derivation is to show that the probabilities of obtaining either one of the relative states $|a_1\rangle$ and $|a_2\rangle$ (identified by Zurek with the "events" of interest to which probabilities are to be assigned [66, p. 12]; see also the discussion in [413]) are equal. Given this result, generalizations to higher-dimensional Hilbert spaces and to the case of unequal absolute values of the Schmidt coefficients in (8.4) can be achieved by means of a counting argument [67].

The result is arrived at in two key steps. First, a few simple and quite natural assumptions (called "facts" by Zurek [67]) are introduced that relate the global quantum state vector (8.4) to properties of the "state of the system $\mathcal{A}$." This is necessary because the global quantum state of the composite system is all that the pure state-vector formalism of quantum mechanics provides for the description of two entangled subsystems. More generally, the relative-state framework presumes nothing besides the global unitarily evolving state vector, which usually contains a high degree of environmental entanglement. There does not exist a quantum state vector that could be assigned to one of the subsystems alone, and (as discussed above) we must not use reduced density matrices to describe the local system. Zurek's "facts" can be stated follows [67]:

(i) The state of $\mathcal{A}$ is completely determined by the global quantum state (8.4).
(ii) The state of $\mathcal{A}$ specifies all measurable properties of $\mathcal{A}$, including probabilities of outcomes of measurements on $\mathcal{A}$.
(iii) Unitary transformations can change the state of $\mathcal{A}$ only if they act on $\mathcal{A}$. That is, when the transformation has the form $\hat{I}_{\mathcal{A}} \otimes \hat{U}_{\mathcal{B}}$, the state of $\mathcal{A}$ remains the same. This amounts to a "no-signaling" assumption, i.e., to the assumption that entanglement cannot be used to send instantaneous "messages" between subsystems (see also [413,414] for discussions of this assumption).

Granted these three assumptions, Zurek shows that measurable properties of $\mathcal{A}$ can depend neither

(1) on the phases φ_i in (8.4), such that we can assume the simplified form

$$|\Psi\rangle = \frac{1}{\sqrt{2}} \left(|a_1\rangle |b_1\rangle + |a_2\rangle |b_2\rangle \right) \tag{8.5}$$

for our purpose of discussing probabilities associated with $\mathcal{A}$;

(2) nor on whether $|a_1\rangle$ is paired with $|b_1\rangle$ or with $|b_2\rangle$, i.e., the unitary transformation acting on $\mathcal{A}$ that changes the quantum state vector

$$|\Psi\rangle = \frac{1}{\sqrt{2}} \left(|a_1\rangle |b_1\rangle + |a_2\rangle |b_2\rangle \right) \tag{8.6}$$

into

$$|\Psi'\rangle = \frac{1}{\sqrt{2}} \left(|a_2\rangle |b_1\rangle + |a_1\rangle |b_2\rangle \right) \tag{8.7}$$

cannot have altered the state of $\mathcal{A}$.

In a way, result (2) already indicates a feature of ignorance about the state of $\mathcal{A}$, since interchanging the potential "outcomes" $|a_i\rangle$ through local operations performed on $\mathcal{A}$ does not change any measurable properties of $\mathcal{A}$. This feature may be viewed as leading to a form of "objective indifference" among the $|a_i\rangle$. It is important to note that this effect is crucially dependent on the feature of entanglement. In a nonentangled pure state of the form $|\psi\rangle = \left(|a_1\rangle + e^{i\varphi} |a_2\rangle\right) / \sqrt{2}$, the phase φ must of course not be ignored (and would be measurable in a suitable interference experiment), and therefore the system described by the "swapped" state vector $|\psi'\rangle = \left(|a_2\rangle + e^{i\varphi} |a_1\rangle\right) / \sqrt{2}$ is clearly physically different from that represented by the original state vector $|\psi\rangle$.

To make the above argument more precise, the second key step of the derivation explicitly connects the notion of probabilities of the outcomes $|a_i\rangle$ in a measurement performed on $\mathcal{A}$ (previously only subsumed under the general heading of "measurable properties of $\mathcal{A}$") to the global state vector via an additional assumption. In [67], Zurek offers three possible choices for this assumption, of which we should quote only one (see also [414]). Namely, it is assumed that the form of the Schmidt product states $|a_i\rangle |b_i\rangle$ appearing in (8.4) implies a perfect correlation between the Schmidt "partners," i.e., that the detection of the Schmidt state $|a_i\rangle$ implies detection of its partner $|b_i\rangle$ with certainty. This leads one to conclude that the probabilities for $|a_i\rangle$ and $|b_i\rangle$ must be equal. Given this assumption and using result (2) above, it can be readily established (see [67, 413]) that the probabilities for $|a_1\rangle$ and $|a_2\rangle$ must be equal, thus completing the derivation.

The need for the final assumption may be considered a reflection of the well-worn phrase that a transition from a nonprobabilistic theory (such as quantum mechanics solely based on deterministically evolving state vectors) to a probabilistic theory (that refers to "probabilities of outcomes of local measurements") requires, at some stage, to "put probabilities in to get probabilities out" [413]. However, in the quantum setting, this introduction of the probability concept relies only on a special case of probability—namely, certainty—and has a far more objective character than in the classical setting. While in the latter case probabilities refer to subjective ignorance in spite of the existence of an underlying well-defined physical state, in the quantum case all that is available, namely, the global entangled quantum state, is perfectly known. The objectivity of ignorance in quantum mechanics can thus be viewed as a consequence of a form of "complementarity" between local and global observables [67] and helps explain the fundamental need for a probabilistic description in the quantum setting despite the deterministic evolution of the global state vector.

It is the great merit of Zurek's proposal to have emphasized this objective character of quantum probabilities arising from the feature of quantum entanglement. On the basis of the above assumptions and the resulting derivation of Born's rule, Zurek [67] has also shown how key elements of the decoherence program, such as the environment-induced superselection of pointer states, can be rederived using the framework of envariance alone, without resorting to a description in terms of reduced density matrices. The importance of such an approach lies in the fact that, as discussed in several places of this book, the usual formalism of decoherence relies fundamentally on reduced density matrices and thus on the usual measurement axioms of quantum mechanics, in particular on Born's rule. It is fair to say that this reliance has been a key challenge to the development of a decoherence-based no-collapse interpretation of quantum mechanics that derives the measurement postulates as emerging in an effective manner from consequences of environmental entanglement alone.

8.2.3 The "Existential Interpretation"

A relative-state interpretation that relies heavily on decoherence has been proposed by Zurek [84, 104] (see also the recent reevaluation in [67]). This approach, termed the "existential interpretation," defines the reality, or objective existence, of a state as the possibility of finding out what the state is and simultaneously leaving it unperturbed, similar to a classical state. Zurek assigns a "relative objective existence" to the environment-superselected robust states. By measuring properties of the system–environment interaction Hamiltonian, the observer could, at least in principle, determine the set of observables that can be measured on the system without perturbing it and thus find out the "objective" state of the system. What actually happens, of course, is that the observer takes advantage of the redundant records of the state of the system encoded in the environment (see Sect. 2.9). By intercepting parts of this environment, the observer can determine the state of the system essentially without perturbing it [16, 66, 110, 321]).

Zurek emphasizes the importance of stable records for observers, i.e., of robust correlations between the environment-selected states and the memory states of the observer. Information must be represented physically [323], and thus the "objective" state of the observer who has detected one of the potential outcomes of a measurement must be physically distinct and objectively different from the state of an observer who has recorded an alternative outcome (since the record states can be determined from the outside without perturbing them—see the previous paragraph). The different objective states of the observer are, via quantum correlations, attached to different branches defined by the environment-selected robust states; they thus ultimately label the different branches of the universal state vector. This is claimed to lead to the perception of classicality. The impossibility of perceiving arbitrary superpositions is explained via the rapid decoherence-induced suppression of interference between different memory states, where each (physically distinct) memory state represents an individual observer identity (see also Chap. 9).

Recently, Zurek has connected the existential interpretation to his envariance program and his derivation of the Born rule (see the previous Sect. 8.2.2) [67]. The derivation can be recast in the framework of the existential interpretation such that probabilities refer explicitly to the future record state of an observer. This concept of probability bears similarities with classical probability theory (for more details on these ideas, see [67]).

8.3 Modal Interpretations

The first type of modal interpretation was suggested by van Fraassen [416, 417] based on his program of "constructive empiricism," which proposes to take only empirical adequacy, but not necessarily "truth," as the goal of science. Since then, a large number of interpretations of quantum mechanics have been suggested that can be considered as modal (for a review and discussion of some of the basic properties and problems of such interpretations, see [418]).

In general, the idea of modal interpretations is to weaken the orthodox eigenvalue–eigenstate link by allowing for the assignment of definite measurement outcomes even if the system is not in an eigenstate of the observable representing the measurement. In this way, one can preserve a purely unitary time evolution *and* account for definite measurement results without the need for an additional collapse postulate. Of course, this immediately raises the question of how physical properties perceived through measurements and measurement results are connected to the quantum state. The general goal of modal interpretations is therefore to specify rules that determine a catalog of possible properties of a system described by the density matrix $\hat{\rho}(t)$. Two different views are typically distinguished, namely, a "semantic approach" that only changes the way of talking about the connection between properties and state, and a "realistic view" that provides a different specification of what the possible properties of a system really are, given the state vector (or the density matrix).

Such an attribution of possible properties must fulfill certain requirements. For instance, probabilities for outcomes of measurements should be consistent with the usual Born probabilities of standard quantum mechanics. It should also be possible to recover our experience of classicality at the level of macroscopic objects. And finally, an explicit time evolution of properties and their probabilities should be definable that is consistent with the results of the Schrödinger equation. As we shall see in the following, decoherence has frequently been used to motivate and define rules for property assignment in modal interpretations. Dieks [419,420] even suggested that one of the central goals of modal approaches is to provide an interpretation of decoherence.

8.3.1 Property Assignment Based on Environment-Induced Superselection

The intrinsic difficulty of modal interpretations is to avoid any *ad hoc* character of the property assignment, yet to find generally applicable rules that lead to a selection of possible properties that include the determinate properties of our experience. To solve this problem, various modal interpretations have embraced the results of the decoherence program. A natural approach would be to employ the environment-induced superselection of preferred bases to define sets of possible quasiclassical properties associated with the correct probabilities. This approach would be based on an entirely physical and very general selection criterion (namely, the stability criterion) and has, for the cases studied, been shown to give results that agree well with our experience, thus matching van Fraassen's goal of empirical adequacy.

Furthermore, since the decoherence program is based solely on Schrödinger dynamics, the task of defining a time evolution of the "property states" and their associated probabilities which is in agreement with the results of unitary quantum mechanics would presumably be easier than in a model of property assignment in which the set of possibilities does not arise dynamically via the Schrödinger equation alone (for a detailed proposal for modal dynamics of the latter type, see [421]). The need for explicit dynamics of property states in modal interpretations is controversial. One can argue that it suffices to show that at each instant of time, the set of possibly possessed properties that can be assigned to the system is empirically adequate, in the sense that it contains the properties of our experience, especially with respect to the properties of macroscopic objects (this is essentially the view of, for example, van Fraassen [416, 417]). On the other hand, this cannot ensure that these properties behave over time in agreement with our experience (for instance, that macroscopic objects which are left undisturbed do not spontaneously change their position in space in an observable manner). In other words, the emergence of classicality is to be tied not only to determinate properties at each instant of time, but also to the existence of quasiclassical "trajectories" in property space. Since decoherence allows one to reidentify components of the decohered density matrix over time, this could be used to derive property states with a continuous, quasiclassical trajectory-like time evolution based on Schrödinger dynamics alone. For discussions of this approach, see [421, 422].

8.3.2 Property Assignment Based on Instantaneous Schmidt Decompositions

Since it is often rather difficult to determine explicitly the robust pointer states in more complicated models (see the predictability-sieve approach described in Sect. 2.8.3), the problem arises of how to specify a simple yet general rule for property assignment based on environment-induced superselection that is easy to apply to concrete cases of interest. To simplify this situation, several modal interpretations have restricted themselves to the orthogonal decomposition of the density matrix to define the set of properties that can be assigned (see, for instance, [68, 423–426]).

For example, the approach of Dieks [425] recognizes, by referring to the decoherence program, the relevance of the environment by considering a composite system–environment state vector and its diagonal Schmidt decomposition, $|\Psi\rangle = \sum_k \sqrt{p_k}\, |s_k\rangle\, |e_k\rangle$, which always exists (see Sect. 2.15.1). Possible properties that can be assigned to the system are then represented by the Schmidt projectors $|s_k\rangle\langle s_k|$. Although all terms are present in the Schmidt expansion (which Dieks calls the "mathematical state"), the "physical state" is postulated to be given by only one of the terms, with probability p_k. A generalization of this approach to a decomposition into any number of subsystems has been described by Vermaas and Dieks [426]. In this sense, the Schmidt decomposition itself is taken to define an interpretation of quantum mechanics. Dieks [427] suggested a physical motivation for the Schmidt decomposition in modal interpretations based on the assumed requirement of a one-to-one correspondence between the properties of the system and its environment. (For a comment on the violation of the property composition principle in such interpretations, see the analysis by Clifton [418].)

However, as discussed in Sects. 2.15.1 and 8.2.1, the states selected by the (instantaneous) orthogonal decomposition of the reduced density matrix will in general differ from the robust environment-superselected states and may have distinctly nonclassical properties. That this will be the case especially when the states selected by the orthogonal decomposition are close to degeneracy (as it is often the case for macroscopic systems with many degrees of freedom) has already been shown in Sect. 2.15.1. This issue has also been explored in more detail in the context of modal interpretations by Bacciagaluppi [428] and Donald [429], who showed that in the case of near-degeneracy, the resulting projectors will be extremely sensitive to the precise form of the state [428]. Clearly such sensitivity is undesired, since the projectors, and thus the properties of the system, will not be well-behaved under the inevitable approximations employed in physics [429].

8.3.3 Property Assignment Based on Decompositions of the Decohered Density Matrix

Other authors have therefore used the orthogonal decomposition of the decohered reduced density matrix (instead of the decomposition of the instantaneous density matrix), which has led to noteworthy results. When the system is represented by an only finite-dimensional Hilbert space, the resulting states were indeed found to be typically close to the robust states selected by the stability criterion, unless again the final composite state was close to degeneracy [430, 431]. Thus, in sufficiently nondegenerate cases, decoherence can ensure that the definite properties selected by modal interpretations of the Dieks type will be reasonably close to the properties corresponding to the ideal pointer states, provided the modal properties are based on the orthogonal decomposition of the decohered reduced density matrix.

On the other hand, Bacciagaluppi [432] showed that, in the case of an infinite-dimensional state space of the system, the predictions of the modal approach [425, 426] and those of decoherence can differ significantly. Using the scattering model of Joos and Zeh [7] described in Chap. 3, it was demonstrated that the definite properties obtained from the orthogonal decomposition of the decohered density matrix were highly delocalized (that is, smeared out over the entire spread of this matrix), although the coherence length of the density matrix itself was shown to be very small, so that decoherence indicated localized properties. Thus, based on these results (and similar ones of Donald [429]), decoherence can be used to argue for the physical inadequacy of the rule for the assignment of definite properties proposed by Dieks [425] and Vermaas and Dieks [426].

8.4 Physical Collapse Theories

The basic idea of physical collapse theories is to introduce an explicit modification of the Schrödinger time evolution to achieve a physical mechanism for state-vector reduction (for an extensive review, see [433]). This is in general motivated by a "realist" interpretation of the state vector, that is, the state vector is directly identified with a physical state. This assumption is then seen as requiring the reduction of a superposition state onto one of its components to establish equivalence to the observed determinate properties of physical states (at least as far as the macroscopic realm is concerned). Physical collapse models are not only motivated by the problem of the nonobservability of macroscopic interference effects, but moreover also by the conceptual goal of resolving a felt "weirdness" in the existing quantum theory [434] as exemplified by Schrödinger's cat paradox. Since such models lead to an objective reduction of the wave function of a system, they allow for the assignment of a pure quantum state, i.e., a definite wave function, to the system at (almost) all times.

The first proposals for theories of this type go back to Pearle [435–437] and Gisin [194], who developed models that modify the unitary dynamics such that a superposition of quantum states evolves continuously into one of its terms (see also the review by Pearle [438]). Typically, terms representing external white noise are added to the Schrödinger equation, causing the squared amplitudes $|c_n(t)|^2$ in the state-vector expansion $|\Psi(t)\rangle = \sum_n c_n(t)\,|\psi_n\rangle$ to fluctuate randomly in time, while maintaining the normalization condition $\sum_n |c_n(t)|^2 = 1$ for all t. Eventually one amplitude $|c_n(t)|^2$ approaches a value of unity, while all other squared coefficients decay to zero (the "gambler's ruin game"), where $|c_n(t)|^2 \longrightarrow 1$ with probability $|c_n(t=0)|^2$ to ensure agreement with the predictions of the Born rule. Such models are known under the heading of *stochastic dynamical reduction.*

These early models exhibit two main difficulties. First, they suffer from the preferred-basis problem. What determines the terms in the state-vector expansion onto which the state vector gets reduced? Why does reduction lead to the distinct macroscopic states of our experience and not superpositions thereof? Second, how can one account for the empirical fact that the effectiveness of the collapse seems to increase from microscopic to macroscopic scales?

These problems motivated *spontaneous localization models*, first proposed by Ghirardi, Rimini, and Weber (henceforth GRW) [439]. Here state-vector reduction is not implemented as a dynamical process (i.e., as a continuous evolution over time), but instead occurs instantaneously and spontaneously, leading to a spatial localization of the wave function. To be precise, the N-particle wave function $\psi(\mathbf{x}_1, \ldots, \mathbf{x}_N)$ is at random intervals multiplied by a Gaussian of the form $\exp\left[-(\mathbf{X} - \mathbf{x}_k)^2/2\Delta^2\right]$ (this process is often called a "hit" or a "jump"), and the resulting product is subsequently normalized. The occurrence of these hits is not explained, but simply postulated as a new fundamental physical mechanism. Both the coordinate $\mathbf{x}_k$ and the center $\mathbf{X}$ of the Gaussian hit are chosen at random, but the probability of a specific $\mathbf{X}$ is postulated to be given by the squared inner product of $\psi(\mathbf{x}_1, \ldots, \mathbf{x}_N)$ with the Gaussian (and therefore hits are more likely to occur where $|\psi|^2$, viewed as a function of $\mathbf{x}_k$ only, is large).

The mean frequency ν of hits for a single microscopic particle is chosen so as to effectively preserve unitary time evolution for microscopic systems, while ensuring that for macroscopic objects composed of a very large number N of particles the localization occurs rapidly (on the order of $N\nu$) in order to preclude the persistence of nonclassical spatial superpositions on timescales shorter than realistic observations could resolve. In their original paper [439], GRW chose $\nu \approx 10^{-16}$ s^{-1}, and thus a macroscopic system containing on the order of 10^{23} particles would undergo localization on average every 10^{-7} s. Inevitable coupling to the environment can in general be expected to lead to a further drastic increase of N and therefore to an even higher localization rate. Note, however, that the localization process itself is independent of any interaction with the environment.

Subsequently, the ideas of stochastic dynamical reduction and the GRW theory were combined into *continuous spontaneous localization models* [440, 441]. Here localization of the GRW type can be shown to emerge from a nonunitary, nonlinear Itô stochastic differential equation, namely, the Schrödinger equation augmented by spatially correlated Brownian motion terms (see also [191, 442]). The particular choice of the stochastic terms determines the preferred basis. Frequently, these terms have been based on the mass density, which yields spatial localization similar to the GRW model [440–442]. Stochastic terms driven by the Hamiltonian, leading to a reduction in the energy basis, have also been studied [443–452].

8.4.1 The Preferred-Basis Problem

Physical reduction theories typically remove wave-function collapse from the restrictive context of the orthodox interpretation (where the external observer arbitrarily selects the measured observable and thus determines the preferred basis). Instead, they understand reduction as a universal mechanism that acts constantly on every state vector regardless of an explicit measurement situation. It is thus particularly important to provide a definition for the states onto which the wave function collapses.

As mentioned before, the original stochastic dynamical reduction models suffered from this preferred-basis problem. Taking into account environment-induced superselection of a preferred basis could help resolve this issue. Since decoherence occurs on extremely short timescales (especially for mesoscopic and macroscopic objects), it would presumably be able to bring about basis selection much faster than the time required for dynamical fluctuations to establish a “winning” expansion coefficient.

By contrast, the GRW theory solves the preferred-basis problem by postulating a mechanism that leads to spatial localization. That is, position is here assumed to be the universal preferred basis. On the one hand, decoherence supplies the physical motivation for this basis selection (see Sect. 2.8.4). On the other hand, however, by restricting the reduction mechanism to spatial localization, GRW bypasses the question of why certain systems are observed to be in robust states which are not necessarily spatially localized. For example, microscopic systems are usually found in energy eigenstates, and SQUIDs are described by states of supercurrents flowing in specific directions (see Sect. 6.3). In the case of microscopic systems, the particular choice of the parameters of the localization process in the GRW model means that these systems remain essentially unaffected by the reduction mechanism. Similarly, in the case of SQUIDs, the GRW mechanism would only result in a small reduction of the supercurrent below the detectable level due to a breaking-up of Cooper pairs, but not in an approximate reduction onto one of the persistent-current states [433, 453, 454].

However, given that these systems are effectively exempt from the GRW collapse, what makes it then so difficult to observe superpositions of energy eigenstates in microscopic systems, or superpositions of supercurrent states in SQUIDs? Of course, one may argue that essentially all our observations must be grounded in a position measurement,[4] and that thus the GRW mechanism

[4]This measurement may ultimately occur only in the brain of the observer; see the objection to the GRW model by Albert and Vaidman [455]. With respect to the general preference for position as the basis of measurements, Bell [401] once said that “in physics the only observations we must consider are position observations, if only the positions of instrument pointers.”

will lead to an indirect reduction of such superpositions through the coupling to measurement devices, which are usually macroscopic and thus susceptible to the localization mechanism. However, such an argument introduces a dependence on measurements for the "world as we perceive it" to emerge. This runs counter to the basic idea of physical collapse theories, namely, that quantum states directly represent a physical, observer-independent reality that does not require measurements at a fundamental level. In comparison, environment-induced superselection constitutes a much more general approach to the problem of the preferred basis, by explaining the emergence of a range of preferred robust observables on the basis of the physical properties of the relevant system–environment interactions.

A similar argument can be made with respect to continuous spontaneous localization models. Here, one essentially preselects a preferred basis through the particular choice of the stochastic terms added to the Schrödinger equation. This allows for a greater range of possible preferred bases, for instance by combining terms driven by the Hamiltonian and by the mass density, leading to a competition between localization in energy and position space (corresponding to the two most frequently observed preferred bases). Nonetheless, any particular choice of terms will again be subject to the charge of possessing an *ad hoc* flavor.

8.4.2 Simultaneous Presence of Decoherence and Spontaneous Localization

Since decoherence will always be present, the assumption that a physical collapse theory holds means that the evolution of a system will be guided by both decoherence effects and the collapse mechanism. Let us first consider the situation in which decoherence and the collapse mechanism act constructively in the same direction, i.e., toward a common preferred basis. This raises the question in which order these two effects influence the evolution of the system [385]. If the collapse occurs on a shorter timescale than the environment-induced superselection of a preferred basis and the suppression of local interference, decoherence will in most cases have very little influence on the evolution of the system, since typically the system will already have evolved into a reduced state. Conversely, if decoherence acts more quickly on the system than the localization mechanism, the interaction with the environment would lead to the preparation of quasiclassical robust states that are subsequently chosen by the localization mechanism. As demonstrated in Chap. 3, decoherence usually occurs on extremely short timescales and can be shown to be typically significantly faster than the action of the reduction mechanism (for such studies related to the GRW model, see [166,456]). This indicates that decoherence will typically play an important role even in the presence of physical wave-function reduction.

The second case corresponds to the situation in which decoherence leads to the selection of a different preferred basis than the basis specified by the collapse mechanism. As remarked by Bacciagaluppi [385] in the context of the GRW theory, one might then imagine the collapse either to occur only at the level of the environment (which would thus serve as an amplifying and recording device with different localization properties than the system under study), or to lead to an explicit competition between decoherence and collapse effects.

8.4.3 The Tails Problem

The clear advantage of physical collapse models over the consideration of decoherence-induced effects alone for a solution to the measurement problem lies in the fact that collapse models achieve an actual quantum-state reduction. Thus one may be tempted to conclude that at the conclusion of the reduction process the system actually is in a determinate state. However, all collapse models lead to an only approximate reduction of the wave function. In the case of dynamical reduction models, the state will always retain small interference terms for finite times. Similarly, in the GRW theory the width Δ of the multiplying Gaussian cannot be made arbitrarily small, and therefore the reduced wave packet cannot become perfectly localized in position space, since this would entail an infinitely large energy gain by the system due to the time–energy uncertainty relation, which would certainly show up experimentally (in their original paper [439], GRW chose $\Delta \approx 10^{-5}$ cm). This need for an only approximate reduction leads to wave function "tails" [457]. That is, in any region in space and at any time $t > 0$, the amplitude of the wave function will remain nonzero if it had been nonzero at $t = 0$ (before the collapse), and thus there will always be a part of the system that is not "here."

In this sense, collapse models are as much "fine for all practical purposes" (to paraphrase Bell [458]) as decoherence is, where perfect orthogonality of the relative states of the environment is only attained as $t \longrightarrow \infty$. The severity of the consequences, however, is not equivalent for the two strategies. Since collapse models directly change the state vector, a single outcome is at least approximately selected, and it only requires a weakening of the eigenvalue–eigenstate link to make this state of affairs correspond to the (objective) existence of a determinate physical property.[5] In the case of decoherence, the lack of an exact disappearance of local interference terms is not the main problem. Even if exact orthogonality of the relative environmental states were ensured at all times (leading to complete decoherence), the resulting reduced density matrix would still represent an improper mixture. We would therefore have to supply some additional interpretive framework to explain our perception of outcomes (see also the comment by GRW [434]).

[5]It should be noted, however, that such "fuzzy" eigenvalue–eigenstate links may in turn lead to difficulties, as the discussion of Lewis's "counting anomaly" has shown [459].

8.4.4 Connecting Decoherence and Collapse Models

It was realized early that there exists a striking formal similarity of the equations of motion that govern the time evolution of density matrices in the GRW approach and in models of decoherence. For example, the GRW equation for a single free mass point in one dimension reads [439]

$$\mathrm{i}\frac{\partial \rho(x,x',t)}{\partial t} = \frac{1}{2m}\left[\frac{\partial^2}{\partial x'^2} - \frac{\partial^2}{\partial x^2}\right]\rho(x,x',t) - \mathrm{i}\Lambda(x-x')^2\rho(x,x',t), \quad (8.8)$$

where the second term on the right-hand side accounts for the destruction of spatial interference terms. This equation is formally identical, for example, to the master equation (3.78) describing the evolution of the reduced density matrix in the presence of environmental scattering.

Joos [460] used this formal similarity to question the need for an explicit reduction-inducing mechanism, at least with respect to achieving the suppression of spatial coherences over macroscopic distances. Provided the value of the GRW localization parameter Λ appearing in (8.8) is chosen in agreement with the value derived from the physical properties of the relevant system–environment interaction, the time evolution of the reduced density matrix describing the dynamics of the system coupled to a scattering environment on the one hand, and the time evolution of the density matrix of the system under the influence of the GRW collapse on the other hand, will be identical.

Of course, as pointed out also by GRW [434] in response to Joos's comment, it is important to bear in mind the fundamental difference between the approximately proper ensembles arising from collapse models and the improper ensembles resulting from decoherence. The latter ensembles are derived from entangled global system–environment states and therefore do not allow for the assignment of a pure quantum state to the system (see Sects. 2.4.6 and 8.1.1). The reduced density matrix arising from decoherence does not describe the state of the system, but only the statistics of local measurements in the context of the usual measurement axioms of quantum mechanics (with these statistics then explaining the nonobservability of macroscopic interference effects). Therefore, despite the formal similarity between the GRW evolution equation (8.8) and the master equation (3.78) for scattering-induced decoherence, these equations describe the dynamics of two types of density matrices whose conceptual, interpretive, and formal underpinnings must be carefully distinguished.

Finally, the formal similarity of the evolution equations may nourish hopes that the postulated reduction mechanisms of collapse models could possibly be derived from novel (nonunitary) interactions with a universal collapse-inducing "environment." For example, Penrose [461, 462] has suggested that quantum gravity might act as such an "environment" (see also [438, 442]).

8.4.5 Experimental Tests of Collapse Models

Collapse theories postulate deviations from Schrödinger dynamics and could thus be tested in experiments. Several proposals for an experimental detection of the GRW collapse and for the demonstration of potential deviations from the predictions of quantum theory when dynamical state-vector reduction is included have been discussed in the literature (see, e.g., [453, 463–466]). Conversely, one may speculate that the simultaneous presence of both decoherence and reduction effects might allow for an experimental disproof of collapse theories by preparing states that differ in an observable manner from the predictions of the reduction models.

If we acknowledge the existence and feasibility of interpretations of quantum mechanics that only appeal to decoherence in explaining the perception of apparent collapses (see, for example, the "existential interpretation" of Zurek [84, 104] described in Sect. 8.2.3), we will not be able to experimentally distinguish between a "true" collapse and a mere suppression of interference due to decoherence. Instead, an experimental situation is required in which a given collapse model predicts a reduction of the wave function, but in which no (significant) suppression of interference through decoherence arises.

The problem in realizing such experiments is the required shielding of the system from decoherence effects. In fact, it would be very difficult to distinguish collapse effects from decoherence, since the large number of atoms required for the collapse mechanism to be effective also leads to strong decoherence [166, 433, 456]. It would therefore be necessary to isolate the system of interest extremely well from its environment, such that decoherence effects can be neglected in comparison with the environment-independent collapse mechanism. Even in this case it might be difficult to exclude the influence of decoherence due to, for example, thermal emission of radiation, as demonstrated in the case of C_{70} interferometry (see Sect. 6.2 and [266, 269]). Based on explicit numerical estimates, Tegmark [166] has shown that decoherence due to scattering of environmental particles such as air molecules or photons will have a much stronger influence on the evolution of the system than the proposed GRW effect of spontaneous localization (see also [433, 456]; for different results for energy-driven reduction models, see [450]).

The increasing size of physical systems for which interference effects have been experimentally observed imposes bounds on the parameters used in collapse models. However, the current experiments demonstrating mesoscopic and macroscopic interferences are still quite far away from disproving the existing collapse theories. For example, even the rather impressive C_{70} diffraction experiments still fall short of ruling out continuous spontaneous localization models (which lead to the strongest deviations from Schrödinger dynamics among all physical collapse theories) by eleven orders of magnitude [467]. A mirror-superposition experiment recently proposed by Marshall et al. [466], which could lead to a spatial superposition involving $O(10^{14})$ atoms, fails to rule out continuous spontaneous localization models by about six orders of magnitude [466,468]. The superpositions observed in coherent quantum tunneling in SQUIDs and other superconducting qubit systems (see Sect. 6.3) also appear to be compatible with dynamical reduction models. As mentioned above, the spatial localization mechanism is ineffective in bringing about a collapse onto the persistent-current states [433,453,454]. However, given the rapid development of experiments demonstrating quantum superpositions and interference effects on increasingly large scales, it may be only a matter of time when it becomes possible to probe the range relevant to a test of physical reduction models.

8.5 Bohmian Mechanics

David Bohm's approach [35–37] is a modification of de Broglie's original "pilot-wave" proposal [34]. In Bohmian mechanics, a system containing N (nonrelativistic) particles is described by a wave function $\psi(t)$ and the configuration $\mathcal{Q}(t) = (\mathbf{q}_1(t), \dots, \mathbf{q}_N(t)) \in \mathbb{R}^{3N}$ of particle positions $\mathbf{q}_i(t)$. Thus the state of the system is represented by $(\psi, \mathcal{Q})$ at each instant t. The evolution of the system is guided by two equations. The wave function $\psi(t)$ is transformed as usual via the standard Schrödinger equation, while the particle positions $\mathbf{q}_i(t)$ of the configuration $\mathcal{Q}(t)$ evolve according to the *guiding equation*

$$\frac{\mathrm{d}\mathbf{q}_i}{\mathrm{d}t} = \mathbf{v}_i^{\psi}(\mathbf{q}_1, \dots, \mathbf{q}_N) \equiv \frac{1}{m_i} \mathrm{Im} \frac{\psi^* \boldsymbol{\nabla}_{\mathbf{q}_i} \psi}{\psi^* \psi}(\mathbf{q}_1, \dots, \mathbf{q}_N), \tag{8.9}$$

where m_i is the mass of the ith particle. The particles follow determinate trajectories described by $\mathcal{Q}(t)$, with the distribution of $\mathcal{Q}(t)$ given by the quantum equilibrium distribution $|\psi|^2$.

In Bohmian mechanics, the wave function plays the role of a "guiding field." That is, through the guiding equation (8.9) the wave function generates a velocity field which the particles follow. Thus the Bohm theory describes the motion of particles along trajectories, similar to Newtonian mechanics. However, instead of the force–acceleration law of classical mechanics, the equation of motion for the particles is now given by (8.9), with the change of the wave function ψ, which enters on the right-hand side of (8.9), determined by the Schrödinger equation.

We can neatly illustrate this formalism in the context of the double-slit experiment [37,470]. Fig. 8.1 shows a set of possible Bohmian trajectories of single particles behind the slits. Each particle follows one of these determinate trajectories and thus passes through either one of the slits. Which particular Bohmian trajectories is taken by an individual particle simply depends on the initial position of the particle. Thus, although each particle at any given time is described by the same wave function ψ, the particles deterministically follow different determinate trajectories. Bohmian mechanics is therefore an

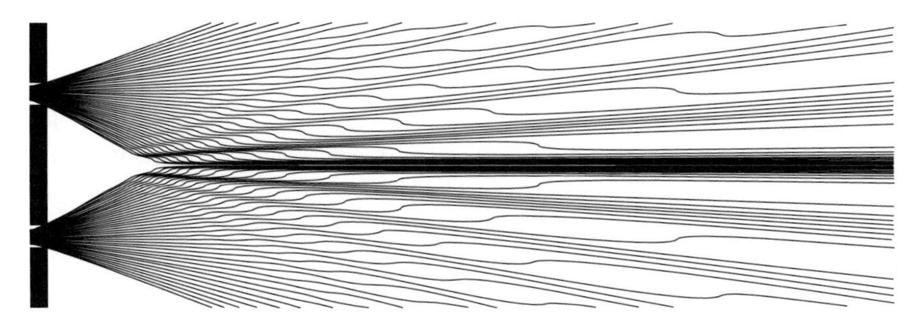

Fig. 8.1. A sample of Bohmian trajectories that particles may follow in the double-slit experiment. Figure reprinted, with kind permission from Springer Science and Business Media, from [469].

example of a hidden-variables theory (see Sect. 2.1.2). That is, the wave function ψ does not constitute a complete description of the physical state of the system, and the always-determinate positions $\mathbf{q}_i(t)$ of the particles play the role of the hidden variables. From Fig. 8.1 we observe that the particles take distinctly non-Newtonian paths. Instead of being straight lines, the particle trajectories become concentrated in areas in which the amplitude of the wave function ψ is large. The familiar maximums and minimums of the interference pattern observed on the screen can therefore be directly related to the "density" of the Bohmian trajectories.

8.5.1 Particles as Fundamental Entities

Bohm's theory has been criticized for attributing a fundamental ontological status to particles. General arguments against particles on a fundamental level of any relativistic quantum theory have been frequently given (see, for instance, [471,472]).[6] Moreover, and this is the point we would like to discuss in this section, it has been argued that the appearance of particles could be derived from the continuous process of decoherence, leading to claims that no fundamental role need be attributed to particles [397,475,476]. Based on the fact that decohered reduced density matrices of mesoscopic and macroscopic systems essentially always represent (improper) ensembles of narrow wave packets in position space, Zeh [397, p. 190] holds that such wave packets can be viewed as representing individual "particle" positions:

> All particle aspects observed in measurements of quantum fields (like spots on a plate, tracks in a bubble chamber, or clicks of a counter)

[6]On the other hand, there are proposals for a "Bohmian mechanics of quantum fields," i.e., a theory that embeds quantum field theory into a Bohmian-style framework [473,474].

can be understood by taking into account this decoherence of the relevant local (i.e., subsystem) density matrix.

Of course, to interpret the improper ensembles of narrow wave packets resulting from decoherence as leading to the perception of individual particles, we must supply an additional interpretive framework that explains why only one of the wave packets is perceived. That is, we need to add some interpretive rule to get from the improper ensemble emerging from decoherence to the perception of individual terms, so decoherence alone does not necessarily make Bohm's particle concept superfluous. But it suggests that the postulate of particles as fundamental entities may well be unnecessary, and taken together with the difficulties in reconciling such a particle theory with a relativistic quantum field theory, Bohm's *a priori* assumption of particles at a fundamental level of the theory appears seriously challenged.

8.5.2 Bohmian Trajectories and Decoherence

A well-known property of Bohmian mechanics is the fact that its trajectories are often highly nonclassical (see, for example, [37, 477, 478]). This poses the serious problem of how Bohm's theory can explain the existence of quasiclassical trajectories at a macroscopic level.

Bohm and Hiley [37] considered the scattering of a beam of environmental particles on a macroscopic system, a process that gives rise to decoherence [7, 17]. The authors demonstrated that this scattering yields quasiclassical trajectories for the system. Furthermore, Appleby [478] showed that for isolated systems, the Bohm theory will typically not give the correct classical limit. It was thus suggested that the inclusion of the environment and the resulting decoherence effects might be helpful in recovering quasiclassical trajectories in Bohmian mechanics [475, 479–483]. The basic idea is then to associate the quasi-Newtonian phase-space trajectories in the improper ensemble created by decoherence (see, e.g., Sect. 5.2.5) with the particle trajectories $\mathcal{Q}(t)$ of the Bohm theory. As pointed out by Bacciagaluppi [385], a great advantage of this strategy lies in the fact that the same approach would allow for a recovery of both quantum and classical phenomena.

However, a careful analysis by Appleby [479] showed that only under certain additional assumptions will processes that lead to decoherence also result in the correct quasiclassical Bohmian trajectories for macroscopic systems (Appleby described the example of the long-time limit of a system that has initially been prepared in an energy eigenstate). Interesting results were also reported by Allori [480], Allori and Zanghì [481], and Allori et al. [482]. These authors demonstrated that decoherence effects can play the role of preserving classical properties of Bohmian trajectories. Furthermore, they showed that, while in standard quantum mechanics it is important to maintain narrow wave packets to account for the emergence of classicality, the Bohmian description of a system by both its wave function and its configuration allows for the derivation of quasiclassical behavior from highly delocalized wave functions.

Sanz and Borondo [483] studied the double-slit experiment in the framework of Bohmian mechanics and in the presence of decoherence. They showed that even when (spatial) coherence is fully lost, and thus interference is absent, nonlocal quantum correlations remain that influence the dynamics of the particles in the Bohm theory. This example demonstrates that in general decoherence does not suffice to ensure the correct classical limit in Bohmian mechanics.

8.6 Summary

We have shown how the environment-induced superselection of preferred states has been incorporated into different interpretations of quantum mechanics in order to achieve a very general and empirically adequate definition (and explanation) of the "determinate quantities" in each interpretation. Thus the decoherence program has made a very significant contribution to solving the infamous preferred-basis problem that had haunted many interpretations for a long time. In some cases, such as in relative-state interpretations, the lack of a clear definition of preferred bases had been at the focus of ongoing criticism directed at such interpretations. In other cases, such as in physical collapse theories or Bohmian mechanics, the particular form of the preferred basis is simply postulated. Here decoherence can provide a physical motivation for the particular choice of the postulated quantity.

We have argued that, within the standard interpretation of quantum mechanics, decoherence cannot solve the problem of outcomes in quantum measurement. We are still left with a multitude of (albeit individually well-localized quasiclassical) components, and we need to supplement or otherwise to interpret this situation in order to explain why and how single outcomes are perceived. Accordingly, we have discussed how environment-induced superselection and the local suppression of interference terms can be put to great use in physically motivating, or potentially disproving, rules and assumptions of alternative interpretive approaches that change (or altogether abandon) the eigenvalue–eigenstate link and/or modify the unitary dynamics in order to account for our perception of outcomes and classicality in general.

For example, to name just a few applications, decoherence can provide a universal criterion for the selection of the branches in relative-state interpretations and a physical argument for the noninterference between these branches from the point of view of an observer. In modal interpretations, decoherence can be used to specify empirically adequate sets of properties that can be assigned to systems. In collapse models, the free parameters (and possibly even the nature of the reduction mechanism itself) might be derivable from environmental interactions. Decoherence can also help select quasiclassical particle trajectories in Bohmian mechanics. Moreover, it has become clear that decoherence is capable of ensuring the empirical adequacy and thus empirical equivalence of different interpretations. This observation has led the physicist Max Tegmark to the suggestion [484, p. 855] that the choice between, for example, the orthodox and the Everett interpretation may become "purely a matter of taste, roughly equivalent to whether one believes mathematical language or human language to be more fundamental."

9 Observations, the Quantum Brain, and Decoherence

Biological systems are rarely analyzed in quantum-mechanical terms. The enormous complexity of such systems makes a quantum-mechanical treatment essentially impossible. Furthermore, structures in biological systems are typically embedded into aqueous environments at room temperatures. It is therefore reasonable to expect that such systems will be subject to very strong decoherence, which will effectively suppress any quantum-coherent behavior. At the same time, if we expect quantum mechanics to be valid on all scales, the behavior of biological structures will be governed by the laws of quantum mechanics.

In this context, the brain is a biological system of particular interest. It constitutes the central organ that processes our perceptions. It is clear that every empirically relevant statement of a physical theory must be related to precisely these perceptions. Therefore, in trying to explain our observations based on what is predicted by the theory, we may need to give an account of the role of the system that delivers these perceptions to us, namely, the brain. Furthermore, as we shall outline below, historically the related question of the role of "consciousness" in quantum theory arose, including proposals that consciousness may induce a collapse of the wave function.

In this chapter, after a few introductory remarks on the general problem of observation in quantum mechanics (Sect. 9.1), in Sect. 9.2 we will explicitly include the observer in the von Neumann measurement scheme (see Sect. 2.5.1) and discuss the resulting consequences. We will then focus on the quantum-mechanical brain (Sect. 9.3). We will describe some explicit modeling studies of the decoherence properties of neuronal (and other) superposition states in the brain, which have resulted in explicit numerical estimates for the relevant decoherence timescales. In Sect. 9.4, we shall also briefly discuss the implications of these results for "subjective" resolutions of the measurement problem.

9.1 The Role of the Observer in Quantum Mechanics

The role of the observer in quantum mechanics has been at the heart of many foundational disputes reaching back to the early days of the theory. Recall that a peculiar feature of quantum mechanics is the fact that, in contrast with classical physics, states are inherently fragile (see Sect. 2.1). The measurement of a particular physical quantity will in general perturb the quantum state of the system and thus the outcome of a subsequent measurement of a different quantity represented by a noncommuting observable.

It thus seems that the physical state of the system "is" what the observer chooses to measure. This feature introduces a flavor of "subjectivity" into our description of nature that is rather unknown in the classical theories. Therefore observations, and consequently observers, seem to play a much more fundamental role in quantum mechanics than in classical physics. We already quoted Heisenberg's radical statement that "the particle trajectory is created by our act of observing it" [42, p. 185]. In this positivist view (which is most apparent in Bohr's writings), the entire formalism of quantum mechanics constitutes, in essence, a recipe for *predicting* what the observer will find if a measurement was to be performed, but it does not make any ontological claims as to "what is really out there." In fact, any statements about the physical state of the system prior to measurement are deemed meaningless by the Copenhagen interpretation.

Even the many people who subscribe to some variant of the Copenhagen interpretation when dealing with practical issues of experimental physics still maintain a belief in the existence of an observer-independent physical reality that quantum mechanics ought to refer to. The overarching question is then how to reconcile the lesson learned from quantum mechanics—that physical systems cannot, in general, be assigned an exhaustive set of premeasurement values of physical quantities—with our intuitively felt need for an "objectively existing" world around us to which we wish the theory to pertain in some way.

At the same time, we are also forced to carefully examine notions that we have come to regard as objective and thus as a required part of a physical theory, despite the fact that they might simply be artifacts of our subjective perception. Von Neumann [60, Chap. VI] reminded us that

> experience makes assertions only of the kind "an observer has made a particular (subjective) observation," but never of the kind "a certain physical quantity has a particular value."[1]

Similarly, d'Espagnat [485, pp. 134–135] cautions us as follows:

> The fact that we perceive such "things" as macroscopic objects lying at distinct places is due, partly at least, to the structure of our sensory and intellectual equipment. We should not, therefore, take it as being part of the body of sure knowledge that we have to take into account for defining a quantum state. (...) In fact, scientists most

[1]The original German text reads: *"Denn die Erfahrung macht nur Aussagen von diesem Typus: ein Beobachter hat eine bestimmte (subjektive) Wahrnehmung gemacht, und nie eine solche: eine physikalische Größe hat einen bestimmten Wert."*

> rightly claim that the purpose of science is to describe human experience, not to describe "what really is"; and as long as we only want to describe human experience, that is, as long as we are content with being able to predict what will be observed in all possible circumstances (...) we need not postulate the existence—in some absolute sense—of unobserved (i.e., not yet observed) objects lying at definite places in ordinary 3-dimensional space.

Von Neumann and d'Espagnat therefore urge us to realize that the often deeply felt commitment to a general objective "definiteness" is only based on our everyday experience of macroscopic systems. If it was possible for us to know, independently of our experience, that definiteness in fact existed in nature, then subjective definiteness would (presumably) be a consequence of the objective definiteness once we have constructed a simple model which connects the "external" physical phenomena with our "internal" perceptual and cognitive apparatus. (Here we may justify the expected simplicity of such a model by referring to the presumed identity of the physical laws governing external and internal processes, an assumption von Neumann referred to as "psycho-physical parallelism" [60].) However, any knowledge is directly or indirectly based on our observations, and therefore there is no way of acquiring knowledge about the objective existence of definiteness without rooting such knowledge in our (subjective) observation of definiteness. At the same time, macroscopic interference and coherence phenomena clearly show the fuzziness of the boundary between the quantum world and the definiteness observed at the level of our direct experience (see Chap. 6).

It may therefore indeed be reasonable to give up the demand that objective definiteness should be an *a priori* part of a satisfactory physical theory—provided, of course, the theory is able to account for all of our subjectively observed definiteness in agreement with our experience. The corresponding "subjective" resolutions of the measurement problem have received an enormous boost from the theory of decoherence. In fact, decoherence makes many of these types of subjective approaches empirically viable in the first place, because decoherence may allow us to derive definiteness from the point of view of local observers without having to enforce definiteness on the global level. We will discuss such strategies in more detail in Sect. 9.4 below.

9.2 Quantum Observers and the Von Neumann Chain

Instead of attributing a particular *a priori* role to measurements and observers, as done in the Copenhagen interpretation (Sect. 8.1) and other interpretations of quantum mechanics, let us now take a different viewpoint. We simply treat the observer as a quantum system interacting with the observed system. We may model this situation using the von Neumann measurement scheme discussed in Sect. 2.5.1, with the observer included as the final link in the *von Neumann chain* of measurement interactions (Fig. 9.1).

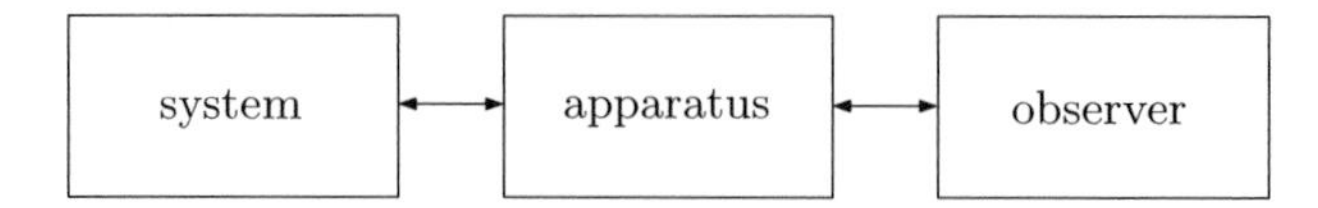

Fig. 9.1. The von Neumann chain with the observer included as the last link.

Then, the quantum states of the observer corresponding to the observation of the different possible outcomes of the measurement become quantum-correlated with the corresponding relative states of the measurement apparatus, which in turn is quantum-correlated with the different components in the superposition state describing the system of interest. We may schematically express this measurement interaction in the usual von Neumann form as [see (2.54)]

$$\left(\sum_n c_n |n\rangle\right) |\text{“apparatus ready”}\rangle \, |\text{“observer ready”}\rangle$$
$$\longrightarrow \sum_n c_n |n\rangle \, |\text{“pointer at } n\text{”}\rangle \, |\text{“observer perceives outcome } n\text{”}\rangle \,, \quad (9.1)$$

where $\sum_n c_n |n\rangle$ is the state of the observed (measured) system. Thus the final quantum state describing measurement in this formalism is an entangled system–apparatus–observer superposition state whose components correspond to the different "outcomes" of the measurement.

However, the inclusion of the observer as in (9.1) does not terminate the von Neumann chain of interacting systems. The addition of further links to this chain—such as an environment, secondary apparatuses or observers, etc.—will again lead to a final entangled state involving states corresponding to all possible outcomes, just as in (9.1). As long as the time evolution of the composite system that includes all these links is strictly unitary, there cannot be any change in the amount of information contained in this state (as measured, for example, by the purity or the von Neumann entropy, see Sect. 2.4.3), and the chain does not terminate. We are thus faced with a problem of infinite regression.

This raises the fundamental question of how and when these different correlated component states may reduce to the single measurement outcome actually experienced by the observer. The problem of whether, how, and where the von Neumann chain becomes terminated is of course, in essence, the problem of outcomes, which we have already discussed in several places in this book (see, for example, Sects. 2.5.4 and 8.1). Von Neumann himself was clearly aware of this problem. To him, the only sure fact was that we, as observers, always perceive definite outcomes at the conclusion of the measurement. Thus the theory must account for such definite perceptions at least at the level of the experience of the observer. The statistical predictions of quantum mechanics are insensitive to where exactly the Heisenberg cut is placed along the observational chain. Thus some form of wave-function collapse may be introduced at any stage between the apparatus and the observer. However, only at the level of the observer an explanation of our perception of manifestly definite outcomes is actually *forced out* by the obvious empirical constraints.

In this spirit, von Neumann postulated a collapse of the wave function within the observer, which he referred to as the "first intervention" (*"erster Eingriff"* in German).[2] This approach stood in stark contrast to the Copenhagen interpretation (see Sect. 8.1). While in the latter the collapse of the wave function was only implicit in the assumption of the existence of intrinsically classical measurement devices that are not to be subjected to further quantum-mechanical analysis, von Neumann did not attribute any special role to the measurement devices themselves. However, unable to resolve the problem of how to explain our perception of definite outcomes from his purely quantum-mechanical formalism of interacting quantum systems, he felt forced to uphold a fundamental quantum–classical boundary, albeit now pushed all the way toward the observer. We may say that von Neumann's ambitious goal of providing a purely quantum-mechanical description of the measurement process was therefore cut short by his *ad hoc* introduction of a collapse at the level of the observer.

Von Neumann himself did not elaborate on the nature of the collapse in the observer, or on what distinguished observers from inanimate objects such as measurement devices. However, by placing the Heisenberg cut within the observer, a new door was opened, namely, the question of the role of "consciousness" in the quantum measurement process and in the (actual or just perceived) collapse of the wave function.

The approach of including rather vague terms such as "mind" and "consciousness" at a fundamental level of a physical theory surely runs counter to the traditional strategy of natural science in general, and physics in particular. At the same time, the peculiar relevance of the observer in the quantum theory—the seemingly inevitable need to depart from the previously unquestioned assumption of an observer-independent reality that is similar in structure to the world directly observed by us—seems to have reintroduced the idea of attributing a distinct (physical) role to the mind of the observer.

Even the Copenhagen interpretation itself, and the positivist attitude embodied in it, implicitly assigns relevance to the concept of a "mind" by regarding quantum mechanics essentially as a set of statements that obtain physical meaning only in the context of verification (i.e., measurements) by an observer. As discussed by d'Espagnat [486], by referring not to what "is" but only to what can be "found out" by the observer, the observer's mind (that verifies, finds out, etc.) constitutes a primitive notion which is prior to

---

[2]Interestingly, the unitary Schrödinger evolution governing the interaction between the system, the apparatus, and the observer was referred to as the "second intervention."

that of scientific reality. Yet, terms such as "mind" and "consciousness" were not an explicit part of the Copenhagen interpretation, and the basic intuition that the knowledge obtained by the observer nonetheless must refer to some underlying physical reality was not truly relinquished by the proponents of this interpretation.

A famous (albeit only temporary) supporter of the idea that consciousness plays a fundamental role in the quantum measurement process was Eugene Wigner. His motivation was grounded in a *Gedankenexperiment* that he devised in the early 1960s and that has since become known as the *problem of Wigner's friend* [487]. This thought experiment is a variation of the Schrödinger-cat setup. The poison is replaced by a measuring device that emits a flash of light when the unstable atom decays, and—crucially—the cat is substituted by a human observer, referred to as "Wigner's friend." Let us call the external observer in the original Schrödinger-cat scenario "Wigner" to distinguish him from his friend.

Thus our setup now contains two human observers: An "inside" observer (Wigner's friend) whose measurement-type interaction with the atom–apparatus system is described quantum-mechanically within the von Neumann scheme, and an "outside" observer (Wigner himself) that later inquires about the result of this interaction. The box of the original cat scenario (housing the cat together with the unstable atom and the poison) corresponds here to the laboratory room containing the atom, the measuring device, and Wigner's friend. Wigner waits outside of the room for some period of time, during which the combined state of the atom, apparatus, and Wigner's friend evolves into the usual cat-type superposition. We may schematically express this evolution in the form

$$
\begin{aligned}
&(\alpha\,|\text{“atom not decayed”}\rangle + \beta\,|\text{“atom decayed”}\rangle) \\
&\qquad \otimes |\text{“device ready”}\rangle\,|\text{“observer ready”}\rangle \\
&\quad \longrightarrow \alpha\,|\text{“atom not decayed”}\rangle\,|\text{“no flash emitted”}\rangle\,|\text{“no flash observed”}\rangle \\
&\qquad\quad + \beta\,|\text{“atom decayed”}\rangle\,|\text{“flash emitted”}\rangle\,|\text{“flash observed”}\rangle\,. \qquad (9.2)
\end{aligned}
$$

Wigner then enters the room to inquire about the outcome of the experiment. He asks his friend whether she has seen a flash of light or not. The friend, evidently, will answer that she either has or has not observed the flash. Depending on the answer, Wigner would conclude that, from his perspective, the cat state describing the combined atom–apparatus–friend system has collapsed onto either one of its two components on the right-hand side of (9.2). But what would Wigner's friend say if Wigner asked her about her experience regarding the flash of light *before* he had entered the laboratory and asked the first question?

In essence, just as in the case of the Schrödinger-cat scenario, the overarching issue brought out by the example of Wigner's friend is the question of when and where the collapse-inducing measurement takes place. In the case of Wigner's friend, however, the difference—and in Wigner's thinking, the crucial difference—to the Schrödinger-cat paradox is that the superposition involves a conscious observer who should be able to provide a verbal statement to Wigner (*after* he has entered the room) about her experience *before* Wigner had gone into the laboratory and had asked her about whether or not she has seen the flash of light.

Wigner himself believed that the superposition of two distinct "states of consciousness" in (9.2)—one corresponding to the observation of a flash, the other to the failure to observe a flash—would need to be regarded as absurd. In his opinion, the conscious observer must always be in only one of these states [487]. This belief led him to conclude that conscious observations cannot be described by the standard linear quantum formalism as in (9.2), i.e., that consciousness must be fundamentally different from material objects, in the sense that it breaks the unitary evolution and induces a collapse of the wave function onto a definite state of the conscious observer.

From a purely empirical point of view, a strong counterargument to the conclusiveness of Wigner's reasoning is that superpositions of different states of consciousness do not necessarily need to be regarded as absurd and thus be excluded, as long as these states do not interfere in any way, i.e., as long these different "quantum versions" of the conscious observer are not "aware" of each other. Decoherence would inevitably also affect the conscious observer, and thus we may conjecture that it would be impossible to empirically confirm the existence of the different "branches of consciousness" in (9.2) from the "inside view" of the observer. We will discuss this point in more detail in Sect. 9.4. In fact, Wigner later abandoned [488] his views on the special role of consciousness in quantum measurement once he became aware of Zeh's paper of 1970 [4], which had introduced some of the basic ideas underlying the decoherence program.

9.3 Decoherence in the Brain: The Brain as a Quantum Computer?

The brain comprises an astonishingly complex network of $O(10^{11})$ neurons interconnected via $O(10^{14})$ synapses. There are convincing arguments that the memory capabilities and dynamical interactions of this neuronal network are the physiological core for the storing and processing of our sensory input. It is not unreasonable to conjecture that ultimately all cognitive processes, including our perception of "consciousness," can be reduced to neuronal activity. Researchers in neurobiology and biophysics typically model the brain in completely classical terms as a massively parallel interconnected web of on–off "switches" (nodes) [489–491]. Each such node represents a single neuron and is turned on or off (corresponding to, respectively, the firing and resting state of a neuron) depending on a particular, often nonlinear, activation function. For obvious reasons, such models are usually referred to as *artificial neuronal networks*.

Broadly speaking, the brain is thus modeled as a classical computer. One of the surprising insights that has been gained from such modeling studies is that even relatively simple networks, based on a fairly small number of nodes and on only very few rules for the activation of each node, can yield an astonishingly complex behavior, not dissimilar to that observed in simple cognitive processes. Yet, despite the success of such classical models, the question of whether quantum effects may play an important role in our cognitive apparatus has been discussed in at least two respects.

First, quantum "uncertainties" and the apparent indeterminism of quantum mechanics have sometimes been invoked to evade the strict determinism mandated by classical physics and to restore a notion of "free will." Such a strategy, however, is fundamentally flawed. Any quantum "uncertainties" would represent wholly uncontrollable effects, completely different from our understanding of the concept of "free will," which is based on our feeling of being able to actively control and decide our actions. For the same reason, the fact that quantum mechanics only predicts probabilities for measurement outcomes and thus seems to reintroduce indeterminism into the description of the physical world cannot be used to infer the existence of a free will. Once again, any actions triggered by such "quantum randomness" would not be controllable by the individual claimed to possess free will. Thus, if anything, invoking quantum theory in these ways renders the old "problem of free will" (if there is any problem at all) even more severe.[3] Finally, we should also note that synaptic transmissions in the brain have a fairly high failure rate due to the complexity of the underlying biological processes [493]. This inevitably leads to a rather high degree of unpredictability on the "everyday level." Any such purely functional unpredictabilities will presumably be far more significant than any quantum-uncertainty effects. Therefore the view of the brain as a deterministic classical computer with a predictable input/output pattern should in fact not be taken too literally.

---

[3] On the subject of free will, Einstein once said in his thoughtful *Credo* of 1932 [492]:

> I do not believe in free will. Schopenhauer's words: "Man can do what he wants, but he cannot will what he wills," accompany me in all situations throughout my life and reconcile me with the actions of others, even if they are rather painful to me. This awareness of the lack of free will keeps me from taking myself and my fellow men too seriously as acting and deciding individuals, and from losing my temper.
> (The original text reads: *"Ich glaube nicht an die Freiheit des Willens. Schopenhauers Wort: "Der Mensch kann wohl tun, was er will, aber er kann nicht wollen, was er will", begleitet mich in allen Lebenslagen und versöhnt mich mit den Handlungen der Menschen, auch wenn sie mir recht schmerzlich sind. Diese Erkenntnis von der Unfreiheit des Willens schützt mich davor, mich selbst und die Mitmenschen als handelnde und urteilende Individuen allzu ernst zu nehmen und den guten Humor zu verlieren."*)

Beyond this "naïve" application of quantum theory to the brain lies the realm of suggestions that the complexity of human cognitive processes generated by the brain may be explainable only if the brain is capable of acting as a quantum computer. In other words, such proposals are based on the idea that at least part of the abilities of the brain may be due to the presence and manipulation of quantum-coherent superposition states of the usual classical states in the brain (such as the resting and firing states of a neuron, for example). If this was true, it would follow that classical artificial neuronal networks would be intrinsically insufficient to fully simulate the brain.

In particular, the question of whether human consciousness may be linked to quantum-coherent processes has frequently been discussed. In the previous Sect. 9.2, we have already alluded to a related but opposite-directed concept, namely, Wigner's early speculation [487] that the collapse of the wave function may be precipitated by the action of consciousness. There, consciousness was invoked to effectively *destroy* the quantum coherence embodied in the global entangled state (9.1) produced by the von Neumann chain of measurement-type interactions. Now, the direction of the problem is reversed: May quantum coherence be associated with the emergence of consciousness? In this context, various macroscopic quantum-coherent processes have been suggested as the origin of consciousness (see, e.g., [494–500]). Most prominently, Penrose [462] and others [501–503] have suggested that so-called microtubules in the brain—dynamically active structures that are a dominant part of the cytoskeleton (i.e., of the internal scaffolding of cells)—may have sufficiently long decoherence times to allow for quantum computations, and, moreover, that such computations could be associated with the emergence of consciousness (see Sect. 9.3.2 below).

Quite independently of issues concerning the rather vague notion of consciousness and its potential role in a fundamental theory of physics, the obvious overarching question that we shall now discuss is then the following. Could the relevant structures in the brain—most notably, neurons and microtubules—sustain quantum coherence long enough to allow for quantum computations in the brain to be carried out? Based on our discussion of decoherence in the previous chapters of this book, we would expect that such decoherence times would be extremely short. After all, neurons and microtubules, while small on a biological scale, are still macroscopic and very complex objects on the typical scales considered in quantum physics. A state of, say, a neuron being in a superposition of firing and resting would fall clearly into the category of superpositions of macroscopically distinct states (see below). Furthermore, these structures are embedded into a macroscopic "warm and wet" environment and interact strongly with this environment.

Our intuition regarding the shortness of decoherence times has been supported by quantitative results reported by Tegmark in 2000 [504]. Tegmark considered models for the interaction of neurons and microtubules with their natural environment within the brain and presented order-of-magnitude estimates for the resulting decoherence timescales of superposition states in these structures. We shall describe these results in the following Sects. 9.3.1 (decoherence in neurons) and 9.3.2 (decoherence in microtubules). Implications of these results for "subjective" resolutions of the measurement problem will be discussed in Sect. 9.4.

9.3.1 Decoherence Timescales for Superposition States in Neurons

Let us first discuss the case of neuronal decoherence as considered by Tegmark [504]. The neuron's key component of interest is here the so-called axon, which can be thought of as a long hollow tube with a diameter of about $d \approx 10\ \mu\text{m}$ (Fig. 9.2). Most of the wall of the axon is coated with myelin, which acts as an insulator. However, there are narrow bands (several nanometers wide) along the axon where this insulation is absent, creating patches of semipermeable membrane (with a thickness of about $h \approx 10$ nm) that allow for the exchange of ions between the inside of the axon and the aqueous medium surrounding the axon. The firing of a neuron is represented by a very rapid flux of sodium ions from the surroundings through these membrane patches into the inside of the axon, mediated by voltage-gated sodium channels located inside the membrane. This firing propagates along the axon at high speeds (up to 100 m/s), causing the sodium channels in every patch to open. The influx of ions only last for about one millisecond, after which the sodium channels close up again, and the neuron returns to its resting state. Using parameters typical for neurons found in the central nervous system, Tegmark estimated that a total of $O(10^6)$ sodium ions traverse the entire length of the axon membrane (combining the contributions from all individual patches) during firing.

This means that during firing there are on the order of $N = 10^6$ more sodium ions in the interior of the axon than during resting (see Fig. 9.3). Thus

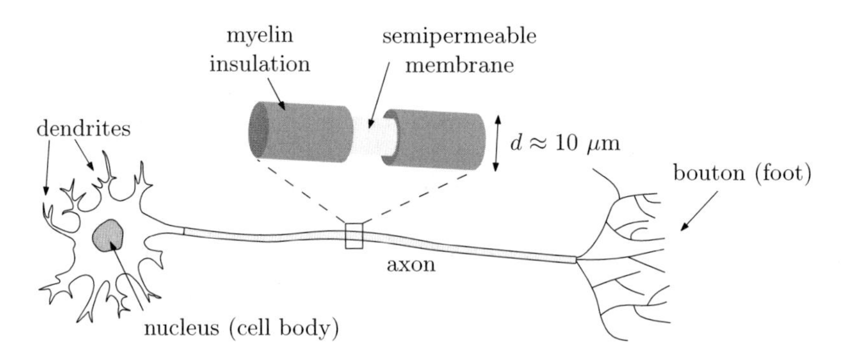

Fig. 9.2. Schematic illustration of a neuron. A segment of the axon is magnified for clarification. For the most part, the axon wall is covered with myelin, which insulates the inside of the axon from the surrounding medium. Small band-shaped areas, however, lack this insulation, and ions can traverse through the resulting semipermeable membrane.

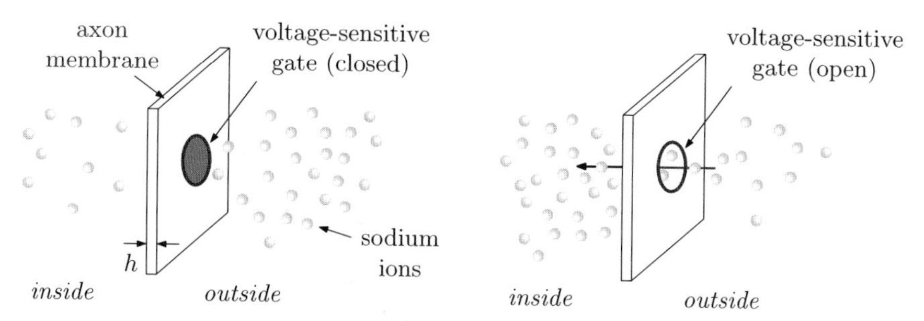

Fig. 9.3. Illustration of the resting (left) and firing (right) states of a neuron. In the firing state, $O(10^6)$ more sodium ions are present on the inside of the axon than in the resting state.

a superposition of firing and resting states corresponds to a superposition of the two states describing a collection of $O(N)$ sodium atoms located on the inside and outside, respectively, of the axon. In other words, we deal with a system composed of $O(N)$ sodium atoms and described by a superposition of two center-of-mass positions a distance $O(h)$ apart.

Tegmark then studied three types of environmental interactions by which the neuronal superposition will lose coherence, namely, collisions of the sodium atoms with other nearby sodium (and potassium) ions, collisions with surrounding water molecules, and electrostatic (Coulomb) interactions with distant ions. All three cases can be treated using the theory for scattering-induced decoherence described in Chap. 3.[4] Let us briefly sketch the reasoning for the first two cases of ion–ion and ion–water collisions. The thermal de Broglie wavelength (2.114) of a sodium (or potassium) ion and of a water molecule at $T = 310$ K (i.e., at human body temperature) are of similar magnitude, namely, $\lambda_{\mathrm{dB}} \approx 0.03$ nm. This value is much smaller than the coherent spatial separation $h \approx 10$ nm. This means that such an ion or water molecule, acting as a scattered environmental particle, can perfectly resolve the separation h. We are therefore entirely within the short-wavelength limit discussed in Sect. 3.3.1.

[4] Note that the model developed in Chap. 3 is based on the assumption that the momentum change of the scatterer, i.e., of the system of interest, is much smaller than the momentum change of the scattered environmental particles, such that recoil of the object can be neglected. Strictly speaking, this approximation does not hold in our case, since the mass of the scatterer (the sodium ion) is similar to the mass of the environmental particle (a sodium or potassium ion, or a water molecule). The following numerical estimates should therefore not be taken too literally. On the other hand, for the present purpose, we shall be content with a few crude estimates. Even if the correct values deviated by several orders of magnitude from our estimates, the general conclusion of extremely short decoherence timescales for such neuronal superposition states, relative to the timescales relevant to biological and cognitive processes, will remain unaffected.

In this limit of maximum decoherence, we have shown that the rate of decay of the off-diagonal elements of the position-space reduced density matrix of the system saturates at a value set by the total scattering rate Γ_{tot} [see (3.48)],

$$\rho(\boldsymbol{x}, \boldsymbol{x}', t) = \rho(\boldsymbol{x}, \boldsymbol{x}', 0)\mathrm{e}^{-\Gamma_{\text{tot}} t}. \tag{9.3}$$

For a single sodium ion in a superposition of being located on the inside and outside of the axon, and using the cross section for Coulomb scattering, Tegmark estimated $\Gamma_{\text{tot}} \approx 10^{14}\ \text{s}^{-1}$ for collisions with sodium and potassium ions, and roughly the same value for the scattering of water molecules. The decoherence timescale τ_{d} for the spatial superposition involving $N = 10^6$ ions is then estimated to be on the order of

$$\tau_{\text{d}} = (N\Gamma_{\text{tot}})^{-1} \approx 10^{-20}\,\text{s}. \tag{9.4}$$

For the third case of decoherence due to Coulomb interactions with more distant ions, Tegmark employed a similar scattering model. Without going into the details here, the resulting estimate for the typical timescale for decoherence of the neuronal superposition turns out to be very similar to that of (9.4), namely, $\tau_{\text{d}} \approx 10^{-19}$ s. All values are summarized in Table 9.1.

How do these values compare to the timescales relevant to biological and cognitive processes in the brain? The duration of a single firing is on the order of 10^{-3} s, which is thus many orders of magnitude larger than τ_{d}. We thus conclude that the process of firing itself is effectively classical by virtue of decoherence. Tegmark estimated the timescale for typical cognitive processes—such as deliberate actions, thoughts, speech, motor response to external stimuli, etc.—to be around 10^{-2}–10^{0} s, which certainly seems reasonable in light of our own experience and of general experimental evidence from neurophysiology. It is clear that such timescales massively exceed the decoherence times of 10^{-19}–10^{-20} s estimated for neuronal superpositions. We can therefore quite safely conclude that such superpositions cannot play a significant role in the above examples of cognitive processes.

We can go even further and also use these results to arrive at insights about the aforementioned "quantum nature of consciousness" suggested by some authors. As we have pointed out, general evidence from neurobiology

Table 9.1. Estimates of decoherence timescales (in seconds) for superpositions of firing and resting neuronal states and of kink-like excitations in microtubules, as reported by Tegmark [504].

Object	Environment	Decoherence timescale
Neuron	Sodium and potassium ions	10^{-20}
Neuron	Water molecules	10^{-20}
Neuron	Distant ions	10^{-19}
Microtubule	Distant ions	10^{-13}

suggests that conscious perceptions are linked to certain neuronal patterns of resting and firing in the brain. As argued by Tegmark [504], this would mean that consciousness cannot have the suggested quantum nature, for the following reason. Suppose that, at some fundamental level, the source of consciousness was indeed due to some subsystem of the brain with a sufficiently long decoherence time, such that one may assume that quantum-coherent processes should be possible. However, to actually create the conscious perception within the observer, this subsystem would need to continuously interact with the neuronal network. Since the neurons are in turn strongly coupled to their environment, these subsystem–neuron interactions would rapidly suppress any quantum coherence within the subsystem, thus effectively destroying the proposed quantum nature of consciousness. This argument is quite independent of the precise form that such a quantum consciousness would take. It only requires the (quite well-motivated) assumption that different conscious perceptions correspond to different neuronal patterns of firing and resting.

Finally, let us note that it is of course no coincidence that the relevant "computational" states of the neurons—i.e., the resting and firing states—correspond to the most ubiquitous environment-superselected states in nature, namely, states that are localized in position space. These neuronal states can be identified with "record states" that are capable of robustly encoding information in spite of environmental interactions [67,104] (see also Zurek's "existential interpretation" outlined in Sect. 8.2.3). If the computational states of neurons corresponded to the spatial superpositions of resting and firing, they would be subject to rapid scattering-induced decoherence and would therefore be hardly useful for encoding information in the brain.

9.3.2 Decoherence Timescales for Superposition States in Microtubules

Tegmark also estimated typical decoherence timescales for superposition states in microtubules. As mentioned above, it has been suggested that such microtubules may allow for quantum-coherent dynamics and may even be the "origin of human consciousness" [462, 501–503]. Microtubules are a key component of neurons in the brain and also play important roles in various cellular processes. They take the shape of hollow cylinders with a diameter of about 24 nm and lengths on the order of micrometers to millimeters (Fig. 9.4). Each cylinder consists of a bundle of 13 individual so-called protofilaments. Each such protofilament is a linear row of tubulin dimers composed of the monomers α-tubulin and β-tubulin. The β subunit carries an extra 18 Ca^{2+} ions, whereas an equal number of negative charges is located at the nearby α subunit. This gives the dimer an electric dipole moment of strength $36e$. Along each protofilament, the tubulin dimers align such that the α subunit of one dimer is in contact with the β subunit of the next dimer. Thus, from a physics point of view, protofilaments are strings of tiny electric dipoles that

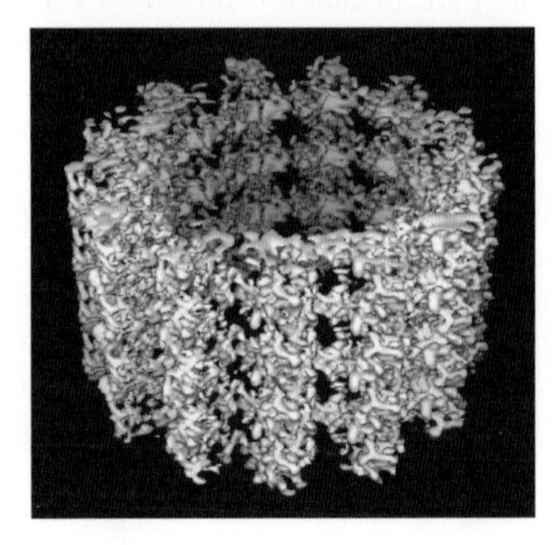

Fig. 9.4. Structure of a microtubule. The image shows a three-dimensional reconstruction based on data obtained from cryoelectron microscopy at a resolution of about 8 Å. Created by the Visualization Group of Ken Downing at the Lawrence Berkeley National Laboratory [505, 506].

are bundled together in parallel so as to create a net electric dipole moment of the microtubule as a whole (Fig. 9.5).

The inside of each dimer contains a single delocalized electron which can be localized in either one of two hydrophobic pockets located toward the side of the α subunit and β subunit, respectively (Fig. 9.6). If the electron is localized toward the β subunit (referred to as the β state of the dimer), the dimer undergoes a conformational distortion of about 30 degrees from

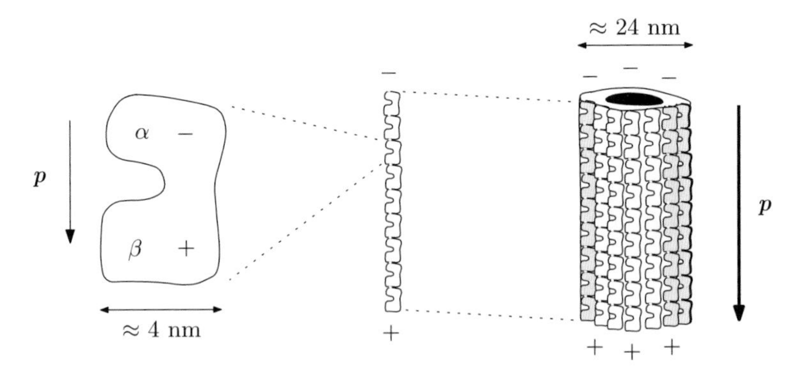

Fig. 9.5. Dipole moment of a microtubule. *Left:* Each tubulin dimer (composed of α and β monomers) acts as a miniature electric dipole. *Middle:* The protofilaments are strings of such electric dipoles. *Right:* Since the protofilaments bundle together in a parallel manner to form the microtubule, the latter possesses a dipole moment that is the sum of the dipole moments of the individual protofilaments.

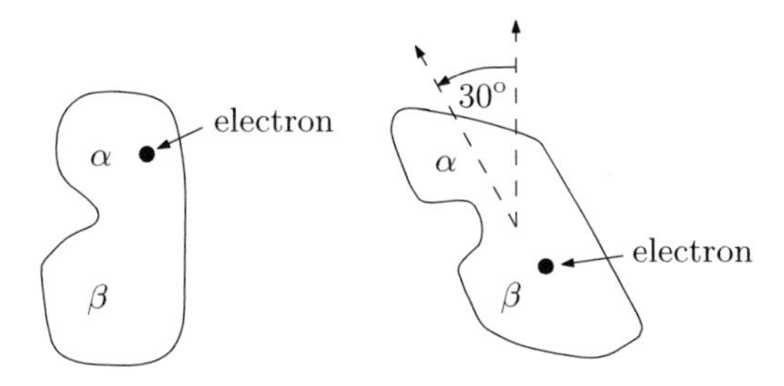

Fig. 9.6. The two conformational states α and β of the tubulin dimer.

the vertical axis associated with the conformation in which the electron is localized toward the α subunit (the α state of the dimer). Thus, the dimer can be described as an effective two-state system, with basis states corresponding to the conformational states α and β. As a consequence of the distortion of the dimer, the direction of the electric dipole moment will be different for the α and β states, and we can associate these two states with two distinct values p_α and p_β of electric dipole moment along the axis of the microtubule. We may thus regard the protofilaments as a set of interacting Ising spin-$\frac{1}{2}$ chains, where the spin states "up" and "down" now correspond to the two possible states α and β of electric dipole moment along the microtubule axis.

Certain processes, such as the supply of energy via a process of hydrolysis, can induce a change between the conformational states α and β of a tubulin dimer located at one end of the microtubule [507]. Mediated by dipole–dipole interactions between the dimers, it has been suggested that the resulting change in the dipole moment associated with the dimer may then propagate rapidly along the axis of the microtubule (see Fig. 9.7). Sataric,

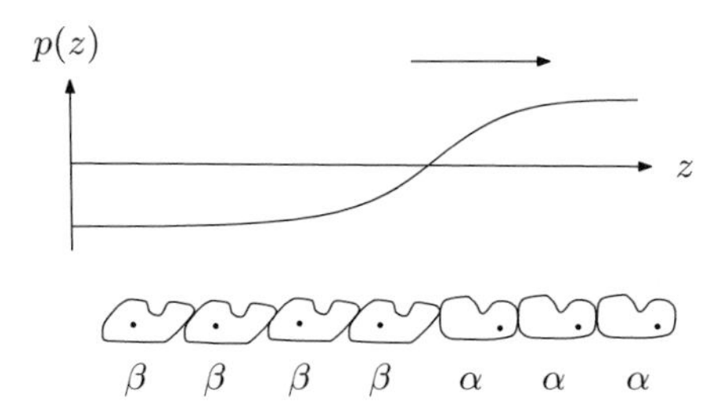

Fig. 9.7. Model for kink-like excitations in microtubules. A change of the local dipole moment at one end of the microtubule, corresponding to a conformational switch between the states α and β of the tubulin dimer, travels along the axis of the microtubule, mediated by dipole–dipole interactions between the dimers. This results in kink-like excitations of the dipole moment $p(z)$ along the axis of the microtubule, with these excitations propagating down the axis of the microtubule at high speeds. (For simplicity, we have associated the states α and β with equal but opposite values of the electric dipole moment along the microtubule axis.)

Tuszynski, and Zakula [508] proposed a detailed theoretical model for such "kink-like" excitations (solitons) to describe the possibility of lossless energy transfer in microtubules.

Tegmark [504], focusing on this particular model, estimated typical decoherence timescales for a superposition of two kink-like excitations separated by a distance of many dimers. He considered Coulomb interactions between the net charge at the location of the kink (taken to be equal to the total charge due to the 18 Ca^{2+} ions in the 13 β subunits contained in the "ring" of dimers formed by a cross section of the microtubule) with distant environmental ions. The corresponding timescale τ_d for the decoherence of a superposition of two well-separated kinks was estimated using a Coulomb-scattering model similar to that employed in the modeling of decoherence of neuronal superpositions due to interactions with distant ions (see Sect. 9.3.1). Tegmark found a value of $\tau_\mathrm{d} \approx 10^{-13}$ s (see also Table 9.1). If we choose to identify the potential nature of quantum-coherent behavior of microtubules with this type of superposition, Tegmark's numerical estimate quite clearly rules out the possibility that microtubules may exhibit quantum-coherent behavior akin to a quantum computer.

However, another possibility has been considered in (rather controversial) models developed by Penrose and Hameroff [462, 501, 502]. Here, instead of the superpositions of spatially separated kink-like excitations studied by Tegmark, the authors considered superpositions of the two conformational states α and β of an individual tubulin dimer. Attempting to challenge Tegmark's claim that decoherence is too strong for microtubules to exhibit quantum-coherent dynamics, Hagan, Hameroff, and Tuszynski [503] suggested that such configurational superpositions involving individual dimers would allow for decoherence times many orders of magnitude longer than the decoherence times for superpositions of solitons estimated by Tegmark.

To calculate the decoherence time of superpositions of the α and β states, Hagan, Hameroff, and Tuszynski [503] used a model similar to that employed by Tegmark but applied it to the dipole moment of a single dimer. However, as pointed out by Rosa and Faber [509], once an unjustified approximation in the calculations presented in [503] is corrected, the resulting decoherence times are of similar magnitude as those originally given by Tegmark for superpositions of spatially separated kink-like excitations. Although not conclusively ruling out the models of the Penrose–Hameroff type, these estimates present a very serious challenge to such theories.

It is fair to say a majority of researchers now uphold the view that biological structures in the brain are most likely too prone to decoherence to allow for any quantum coherence to persist over timescales relevant to cognitive and conscious processes [510]. Therefore classical models of the brain remain largely unchallenged to date. Based on Tegmark's numerical results and on general intuitions about decoherence on macroscopic scales, it is unlikely that this situation will change any time soon.

9.4 "Subjective" Resolutions of the Measurement Problem

The final state of the von Neumann chain (9.1) describes different "mental states" of the observer (corresponding to the perception of the different outcomes labeled by n) quantum-correlated with the relative states of the apparatus, system, etc. Approaches toward a *subjective* resolution of the measurement problem then try to explain how the observer's perception of only a single one of these possible outcomes may arise, in spite of the persistence of coherence in the global state. In this case the "physical reality out there" (whatever it might be) may well continue to be exhaustively described by the global superposition state (9.1), without the (empirical) need to single out any of the component states. As we have discussed in Sect. 9.1, there is no fundamental reason that would prevent us from regarding such a subjective resolution of the measurement problem as satisfactory and sufficient, provided the theory is capable of correctly accounting for all possible observations (see also [1, 320, 485]).

We mentioned above that current scientific evidence suggests that most, if not all, cognitive activity—and thus, presumably, the "states of mind," or "mental states" of an observer[5]—can be associated with certain resting/firing patterns of the network of neurons in the brain (see [493, 511] for more precise definitions of the relationship between perceptions and the neuronal patterns). This would mean that, broadly speaking, each of the abstract states "observer perceives outcome n" on the right-hand side of (9.1) corresponds to a certain resting/firing pattern of a collection of neurons. As we have seen above, superpositions of firing and resting states will be rapidly decohered by the interaction with the environment. It is therefore reasonable to conclude that this decoherence will result in a practically irreversible dynamical decoupling of the "branches" on the right-hand side of (9.1), which correspond to the distinct outcomes labeled by the index n.

One may now ask the question of why the robust neuronal pointer states (i.e., the resting and firing states) become quantum-correlated precisely with the stable quasiclassical pointer states in the world around us, i.e., why the "outcome states" in (9.1) correspond to the familiar states of our experience. For example, why does an environment-superselected resting or firing state of a neuron get correlated with spatially well-localized states of macroscopic objects, instead of with superpositions of macroscopically separated states? Apart from the obvious reason of a limited set of available interaction Hamiltonians and thus of a heavily constrained set of detectable observables, Zurek [13] has pointed out another important reason:

[5]The frequent use of quotation marks in the following paragraphs inevitably reflects the rather vague nature of notions such as "mind," "consciousness," and "mental states."

> Our senses did not evolve for the purpose of verifying quantum mechanics. Rather, they have developed in the process in which survival of the fittest played a central role. There is no evolutionary reason for perception when nothing can be gained from prediction. And, as the predictability sieve illustrates, only quantum states that are robust in spite of decoherence, and hence, effectively classical, have predictable consequences. Indeed, classical reality can be regarded as nearly synonymous with predictability.

We also need to show that the branches defined by the (objective) physical interactions between the subsystems indeed correspond to the (subjective) conscious experiences of the different individual measurement outcomes. That is, we ought to demonstrate that the classical neuronal resting/firing patterns associated with each individual branch indeed represent the relevant collective memory (or "record") states in the brain associated with the conscious perception of the different possible outcomes (corresponding to von Neumann's aforementioned principle of a "psycho-physical parallelism" [60]).

Needless to say, quantum mechanics itself does not allow us to derive a relationship between subjective experience and its physical correlates. This fact has led some to the conclusion that the question of this connection can only be fully answered through the introduction of new physical laws [493]. Other authors, such as Zeh [39, 512], have suggested that the empirical fact of decohering wave-function components in neuronal processes constitutes sufficient (if not compelling) grounds for postulating, within the quantum-mechanical formalism, the "existence of consciousness." In this picture, dynamically autonomous conscious observers are then associated with the robust components of the global wave function of the type (9.1), with each of the components labeled by the decohered neuronal states corresponding to definite resting/firing patterns. We thus obtain a *multitude* of classical "worlds" (defined by the robust branches) within a *single* quantum universe (described by the unitarily evolving global quantum state vector).

Given this setting, it may indeed not be far-fetched to conclude that a multitude of such conscious perceptions of the observer can co-"exist," while at the same time this co-"existence" could never be explicitly empirically confirmed (which is why we have put the terms "exist" and "existence" in quotation marks). In view of this nonobservability, Zeh called the "existence" of different conscious versions of the observer[6] a "heuristic fiction" [512], since the "existence" of multiple branches representing different conscious perceptions arises simply as a consequence of the assumption of universally valid unitary dynamics (together with a "realist" interpretation of the global quantum state).

---

[6]Since the identity of an observer may be regarded as determined by the observer's particular conscious perceptions, it may be difficult to argue that this situation would represent different conscious versions of the *same* observer.

The conjecture is then that, because the different conscious versions of the observer would not be "aware" of each other, from the inside perspective of the observer one should be able to account for the empirically required perception of definite measurement outcomes, without relinquishing the assumption that quantum states describe (some form of) physical reality and that unitary dynamics is universally valid.

Not surprisingly, the resulting "many-minds" interpretation of quantum mechanics (e.g., [4, 5, 16, 39, 67, 104, 388, 512]; see also Sect. 8.2) sits uncomfortably with many people. It is not so much a concern about the empirical adequacy (i.e., the possibility of experimental disproof) of such a view that creates the discomfort. After all, decoherence, together with an additional assumption about the connection between neuronal resting/firing patterns and the emergence of our conscious perceptions (an assumption that in turn could find its physical justification in decoherence effects), may indeed suffice to ensure the perception of definite outcomes at the level of the observer.

Rather, the discomfort with many-minds interpretations (as well as with "many-worlds" interpretations; see Sect. 8.2) is rooted in our feeling that such interpretations run counter to a deeply ingrained philosophical intuition of ours, namely, that the subjective observations of the world around us are a reasonably good mirror of the "objective external physical reality." In the many-minds picture, the quantum universe described by the global quantum state would appear "to be" of a radically different form than the observed structure of the world (given by a single "classical" branch defined by the locality of observers and by decoherence processes), even though we would presumably never be able to empirically verify the "existence" of this global quantum universe. Of course, ultimately this amounts to a purely philosophical discussion about the ontological status of the "other" (unobservable) branches. After all, terminology such as "physical reality" and "to be" can only refer to existence as confirmed (or at least verifiable) by our subjective observations. Zeh [397] has put it as follows:

> [A]fter an observation one need not necessarily conclude that only one component now *exists* but only that only one component *is observed*. (...) Superposed world components describing the registration of different macroscopic properties by the "same" observer are dynamically entirely independent of one another: they describe different observers. (...) He who considers this conclusion of an indeterminism or splitting of the observer's identity, derived from the Schrödinger equation in the form of dynamically decoupling ("branching") wave packets on a fundamental global configuration space, as unacceptable or "extravagant" may instead dynamically formalize the superfluous hypothesis of a disappearance of the "other" components by whatever method he prefers, but he should be aware that he may thereby also create his own problems: Any deviation from the global Schrödinger equation

must in principle lead to observable effects, and it should be recalled that none have ever been discovered.

Regardless of one's personal interpretive preferences, one of the main take-home points of this section is the insight that decoherence constitutes the core mechanism for ensuring the empirical adequacy of interpretations such as the many-minds view (see also Sect. 8.2 on this issue). Indeed, without decoherence the many-minds (and many-worlds) interpretations would stand no chance. Decoherence, or, more precisely, the physical interactions between subsystems (including the observer) and the resulting entanglement, account for the emergence of locally preferred states, thus dynamically defining the structure of the branching. Decoherence also ensures the nonobservability of interference effects between these preferred states from the point of local observers, leading to effectively independent (i.e., dynamically decoupled) "classical" branches.

The question of how exactly these branches (containing, among other things, certain patterns of neuronal states) may be linked to our actual conscious experiences is of course inherently difficult to answer, as it touches upon many aspects in the physical, chemical, biological, and psychological sciences. However, the important point is that, while the practical details of this modern "psycho-physical parallelism" are presently only poorly understood, decoherence provides the fundamental ingredients necessary to *in principle* describe observers and observations in a purely quantum-mechanical and empirically adequate manner. For example, the robust, temporally extended branches "created" by decoherence contain all of the essential properties associated with observers, such as well-defined stable measurement records that are correlated with the familiar determinate quantities of our experience and that are also redundantly imprinted in the environment. In contrast with the extraneous introduction of a collapse postulate, or with the assumption of the existence of intrinsically classical measurement devices or a collapse-inducing consciousness outside of the laws of quantum mechanics, this description is based solely on interacting physical systems governed by a universally valid quantum theory and is open to precise quantitative theoretical and experimental analysis.

Appendix: The Interaction Picture

The technique of expressing operators and quantum states in the so-called *interaction picture* is of great importance in treating quantum-mechanical problems in a perturbative fashion. It plays an important role, for example, in the derivation of the Born–Markov master equation for decoherence detailed in Sect. 4.2.2. We shall review the basics of the interaction-picture approach in the following.

We begin by considering a Hamiltonian of the form

$$\hat{H} = \hat{H}_0 + \hat{V}. \tag{A.1}$$

Here $\hat{H}_0$ denotes the part of the Hamiltonian that describes the free (unperturbed) evolution of a system $\mathcal{S}$, whereas $\hat{V}$ is some added external perturbation. In *applications* of the interaction-picture formalism, $\hat{V}$ is typically assumed to be weak in comparison with $\hat{H}_0$, and the idea is then to determine the approximate dynamics of the system given this assumption. In the following, however, we shall proceed with exact calculations only and make no assumption about the relative strengths of the two components $\hat{H}_0$ and $\hat{V}$.

From standard quantum theory, we know that the expectation value of an operator observable $\hat{A}(t)$ is given by the trace rule [see (2.17)],

$$\langle \hat{A}(t) \rangle = \mathrm{Tr}\left[\hat{A}(t)\hat{\rho}(t)\right] = \mathrm{Tr}\left[\hat{A}(t)\mathrm{e}^{-\mathrm{i}\hat{H}t}\hat{\rho}(0)\mathrm{e}^{\mathrm{i}\hat{H}t}\right]. \tag{A.2}$$

For reasons that will immediately become obvious, let us rewrite this expression as

$$\langle \hat{A}(t) \rangle = \mathrm{Tr}\left[\left(\mathrm{e}^{\mathrm{i}\hat{H}_0 t}\hat{A}(t)\mathrm{e}^{-\mathrm{i}\hat{H}_0 t}\right)\left(\mathrm{e}^{\mathrm{i}\hat{H}_0 t}\mathrm{e}^{-\mathrm{i}\hat{H}t}\hat{\rho}(0)\mathrm{e}^{\mathrm{i}\hat{H}t}\mathrm{e}^{-\mathrm{i}\hat{H}_0 t}\right)\right]. \tag{A.3}$$

In inserting the time-evolution operators $\mathrm{e}^{\pm\mathrm{i}\hat{H}_0 t}$ at the beginning and the end of the expression in square brackets, we have made use of the fact that the trace is cyclic under permutations of the arguments, i.e., that $\mathrm{Tr}\left(\hat{A}\hat{B}\hat{C}\cdots\right) = \mathrm{Tr}\left(\hat{B}\hat{C}\cdots\hat{A}\right)$, etc. Thus we could move the first factor $\mathrm{e}^{\mathrm{i}\hat{H}_0 t}$ to the end of the argument of the trace, canceling out the extra factor $\mathrm{e}^{-\mathrm{i}\hat{H}_0 t}$ at that position.

Equation (A.3) motivates us to now introduce the *interaction-picture* form of general operators $\hat{A}(t)$ and density matrices $\hat{\rho}(t)$ as

$$\hat{A}^{(I)}(t) = \mathrm{e}^{\mathrm{i}\hat{H}_0 t}\hat{A}(t)\mathrm{e}^{-\mathrm{i}\hat{H}_0 t}, \tag{A.4}$$

$$\begin{aligned} \hat{\rho}^{(I)}(t) &= \mathrm{e}^{\mathrm{i}\hat{H}_0 t}\hat{\rho}(t)\mathrm{e}^{-\mathrm{i}\hat{H}_0 t} \\ &= \mathrm{e}^{\mathrm{i}\hat{H}_0 t}\mathrm{e}^{-\mathrm{i}\hat{H}t}\hat{\rho}(0)\mathrm{e}^{\mathrm{i}\hat{H}t}\mathrm{e}^{-\mathrm{i}\hat{H}_0 t}. \end{aligned} \tag{A.5}$$

Here the superscript (I) is used to denote interaction-picture operators. Note that the dynamics of the interaction-picture operators $\hat{A}^{(I)}(t)$ are fully determined by the unperturbed (free) Hamiltonian $\hat{H}_0$ rather than by the total Hamiltonian $\hat{H}$.

With the new definitions (A.4) and (A.5), our expectation-value equation (A.3) can then be written in the compact form

$$\langle \hat{A}(t) \rangle = \mathrm{Tr}\left[\hat{A}^{(I)}(t)\hat{\rho}^{(I)}(t)\right]. \tag{A.6}$$

Let us now determine an evolution equation for the interaction-picture density matrix $\hat{\rho}^{(I)}(t)$. Using the standard Liouville–von Neumann equation for the density matrix $\hat{\rho}(t)$ in the Schrödinger picture,

$$\frac{\mathrm{d}}{\mathrm{d}t}\hat{\rho}(t) = -\mathrm{i}\left[\hat{H}, \hat{\rho}(t)\right], \tag{A.7}$$

we obtain

$$\begin{aligned} \frac{\mathrm{d}}{\mathrm{d}t}\hat{\rho}^{(I)}(t) &= \mathrm{i}\left[\hat{H}_0, \hat{\rho}^{(I)}(t)\right] + \mathrm{e}^{\mathrm{i}\hat{H}_0 t}\left(\frac{\mathrm{d}}{\mathrm{d}t}\hat{\rho}(t)\right)\mathrm{e}^{-\mathrm{i}\hat{H}_0 t} \\ &\overset{(A.7)}{=} \mathrm{i}\left[\hat{H}_0, \hat{\rho}^{(I)}(t)\right] - \mathrm{i}\mathrm{e}^{\mathrm{i}\hat{H}_0 t}\left[\hat{H}, \hat{\rho}(t)\right]\mathrm{e}^{-\mathrm{i}\hat{H}_0 t} \\ &\overset{(A.1)}{=} \mathrm{i}\left[\hat{H}_0, \hat{\rho}^{(I)}(t)\right] - \mathrm{i}\mathrm{e}^{\mathrm{i}\hat{H}_0 t}\left[\hat{H}_0 + \hat{V}, \hat{\rho}(t)\right]\mathrm{e}^{-\mathrm{i}\hat{H}_0 t}. \end{aligned} \tag{A.8}$$

From (A.4) and (A.5) we see that the application of the time-evolution operators $\mathrm{e}^{\pm\mathrm{i}\hat{H}_0 t}$ to the quantities in the second commutator amounts to a transformation of these quantities to the interaction picture. Equation (A.8) therefore becomes

$$\begin{aligned} \frac{\mathrm{d}}{\mathrm{d}t}\hat{\rho}^{(I)}(t) &= \mathrm{i}\left[\hat{H}_0, \hat{\rho}^{(I)}(t)\right] - \mathrm{i}\left[\hat{H}_0, \hat{\rho}^{(I)}(t)\right] - \mathrm{i}\left[\hat{V}^{(I)}(t), \hat{\rho}^{(I)}(t)\right] \\ &= -\mathrm{i}\left[\hat{V}^{(I)}(t), \hat{\rho}^{(I)}(t)\right]. \end{aligned} \tag{A.9}$$

This establishes our first main result: The time evolution of the interaction-picture density operator is given by an equation of the Liouville–von Neumann type, but with the interaction-picture perturbation $\hat{V}^{(I)}(t)$ instead of the full Hamiltonian $\hat{H}$. Appropriately, (A.9) is called the *interaction-picture Liouville–von Neumann equation.*

Up to now, we have considered the case of a single system $\mathcal{S}$ subject to some perturbation $\hat{V}$ whose origin we have not specified further. Let us now enter the setting relevant to decoherence: We suppose that the perturbation is due to the interaction with some external environment $\mathcal{E}$ described by an interaction Hamiltonian $\hat{H}_{\text{int}}$, i.e., $\hat{V} \equiv \hat{H}_{\text{int}}$. If we denote the self-Hamiltonians of the system and the environment by $\hat{H}_{\mathcal{S}}$ and $\hat{H}_{\mathcal{E}}$, respectively, we may write the total Hamiltonian of the composite system $\mathcal{SE}$ in the form

$$\hat{H} = \hat{H}_0 + \hat{V} \equiv \underbrace{\hat{H}_{\mathcal{S}} + \hat{H}_{\mathcal{E}}}_{\equiv \hat{H}_0} + \underbrace{\hat{H}_{\text{int}}}_{\equiv \hat{V}} . \tag{A.10}$$

Following the usual strategy, we are interested in determining the time evolution of the reduced density operator $\hat{\rho}_{\mathcal{S}}(t)$. In our case, we would like to obtain the reduced interaction-picture density operator

$$\hat{\rho}_{\mathcal{S}}^{(I)}(t) \equiv \text{Tr}_{\mathcal{E}} \left[\hat{\rho}^{(I)}(t) \right] . \tag{A.11}$$

We will now address the following two questions:

1. Given the reduced density operator $\hat{\rho}_{\mathcal{S}}(t)$ in the Schrödinger picture, what is the corresponding operator $\hat{\rho}_{\mathcal{S}}^{(I)}(t)$ in the interaction picture?
2. What is the appropriate time-evolution equation for $\hat{\rho}_{\mathcal{S}}^{(I)}(t)$?

Let us answer the first question. Evaluating the trace in (A.11) gives

$$\begin{aligned} \text{Tr}_{\mathcal{E}} \left[\hat{\rho}^{(I)}(t) \right] &\overset{(A.5)}{=} \text{Tr}_{\mathcal{E}} \left[e^{i\hat{H}_0 t} \hat{\rho}(t) e^{-i\hat{H}_0 t} \right] \\ &\overset{(A.10)}{=} e^{i\hat{H}_{\mathcal{S}} t} \left\{ \text{Tr}_{\mathcal{E}} \left[e^{i\hat{H}_{\mathcal{E}} t} \hat{\rho}(t) e^{-i\hat{H}_{\mathcal{E}} t} \right] \right\} e^{-i\hat{H}_{\mathcal{S}} t} \\ &= e^{i\hat{H}_{\mathcal{S}} t} \left\{ \text{Tr}_{\mathcal{E}} \left[\hat{\rho}(t) \right] \right\} e^{-i\hat{H}_{\mathcal{S}} t} \\ &= e^{i\hat{H}_{\mathcal{S}} t} \hat{\rho}_{\mathcal{S}}(t) e^{-i\hat{H}_{\mathcal{S}} t} \\ &\overset{(A.11)}{\equiv} \hat{\rho}_{\mathcal{S}}^{(I)}(t), \end{aligned} \tag{A.12}$$

where in the second-to-last line we have again used the invariance of the trace operation under cyclic permutations of the arguments (or, put slightly differently, the invariance of the trace under unitary transformations of the argument).

Equation (A.12) shows that the reduced density operator in the interaction picture is obtained by a unitary transformation of the reduced density operator in the Schrödinger picture involving the free *system* Hamiltonian $\hat{H}_{\mathcal{S}}$ only, i.e.,

$$\hat{\rho}_{\mathcal{S}}^{(I)}(t) = e^{i\hat{H}_{\mathcal{S}} t} \hat{\rho}_{\mathcal{S}}(t) e^{-i\hat{H}_{\mathcal{S}} t} . \tag{A.13}$$

The analogy of (A.13) to the corresponding transformation for the total interaction-picture density operator $\hat{\rho}^{(I)}(t)$, see (A.5), should now be clear. In each case, the transformation involves only the *free* Hamiltonian, namely, the Hamiltonian $\hat{H}_0$ for $\hat{\rho}^{(I)}(t)$ and the system Hamiltonian $\hat{H}_{\mathcal{S}}$ for the reduced interaction-picture density operator $\hat{\rho}_{\mathcal{S}}^{(I)}(t)$.

We note that the definition (A.13) also ensures, as desired, that expectation values of system observables $\hat{A}_{\mathcal{S}}(t)$ automatically agree in both the Schrödinger and interaction pictures, i.e., that

$$\langle \hat{A}_{\mathcal{S}}(t) \rangle = \mathrm{Tr}_{\mathcal{S}} \left[\hat{\rho}_{\mathcal{S}}(t) \hat{A}_{\mathcal{S}}(t) \right] = \mathrm{Tr}_{\mathcal{S}} \left[\hat{\rho}_{\mathcal{S}}^{(I)}(t) \hat{A}_{\mathcal{S}}^{(I)}(t) \right], \tag{A.14}$$

where

$$\hat{A}_{\mathcal{S}}^{(I)}(t) \stackrel{(\mathrm{A.4})}{=} \mathrm{e}^{\mathrm{i}\hat{H}_0 t} \hat{A}_{\mathcal{S}}(t) \mathrm{e}^{-\mathrm{i}\hat{H}_0 t} \stackrel{(\mathrm{A.10})}{=} \mathrm{e}^{\mathrm{i}\hat{H}_{\mathcal{S}} t} \hat{A}_{\mathcal{S}}(t) \mathrm{e}^{-\mathrm{i}\hat{H}_{\mathcal{S}} t}. \tag{A.15}$$

The proof of (A.14) is immediate:

$$\begin{aligned} \mathrm{Tr}_{\mathcal{S}} \left[\hat{\rho}_{\mathcal{S}}^{(I)}(t) \hat{A}_{\mathcal{S}}^{(I)}(t) \right] &= \mathrm{Tr}_{\mathcal{S}} \left[\mathrm{e}^{\mathrm{i}\hat{H}_{\mathcal{S}} t} \hat{\rho}_{\mathcal{S}}(t) \hat{A}_{\mathcal{S}}(t) \mathrm{e}^{-\mathrm{i}\hat{H}_{\mathcal{S}} t} \right] \\ &= \mathrm{Tr}_{\mathcal{S}} \left[\hat{\rho}_{\mathcal{S}}(t) \hat{A}_{\mathcal{S}}(t) \right]. \end{aligned} \tag{A.16}$$

Finally, let us now answer the second question posed above, i.e., let us find an equation of motion for the reduced interaction-picture density operator $\hat{\rho}_{\mathcal{S}}^{(I)}(t)$. Taking the trace over the environment on both sides of our evolution equation (A.9) for the full interaction-picture density operator $\hat{\rho}^{(I)}(t)$, with $\hat{V}^{(I)}(t) \equiv \hat{H}_{\mathrm{int}}^{(I)}(t)$, yields

$$\frac{\mathrm{d}}{\mathrm{d}t} \hat{\rho}_{\mathcal{S}}^{(I)}(t) = -\mathrm{i}\, \mathrm{Tr}_{\mathcal{E}} \left[\hat{H}_{\mathrm{int}}^{(I)}(t), \hat{\rho}^{(I)}(t) \right]. \tag{A.17}$$

A closer inspection reveals that this equation is *not* of the standard Liouville–von Neumann form, since the right-hand side depends on the *full* density operator $\hat{\rho}^{(I)}(t)$ rather than on the reduced density operator $\hat{\rho}_{\mathcal{S}}^{(I)}(t)$. This fact should not be too surprising to the reader. After all, the state of the environment will generally influence the evolution of the reduced density operator.

Thus, in general, we must consider the full interacting system–environment combination to determine the reduced dynamics. (This observation is, of course, independent of whether we work in the Schrödinger or the interaction picture.) Only once certain approximations are imposed (for instance, the assumption of weak system–environment coupling), we may obtain an evolution equation for the reduced density operator of the system that does not explicitly depend on the time-dependent total (system–environment) density operator. Such *master equations* are derived in Chaps. 4 and 5.

References

1. M. Schlosshauer, Decoherence, the measurement problem, and interpretations of quantum mechanics, *Rev. Mod. Phys.* **76**, 1267–1305 (2004).
2. E. Schrödinger, Die gegenwärtige Situation in der Quantenmechanik, *Naturwissenschaften* **23**, 807–812, 823–828, 844–849 (1935).
3. A. Pais, Einstein and the quantum theory, *Rev. Mod. Phys.* **51**, 863–914 (1979).
4. H. D. Zeh, On the interpretation of measurement in quantum theory, *Found. Phys.* **1**, 69–76 (1970).
5. H. D. Zeh, Toward a quantum theory of observation, *Found. Phys.* **3**, 109–116 (1973).
6. O. Kübler, H. D. Zeh, Dynamics of quantum correlations, *Ann. Phys. (N.Y.)* **76**, 405–418 (1973).
7. E. Joos, H. D. Zeh, The emergence of classical properties through interaction with the environment, *Z. Phys. B: Condens. Matter* **59**, 223–243 (1985).
8. W. H. Zurek, Pointer basis of quantum apparatus: Into what mixture does the wave packet collapse?, *Phys. Rev. D* **24**, 1516–1525 (1981).
9. W. H. Zurek, Environment-induced superselection rules, *Phys. Rev. D* **26**, 1862–1880 (1982).
10. A. Einstein, B. Podolsky, N. Rosen, Can quantum-mechanical description of physical reality be considered complete?, *Phys. Rev.* **47**, 777–780 (1935).
11. H. D. Zeh, Roots and fruits of decoherence, in: B. Duplantier, J.-M. Raimond, V. Rivasseau (Eds.), *Quantum Decoherence*, Birkhäuser, 2006, pp. 151–175.
12. W. H. Zurek, Reduction of the wavepacket: How long does it take?, in: G. T. Moore, M. O. Scully (Eds.), *Frontiers of Nonequilibrium Statistical Mechanics*, Plenum Press, New York, 1986, pp. 145–149, first published in 1984 as Los Alamos report LAUR 84-2750.
13. W. H. Zurek, Decoherence and the transition from quantum to classical, *Phys. Today* **44**, 36–44 (1991). See also the updated version available as eprint quant-ph/0306072.
14. E. Joos, Elements of environmental decoherence, in: P. Blanchard, D. Giulini, E. Joos, C. Kiefer, I.-O. Stamatescu (Eds.), *Decoherence: Theoretical, Experimental, and Conceptual Problems*, Springer, Berlin, 2000, pp. 1–17.
15. J. P. Paz, W. H. Zurek, Environment-induced decoherence and the transition from quantum to classical, in: R. Kaiser, C. Westbrook, F. David (Eds.), *Coherent Atomic Matter Waves, Les Houches Session LXXII*, Vol. 72 of *Les Houches Summer School Series*, Springer, Berlin, 2001, pp. 533–614.
16. W. H. Zurek, Decoherence, einselection, and the quantum origins of the classical, *Rev. Mod. Phys.* **75**, 715–775 (2003).

17. E. Joos, H. D. Zeh, C. Kiefer, D. Giulini, J. Kupsch, I.-O. Stamatescu, *Decoherence and the Appearance of a Classical World in Quantum Theory*, 2nd Edition, Springer, New York, 2003.
18. H.-P. Breuer, F. Petruccione, *The Theory of Open Quantum Systems*, Oxford University Press, Oxford, 2002.
19. M. Born, Zur Quantenmechanik der Stoßvorgänge, *Z. Phys.* **37**, 863–867 (1926). Reprinted and translated in [40], pp. 52–55.
20. W. Wootters, W. Zurek, A single quantum cannot be cloned, *Nature* **299**, 802–803 (1982).
21. D. Dieks, Communication by EPR devices, *Phys. Lett. A* **92**, 271–272 (1982).
22. M. Alford, S. Coleman, J. March-Russel, Disentangling nonabelian discrete quantum hair, *Nucl. Phys. B* **351**, 735–748 (1991).
23. D. Dieks, Overlap and distinguishability of quantum states, *Phys. Lett. A* **126**, 303–307 (1988).
24. A. Peres, How to differentiate between non-orthogonal states, *Phys. Lett. A* **128**, 19 (1988).
25. G. Jaeger, A. Shimony, Optimal distinction between two non-orthogonal quantum states, *Phys. Lett. A* **197**, 83–87 (1995).
26. P. Busch, Is the quantum state (an) observable?, in: R. S. Cohen, M. A. Horne, J. Stachel (Eds.), *Potentiality, Entanglement and Passion-at-a-Distance: Quantum Mechanical Studies for Abner Shimony*, Kluwer, Dordrecht, 1997, pp. 61–70.
27. J. S. Bell, *Speakable and Unspeakable in Quantum Mechanics*, Cambridge University Press, Cambridge, England, 1987.
28. M. Daumer, D. Dürr, S. Goldstein, N. Zanghì, Naive realism about operators, *Erkenntnis* **45**, 379–398 (1996).
29. A. Fine, *The Shaky Game: Einstein, Realism and the Quantum Theory*, 2nd Edition, *Science and its Conceptual Foundations Series*, University of Chicago Press, Chicago/London, 1996.
30. J. S. Bell, On the Einstein–Podolsky–Rosen paradox, *Physics* **1**, 195–200 (1964).
31. J. S. Bell, On the problem of hidden variables in quantum mechanics, *Rev. Mod. Phys.* **38**, 447–452 (1966).
32. L. E. Szabo, A. Fine, A local hidden variable theory for the GHZ experiment, *Phys. Lett. A* **295**, 229–240 (2002).
33. G. Weihs, T. Jennewein, C. Simon, H. Weinfurter, A. Zeilinger, Violation of Bell's inequality under strict Einstein locality condition, *Phys. Rev. Lett.* **58**, 5039–5043 (1998).
34. L. de Broglie, *An Introduction to the Study of Wave Mechanics*, E. P. Dutton and Co., New York, 1930.
35. D. Bohm, A suggested interpretation of the quantum theory in terms of "hidden variables", I and II, *Phys. Rev.* **85**, 166–193 (1952).
36. D. Bohm, J. Bub, A proposed solution of the measurement problem in quantum mechanics by a hidden variable theory, *Rev. Mod. Phys.* **38**, 453–469 (1966).
37. D. Bohm, B. Hiley, *The Undivided Universe*, Routledge, London, 1993.
38. S. Kochen, E. Specker, The problem of hidden variables in quantum mechanics, *J. Math. Mech.* **17**, 59–87 (1967).

39. H. D. Zeh, The wave function: It or bit?, in: J. D. Barrow, P. C. W. Davies, C. L. Harper Jr. (Eds.), *Science and Ultimate Reality: Quantum Theory, Cosmology and Complexity*, Cambridge University Press, 2004, pp. 103–120.
40. J. A. Wheeler, W. H. Zurek (Eds.), *Quantum Theory and Measurement*, Princeton University, Princeton, 1983.
41. J. A. Wheeler, The "past" and the "delayed-choice" double-slit experiment, in: A. R. Marlow (Ed.), *Mathematical Foundations of Quantum Theory*, Academic Press, New York, 1978, pp. 9–48.
42. W. Heisenberg, Über die Grundprinzipien der "Quantenmechanik", *Forschungen und Fortschritte* **3**, 83 (1927). Reprinted and translated in [40], pp. 62–84.
43. A. Einstein, H. Born, M. Born, *Briefwechsel: 1916–1955*, Edition Erbrich, Frankfurt am Main, 1982.
44. C. A. Fuchs, A. Peres, Quantum theory needs no "interpretation", *Phys. Today* **53**, 70–71 (2000).
45. L. E. Ballentine, The statistical interpretation of quantum mechanics, *Rev. Mod. Phys.* **42**, 358–381 (1970).
46. L. E. Ballentine, *Quantum Mechanics: A Modern Development*, 2nd Edition, World Scientific Publishing, New Jersey, 1998.
47. B. d'Espagnat, Two remarks on the theory of measurement, *Nuovo Cimento Suppl.* **1**, 828–838 (1966).
48. B. d'Espagnat, *Conceptual Foundations of Quantum Mechanics*, 2nd Edition, Benjamin, Reading, Massachusetts, 1976.
49. B. d'Espagnat, *Veiled Reality, an Analysis of Present-Day Quantum Mechanical Concepts*, Addison-Wesley, Reading, Massachusetts, 1995.
50. Y. Aharonov, L. Vaidman, Complete description of a quantum system at a given time, *J. Phys. A: Math. Gen.* **24**, 2315–2328 (1991).
51. N. F. Ramsey, A molecular beam resonance method with separated oscillating fields, *Phys. Rev.* **78**, 695–699 (1950).
52. E. Schrödinger, Discussion of probability relations between separated systems, *Proc. Cambridge Philos. Soc.* **31**, 555–563 (1935).
53. E. Schrödinger, Discussion of probability relations between separated systems, *Proc. Cambridge Philos. Soc.* **36**, 446–451 (1936).
54. R. Kaltenbaek, M. Aspelmeyer, T. Jennewein, C. Brukner, A. Zeilinger, M. Pfennigbauer, W. R. Leeb, Proof-of-concept experiments for quantum physics in space, in: *SPIE Proceedings on Quantum Communications and Quantum Imaging*, International Society for Optical Engineering, 2003, pp. 252–268.
55. J. P. Dowling, G. J. Milburn, Quantum technology: The second quantum revolution, *Proc. R. Soc. Lond. A* **361**, 1655–1674 (2003).
56. M. A. Nielsen, I. L. Chuang, *Quantum Computation and Quantum Information*, Cambridge University Press, Cambridge, 2000.
57. E. Schrödinger, The present situation in quantum mechanics: A translation of Schrödinger's "cat paradox" paper, *Proc. Am. Philos. Soc.* **124**, 323–338 (1980). Translated by John D. Trimmer.
58. J. J. Sakurai, *Modern Quantum Mechanics*, Addison-Wesley, Reading, Massachusetts, 1994.
59. J. von Neumann, Thermodynamik quantummechanischer Gesamheiten, *Gött. Nach.* **1**, 273–291 (1927).
60. J. von Neumann, *Mathematische Grundlagen der Quantenmechanik*, Springer, Berlin, 1932.

61. D. Petz, Entropy, von Neumann and the von Neumann entropy, in: M. Rédei, M. Stöltzner (Eds.), *John von Neumann and the Foundations of Quantum Physics*, Kluwer, Dordrecht, 2001, pp. 83–96.
62. L. D. Landau, The damping problem in wave mechanics, *Z. Phys.* **45**, 430–441 (1927).
63. W. H. Furry, Note on the quantum mechanical theory of measurement, *Phys. Rev.* **49**, 393–399 (1936).
64. O. Pessoa, Can the decoherence approach help to solve the measurement problem?, *Synthese* **113**, 323–346 (1998).
65. W. H. Zurek, Environment-assisted invariance, entanglement, and probabilities in quantum physics, *Phys. Rev. Lett.* **90**, 120404 (2003).
66. W. H. Zurek, Quantum Darwinism and envariance, in: J. D. Barrow, P. C. W. Davies, C. H. Harper (Eds.), *Science and Ultimate Reality*, Cambridge University Press, Cambridge, England, 2004, pp. 121–137.
67. W. H. Zurek, Probabilities from entanglement, Born's rule $p_k = |\psi_k|^2$ from envariance, *Phys. Rev. A* **71**, 052105 (2005).
68. J. Bub, *Interpreting the Quantum World*, 1st Edition, Cambridge University Press, Cambridge, England, 1997.
69. V. B. Braginsky, F. Y. Khalili, Quantum nondemolition measurements: the route from toys to tools, *Rev. Mod. Phys.* **68**, 1–11 (1996).
70. N. Bohr, The quantum postulate and the recent development of atomic theory, in: E. Rüdinger, J. Kolckar (Eds.), *Collected Works, Vol. 6: Foundations of Quantum Physics I.*, North-Holland, Amsterdam, 1985, pp. 109–136.
71. M. Jammer, *The Philosophy of Quantum Mechanics*, 1st Edition, John Wiley & Sons, New York, 1974.
72. N. Bohr, Discussions with Einstein on epistemological problems in atomic physics, in: P. A. Schilpp (Ed.), *Albert Einstein: Philosopher–Scientist*, Library of Living Philosophers, Evanston, Illinois, 1949, pp. 200–241, reprinted in [40], pp. 9–49.
73. P. Bertet, S. Osnaghi, A. Rauschenbeutel, G. Nogues, A. Auffeves, M. Brune, J. M. Raimond, S. Haroche, A complementarity experiment with an interferometer at the quantum–classical boundary, *Nature* **411**, 166–170 (2001).
74. W. K. Wootters, W. H. Zurek, Complementarity in the double-slit experiment: Quantum nonseparability and a quantitative statement of Bohr's principle, *Phys. Rev. D* **19**, 473–484 (1979).
75. M. O. Scully, K. Drühl, Quantum eraser: A proposed photon correlation experiment concerning observation and "delayed choice" in quantum mechanics, *Phys. Rev. A* **25**, 2208–2213 (1982).
76. W. H. Zurek, Information transfer in quantum measurements: Irreversibility and amplification, in: P. Meystre, M. O. Scully (Eds.), *Quantum Optics, Experimental Gravitation, and Measurement Theory*, Plenum Press, 1983, pp. 87–116, lecture notes from the NATO Advanced Study Institute held in Bad Windsheim, Germany, August 17–30, 1981.
77. C. Kiefer, E. Joos, Decoherence: Concepts and examples, in: P. Blanchard, A. Jadczyk (Eds.), *Quantum Future: From Volta and Como to the Present and Beyond (Proceedings of the Xth Max Born Symposium Held in Przesieka, Poland, 24–27 September 1997)*, Springer, Berlin, 1999, p. 105.
78. G. C. Wick, A. S. Wightman, E. P. Wigner, The intrinsic parity of elementary particles, *Phys. Rev.* **88**, 101–105 (1952).

79. G. C. Wick, A. S. Wightman, E. P. Wigner, Superselection rule for charge, *Phys. Rev. D* **1**, 3267–3269 (1970).
80. A. Galindo, A. Morales, R. Núñez-Lagos, Superselection principle and pure states of n-identical particles, *J. Math. Phys.* **3**, 324–328 (1962).
81. A. S. Wightman, Superselection rules: old and new, *Nuovo Cimento B* **110**, 751–769 (1995).
82. C. Cisnerosy, R. P. Martinez-y-Romeroz, H. N. Núñez-Yépe, A. L. Salas-Brito, Limitations to the superposition principle: Superselection rules in non-relativistic quantum mechanics, *Eur. J. Phys.* **19**, 237–243 (1998).
83. D. Giulini, Decoherence: A dynamical approach to superselection rules?, *Lect. Notes Phys.* **559**, 67–92 (2000).
84. W. H. Zurek, Preferred states, predictabilty, classicality, and the environment-induced decoherence, *Prog. Theor. Phys.* **89**, 281–312 (1993).
85. D. Giulini, C. Kiefer, H. D. Zeh, Symmetries, superselection rules, and decoherence, *Phys. Lett. A* **199**, 291–298 (1995).
86. W. H. Zurek, S. Habib, J. P. Paz, Coherent states via decoherence, *Phys. Rev. Lett.* **70**, 1187–1190 (1993).
87. E. P. Wigner, Die Messung quantenmechanischer Operatoren, *Z. Phys.* **133**, 101–108 (1952).
88. H. Araki, M. M. Yanase, Measurement of quantum mechanical operators, *Phys. Rev.* **120**, 622–626 (1960).
89. G. M. Palma, K.-A. Suominen, A. K. Ekert, Quantum computers and dissipation, *Proc. R. Soc. Lond. A* **452**, 567–584 (1996).
90. D. A. Lidar, I. L. Chuang, K. B. Whaley, Decoherence-free subspaces for quantum computation, *Phys. Rev. Lett.* **81**, 2594–2597 (1998).
91. P. Zanardi, M. Rasetti, Noiseless quantum codes, *Phys. Rev. Lett.* **79**, 3306–3309 (1997).
92. P. Zanardi, M. Rasetti, Error avoiding quantum codes, *Mod. Phys. Lett. B* **11**, 1085–1093 (1997).
93. P. Zanardi, Dissipation and decoherence in a quantum register, *Phys. Rev. A* **57**, 3276–3284 (1998).
94. D. A. Lidar, D. Bacon, K. B. Whaley, Concatenating decoherence-free subspaces with quantum error correcting codes, *Phys. Rev. Lett.* **82**, 4556–4559 (1999).
95. D. Bacon, J. Kempe, D. A. Lidar, K. B. Whaley, Universal fault-tolerant quantum computation on decoherence-free subspaces, *Phys. Rev. Lett.* **85**, 1758–1761 (2000).
96. L.-M. Duan, G.-C. Guo, Reducing decoherence in quantum-computer memory with all quantum bits coupling to the same environment, *Phys. Rev. A* **57**, 737–741 (1998).
97. P. Zanardi, Stabilizing quantum information, *Phys. Rev. A* **63**, 012301 (2001).
98. E. Knill, R. Laflamme, L. Viola, Theory of quantum error correction for general noise, *Phys. Rev. Lett.* **82**, 2525–2528 (2000).
99. D. Bacon, D. A. Lidar, K. B. Whaley, Robustness of decoherence-free subspaces for quantum computation, *Phys. Rev. A* **60**, 1944–1955 (1999).
100. J. Kempe, D. Bacon, D. A. Lidar, K. B. Whaley, Theory of decoherence-free fault-tolerant universal quantum computation, *Phys. Rev. A* **63**, 042307 (2001).

101. D. A. Lidar, K. B. Whaley, Decoherence-free subspaces and subsystems, in: F. Benatti, R. Floreanini (Eds.), *Irreversible Quantum Dynamics*, Vol. 622 of *Springer Lecture Notes in Physics*, Springer, Berlin, 2003, pp. 83–120, also available as eprint quant-ph/0301032.
102. D. A. R. Dalvit, J. Dziarmaga, W. H. Zurek, Decoherence in Bose–Einstein condensates: Towards bigger and better Schrödinger cats, *Phys. Rev. A* **62**, 013607 (2000).
103. J. P. Paz, W. H. Zurek, Quantum limit of decoherence: Environment induced superselection of energy eigenstates, *Phys. Rev. Lett.* **82**, 5181–5185 (1999).
104. W. H. Zurek, Decoherence, einselection, and the existential interpretation (The Rough Guide), *Philos. Trans. R. Soc. London, Ser. A* **356**, 1793–1821 (1998).
105. L. Diósi, C. Kiefer, Robustness and diffusion of pointer states, *Phys. Rev. Lett.* **85**, 3552–3555 (2000).
106. J. Eisert, Exact decoherence to pointer states in free open quantum systems is universal, *Phys. Rev. Lett.* **92**, 210401 (2004).
107. R. A. Harris, L. Stodolsky, On the time dependence of optical activity, *J. Chem. Phys.* **74**, 2145–2155 (1981).
108. H. D. Zeh, The meaning of decoherence, in: P. Blanchard, D. Giulini, E. Joos, C. Kiefer, I. Stamatescu (Eds.), *Decoherence: Theoretical, Experimental, and Conceptual Problems, Lecture Notes in Physics No. 538*, Springer, Berlin, 2000, pp. 19–42.
109. W. G. Unruh, W. H. Zurek, Reduction of a wavepacket in quantum Brownian motion, *Phys. Rev. D* **40**, 1071–1094 (1989).
110. H. Ollivier, D. Poulin, W. H. Zurek, Emergence of objective properties from subjective quantum states: Environment as a witness, *Phys. Rev. Lett.* **93**, 220401 (2004).
111. H. Ollivier, D. Poulin, W. H. Zurek, Environment as a witness: selective proliferation of information and emergence of objectivity, *Phys. Rev. A* **72**, 042113 (2005).
112. R. Blume-Kohout, W. H. Zurek, A simple example of "Quantum Darwinism": Redundant information storage in many-spin environments, *Found. Phys.* **35**, 1857–1876 (2005).
113. R. Blume-Kohout, W. H. Zurek, Quantum Darwinism: Entanglement, branches, and the emergent classicality of redundantly stored quantum information, *Phys. Rev. A* **73**, 062310 (2006).
114. C. E. Shannon, A mathematical theory of communication, *Bell Syst. Tech. J.* **27**, 379–423, 623–656 (1948).
115. C. E. Shannon, Communication theory of secrecy systems, *Bell Syst. Tech. J.* **28**, 656–715 (1949).
116. A. Galindo, M. A. Martín-Delgado, Information and computation: Classical and quantum aspects, *Rev. Mod. Phys.* **74**, 347–423 (2002).
117. H. Ollivier, W. H. Zurek, Quantum discord: A measure of the quantumness of correlations, *Phys. Rev. Lett.* **88**, 017901 (2001).
118. F. M. Cucchietti, J. P. Paz, W. H. Zurek, Gaussian decoherence from random spin environments, *Phys. Rev. A* **72**, 052113 (2005).
119. F. M. Cucchietti, H. M. Pastawski, D. A. Wisniacki, Decoherence as decay of the Loschmidt echo in a Lorentz gas, *Phys. Rev. E* **65**, 045206 (2002).
120. F. M. Cucchietti, D. A. R. Dalvit, J. P. Paz, W. H. Zurek, Decoherence and the Loschmidt echo, *Phys. Rev. Lett.* **91**, 210403 (2003).

121. J. Allinger, U. Weiss, Nonuniversality of dephasing in quantum transport, *Z. Physik B* **98**, 289–296 (1995).
122. M. Schlosshauer, Self-induced decoherence approach: Strong limitations on its validity in a simple spin bath model and on its general physical relevance, *Phys. Rev. A* **72**, 012109 (2005).
123. A. M. Jayannavar, Comment on some theories of state reduction, *Phys. Lett. A* **167**, 433–434 (1992).
124. A. Venugopalan, D. Kumar, R. Ghosh, Environment-induced decoherence II. Effect of decoherence on Bell's inequality for an EPR pair, *Physica A* **220**, 576–584 (1995).
125. C. J. Myatt, B. E. King, Q. A. Turchette, C. A. Sackett, D. Kielpinski, W. M. Itano, C. Monroe, D. J. Wineland, Decoherence of quantum superpositions through coupling to engineered reservoirs, *Nature* **403**, 269–273 (2000).
126. H. Wakita, Measurement in quantum mechanics. II — Reduction of a wave packet, *Prog. Theor. Phys.* **27**, 139–144 (1962).
127. E. P. Wigner, The problem of measurement, *Am. J. Phys.* **31**, 6–15 (1963).
128. E. Jaynes, Quantum beats, in: A. O. Barut (Ed.), *Foundations of Radiation Theory and Quantum Electrodynamics*, Plenum Press, New York, 1980, pp. 37–43.
129. A. Peres, Can we undo quantum measurements?, *Phys. Rev. D* **22**, 879–883 (1980).
130. M. O. Scully, B. G. Englert, H. Walther, Quantum optical tests of complementarity, *Nature (London)* **351**, 111–116 (1991).
131. Y.-H. Kim, R. Yu, S. P. Kulik, Y. Shih, M. O. Scully, Delayed "choice" quantum eraser, *Phys. Rev. Lett.* **84**, 1–5 (2000).
132. P. G. Kwiat, A. M. Steinberg, R. Y. Chiao, Observation of a "quantum eraser": A revival of coherence in a two-photon interference experiment, *Phys. Rev. A* **45**, 7729–7739 (1992).
133. A. G. Zajonc, L. J. Wand, X. Y. Zou, L. Mandel, Quantum eraser, *Nature* **353**, 507–508 (1991).
134. T. J. Herzog, P. G. Kwiat, H. Weinfurter, A. Zeilinger, Complementarity and the quantum eraser, *Phys. Rev. Lett.* **75**, 3034–3037 (1995).
135. S. P. Walborn, M. O. T. Cunha, S. Pádua, C. H. Monken, Double-slit quantum eraser, *Phys. Rev. A* **65**, 033818 (2002).
136. C. H. Bennett, Demons, engines, and the second law, *Sci. Amer.* **257**, 108–117 (1987).
137. H. D. Zeh, *The Physical Basis of the Direction of Time*, 5th Edition, Springer, Berlin, 2007.
138. G. Auletta, *Foundations and Interpretation of Quantum Mechanics in the Light of a Critical-Historical Analysis of the Problems and of a Synthesis of the Results*, World Scientific, Singapore, 2000.
139. N. P. Landsman, Observation and superselection in quantum mechanics, *Stud. Hist. Philos. Mod. Phys.* **26**, 45–73 (1995).
140. H. Everett, "Relative state" formulation of quantum mechanics, *Rev. Mod. Phys.* **29**, 454–462 (1957).
141. N. D. Mermin, The Ithaca interpretation of quantum mechanics, *Pramana* **51**, 549–565 (1998).
142. N. D. Mermin, What is quantum mechanics trying to tell us?, *Am. J. Phys.* **66**, 753–767 (1998).

143. C. Rovelli, Relational quantum mechanics, *Int. J. Theor. Phys.* **35**, 1637–1678 (1996).
144. E. Schmidt, Zur Theorie der linearen und nichtlinearen Integralgleichungen, *Math. Annalen* **63**, 433–476 (1907).
145. A. Albrecht, Investigating decoherence in a simple system, *Phys. Rev. D* **46**, 5504–5520 (1992).
146. A. Albrecht, Following a "collapsing" wave function, *Phys. Rev. D* **48**, 3768–3778 (1993).
147. A. O. Barvinsky, A. Y. Kamenshchik, Preferred basis in quantum theory and the problem of classicalization of the quantum Universe, *Phys. Rev. D* **52**, 743–757 (1995).
148. A. Kent, J. McElwaine, Quantum prediction algorithms, *Phys. Rev.* **55**, 1703–1720 (1997).
149. E. Wigner, On the quantum correction for thermodynamic equilibrium, *Phys. Rev.* **40**, 749–759 (1932).
150. M. Hillery, R. F. O'Connell, M. O. Scully, E. P. Wigner, Distribution functions in physics: Fundamentals, *Phys. Rep.* **106**, 121–167 (1984).
151. R. L. Hudson, When is the Wigner quasi-probability density non-negative?, *Rep. Math. Phys.* **6**, 249–252 (1974).
152. K. Kraus, *States, Effects, and Operations*, Springer, Berlin, 1983.
153. L. de Broglie, Sur la possibilité de relier les phénomènes d'interférences et de diffraction à la théorie des quanta de lumière, *Comptes Rendus Hebdomadaires des Séances de l'Académie des Sciences (Paris)* **183**, 447–448 (1926).
154. L. de Broglie, La mécanique ondulatoire et la structure atomique de la matière et du rayonnement, *Le Journal de Physique et le Radium* **8**, 225–241 (1927).
155. L. de Broglie, La structure atomique de la matière et du rayonnement et la mécanique ondulatoire, *Comptes Rendus Hebdomadaires des Séances de l'Académie des Sciences (Paris)* **184**, 273–274 (1927).
156. L. de Broglie, *Recherches Sur La Théorie Des Quanta*, Ph.D. thesis, Faculty of Sciences at Paris University (1924).
157. E. Schrödinger, Der stetige Übergang von der Micro- zur Macromechanik, *Naturwissenschaften* **14**, 664–666 (1926).
158. W. H. Zurek, Decoherence, chaos, quantum–classical correspondence, and the algorithmic arrow of time, *Phys. Scri. T* **76**, 186–198 (1998).
159. L. Diósi, Quantum master equation of a particle in a gas environment, *Europhys. Lett.* **30**, 63–68 (1995).
160. K. Hornberger, J. E. Sipe, Collisional decoherence reexamined, *Phys. Rev. A* **68**, 012105 (2003).
161. S. L. Adler, Normalization of collisional decoherence: Squaring the delta function, and an independent cross-check, *J. Phys. A: Math. Gen.* **39**, 14067–14074 (2006).
162. M. R. Gallis, G. N. Fleming, Environmental and spontaneous localization, *Phys. Rev. A* **42**, 38–48 (1990).
163. J. J. Halliwell, Two derivations of the master equation of quantum Brownian motion, *J. Phys. A: Math. Theor.* **40**, 3067–3080 (2007).
164. K. Hornberger, Monitoring approach to open quantum dynamics using scattering theory, *EPL* **77**, 50007 (2006).
165. K. Hornberger, Master equation for a quantum particle in a gas, *Phys. Rev. Lett.* **97**, 060601 (2006).

166. M. Tegmark, Apparent wave function collapse caused by scattering, *Found. Phys. Lett.* **6**, 571–590 (1993).
167. R. Shankar, *Principles of Quantum Mechanics*, 2nd Edition, Plenum Press, New York, 1994.
168. J. D. Jackson, *Classical Electrodynamics*, John Wiley, New York, 1999.
169. P. Zitzewitz, R. Neff, *Physics*, Glencoe, New York, 1995.
170. D. W. Keith, C. R. Ekstrom, Q. A. Turchette, D. E. Pritchard, An interferometer for atoms, *Phys. Rev. Lett.* **66**, 2693–2696 (1991).
171. D. A. Kokorowski, A. D. Cronin, T. D. Roberts, D. E. Pritchard, From single- to multiple-photon decoherence in an atom interferometer, *Phys. Rev. Lett.* **86**, 2191–2195 (2001).
172. H. Uys, J. D. Perreault, A. D. Cronin, Matter-wave decoherence due to a gas environment in an atom interferometer, *Phys. Rev. Lett.* **95**, 150403 (2005).
173. M. Morikawa, Quantum decoherence and classical correlation in quantum mechanics, *Phys. Rev. D* **42**, 2929–2932 (1990).
174. A. G. Redfield, On the theory of relaxation processes, *IBM J. Res. Develop.* **1**, 19–31 (1957).
175. K. Blum, *Density Matrix Theory and Applications*, Plenum Press, New York, London, 1981.
176. V. Gorini, A. Kossakowski, E. C. G. Sudarshan, Completely positive dynamical semigroups of N-level systems, *J. Math. Phys.* **17**, 821–825 (1976).
177. G. Lindblad, On the generators of quantum dynamical semigroups, *Commun. Math. Phys.* **48**, 119–130 (1976).
178. F. Haake, Statistical treatment of open systems by generalized master equations, in: G. Höhler (Ed.), *Springer Tracts in Modern Physics*, Vol. 66, Springer, Berlin, 1973, pp. 98–168.
179. V. Gorini, A. Frigerio, M. Verri, A. Kossakowski, E. C. G. Sudarshan, Properties of quantum Markovian master equations, *Rep. Math. Phys.* **13**, 149–173 (1978).
180. E. B. Davies, Markovian master equations, *Commun. Math. Phys.* **39**, 91–110 (1974).
181. A. Kossakowski, On quantum statistical mechanics of non-Hamiltonian systems, *Rep. Math. Phys.* **3**, 247–288 (1972).
182. R. Dümcke, H. Spohn, The proper form of the generator in the weak-coupling limit, *Z. Phys. B* **34**, 419–422 (1979).
183. R. Karrlein, H. Grabert, Exact time evolution and master equations for the damped harmonic oscillator, *Phys. Rev. A* **55**, 153–164 (1997).
184. J. P. Paz, S. Habib, W. H. Zurek, Reduction of the wave packet: Preferred observable and decoherence time scale, *Phys. Rev. D* **47**, 488–501 (1993).
185. L. D. Romero, J. P. Paz, Decoherence and initial correlations in quantum Brownian motion, *Phys. Rev. A* **55**, 4070–4083 (1997).
186. A. Barchielli, V. P. Belavkin, Measurements continuous in time and a posteriori states in quantum mechanics, *J. Phys. A: Math. Gen.* **24**, 1495–1514 (1991).
187. V. P. Belavkin, Non-demolition measurements, nonlinear filtering and dynamic programming of quantum stochastic processes, in: *Lecture notes in Control and Information Sciences*, Vol. 121, Springer, Berlin, 1989, pp. 245–265.
188. V. P. Belavkin, A continuous counting observation and posterior quantum dynamics, *J. Phys. A: Math. Gen.* **22**, L1109–L1114 (1989).

189. V. P. Belavkin, A new wave equation for a continuous non-demolition measurement, *Phys. Lett. A* **140**, 355–358 (1989).
190. V. P. Belavkin, The interplay of classical and quantum stochastics: Diffusion, measurement and filtering, in: P. Garbaczewksi, M. Wolf, A. Veron (Eds.), *Chaos: The Interplay Between Stochastic and Deterministic Behaviour*, *Lecture Notes in Physics*, Springer, 1995, pp. 21–41.
191. L. Diósi, Continuous quantum measurement and Itô formalism, *Phys. Lett. A* **129**, 419–423 (1988).
192. L. Diósi, Localized solution of a simple nonlinear quantum Langevin equation, *Phys. Lett. A* **132**, 233–236 (1988).
193. L. Diósi, Quantum stochastic processes as models for state vector reduction, *J. Phys. A* **21**, 2885–2898 (1988).
194. N. Gisin, Quantum measurements and stochastic processes, *Phys. Rev. Lett.* **52**, 1657–1660 (1984).
195. N. Gisin, Stochastic quantum dynamics and relativity, *Helv. Phys. Acta* **62**, 363–371 (1989).
196. H. M. Wiseman, Quantum theory of continuous feedback, *Phys. Rev. A* **49**, 2133–2150 (1994).
197. H.-S. Goan, G. J. Milburn, H. M. Wiseman, H. B. Sun, Continuous quantum measurement of two coupled quantum dots using a point contact: A quantum trajectory approach, *Phys. Rev. B* **63**, 125326 (2001).
198. M. B. Plenio, P. L. Knight, The quantum-jump approach to dissipative dynamics in quantum optics, *Rev. Mod. Phys.* **70**, 101–144 (1998).
199. N. V. Prokof'ev, P. C. E. Stamp, Theory of the spin bath, *Rep. Prog. Phys.* **63**, 669–726 (2000).
200. M. Dubé, P. C. E. Stamp, Mechanisms of decoherence at low temperatures, *Chem. Phys.* **268**, 257–272 (2001).
201. S. Nakajima, On quantum theory of transport phenomena, *Prog. Theor. Phys.* **20**, 948–959 (1958).
202. R. Zwanzig, Statistical mechanics of irreversibility, in: *Boulder Lecture Notes in Theoretical Physics*, Vol. III, Interscience, New York, 1960, pp. 106–141.
203. R. Zwanzig, Ensemble method in the theory of irreversibility, *J. Chem. Phys.* **33**, 1338–1341 (1960).
204. S. Chaturvedi, F. Shibata, Time-convolutionsless projection operator formalism for elimination of fast variables. Applications to Brownian motion, *Z. Phys. B* **35**, 297–308 (1979).
205. F. Shibata, T. Arimitsu, Expansion formulas and nonequilibrium statistical mechanics, *J. Phys. Soc. Jpn.* **49**, 891–897 (1980).
206. A. Royer, Cumulant expansions and pressure broadening as an example of relaxation, *Phys. Rev. A* **6**, 1741–1760 (1972).
207. A. Royer, Combining projection superoperators and cumulant expansions in open quantum dynamics with initial correlations and fluctuating Hamiltonians and environments, *Phys. Lett. A* **315**, 335–351 (2003).
208. A. O. Caldeira, A. H. Castro Neto, T. O. de Carvalho, Dissipative quantum systems modeled by a two-level-reservoir coupling, *Phys. Rev. B* **48**, 13974–13976 (1993).
209. A. J. Leggett, S. Chakravarty, A. T. Dorsey, M. P. A. Fisher, A. Garg, Dynamics of the dissipative two-state system, *Rev. Mod. Phys.* **59**, 1–85 (1987).
210. N. V. Prokof'ev, P. C. E. Stamp, Giant spins and topological decoherence: a Hamiltonian approach, *J. Phys. Chem. Lett.* **5**, L663–L670 (1993).

211. R. Feynman, F. L. Vernon, The theory of a general quantum system interacting with a linear dissipative system, *Ann. Phys. (N.Y.)* **24**, 118–173 (1963).
212. A. Caldeira, A. Leggett, Quantum tunneling in a dissipative system, *Ann. Phys. (N.Y.)* **149**, 374–456 (1983).
213. O. V. Lounasmaa, *Experimental Principles and Methods below 1 K*, Academic Press, New York, 1974.
214. N. V. Prokof'ev, P. C. E. Stamp, Decoherence in the quantum dynamics of a "central spin" coupled to a spin environment, eprint cond-mat/9511011.
215. N. V. Prokof'ev, P. C. E. Stamp, Quantum relaxation of magnetisation in magnetic particles, *J. Low Temp. Phys.* **104**, 143–210 (1996).
216. P. C. E. Stamp, Quantum environments: Spin baths, oscillator baths, and applications to quantum magnetism, in: S. Tomsovic (Ed.), *Tunnelling in Complex Systems*, World Scientific, Singapore, 1998, pp. 101–197.
217. N. D. Mermin, Can a phase transition make quantum mechanics less embarassing?, *Physica (Amsterdam)* **177A**, 561–566 (1991).
218. U. Weiss, *Quantum Dissipative Systems*, World Scientific, Singapore, 1999.
219. B. L. Hu, J. P. Paz, Y. Zang, Quantum Brownian motion in a general environment: Exact master equation with nonlocal dissipation and colored noise, *Phys. Rev. D* **45**, 2843–2861 (1992).
220. F. C. Lombardo, P. I. Villar, Decoherence induced by zero point fluctuations in quantum Brownian motion, *Phys. Lett. A* **336**, 16–24 (2005).
221. A. O. Caldeira, A. J. Leggett, Path integral approach to quantum Brownian motion, *Physica A* **121**, 587–616 (1983).
222. D. F. Walls, M. J. Collett, G. J. Milburn, Analysis of quantum measurement, *Phys. Rev. D* **32**, 3208–3215 (1985).
223. M. Tegmark, H. S. Shapiro, Decoherence produces coherent states: An explicit proof for harmonic chains, *Phys. Rev. E* **50**, 2538–2547 (1994).
224. M. R. Gallis, Emergence of classicality via decoherence described by Lindblad operators, *Phys. Rev. A* **53**, 655–660 (1996).
225. H. M. Wiseman, J. A. Vaccaro, Maximally robust unravelings of quantum master equations, *Phys. Lett. A* **250**, 241–248 (1998).
226. G.-S. Paraoanu, Selection of squeezed states via decoherence, *Europhys. Lett.* **47**, 279–284 (1999).
227. F. Haake, D. F. Walls, Overdamped and amplifying meters in the quantum theory of measurement, *Phys. Rev. A* **36**, 730–739 (1987).
228. M. R. Gallis, Spatial correlations of random potentials and the dynamics of quantum coherence, *Phys. Rev. A* **45**, 47–53 (1992).
229. J. R. Anglin, J. P. Paz, W. H. Zurek, Deconstructing decoherence, *Phys. Rev. A* **55**, 4041–4053 (1997).
230. A. O. Caldeira, A. J. Leggett, Influence of damping on quantum interference: An exactly soluble model, *Phys. Rev. A* **31**, 1059–1066 (1985).
231. F. Haake, R. Reibold, Strong damping and low-temperature anomalies for the harmonic oscillator, *Phys. Rev. A* **32**, 2462–2475 (1985).
232. H. Grabert, P. Schramm, G.-L. Ingold, Quantum Brownian motion: The functional integral approach, *Phys. Rep.* **168**, 115–207 (1988).
233. B. M. Garraway, Decay of an atom coupled strongly to a reservoir, *Phys. Rev. A* **55**, 4636–4639 (1997).
234. B. M. Garraway, Nonperturbative decay of an atomic system in a cavity, *Phys. Rev. A* **55**, 2290–2303 (1997).

235. J. Gilmore, R. H. McKenzie, Criteria for quantum coherent transfer of excitons between chromophores in a polar solvent, *Chem. Phys. Lett.* **421**, 266–271 (2006).
236. W. G. Unruh, Maintaining coherence in quantum computers, *Phys. Rev. A* **51**, 992–997 (1995).
237. J. H. Reina, L. Quiroga, N. F. Johnson, Decoherence of quantum registers, *Phys. Rev. A 65* **65**, 032326 (2002).
238. R. J. Glauber, Photon correlations, *Phys. Rev. Lett.* **10**, 84–86 (1963).
239. R. J. Glauber, The quantum theory of optical coherence, *Phys. Rev.* **130**, 2529–2539 (1963).
240. R. J. Glauber, Coherent and incoherent states of the radiation field, *Phys. Rev.* **131**, 2766–2788 (1963).
241. D. F. Walls, G. J. Milburn, *Quantum Optics*, Springer, Berlin, 1994.
242. V. V. Dobrovitski, H. A. D. Raedt, M. I. Katsnelson, B. N. Harmon, Quantum oscillations without quantum coherence, *Phys. Rev. Lett.* **90**, 210401 (2003).
243. B. Golding, J. E. Graebner, R. J. Schutz, Intrinsic decay lengths of quasi-monochromatic phonons in a glass below 1 K, *Phys. Rev. B* **14**, 1660–1662 (1976).
244. W. A. Phillips, Two-level states in glasses, *Rep. Prog. Phys.* **50**, 1657–1708 (1987).
245. C. Monroe, D. M. Meekhof, B. E. King, D. J. A. Wineland, "Schrödinger cat" superposition state of an atom, *Science* **272**, 1131–1136 (1996).
246. M. Brune, E. Hagley, J. Dreyer, X. Maître, A. Maali, C. Wunderlich, J. M. Raimond, S. Haroche, Observing the progressive decoherence of the "meter" in a quantum measurement, *Phys. Rev. Lett.* **77**, 4887–4890 (1996).
247. J. M. Raimond, M. Brune, S. Haroche, Manipulating quantum entanglement with atoms and photons in a cavity, *Rev. Mod. Phys.* **73**, 565–582 (2001).
248. J. I. Kim, K. M. F. Romero, A. M. Horiguti, L. Davidovich, M. C. Nemes, A. F. R. de Toledo Piza, Classical behavior with small quantum numbers: The physics of Ramsey interferometry of Rydberg atoms, *Phys. Rev. Lett.* **82**, 4737–4740 (1999).
249. L. Davidovich, M. Brune, J. M. Raimond, S. Haroche, Mesoscopic quantum coherences in cavity QED: Preparation and decoherence monitoring schemes, *Phys. Rev. A* **53**, 1295–1309 (1996).
250. X. Maître, E. Hagley, J. Dreyer, A. Maali, C. W. M. Brune, J. M. Raimond, S. Haroche, An experimental study of a Schrödinger cat decoherence with atoms and cavities, *J. Mod. Opt.* **44**, 2023–2032 (1997).
251. A. Auffeves, P. Maioli, T. Meunier, S. Gleyzes, G. Nogues, M. Brune, J. M. Raimond, S. Haroche, Entanglement of a mesoscopic field with an atom induced by photon graininess in a cavity, *Phys. Rev. Lett.* **91**, 230405 (2003).
252. S. M. Dutra, *Cavity Quantum Electrodynamics: The Strange Theory of Light in a Box*, John Wiley & Sons, New York, 2004.
253. J. M. Raimond, M. Brune, S. Haroche, Reversible decoherence of a mesoscopic superposition of field states, *Phys. Rev. Lett.* **79**, 1964–1967 (1997).
254. S. G. Mokarzel, A. N. Salgueiro, M. C. Nemes, Modeling the reversible decoherence of mesoscopic superpositions in dissipative environments, *Phys. Rev. A* **65**, 044101 (2002).
255. R. P. Crease, *The Prism and the Pendulum: The Ten Most Beautiful Experiments in Science*, Random House, New York, 2003.

256. R. P. Feynman, R. B. Leighton, M. Sands, *The Feynman Lectures on Physics*, Vol. I, Addison–Wesley, New York, 1963.
257. C. Jönsson, Elektroneninterferenzen an mehreren künstlich hergestellten Feinspalten, *Z. Physik* **161**, 454–474 (1961).
258. C. Jönsson, Electron diffraction at multiple slits, *Am. J. Phys.* **42**, 4–11 (1974).
259. T. Young, Experimental demonstration of the general law of the interference of light, in: M. Shamos (Ed.), *Great Experiments in Physics*, Holt Reinhart and Winston, New York, 1959, pp. 96–101.
260. A. Tonomura, J. Endo, T. Matsuda, T. Kawasaki, H. Ezawa, Demonstration of single-electron buildup of an interference pattern, *Am. J. Phys.* **57**, 117–120 (1989).
261. P. Sonnentag, F. Hasselbach, Measurement of decoherence of electron waves and visualization of the quantum-classical transition, *Phys. Rev. Lett.* **98**, 200402 (2007).
262. M. Arndt, O. Nairz, J. Vos-Andreae, C. Keller, G. van der Zouw, A. Zeilinger, Wave–particle duality of C_{60} molecules, *Nature* **401**, 680–682 (1999).
263. B. Brezger, L. Hackermüller, S. Uttenthaler, J. Petschinka, M. Arndt, A. Zeilinger, Matter-wave interferometer for large molecules, *Phys. Rev. Lett.* **88**, 100404 (2002).
264. K. Hornberger, S. Uttenthaler, B. Brezger, L. Hackermüller, M. Arndt, A. Zeilinger, Collisional decoherence observed in matter wave interferometry, *Phys. Rev. Lett.* **90**, 160401 (2003).
265. L. Hackermüller, S. Uttenthaler, K. Hornberger, E. Reiger, B. Brezger, A. Zeilinger, M. Arndt, Wave nature of biomolecules and fluorofullerenes, *Phys. Rev. Lett.* **91**, 090408 (2003).
266. L. Hackermüller, K. Hornberger, B. Brezger, A. Zeilinger, M. Arndt, Decoherence of matter waves by thermal emission of radiation, *Nature* **427**, 711–714 (2004).
267. L. Hackermüller, K. Hornberger, B. Brezger, A. Zeilinger, M. Arndt, Decoherence in a Talbot–Lau interferometer: the influence of molecular scattering, *Appl. Phys. B* **77**, 781–787 (2003).
268. K. Hornberger, J. E. Sipe, M. Arndt, Theory of decoherence in a matter wave Talbot-Lau interferometer, *Phys. Rev. A* **70**, 053608 (2004).
269. K. Hornberger, L. Hackermüller, M. Arndt, Influence of molecular temperature on the coherence of fullerenes in a near-field interferometer, *Phys. Rev. A* **71**, 023601 (2005).
270. K. Hornberger, Thermal limitation of far-field matter-wave interference, *Phys. Rev. A* **73**, 052102 (2006).
271. M. Arndt, O. Nairz, A. Zeilinger, Interferometry with macromolecules: Quantum paradigms tested in the mesoscopic world, in: R. A. Bertlmann, A. Zeilinger (Eds.), *Quantum [Un]Speakables: From Bell to Quantum Information*, Springer, Berlin, 2002, pp. 333–351.
272. A. Stibor, K. Hornberger, L. Hackermüller, A. Zeilinger, M. Arndt, Talbot–Lau interferometry with fullerenes: Sensitivity to inertial forces and vibrational dephasing, *Laser Phys.* **15**, 10–17 (2005).
273. M. F. Bocko, A. M. Herr, M. J. Feldman, Prospects for quantum coherent computation using superconducting electronics, *IEEE Trans. Appl. Supercond.* **7**, 3638–3641 (1997).
274. A. J. Leggett, Macroscopic quantum systems and the quantum theory of measurement, *Suppl. Prog. Theor. Phys.* **69**, 80–100 (1980).

275. J. R. Friedman, V. Patel, W. Chen, S. K. Yolpygo, J. E. Lukens, Quantum superposition of distinct macroscopic states, *Nature* **406**, 43–46 (2000).
276. C. H. van der Wal, A. C. J. ter Haar, F. K. Wilhelm, R. N. Schouten, C. J. P. M. Harmans, T. P. Orlando, S. Lloyd, J. E. Mooij, Quantum superposition of macroscopic persistent-current states, *Science* **290**, 773–777 (2000).
277. H. K. Onnes, On the sudden rate at which the resistance of mercury disappears, *Akad. van Wetenschappen* **14**, 818–821 (1911).
278. L. N. C. J. Bardeen, J. R. Schrieffer, Theory of superconductivity, *Phys. Rev.* **108**, 1175–1204 (1957).
279. K. K. Likharev, Superconducting weak links, *Rev. Mod. Phys.* **51**, 101–159 (1979).
280. Y. Makhlin, G. Schön, A. Shnirman, Quantum-state engineering with Josephson-junction devices, *Rev. Mod. Phys.* **73**, 357–400 (2001).
281. P. Silvestrini, B. Ruggiero, Y. N. Ovchinnikov, Resonant macroscopic quantum tunneling in SQUID systems, *Phys. Rev. B* **54**, 1246–1250 (1996).
282. R. Rouse, S. Han, J. E. Lukens, Resonant tunneling between macroscopically distinct levels of a SQUID, in: G. D. Palazzi, C. Cosmelli, L. Zanello (Eds.), *Phenomenology of Unification from Present to Future*, World Scientic, Singapore, 1998, pp. 207–209.
283. I. Chiorescu, Y. Nakamura, C. J. P. M. Harmans, J. E. Mooij, Coherent quantum dynamics of a superconducting flux qubit, *Science* **21**, 1869–1871 (2003).
284. E. Il'ichev, N. Oukhanski, A. Izmalkov, T. Wagner, M. Grajcar, H.-G. Meyer, A. Y. Smirnov, A. M. van den Brink, M. H. S. Amin, A. M. Zagoskin, Continuous monitoring of Rabi oscillations in a Josephson flux qubit, *Phys. Rev. Lett.* **91**, 097906 (2003).
285. A. J. Leggett, Testing the limits of quantum mechanics: motivation, state of play, prospects, *J. Phys.: Condens. Matter* **14**, R415–R451 (2002).
286. P. Bertet, I. Chiorescu, G. Burkard, K. Semba, C. J. P. M. Harmans, D. P. DiVincenzo, J. E. Mooij, Dephasing of a superconducting qubit induced by photon noise, *Phys. Rev. Lett.* **95**, 257002 (2005).
287. Y. Nakamura, Y. A. Pashkin, J. S. Tsai, Coherent control of macroscopic quantum states in a single-Cooper-pair box, *Nature* **398**, 786–788 (1999).
288. D. Vion, A. Aassime, A. Cottet, P. Joyez, H. Pothier, C. Urbina, D. Esteve, M. H. Devoret, Manipulating the quantum state of an electrical circuit, *Science* **296**, 886–889 (2002).
289. Y. Yu, S. Han, X. Chu, S.-I. Chu, Z. Wang, Coherent temporal oscillations of macroscopic quantum states in a Josephson junction, *Science* **296**, 889–892 (2002).
290. J. M. Martinis, S. Nam, J. Aumentado, C. Urbina, Rabi oscillations in a large Josephson-junction qubit, *Phys. Rev. Lett.* **89**, 117901 (2002).
291. J. M. Martinis, K. B. Cooper, R. McDermott, M. Steffen, M. Ansmann, K. Osborn, K. Cicak, S. Oh, D. P. Pappas, R. W. Simmonds, C. C. Yu, Decoherence in Josephson qubits from dielectric loss, *Phys. Rev. Lett.* **95**, 210503 (2005).
292. J. M. Martinis, S. Nam, J. Aumentado, K. M. Lang, C. Urbina, Decoherence of a superconducting qubit due to bias noise, *Phys. Rev. B* **67**, 094510 (2003).
293. M. R. Andrews, C. G. Townsend, H.-J. Miesner, D. S. Durfee, D. M. Kurn, W. Ketterle, Observation of interference between two Bose condensates, *Science* **275**, 637–641 (1997).

294. Y. Shin, M. Saba, T. A. Pasquini, W. Ketterle, D. E. Pritchard, A. E. Leanhardt, Atom interferometry with Bose–Einstein condensates in a double-well potential, *Phys. Rev. Lett.* **92**, 050405 (2004).
295. J. Javanainen, S. M. Yoo, Quantum phase of a Bose–Einstein condensate with an arbitrary number of atoms, *Phys. Rev. Lett.* **76**, 161–164 (1996).
296. A. Röhrl, M. Naraschewski, A. Schenzle, H. Wallis, Transition from phase locking to the interference of independent Bose condensates: Theory versus experiment, *Phys. Rev. Lett* **78**, 4143–4146 (1997).
297. J. Javanainen, Bose–Einstein condensates interfere and survive, *Science* **307**, 1883–1885 (2005).
298. M. Saba, T. A. Pasquini, C. Sanner, Y. Shin, W. Ketterle, D. E. Pritchard, Light scattering to determine the relative phase of two Bose–Einstein condensates, *Science* **307**, 1945–1948 (2005).
299. J. I. Cirac, M. Lewenstein, K. Mølmer, P. Zoller, Quantum superposition states of Bose–Einstein condensates, *Phys. Rev. A* **57**, 1208–1218 (1998).
300. J. Ruostekoski, M. J. Collett, R. Graham, D. F. Walls, Macroscopic superpositions of Bose–Einstein condensates, *Phys. Rev. A* **57**, 511–517 (1998).
301. D. Gordon, C. M. Savage, Creating macroscopic quantum superpositions with Bose–Einstein condensates, *Phys. Rev. A* **59**, 4623–4629 (1999).
302. J. A. Dunningham, K. Burnett, Proposals for creating Schrödinger cat states in Bose–Einstein condensates, *J. Mod. Opt.* **48**, 1837–1853 (2001).
303. J. Calsamiglia, M. Mackie, K.-A. Suominen, Superposition of macroscopic numbers of atoms and molecules, *Phys. Rev. Lett.* **87**, 160403 (2001).
304. P. J. Y. Louis, P. M. R. Brydon, C. M. Savage, Macroscopic quantum superposition states in Bose–Einstein condensates: Decoherence and many modes, *Phys. Rev. A* **64**, 053613 (2001).
305. A. Micheli, D. Jaksch, J. I. Cirac, P. Zoller, Many particle entanglement in two-component Bose–Einstein condensates, *Phys. Rev. A* **67**, 013601 (2003).
306. M. Blencowe, Quantum electromechanical systems, *Phys. Rep.* **395**, 159–222 (2004).
307. K. C. Schwab, M. L. Roukes, Putting mechanics into quantum mechanics, *Phys. Today*, 36–42 (July 2005).
308. M. L. Roukes, Nanoelectromechanical systems face the future, *Phys. World* **14**, 25–31 (2001).
309. M. D. LaHaye, O. Buu, B. Camarota, K. C. Schwab, Approaching the quantum limit of a nanomechanical resonator, *Science* **304**, 74–77 (2004).
310. X. M. H. Huang, C. A. Zorman, M. Mehregany, M. L. Roukes, Nanoelectromechanical systems: Nanodevice motion at microwave frequencies, *Nature* **421**, 496 (2003).
311. A. D. Armour, M. P. Blencowe, K. C. Schwab, Entanglement and decoherence of a micromechanical resonator via coupling to a Cooper-pair box, *Phys. Rev. Lett.* **88**, 148301 (2002).
312. K.-H. Ahn, P. Mohanty, Quantum friction of micromechanical resonators at low temperatures, *Phys. Rev. Lett.* **90**, 085504 (2003).
313. C. Seoanez, F. Guinea, A. H. C. Neto, Dissipation due to two-level systems in nano-mechanical devices, eprint cond-mat/0611153.
314. J. Eisert, M. B. Plenio, S. Bose, J. Hartley, Towards quantum entanglement in nanoelectromechanical devices, *Phys. Rev. Lett.* **93**, 190402 (2004).
315. S. Savel'ev, X. Hu, F. Nori, Quantum electromechanics: Qubits from buckling nanobars, *New J. Phys.* **8**, 105 (2006).

316. P. Mohanty, D. A. Harrington, K. L. Ekinci, Y. T. Yang, M. J. Murphy, M. L. Roukes, Intrinsic dissipation in high-frequency micromechanical resonators, *Phys. Rev. B* **66**, 085416 (2002).
317. G. Zolfagharkhani, A. Gaidarzhy, S.-B. Shim, R. L. Badzey, P. Mohanty, Quantum friction in nanomechanical oscillators at millikelvin temperatures, *Phys. Rev. B* **72**.
318. R. N. Kleiman, G. Agnolet, D. J. Bishop, Two-level systems observed in the mechanical properties of single-crystal silicon at low temperatures, *Phys. Rev. Lett.*, 2079–2082 (1987).
319. R. E. Mihailovich, J. M. Parpia, Low temperature mechanical properties of boron-doped silicon, *Phys. Rev. Lett.* **68**, 3052–3055 (1992).
320. M. Schlosshauer, Experimental motivation and empirical consistency in minimal no-collapse quantum mechanics, *Ann. Phys.* **321**, 112–149 (2006).
321. W. H. Zurek, Einselection and decoherence from an information theory perspective, *Ann. Phys. (Leipzig)* **9**, 855–864 (2000).
322. G. Moore, An update on Moore's law, Speech given at the Intel Developer's Forum 1997.
323. R. Landauer, Information is inevitably physical, in: A. J. G. Hey (Ed.), *Feynman and Computation: Exploring the Limits of Computers*, Perseus, 1999, pp. 77–92.
324. R. Raussendorf, H. J. Briegel, A one-way quantum computer, *Phys. Rev. Lett.* **86**, 5188–5191 (2001).
325. M. A. Nielsen, Quantum computation by measurement and quantum memory, *Phys. Lett. A* **308**, 96–100 (2003).
326. R. Raussendorf, D. E. Browne, H. J. Briegel, Measurement-based quantum computation on cluster states, *Phys. Rev. A* **68**, 022312 (2003).
327. M. A. Nielsen, Cluster-state quantum computation, *Rep. Math. Phys.* **57**, 147–161 (2006).
328. W. H. Zurek, Reversibility and stability of information processing systems, *Phys. Rev. Lett.* **53**, 391–394 (1984).
329. D. Deutsch, A. Barenco, A. Ekert, Universality in quantum computation, *Proc. R. Soc. Lond. A* **449**, 669–677 (1995).
330. A. Barenco, C. H. Bennett, R. Cleve, D. P. DiVincenzo, N. Margolus, P. Shor, T. Sleator, J. A. Smolin, H. Weinfurter, Elementary gates for quantum computation, *Phys. Rev. A* **52**, 3457–3467 (1995).
331. S. Lloyd, Almost any quantum logic gate is universal, *Phys. Rev. Lett.* **75**, 346–349 (1995).
332. D. P. DiVincenzo, Two-bit gates are universal for quantum computation, *Phys. Rev. A* **51**, 1015–1022 (1995).
333. R. P. Feynman, Simulating physics with computers, *Int. J. Theor. Phys.* **21**, 467–488 (1982).
334. A. Steane, Quantum computing, *Rep. Prog. Phys.* **61**, 117–173 (1998).
335. D. Deutsch, Quantum theory, the Church–Turing principle and the universal quantum computer, *Proc. R. Soc. Lond. A* **400**, 97–117 (1985).
336. S. Lloyd, Universal quantum simulators, *Science* **273**, 1073–1078 (1995).
337. D. Deutsch, R. Jozsa, Rapid solution of problems by quantum computation, *Proc. R. Soc. Lond. A* **439**, 553–558 (1992).
338. A. Berthiaume, G. Brassard, Oracle quantum computing, in: *Proc. Workshop on Physics of Computation: PhysComp '92*, IEEE Computer Society Press, Los Alamitos, CA, 1992, pp. 60–62.

339. A. Berthiaume, G. Brassard, The quantum challenge to structural complexity theory, in: *Proc. 7th Annual Structure in Complexity Theory Conf.*, IEEE Computer Society Press, Los Alamitos, CA, 1992, pp. 132–137.
340. E. Bernstein, U. Vazirani, Quantum complexity theory, in: *Proceedings of the 25th Annual ACM Symposium on Theory of Computing*, ACM, New York, 1993, pp. 11–20.
341. D. Simon, On the power of quantum computation, in: *Proc. 35th Annual Symp. on Foundations of Computer Science*, IEEE Computer Society Press, Los Alamitos, CA, 1994, pp. 124–134.
342. P. W. Shor, Polynomial-time algorithms for prime factorization and discrete logarithms on a quantum computer, in: *Proceedings of the 35th Annual Symposium on the Foundations of Computer Science*, IEEE Computer Society Press, Los Alamitos, CA, 1994, pp. 124–134.
343. P. W. Shor, Polynomial-time algorithms for prime factorization and discrete logarithms on a quantum computer, *SIAM J. Comp.* **26**, 1484–1509 (1997).
344. L. K. Grover, A fast quantum mechanical algorithm for database search, in: *Proc. of the 28th Annual ACM Symposium on the Theory of Computing*, ACM, New York, 1996, pp. 212–219.
345. L. K. Grover, Quantum mechanics helps in search for a needle in a haystack, *Phys. Rev. Lett.* **79**, 325–328 (1997).
346. E. Knill, R. Laflamme, G. J. Milburn, A scheme for efficient quantum computation with linear optics, *Nature* **409**, 46–52 (2001).
347. S. Schneider, G. J. Milburn, Decoherence in ion traps due to laser intensity and phase fluctuations, *Phys. Rev. A* **57**, 3748–3752 (1998).
348. C. Miquel, J. P. Paz, W. H. Zurek, Quantum computation with phase drift errors, *Phys. Rev. Lett.* **78**, 3971–3974 (1997).
349. L. M. K. Vandersypen, I. L. Chuang, NMR techniques for quantum control and computation, *Rev. Mod. Phys.* **76**, 1037–1069 (2004).
350. S. Schneider, G. J. Milburn, Decoherence and fidelity in ion traps with fluctuating trap parameters, *Phys. Rev. A* **59**, 3766–3774 (1999).
351. Q. A. Turchette, C. J. Myatt, B. E. King, C. A. Sackett, D. Kielpinski, W. M. Itano, C. Monroe, D. J. Wineland, Decoherence and decay of motional quantum states of a trapped atom coupled to engineered reservoirs, *Phys. Rev. A* **62**, 053807 (2000).
352. A. M. Steane, Error correcting codes in quantum theory, *Phys. Rev. Lett.* **77**, 793–797 (1996).
353. P. W. Shor, Scheme for reducing decoherence in quantum computer memory, *Phys. Rev. A* **52**, R2493–R2496 (1995).
354. A. M. Steane, Quantum computing and error correction, in: P. Turchi, A. Gonis (Eds.), *Decoherence and its implications in quantum computation and information transfer*, IOS Press, Amsterdam, 2001, pp. 284–298, also available as eprint quant-ph/0304016.
355. E. Knill, R. Laflamme, A. Ashikhmin, H. Barnum, L. Viola, W. Zurek, Introduction to quantum error correction, *LA Science* **27**, 188–225 (2002).
356. C. Miquel, J. P. Paz, R. Perazzo, Factoring in a dissipative quantum computer, *Phys. Rev. A* **54**, 2605–2613 (1996).
357. A. Jamiolkowski, Linear transformations which preserve trace and positive semidefiniteness of operators, *Rep. Math. Phys.* **3**, 275–278 (1972).
358. J. Preskill, Reliable quantum computers, *Proc. R. Soc. Lond. A* **454**, 385–410 (1998).

359. D. G. Cory, M. D. Price, W. Maas, E. Knill, R. Laflamme, W. H. Zurek, T. F. Havel, S. S. Somaroo, Experimental quantum error correction, *Phys. Rev. Lett.* **81**, 2152–2155 (1998).
360. P. W. Shor, Fault-tolerant quantum computation, in: *Proceedings of the 37th Symposium on the Foundations of Computer Science*, IEEE Press, Los Alamitos, 1996, pp. 56–65, also available as eprint quant-ph/9605011.
361. D. P. DiVincenzo, P. W. Shor, Fault-tolerant error correction with efficient quantum codes, *Phys. Rev. Lett.* **77**, 3260–3263 (1996).
362. D. Gottesman, A theory of fault-tolerant quantum computation, *Phys. Rev. A* **57**, 127–137 (1998).
363. D. Gottesman, I. L. Chuang, Quantum teleportation is a universal computational primitive, *Nature* **402**, 390–393 (1999).
364. A. M. Steane, Active stabilization, quantum computation, and quantum state synthesis, *Phys. Rev. Lett.* **78**, 2252–2255 (1997).
365. J. Preskill, Battling decoherence: the fault-tolerant quantum computer, *Phys. Today*, 24–30 (June 1999).
366. E. Knill, R. Laflamme, W. H. Zurek, Resilient quantum computation, *Science* **279**, 342–345 (1998).
367. E. Knill, R. Laflamme, W. H. Zurek, Resilient quantum computation: error models and thresholds, *Proc. R. Soc. Lond. A* **454**, 365–384 (1998).
368. A. Y. Kitaev, Quantum error correction with imperfect gates, in: *Quantum Communication, Computing and Measurement (Proc. 3rd Int. Conf. of Quantum Communication and Measurement)*, Plenum Press, New York, 1997, pp. 181–188.
369. D. Aharonov, M. Ben-Or, Fault-tolerant quantum computation with constant error, in: *Proceedings of the 29th Annual ACM Symposium on the Theory of Computing*, ACM, New York, 1996, pp. 176–188, also available as eprint quant-ph/9611025.
370. K. M. Svore, A. W. Cross, I. L. Chuang, A. V. Aho, Pseudothreshold or threshold? — More realistic threshold estimates for fault-tolerant quantum computing, eprint quant-ph/0508176.
371. D. A. Lidar, D. Bacon, J. Kempe, K. B. Whaley, Decoherence-free subspaces for multiple-qubit errors. II. Universal, fault-tolerant quantum computation, *Phys. Rev. A* **63**, 022307 (2001).
372. P. G. Kwiat, A. J. Berglund, J. B. Altepeter, A. G. White, Experimental verification of decoherence-free spaces, *Science* **290**, 498–501 (2000).
373. D. Kielpinski, V. Meyer, M. A. Rowe, C. A. Sackett, W. M. Itano, C. Monroe, D. J. Wineland, A decoherence-free quantum memory using trapped ions, *Science* **291**, 1013–1015 (2001).
374. L. Viola, E. M. Fortunato, M. A. Pravia, E. Knill, R. Laflamme, D. G. Cory, Experimental realization of noiseless subsystems for quantum information processing, *Science* **293**, 2059–2063 (2001).
375. J. F. Poyatos, J. I. Cirac, P. Zoller, Quantum reservoir engineering with laser cooled trapped ions, *Phys. Rev. Lett.* **77**, 4728–4731 (1996).
376. A. R. R. Carvalho, P. Milman, R. L. de Matos Filho, L. Davidovich, Decoherence, pointer engineering, and quantum state protection, *Phys. Rev. Lett.* **86**, 4988–4991 (2001).
377. L. Viola, S. Lloyd, Dynamical suppression of decoherence in two-state quantum systems, *Phys. Rev. A* **58**, 2733–2744 (1998).

378. L. Viola, E. Knill, S. Lloyd, Dynamical decoupling of open quantum systems, *Phys. Rev. Lett.* **82**, 2417–2421 (1999).
379. P. Zanardi, Symmetrizing evolutions, *Phys. Lett. A* **258**, 77–91 (1999).
380. L. Viola, E. Knill, S. Lloyd, Dynamical generation of noiseless quantum subsystems, *Phys. Rev. Lett.* **85**, 3520–3523 (2000).
381. L.-A. Wu, D. A. Lidar, Creating decoherence-free subspaces using strong and fast pulses, *Phys. Rev. Lett.* **88**, 207902 (2002).
382. L.-A. Wu, M. S. Byrd, D. A. Lidar, Efficient universal leakage elimination for physical and encoded qubits, *Phys. Rev. Lett.* **89**, 127901 (2002).
383. G. Bacciagaluppi, A. Valentini, *Quantum Theory at the Crossroads: Reconsidering the 1927 Solvay Conference*, Cambridge University Press, Cambridge, 2007, also available as eprint quant-ph/0609184.
384. N. D. Mermin, What's wrong with this pillow?, *Phys. Today* **42**, 9 (April 1989).
385. G. Bacciagaluppi, The role of decoherence in quantum mechanics, in: E. N. Zalta (Ed.), *The Stanford Encyclopedia of Philosophy*, 2003, online at http://plato.stanford.edu/archives/win2003/entries/qm-decoherence.
386. B. S. DeWitt, Quantum mechanics and reality, *Phys. Today* **23**, 30–35 (September 1970).
387. D. Deutsch, Quantum theory as a universal physical theory, *Int. J. Theor. Phys.* **24**, 1–41 (1985).
388. M. Lockwood, 'Many minds' interpretations of quantum mechanics, *Br. J. Philos. Sci.* **47**, 159–188 (1996).
389. S. Saunders, Time, quantum mechanics, and decoherence, *Synthese* **102**, 235–266 (1995).
390. S. Saunders, Naturalizing metaphysics, *The Monist* **80**, 44–69 (1997).
391. S. Saunders, Time, quantum mechanics, and probability, *Synthese* **114**, 373–404 (1998).
392. D. Deutsch, Comment on Lockwood, *Br. J. Philos. Sci.* **47**, 222–228 (1996).
393. D. Deutsch, The structure of the multiverse, *Proc. R. Soc. Lond. A* **458**, 2911–2923 (2002).
394. L. Vaidman, On schizophrenic experiences of the neutron or why we should believe in the many-worlds interpretation of quantum theory, *Int. Stud. Philos. Sci.* **12**, 245–261 (1998).
395. D. Wallace, Worlds in the Everett interpretation, *Stud. Hist. Philos. Mod. Phys.* **33**, 637–661 (2002).
396. D. Wallace, Everett and structure, *Stud. Hist. Philos. Mod. Phys.* **34**, 87–105 (2003).
397. H. D. Zeh, There are no quantum jumps, nor are there particles!, *Phys. Lett. A* **172**, 189–192 (1993).
398. H. P. Stapp, The basis problem in many-worlds theories, *Can. J. Phys.* **80**, 1043–1052 (2002).
399. J. N. Butterfield, Some worlds of quantum theory, in: R. J. Russell, P. Clayton, K. Wegter-McNelly, J. Polkinghorne (Eds.), *Quantum Mechanics: Scientific Perspectives on Divine Action*, Vatican Observatory and The Center for Theology and the Natural Sciences, Vatican City State, 2001, pp. 111–140, also available as an eprint from the Pittsburgh Philosophy of Science Archive at http://philsci-archive.pitt.edu/archive/00000203.
400. A. Kent, Against many worlds interpretations, *Int. J. Mod. Phys. A* **5**, 1745 (1990).

401. J. S. Bell, On the impossible pilot wave, *Found. Phys.* **12**, 989–999 (1982).
402. B. S. DeWitt, The many-universes interpretation of quantum mechanics, in: B. d'Espagnat (Ed.), *Foundations of Quantum Mechanics*, Academic Press, New York, 1971, pp. 211–262.
403. N. Graham, *The Everett Interpretation of Quantum Mechanics*, University of North Carolina Press, Chapel Hill, 1970.
404. N. Graham, The measurement of relative frequency, in: B. S. DeWitt, N. Graham (Eds.), *The Many-Worlds Interpretation of Quantum Mechanics*, Princeton University Press, Princeton, 1973, pp. 229–253.
405. H. Stein, The Everett interpretation of quantum mechanics: Many worlds or none?, *Nôus* **18**, 635–652 (1984).
406. E. J. Squires, On an alleged "proof" of the quantum probability law, *Phys. Lett. A* **145**, 67–68 (1990).
407. D. Deutsch, Quantum theory of probability and decisions, *Proc. R. Soc. Lond. A* **455**, 3129–3197 (1999).
408. H. D. Zeh, What is achieved by decoherence?, in: M. Ferrero, A. van der Merwe (Eds.), *New Developments on Fundamental Problems in Quantum Physics (Oviedo II)*, Kluwer, Dordrecht, 1997, pp. 441–452.
409. H. Barnum, C. M. Caves, J. Finkelstein, C. A. Fuchs, R. Schack, Quantum probability from decision theory?, *Proc. R. Soc. Lond. A* **456**, 1175–1182 (2000).
410. R. D. Gill, On an argument of David Deutsch, in: M. Schürmann, U. Franz (Eds.), *Quantum Probability and Infinite Dimensional Analysis: From Foundations to Applications*, Vol. 18 of *QP–PQ: Quantum Probability and White Noise Analysis*, World Scientific, Singapore, 2005, pp. 277–292.
411. D. Wallace, Everettian rationality: defending Deutsch's approach to probability in the Everett interpretation, *Stud. Hist. Philos. Mod. Phys.* **34**, 415–439 (2003).
412. S. Saunders, Derivation of the Born rule from operational assumptions, *Proc. R. Soc. A: Math. Phys. Eng. Sci.* **460**, 1771–1788 (2004).
413. M. Schlosshauer, A. Fine, On Zurek's derivation of the Born rule, *Found. Phys.* **35**, 197–213 (2005).
414. H. Barnum, No-signalling-based version of Zurek's derivation of quantum probabilities: A note on 'Environment-assisted invariance, entanglement, and probabilities in quantum physics', eprint quant-ph/0312150.
415. U. Mohrhoff, Probabilities from envariance?, *Int. J. Quantum Inf.* **2**, 221–230 (2004).
416. B. van Fraassen, Semantic analysis of quantum logic, in: C. A. Hooker (Ed.), *Contemporary Research in the Foundations and Philosophy of Quantum Theory*, Reidel, Dordrecht, 1973, pp. 180–213.
417. B. van Fraassen, *Quantum Mechanics: An Empiricist View*, Clarendon, Oxford, 1991.
418. R. Clifton, The properties of modal interpretations of quantum mechanics, *Br. J. Philos. Sci.* **47**, 371–398 (1996).
419. D. Dieks, Modal interpretation of quantum mechanics, measurements, and macroscopic behavior, *Phys. Rev. A* **49**, 2290–2300 (1994).
420. D. Dieks, Objectification, measurement and classical limit according to the modal interpretation of quantum mechanics, in: I. Busch, P. Lahti, P. Mittelstaedt (Eds.), *Proceedings of the Symposium on the Foundations of Modern Physics*, World Scientific, Singapore, 1994, pp. 160–167.

421. G. Bacciagaluppi, M. Dickson, Dynamics for modal interpretations, *Found. Phys.* **29**, 1165–1201 (1999).
422. M. Hemmo, *Quantum Mechanics Without Collapse: Modal Interpretations, Histories and Many Worlds*, Ph.D. thesis, University of Cambridge (1996).
423. S. Kochen, A new interpretation of quantum mechanics, in: P. Lahti, P. Mittelstaedt (Eds.), *Symposium on the Foundations of Modern Physics: 50 Years of the Einstein–Podolsky–Rosen Experiment (Joensuu, Finland, 1985)*, World Scientific, Singapore, 1985, pp. 151–169.
424. R. A. Healey, *The Philosophy of Quantum Mechanics: An Interactive Interpretation*, Cambridge University Press, Cambridge, England/New York, 1989.
425. D. Dieks, Resolution of the measurement problem through decoherence of the quantum state, *Phys. Lett. A* **142**, 439–446 (1989).
426. P. E. Vermaas, D. Dieks, The modal interpretation of quantum mechanics and its generalization to density operators, *Found. Phys.* **25**, 145–158 (1995).
427. D. Dieks, Physical motivation of the modal interpretation of quantum mechanics, *Phys. Lett. A* **197**, 367–371 (1995).
428. G. Bacciagaluppi, M. J. Donald, P. E. Vermaas, Continuity and discontinuity of definite properties in the modal interpretation, *Helv. Phys. Acta* **68**, 679–704 (1995).
429. M. Donald, Discontinuity and continuity of definite properties in the modal interpretation, in: D. Dieks, P. Vermaas (Eds.), *The Modal Interpretation of Quantum Mechanics*, Kluwer, Dordrecht, 1998, pp. 213–222.
430. G. Bacciagaluppi, M. Hemmo, Modal interpretations, decoherence and measurements, *Stud. Hist. Philos. Mod. Phys.* **27**, 239–277 (1996).
431. G. Bene, Quantum origin of classical properties within the modal interpretations, eprint quant-ph/0104112.
432. G. Bacciagaluppi, Delocalized properties in the modal interpretation of a continuous model of decoherence, *Found. Phys.* **30**, 1431–1444 (2000).
433. A. Bassi, G. C. Ghirardi, Dynamical reduction models, *Phys. Rep.* **379**, 257–426 (2003).
434. G. C. Ghirardi, A. Rimini, T. Weber, Disentanglement of quantum wave functions: Answer to "Comment on 'Unified dynamics for microscopic and macroscopic systems'", *Phys. Rev. D* **36**, 3287–3289 (1987).
435. P. Pearle, Reduction of the state vector by a nonlinear Schrödinger equation, *Phys. Rev. D* **13**, 857–868 (1976).
436. P. M. Pearle, Toward explaining why events occur, *Int. J. Theor. Phys.* **48**, 489–518 (1979).
437. P. Pearle, Might God toss coins?, *Found. Phys.* **12**, 249–263 (1982).
438. P. M. Pearle, Collapse models, in: H.-P. Breuer, F. Petruccione (Eds.), *Open Systems and Measurement in Relativistic Quantum Theory, Proceedings of the workshop held at the Instituto Italiano per gli Studi Filosofici*, Springer, Berlin, 1999, pp. 31–65.
439. G. C. Ghirardi, A. Rimini, T. Weber, Unified dynamics for microscopic and macroscopic systems, *Phys. Rev. D* **34**, 470–491 (1986).
440. P. Pearle, Combining stochastic dynamical state-vector reduction with spontaneous localization, *Phys. Rev. A* **39**, 2277–2289 (1989).
441. G. C. Ghirardi, P. Pearle, A. Rimini, Markov processes in Hilbert space and continuous spontaneous localization of systems of identical particles, *Phys. Rev. A* **42**, 78–89 (1990).

442. L. Diósi, Models for universal reduction of macroscopic quantum fluctuations, *Phys. Rev. A* **40**, 1165–1174 (1989).
443. D. Bedford, D. Wang, Toward an objective interpretation of quantum mechanics, *Nuovo Cimento Soc. Ital. Fis., A* **26**, 313–325 (1975).
444. D. Bedford, D. Wang, A criterion for spontaneous state reduction, *Nuovo Cimento Soc. Ital. Fis., A* **37**, 55–62 (1977).
445. G. J. Milburn, Intrinsic decoherence in quantum mechanics, *Phys. Rev. A* **44**, 5401–5406 (1991).
446. I. Percival, Quantum spacetime fluctuations and primary state diffusion, *Proc. R. Soc. Lond. A* **451**, 503–513 (1995).
447. I. Percival, *Quantum State Diffusion*, Cambridge University Press, Cambridge, England, 1998.
448. L. P. Hughston, Geometry of stochastic state vector reduction, *Proc. R. Soc. Lond. A* **452**, 953–979 (1996).
449. D. I. Fivel, An indication from the magnitude of CP violations that gravitation is a possible cause of wave-function collapse, eprint quant-ph/9710042.
450. S. L. Adler, Environmental influence on the measurement process in stochastic reduction models, *J. Phys. A* **35**, 841–858 (2002).
451. S. L. Adler, D. C. Brody, T. A. Brun, L. P. Hughston, Martingale models for quantum state reduction, *J. Phys. A* **34**, 8795–8820 (2001).
452. S. L. Adler, L. P. Horwitz, Structure and properties of Hughston's stochastic extension of the Schrödinger equation, *J. Math. Phys.* **41**, 2485–2499 (2000).
453. A. I. M. Rae, Can GRW theory be tested by experiments on SQUIDs?, *J. Phys. A* **23**, L57–L60 (1990).
454. M. Buffa, O. Nicrosini, A. Rimini, Dissipation and reduction effects of spontaneous localization on superconducting states, *Found. Phys. Lett.* **8**, 105–125 (1995).
455. D. Z. Albert, L. Vaidman, On a proposed postulate of state-reduction, *Phys. Lett. A* **129**, 1–4 (1989).
456. F. Benatti, G. C. Ghirardi, R. Grassi, Quantum mechanics with spontaneous localization and experiments, in: E. G. Beltrametti, J.-M. Lévy-Leblond (Eds.), *Advances in Quantum Phenomena*, Plenum Press, New York, 1995, pp. 263–279.
457. D. Albert, B. Loewer, Tails of Schrödinger's cat, in: R. Clifton (Ed.), *Perspectives on Quantum Reality*, Kluwer, Dordrecht, The Netherlands, 1996, pp. 81–91.
458. J. S. Bell, Against "measurement", in: A. I. Miller (Ed.), *Sixty-Two Years of Uncertainty*, Plenum Press, New York, 1990, pp. 17–31.
459. P. Lewis, Quantum mechanics, orthogonality, and counting, *Br. J. Philos. Sci.* **48**, 313–328 (1997).
460. E. Joos, Comment on "Unified dynamics for microscopic and macroscopic systems", *Phys. Rev. D* **36**, 3285–3286 (1987).
461. R. Penrose, *The Emperor's New Mind*, Oxford University Press, Oxford, 1989.
462. R. Penrose, *Shadows of the Mind*, Oxford University Press, New York, 1994.
463. E. J. Squires, Wavefunction collapse and ultraviolet photons, *Phys. Lett. A* **158**, 431–432 (1991).
464. P. Pearle, Experimental tests of dynamical state-vector reduction, *Phys. Rev. D* **29**, 235–240 (1984).
465. P. Pearle, Stochastic dynamical reduction theories and superluminal communication, *Phys. Rev. D* **33**, 2240–2252 (1986).

466. W. Marshall, C. Simon, R. Penrose, D. Bouwmeester, Towards quantum superpositions of a mirror, *Phys. Rev. Lett* **91**, 130401 (2003).
467. S. L. Adler, *Quantum Theory as an Emergent Phenomenon*, Cambridge University Press, Cambridge, England, 2004, Ch. 6.5.
468. A. Bassi, E. Ippoliti, S. L. Adler, Towards quantum superpositions of a mirror: An exact open systems analysis, *Phys. Rev. Lett.* **94**, 030401 (2005).
469. D. Dürr, *Bohmsche Mechanik als Grundlage der Quantenmechanik*, Springer, Heidelberg, 2001.
470. C. Philippidis, C. Dewdney, B. J. Hiley, Quantum interference and the quantum potential, *Nuovo Cimento* **52B**, 15–28 (1979).
471. D. B. Malament, In defense of dogma: Why there cannot be a relativistic quantum mechanics of (localizable) particles, in: R. Clifton (Ed.), *Perspectives on Quantum Reality*, 1st Edition, Kluwer, Boston, 1996, pp. 1–10.
472. H. Halvorson, R. Clifton, No place for particles in relativistic quantum theories?, *Philos. Sci.* **69**, 1–28 (2002).
473. D. Dürr, S. Goldstein, R. Tumulka, N. Zanghì, Bohmian mechanics and quantum field theory, *Phys. Rev. Lett.* **93**, 090402 (2004).
474. D. Dürr, S. Goldstein, R. Tumulka, N. Zanghì, Trajectories and particle creation and annihilation in quantum field theory, *J. Phys. A* **36**, 4143–4150 (2003).
475. H. D. Zeh, Why Bohm's quantum theory?, *Found. Phys. Lett.* **12**, 197–200 (1999).
476. H. D. Zeh, There is no "first" quantization, *Phys. Lett. A* **309**, 329–334 (2003).
477. P. R. Holland, *The Quantum Theory of Motion*, Cambridge University Press, Cambridge, England, 1993.
478. D. M. Appleby, Generic Bohmian trajectories of an isolated particle, *Found. Phys.* **29**, 1863–1884 (1999).
479. D. M. Appleby, Bohmian trajectories post-decoherence, *Found. Phys.* **29**, 1885–1916 (1999).
480. V. Allori, *Decoherence and the Classical Limit of Quantum Mechanics*, Ph.D. thesis, Physics Department, University of Genova (2001).
481. V. Allori, N. Zanghì, On the classical limit of quantum mechanics, *Int. J. Theor. Phys.* **43**, 1743–1755 (2004).
482. V. Allori, D. Dürr, S. Goldstein, N. Zanghì, Seven steps towards the classical world, *J. Opt. B: Quantum Semiclass. Opt.* **4**, S482–S488 (2002).
483. A. S. Sanz, F. Borondo, A Bohmian view on quantum decoherence, eprint quant-ph/0310096.
484. M. Tegmark, The interpretation of quantum mechanics: Many worlds or many words?, *Fortschr. Phys.* **46**, 855–862 (1998).
485. B. d'Espagnat, A note on measurement, *Phys. Lett. A* **282**, 133–137 (2000).
486. B. d'Espagnat, Consciousness and the Wigner's friend problem, *Found. Phys.* **35**, 1943–1966 (2005).
487. E. P. Wigner, Remarks on the mind-body question, in: I. J. Good (Ed.), *The Scientist Speculates: An Anthology of Partly-Baked Ideas*, Heinemann, London, 1961, pp. 284–302.
488. J. Mehra, A. S. Wightman (Eds.), *The Collected Works of E. P. Wigner*, Vol. VI, Springer, Berlin, 1995, p. 271.
489. F. H. Eeckman, J. M. Bower, *Computation and Neural Systems*, Kluwer, Boston, 1993.

490. R. L. Harvey, *Neural Network Principles*, Prentice Hall, Englewood Cliffs, 1994.
491. D. J. Amit, *Modeling Brain Functions*, Cambridge University Press, Cambridge, 1989.
492. A. Einstein, Einstein's Credo, courtesy of the Albert Einstein Archives, Hebrew University of Jerusalem, Israel. Speech recorded in 1932 for the German League for Human Rights (Deutsche Liga für Menschenrechte).
493. M. J. Donald, Neural unpredictability, the interpretation of quantum theory, and the mind-body problem, eprint quant-ph/0208033.
494. L. H. Domash, The transcendental meditation technique and quantum physics: Is pure consciousness a macroscopic quantum state in the brain?, in: D. W. Orme-Johnson, J. T. Farrow (Eds.), *Scientific Research on the Transcendental Meditation Program: Collected Papers*, Vol. 1, M.E.R.U. Press, New York, 1977, pp. 652–670.
495. E. H. Walker, The nature of consciousness, *Math. Biosci.* **7**, 131–178 (1970).
496. H. P. Stapp, Exact solution of the infrared problem, *Phys. Rev. D* **28**, 1386–1418 (1983).
497. D. Zohar, *The Quantum Self*, William Morrow, New York, 1990.
498. H. C. Rosu, Essay on mesoscopic and quantum brain, *Metaphysical Review* **3**, 1–12 (1997).
499. L. M. Ricciardi, H. Umezawa, Brain physics and many-body problems, *Kibernetik* **4**, 44–48 (1967).
500. A. Vitiello, Dissipation and memory capacity in the quantum brain model, *Int. J. Mod. Phys. B* **9**, 973–989 (1996).
501. S. R. Hameroff, R. Penrose, Orchestrated reduction of quantum coherence in brain microtubules: a model for consciousness?, in: S. R. Hameroff, A. K. Kasszniak, A. C. Scott (Eds.), *Toward a Science of Consciousness — The First Tucson Discussions and Debates*, MIT Press, Cambridge, 1996, pp. 507–540.
502. S. Hameroff, R. Penrose, Conscious events as orchestrated space-time selections, *J. Conscious. Stud.* **3**, 36–53 (1996).
503. S. Hagan, S. R. Hameroff, J. A. Tuszynski, Quantum computation in brain microtubules: Decoherence and biological feasibility, *Phys. Rev. E* **65**, 061901 (2002).
504. M. Tegmark, Importance of quantum decoherence in brain processes, *Phys. Rev. E* **61**, 4194–4206 (2000).
505. E. Nogales, M. Whittaker, R. A. Milligan, K. H. Downing, High resolution model of the microtubule, *Cell* **96**, 79–88 (1999).
506. E. Nogales, S. G. Wolf, K. H. Downing, Structure of the $\alpha\beta$ tubulin dimer by electron crystallography, *Nature* **391**, 199–203 (1998).
507. R. Melki, M.-F. Carlier, D. Pantaloni, S. N. Timasheff, Cold depolymerization of microtubules to double rings: geometric stabilization of assemblies, *Biochemistry* **28**, 9143–9152 (1989).
508. M. V. Sataric, J. A. Tuszynski, R. B. Zakula, Kinklike excitations as an energy-transfer mechanism in microtubules, *Phys. Rev. E* **48**, 589–597 (1993).
509. L. P. Rosa, J. Faber, Quantum models of the mind: Are they compatible with environment decoherence?, *Phys. Rev. E* **70** (3), 31902 (2004).
510. C. Seife, Cold numbers unmake the quantum mind, *Science* **287**, 791 (2000).
511. M. J. Donald, A mathematical characterization of the physical structure of observers, *Found. Phys.* **25**, 529–571 (1995).

512. H. D. Zeh, The problem of conscious observation in quantum mechanical description, *Found. Phys. Lett.* **13**, 221–233 (2000).

Index

actualization by observation, 19
ancilla, 311–320
anomalous-diffusion coefficient, 186, 191, 200, 205, 220
apparatus, 32, 49, 51, 320–321
artificial neuronal networks, 365
asymmetry energy, 174
axon, 368

bath, 171
beables, 16
Bell inequalities, 17
Bell states, 29, 32
bilinear coupling, 179
bit-flip error, 309
black-box view of measurements, 334
Bloch sphere, 295
Bohmian mechanics, 18, 354–357
  and decoherence, 356–357
  particle concept in, 355–356
  trajectories in, 354
Born approximation, 156, 160–161
Born probabilities, 35, 333
Born rule, 15, 35, 331
  and envariance, 340–343
  Zurek's derivation of, 340–343
Born–Markov master equation, 155–165
  assumptions entering into, 156
  derivation of, 158–165
  structure of, 156–158
Born–Pauli interpretation, 19
Bose–Einstein condensates, 282–283
brain, 359–378
branch, 336, 375

C_{70} interference experiments, 258–270
  emission of thermal radiation in, 265–267
  interference pattern in, 262–263
  setup for, 259–262
  which-path information in, 263–265
Caldeira–Leggett master equation, 191–194
canonical models, 171–241
cavity quantum electrodynamics, 255
charge qubits, 281
chiral molecules, 83
classical ensemble, 37
CNOT gate, 306
coarse-graining, 37, 59
code subspace, 315
coherence, 5, 20
  delocalization of, 8
  recurrence of, 92, 213
  relocalization of, 99, 255–258
  revival of, 98–101, 255–258
coherence length, 140–147
coherent oscillations, 275–279
coherent spreading, 116, 144, 145
coherent states, 117, 202, 249
collapse postulate, 3, 15, 330
commutativity criterion, 77
complementarity principle, 61–65
conditional quantum dynamics, 306
conjugate basis, 23
consciousness, 363–367, 370–371, 376–378
conservation laws, 30
contextuality, 18
continuous spontaneous localization, 348–354
controllability (in quantum computers), 301–302
Cooper pairs, 271
Copenhagen interpretation, 330–336, 360, 363

correlations
  classical, 30
  quantum, 29
  quantum vs. classical, 30–32

dc-SQUID, 272
de Broglie wavelength (matter), 116
de Broglie wavelength (thermal), 94
decoherence
  and quantum error correction, 311–321
  and resolution into subsystems, 101–103
  apparatus-induced, 320–321
  canonical models for, 171–241
  collective, 322–323
  due to environmental scattering, 115–151
  dynamics of scattering-induced, 139–150
  experiments on, 243–291
  factor, 91, 122, 212
  fake, 96
  history of, 10–12
  in Bose–Einstein condensates, 282–283
  in cavity QED, 251–258
  in energy, 81, 228
  in interpretations of quantum mechanics, 329–358
  in momentum, 194
  in neurons, 368–371
  in phase space, 173, 194–203
  in quantum Brownian motion, 178–206
  in quantum computers, 293–328
  in quantum-electromechanical systems, 287–288
  in SQUIDs, 279–282
  in the brain, 365–374
  independent, 322–323
  irreversibility of, 5, 69
  on dynamical timescales, 197
  rate, *see separate entry*
  reversal of, 98–101, 255–258
  timescale (general), 94
  timescale (in quantum Brownian motion), 197
  timescale (scattering), 127, 131, 134
  virtual, 98–101
  vs. classical noise, 95–98, 302–304
  vs. controllability, 301–302
  vs. dissipation, 93–95
decoherence rate, 70, 94
  in environmental scattering, 131, 134
  in microtubules, 371–374
  in neurons, 368–371
  in the Caldeira–Leggett model, 193
  in the spin–boson model, 216
  saturation of, 132, 203
decoherence timescale, *see* decoherence rate
decoherence-free subspace, 79–80, 321–327
  experimental realizations of, 325–326
  structure and size of, 322–325
delayed-choice experiments, 100
delocalization of phase relations, 69, 98–101
delocalized modes, 175
density matrices, 33–49
  basis ambiguity of, 41–42
  interpretation of, 36, 39, 48–49
  mixed-state, 36–41
  pure-state, 34–36
  purity of, 40
  reduced, 44–49
  von Neumann entropy of, 41
density operator, *see* density matrices
dephasing, 97, 270, 282
differential cross section, 123
direct peaks, 107
dissipation, 93–95, 188–191
dissipation kernel, 181
distinguishability, 32–33, 47–48
double-slit experiment, 23, 60, 258–259, 354
double-well potential, 174, 273
dynamical decoupling, 10, 326–327, 375
dynamical symmetry, 80
dynamically decoupled subspace, 327
Dyson series, 210

eigenvalue–eigenstate link, 59, 331
einselection, *see* environment-induced superselection
emission of thermal radiation, 115, 265–267

ensemble
  average, 43, 96
  classical (proper), 20
  dephasing, 97, 270, 282
  improper, 49
  interpretation, 19
  physical, 43–44, 96, 303
  width, 141–147
entanglement, 4, 28–33
  and distinguishability, 32–33
  and which-path information, 60–68
  as a resource, 7
  quantification of, 32–33
entropy, 41, 313
  Shannon, 86
  thermodynamic, 41
  von Neumann, 41
envariance, 340–343
environment as a witness, 86
environment engineering, 326–327
environment self-correlation functions, 157, 181
environment-assisted invariance, 340–343
environment-induced superselection, 6, 71–85
  and implications for interpretations, 329–358
  and predictability sieve, 82–83
  commutativity criterion for, 77
  general case of, 81–83
  in Bohmian mechanics, 355–357
  in modal interpretations, 345
  in physical-collapse theories, 349–351
  in relative-state interpretations, 337–344
  in the Copenhagen interpretation, 333–334
  in the quantum limit of decoherence, 81
  in the quantum-measurement limit, 76–80
  of charge, 84
  of position, 83
  stability criterion for, 73
environmental monitoring, 65–68, 78
environmental scattering, 65, 115–151
environmental self-correlations, 156, 163
epistemic view, 19
EPR, 6, 31
existential interpretation, 343–344

faithful measurement, 75
fake decoherence, 96
fault tolerance, 320–321
first intervention, 58, 363
flux quantum, 272
Fokker–Planck equation, 184
free will, 366

gate operation, 296
Gaussian density matrix, 140
Gaussian wave packet, 116
Grover's search algorithm, 300
GRW theory, *see* physical collapse theories
guiding equation, 354

Hadamard rotation, 317
Hamiltonian
  interaction, 76
  Lamb-shifted, 155
  open-system decomposition, 76
  self, 76
harmonic oscillator, 117, 171
Heisenberg cut, 27, 335
heuristic fiction, 376
hidden-variables theories, 17–18, 355
Hilbert space, 15
Hyperion, 118

ideal quantum measurement, 52
ignorance-interpretable mixture, 42
improper mixture, 49, 69, 332
infinite regression, 362
information is physical, 294, 343
interaction Hamiltonian, 76
  diagonal decomposition of, 78
interaction picture, 379–382
interaction-picture Liouville–von Neumann equation, 380
interference, 47–48
  and distinguishability, 47–48
  experiments, 23–26, 243–291

in Bose–Einstein condensates, 282–283
in C_{70} diffraction, 258–270
in SQUIDs, 275–279
local damping of, 68–71
nonobservability of, 50, 55–57, 333
pattern, 23, 24, 55, 60
terms, 34
internal environment, 176
interpretation basis, 338
ion trap, 301, 303, 305, 326
irreversibility
and information loss, 101
and quantum erasure, 100
of collapse, 63
of decoherence, 69, 93, 98, 304
of measurements, 69, 100
isolated systems, 1–8
Itô equation, 348

Josephson effect, 271
Josephson junction, 272

kink-like excitations in microtubules, 374
Kochen–Specker theorem, 18
Kraus-operator formalism, 110–112

Lamb-shifted Hamiltonian, 155
Lindblad master equations, 165–169
Lindblad operator, 167
Liouville–von Neumann equation, 139
interaction-picture, 380
local theory, 4
locality of states, 4
localization, 115–151
localized modes, 175
long-wavelength limit
in environmental scattering, 130–132
in quantum Brownian motion, 204
Loschmidt echo, 88
low-temperature environment, 175

many-minds interpretation, 336, 377
many-worlds interpretation, 336
mapping, 173–178
and oscillator environments vs. spin environments, 176–178
of central systems, 173–174
of oscillator environments, 175
of phase-space particles, 173
of spin environments, 175–176
of spin environments onto oscillator environments, 228–237
of two-level systems, 173–174
Markov approximation, 156, 163
master equations, 153–170
Born–Markov form of, 155–165
general formalism of, 154–155
in Caldeira–Leggett form, 191–194
in Lindblad form, 165–169
non-Markovian, 169–170
measurement problem, 49–60
subjective resolutions of, 375–378
measurement-like interaction, 49, 67
measurements, 15
actualization in, 19
and postselection, 21
Born–Pauli interpretation of, 19
collapse in, 15
contextuality in, 18
interference, 21–26
irreversible, 69, 100
local, 44
noncommutativity in, 15
nondemolition, 52, 74
observables and, 15
of superpositions, 21–26
projective, 15
virtual, 101
memory effects, 156, 163, 169
microtubules, 367, 371–374
mixed states, 19, 36–41
mixture
ignorance-interpretable, 42
improper, 49, 69, 332
proper, 20, 42
modal interpretations, 344–347
and environment-induced superselection, 345
Schmidt decomposition in, 345–346
models
canonical, 171–241
for quantum Brownian motion, 178–206
spin–boson, 207–222
spin-environment, 88–93, 222–237

mutual information, 86
myelin, 368

neurons, 365–371, 375–378
NMR, 88, 97, 301
no-cloning theorem, 16, 306
no-signaling theorem, 31
noise kernel, 181
non-Hermitian observables, 53
non-Markovian dynamics, 169–170
noncommutativity of observables, 15, 54
nondemolition measurement, 52
  by the environment, 74
nonlocality, 4, 6, 32
nonunitary dynamics, 10, 28, 69, 111, 155, 290, 330, 347–354
normal-diffusion coefficient, 185, 189, 191, 200
nuclear magnetic resonance, 88, 97, 301

objectification through redundancy, 88
observables, 15
observation
  in quantum mechanics, 359–361
  indirect, 85
off-diagonal terms, 34
openness of quantum systems, 3–4, 69, 74
operator-sum formalism, 110–112
optical quantum computing, 301
orthodox interpretation, 330–336
oscillator environments, 175
  in the weak-coupling limit, 176
outcomes, 16
  in hidden-variables theories, 17
  problem of, 50, 57–60, 331–333, 362
overcomplete set of states, 53, 75

partial trace, 44, 333
  and decoherence, 45
  interpretation of, 44, 333
phase qubits, 281
phase relations, 69
  delocalization of, 69, 98–101
phase-flip error, 309
physical collapse theories, 347–354
  competition with decoherence in, 350–351
  experimental tests of, 353–354
  preferred-basis problem in, 349–350
  tails problem in, 351
physical ensembles, 43–44, 96, 303
physical reality, 15, 360, 375
Poincaré recurrence, 93, 213
pointer (of apparatus), 51
pointer basis, 75–85
pointer observable, 77–85
pointer states, 74–85
  and implications for interpretations, 329–358
  and predictability sieve, 82
  commutativity criterion for, 77
  general selection methods for, 81–83
  in charge space, 84
  in modal interpretations, 345
  in phase space, 194–203
  in position space, 83
  in relative-state interpretations, 337–344
  in the Copenhagen interpretation, 333–334
  in the quantum limit of decoherence, 81
  in the quantum-measurement limit, 76–80
  predictability of, 82
  redundant encoding of, 87
  stability criterion for, 73
pointer subspace, *see* decoherence-free subspace
positivist view, 67, 360
postselection, 21
predictability, 82, 376
predictability sieve, 82–83
preferred basis, 55, 75–85
preferred observable, 55, 74–85
preferred states, *see* pointer states
preferred-basis problem, 50, 53–55, 71, 83–85
  in modal interpretations, 344–347
  in physical collapse theories, 349–350
  in relative-state interpretations, 337–339
  in the Copenhagen interpretation, 333–334
premeasurement, 53

probabilities
  and envariance, 340–343
  Born, 35
  classical, 37
  epistemic, 17
  in mixed states, 36
  in relative-state interpretations, 339–343
  quantum, 15–18, 331
probability amplitude, 18
problem of outcomes, 50, 57–60, 331–333, 362
problem of the nonobservability of interference, 50, 55–57, 69, 333
proper mixture, 20, 42
protofilaments, 371
psycho-physical parallelism, 361, 376–378
pure states, 19
purification of the environment, 109
purity, 40, 147, 201

quantum algorithms, 300
quantum bit, 88, 270–282, 295–297
quantum brain, 359–378
quantum Brownian motion, 178–206
  and oscillator–spin models, 234–237
  Born–Markov master equation for, 178–182
  decoherence rate in, 193
  dynamics of, 194–203
  exact master equation for, 206
  harmonic-oscillator master equation for, 182–187
  high-temperature limit of, 191–194
  in the Wigner representation, 184, 195–197
  limitations of models for, 203–205
  phase-space decoherence in, 194–203
  pre-Markovian form of the master equation for, 186
quantum computers, 30, 293–328
  and the brain, 365–374
  decoherence vs. classical noise in, 302–304
  decoherence vs. controllability in, 301–302
  error correction in, 304–321
  fault tolerance in, 320–321
  ion-trap, 301, 303, 305, 326
  NMR, 301
  optical, 301
  physical realizations of, 300–301
  power of, 294–297
  read-out of, 297–298
  redundant encoding in, 305–307, 315–320
  storage capacity of, 296
quantum computing, *see* quantum computers
quantum correlations, 4, 29
quantum cryptography, 30
quantum Darwinism, 86
quantum discord, 87
quantum erasure, 98–101, 304, 316
quantum error correction, 304–321
  classical vs. quantum, 305–307
  discretization of errors in, 307–311
  fault-tolerant, 320–321
  three-bit code for, 315–320
quantum holism, 6, 33, 297
quantum information theory, 293–328
quantum limit of decoherence, 81, 228
quantum Monte Carlo, 297
quantum observer, 361–365
quantum parallelism, 297
quantum states
  classical vs. quantum, 14–16
  completeness of, 16
  concept and interpretation of, 14–20
  fragility of, 16, 74
  global, 29
  in Bell form, 29
  mixed, 19
  ontological status of, 18–20
  preparation procedure for, 36
  probabilistic nature of, 16–18
  pure, 19
  realist interpretation of, 19, 347, 376
  relative, 29
quantum-electromechanical systems, 234, 284–288, 335
  decoherence in, 287–288
  superposition states in, 285–287
quantum-jump approach, 168
quantum-measurement limit, 76–80, 228

quantum-to-classical transition, 2, 49–60
quantum-trajectory approach, 168
qubit, 88, 270–282, 295–297

Rabi frequency, 246
Rabi oscillations, 25, 246–247, 276
Ramsey interferometry, 25–26, 279
recoherence, 99, 255–258
recoiling slit, 61
recurrence of coherence, 92, 213
Redfield equation, 163, 187
reduced density matrices, 44–49, 64, 68
  and decoherence, 45
  derivation of, 45–46
  interpretation of, 48–49, 333
  positivity of, 165
  vs. ensembles, 48–49
redundancy, 85–88, 305, 335, 343, 378
redundant encoding of information
  in classical computers, 305–307
  in quantum computers, 315–320
  in the environment, 85–88
relative states, 29, 47
relative-state interpretations, 336–344
  preferred-basis problem in, 337–339
  probabilities in, 339–343
  Schmidt decomposition in, 338
relaxation timescale, 93
reset, 101
resolution into subsystems, 101–103
resonant $\pi/2$ pulse, 247
resting/firing of neurons, 365–371
reversible decoherence, 98–101, 255–258
revival of coherence, 98–101, 255–258
rf-SQUID, 272

S-matrix, 119
scattering, 7, 45, 65, 115–151
  master equation for, 140
  of air molecules, 136–138
  of photons, 132–136
scattering constant, 131
Schmidt basis, *see* Schmidt decomposition
Schmidt decomposition, 104–106
  and classicality, 105–106, 338
  and decoherence, 105–106
  and pointer states, 105–106
  degeneracy in, 105, 346
  for density matrices, 104
  in modal interpretations, 345–346
  in relative-state interpretations, 338
  uniqueness of, 104
Schmidt states, *see* Schmidt decomposition
Schrödinger equation, 21
  augmented, 348
  stochastic, 169, 347
Schrödinger's cat, 2–3, 57–58, 364
second intervention, 363
selection of quasiclassical properties, 83–85
self-Hamiltonian, 76
Shannon entropy, 86
Shor's factoring algorithm, 300
short-wavelength limit, 128–129
shut-up-and-calculate interpretation, 329
Solvay conference, 61, 329
spatial localization, 74, 83, 115–151, 265, 348–351
special relativity, 31
spectral density, 181
  cutoff in, 188
  effective, 233
  in Lorentz–Drude form, 188
  ohmic, 188
spin echo, 98
spin environments, 175–176, 222–237
spin–boson model, 207–222
  and spin-environment models, 233–234
  Born–Markov master equation for, 218–222
  simplified version of, 208–218
spin-environment models, 222–237
  in the quantum limit of decoherence, 228
  in the quantum-measurement limit, 228
  in the weak-coupling limit, 228–237
  non-Markovian, 237
  simple example of, 223–228
splitting (of branches), 336, 375–378
spontaneous localization models, 348–354

spooky action at a distance, 31
SQUID, 176, 270–282
  basic principles of, 272–275
  observation of decoherence in, 279–282
  physical collapse theories and, 349–350, 354
  Rabi oscillations in, 276
  superposition states in, 275–279
stability criterion, 73
standard interpretation, 330–336
Stern–Gerlach experiment, 21–23, 37
  reversible, 99
stochastic dynamical reduction, 348–354
subjectivity, 360
subsystem, 28, 101–103
superconducting qubits, 270–282
superconductivity, 271–272
supercurrent, 271–272
superposition principle, 20–28
  interpretation of, 20–21
  scope of, 26–28
superpositions, 20–28
  amplification of, 52
  and interference experiments, 23–26
  coherent, 20
  direct measurements of, 21–23
  experimental verification of, 21–26
  of Gaussians, 194
  of radiation fields, 245
  of supercurrents, 275–279
  qubit state in, 295
  relocalization of, 99, 255–258
  vs. classical ensembles, 20
superselection rules, 6, 73
symmetrization of the environment, 326
system–observer duality, 336

T-matrix, 122, 123
tails problem, 351
Talbot–Lau effect, 260–261
thermal bath, 171
three-bit code, 315–320
threshold theorem, 321
timescale
  for coherent spreading, 117
  for decoherence (general), 70, 94
  for decoherence in quantum Brownian motion, 197
  for decoherence in the Caldeira–Leggett model, 185
  for decoherence in the spin–boson model, 216
  for recurrence of coherence, 93
  for relaxation, 93
  for scattering-induced decoherence, 127, 131, 134
trace operation, 34–36
  and decoherence, 45
  interpretation of, 36, 44, 333
  partial, 44, 333
trace rule, 36, 46
  for mixed states, 38
tubulin, 371
tunneling, 174, 271, 277
tunneling matrix element, 174

uncertainty principle, 62, 202, 366

virtual decoherence, 98–101
virtual measurement, 101
von Neumann chain, 361–365
von Neumann entropy, 41
von Neumann measurement scheme, 50–53, 361–362

wave packets, 18
  coherent spreading of, 116
  Gaussian, 116
weak-coupling limit, 176
weight (in quantum error correction), 310
which-path information, 60–68, 263–265
  and entanglement, 60–68
  and environmental monitoring, 65–68
Wigner representation, 106–109
  direct peaks in, 107
  in quantum Brownian motion, 184, 195–197
  oscillatory pattern in, 107
Wigner's friend, 364–365

THE FRONTIERS COLLECTION

Information and Its Role in Nature
By J. G. Roederer

Relativity and the Nature of Spacetime
By V. Petkov

Quo Vadis Quantum Mechanics?
Edited by A. C. Elitzur, S. Dolev, N. Kolenda

Life – As a Matter of Fat
The Emerging Science of Lipidomics
By O. G. Mouritsen

Quantum–Classical Analogies
By D. Dragoman and M. Dragoman

Knowledge and the World
Challenges Beyond the Science Wars
Edited by M. Carrier, J. Roggenhofer, G. Küppers, P. Blanchard

Quantum–Classical Correspondence
By A. O. Bolivar

Mind, Matter and Quantum Mechanics
By H. Stapp

Quantum Mechanics and Gravity
By M. Sachs

Extreme Events in Nature and Society
Edited by S. Albeverio, V. Jentsch, H. Kantz

The Thermodynamic Machinery of Life
By M. Kurzynski

The Emerging Physics of Consciousness
Edited by J. A. Tuszynski

Weak Links
Stabilizers of Complex Systems from Proteins to Social Networks
By P. Csermely

Mind, Matter and the Implicate Order
By P.T.I. Pylkkänen

Quantum Mechanics at the Crossroads
New Perspectives from History, Philosophy and Physics
Edited by J. Evans, A.S. Thorndike

Particle Metaphysics
A Critical Account of Subatomic Reality
By B. Falkenburg

The Physical Basis of the Direction of Time
By H.D. Zeh

Asymmetry: The Foundation of Information
By S.J. Muller

Mindful Universe
Quantum Mechanics and the Participating Observer
By H. Stapp

Decoherence and the Quantum-To-Classical Transition
By M. Schlosshauer

Printing: Krips bv, Meppel
Binding: Stürtz, Würzburg